Progress in Probability
Volume 44

Perplexing Problems in Probability

Festschrift in Honor of Harry Kesten

Maury Bramson
Rick Durrett
Editors

Birkhäuser
Boston • Basel • Berlin

Maury Bramson
School of Mathematics
University of Minnesota
Minneapolis, MN 55455

Rick Durrett
Department of Mathematics
Cornell University
Ithaca, NY 14853

Library of Congress Cataloging-in-Publication Data

Perplexing problems in probability : festschrift in honor of Harry
Kesten / Maury Bramson, Rick T. Durrett, editors.
p. cm. — (Progress in probability ; v. 44)
ISBN 0-8176-4093-2 (alk. paper). — ISBN 3-7643-4093-2 (alk.
paper)
1. Probabilities. I. Kesten, Harry, 1931- . II. Bramson,
Maury, 1951- . III. Durrett, Richard, 1951- . IV. Series:
Progress in probability ; 44.
QA273.18.P47 1999
519.2—dc21 99-14285
CIP

AMS Subject Classifications: 60F05, 60F10, 60E07, 60G25, 60G07, 60J10, 60J15, 60K25, 60K35, 62G30, 52A10, 82B20

Printed on acid-free paper.

Birkhäuser

ISBN 0-8176-4093-2
ISBN 3-7643-4093-2

Formatted from authors' files by TEXniques, Inc., Cambridge, MA.
Printed and bound by Hamilton Printing Company, Rensselaer, NY.
Printed in the United States of America.

9 8 7 6 5 4 3 2 1

Harry Kesten

Dedicated to Harry Kesten
In honor of 40 years of impressive research accomplishments in probability

Contents

1

Harry Kesten's Publications

A Personal Perspective

Rick Durrett

Writing about Harry Kesten's life work is a daunting task. At of the writing of this paper, he has published almost 150 papers totaling more than 5000 pages. The topics range from refined results for the classical topics of random walks, renewal theory, Lévy processes, and branching processes to questions of interest in statistical mechanics: first passage percolation, percolation, DLA, and the models named after Ising, Potts, and Heisenberg. In most cases Harry has solved other people's problems, so his publication list makes excursions into dozens of other topics from the local times for Markov processes [25, 43] to existence and uniqueness of Markov random fields [54]; from the speed of convergence of martingales [59] and properties of positive harmonic functions [61] to Chung-type laws of the iterated logarithm [143].

When I was born (i.e., emerged from graduate school) in 1976, the definitive work of the dynamic duo of Kesten and Spitzer on random walks [17, 20, 26] was legendary and, together with Kesten's individual work on random walks [21, 32, 34, 35] and his joint work with Stigum [27, 28, 33] and with Ney and Spitzer [31] on branching processes, was an important part of one's graduate education. In particular, the Kesten and Stigum result that $E(Z_1 \log^+ Z_1) < \infty$ is necessary and sufficient for the convergence of mean normalized branching processes Z_n/m^n to a limit with mean EZ_0 is something that is usually mentioned (but not proved) when branching processes are discussed. These and other results of that era [41, 45, 49, 51, 58] showed us that Kesten was someone who could prove results under a minimal number of assumptions and who could disentangle the mysteries of random walks with infinite mean.

However, most of all we knew Kesten as a problem solver. A classical and well-known example is Kesten's impressive work calculating "Hitting probabilities of single points for a process with independent increments" which appeared as Memoirs of the AMS, No. 93 [37] and was the subject of his 1970 address at the International Congress of Mathematicians in Nice [39]. A more personal bit of data is the following testimonial from `G.R.Grimmett@statslab.cam.ac.uk`: "Rick: I was perhaps understandably impressed by 'Supercritical branching processes with countably many types and the size of random Cantor sets' [106] which was inspired by Dekking/Grimmett. Harry was just able to blast through a version of the problem doing substantially better than anyone else. I understood then how well he grasped branching processes." Indeed, one does not really understand Kesten's prowess until you have seen him demolish a problem that you have worked on.

The first two decades of Kesten's work, while containing important contributions, are difficult for me to properly put in context since the work was already complete when I started learning probability. Because of this I have decided to tell the story of his work as I experienced it. Thus, following the style of some paperback novelists, I will begin in the middle of the story with some exciting events to grab the reader's attention. Then after the story line is established, I will go back and fill in earlier developments.

I spent the 1980–81 academic year at Cornell. At this time, Dynkin's Russian style seminar, held Wednesdays 7–9 PM, was a lively affair, with Avi Mandelbaum, Bob Vanderbei and Patrick Sheppard as students. Kesten had recently proved that "The critical probability for bond percolation on the square lattice equals 1/2" [67] and followed this up with "power estimates of functions in percolation theory" [71]. The title of the last paper is a double entendre. The results concern the power law behavior of functions near the critical value but introduced powerful new rigorous renormalization arguments. To facilitate writing his book [72], Kesten taught a graduate seminar on percolation and first passage percolation. Harry prepared for class while swimming laps in the pool at Teagle Hall, so in his lectures you got to see how he thought. He would start with the main idea of the proof, but then often would have to go back and insert a technicality at an angle on the margin of the board. This made it difficult for the students to get good notes, but for me it provided valuable insights about why things are true and how he went about solving problems.

In this brief article there is not enough space to discuss why things are true, so we will only discuss what Kesten (and others) have done. Our first two topics, percolation and first passage percolation (which we will interpret very broadly) are those of Harry's course in 1980 and of his Wald Lectures in 1986 (see [95]). In each case we will take the subject from the 80's up to today. For the third section of the paper we will go back to Kesten's Ph.D. thesis and follow his work on random walks and related topics up to the present. In these three forays we will touch on much, but by no means all, of Kesten's best work. Like a one week bus tour of Europe, there is only time to drive past the outside of some of the most important landmarks. We apologize in advance for the fact that in order to say things quickly, we will not always be able to say things carefully. We will never intentionally lie about what is true, but sometimes we will not take the time to sort out all the details of who did what when.

1.1 Percolation

Broadbent and Hammersley (1957), and Hammersley (1959) introduced percolation as a model for the spread of a fluid or gas through a random medium. To formulate the bond percolation model in d dimensions, we make the d-dimensional integer lattice $\mathbf{Z}^d$ into a graph by drawing edges connecting adjacent sites. We imagine that the edges are channels and that fluid will move through a channel if and only if the channel is wide enough. We declare that the edges are independently designated as open (wide enough) or closed with probabilities p and $1 - p$ respectively, and let P_p denote the resulting probability measure on the configurations of open and closed edges. We will also sometimes consider site percolation in which the sites are independently open with probability p or closed with probability $1 - p$, but for this article the default process is bond percolation.

With that set-up it is natural to ask about the set of sites C_0 that can be reached from the origin by a path of open edges. The first papers mentioned above showed that if p is small, then the number of points in C_0, $|C_0|$, is always finite, while if p is close enough to 1, then

$$\theta(p) = P_p(|C_0| = \infty) > 0. \tag{1.1.1}$$

This and an obvious monotonicity establishes the existence of a critical value $p_c = \inf\{p : \theta(p) > 0\}$ but does not give much information about its value. The first step in that direction for two dimensional bond percolation was taken by Harris (1960). He noticed that when $p = 1/2$, symmetry dictates that the probability of a left to right crossing of a "sponge," an $n \times (n + 1)$ piece of the square lattice, is 1/2, and used this observation to show that at $p = 1/2$ the origin is surrounded by infinitely many cut sets of vacant edges, so $p_c \geq 1/2$.

The next step in this direction was taken by Sykes and Essam (1964) who introduced a quantity they called the *free energy*:

$$\Delta(p) = E_p(1/|C_0|; |C_0| > 0). \tag{1.1.2}$$

Probabilistically, this is the limiting value of the number of clusters per unit volume. By analogy with the Ising model, Sykes and Essam argued that the phase transition in percolation must be manifest in a "singularity" at p_c. Their calculations showed that the square lattice, $\Delta(p) - \Delta(1 - p)$ is a polynomial in p, so assuming that such a singularity was unique, they arrived at $p_c = 1/2$ for the square lattice. In addition, Sykes and Essam used variations of this argument to show that the critical value of site percolation on the triangular lattice is 1/2 and supplemented this with the star-triangle transformation to show that the critical values for bond percolation on the triangular and hexagonal lattice are ρ and $1 - \rho$ where $\rho = 2\sin(\pi/18)$ is the unique root of $3\rho - \rho^3 = 1$ in $(0, 1)$.

Not much beyond Harris' result was rigorously proved about percolation until 1978, when Russo (1978) and Seymour and Welsh (1978) provided two valuable steps. The first is that if sponge crossing probabilities are large enough, then percolation occurs. To state the second, we need to define the dual of a planar graph, which is constructed by putting sites in each component of the complement of the graph and connecting two sites by an edge if the boundaries of their associated components share an edge. Denoting the dual by a star and defining

$$p_T = \sup\{p : E_p(|C_0|) = \infty\}, \tag{1.1.3}$$

the second fact is $p_c + p_T^* = 1$. The third crucial ingredient was provided by Kesten, who showed that if n is large, then the sponge crossing probability for two dimensional bond percolation is a very steep function of p near $p = 1/2$. Since small sponge crossing probabilities imply that the cluster size is finite, this completed the proof of $p_c = 1/2$.

The computation of the critical value for the square lattice was soon generalized to the other graphs Sykes and Essam considered, see [72, Chapters 1–3], and a flood of new results followed. Some of the results were proved for general d initially and most are known in that generality now, but to make the storytelling simple we will restrict our attention here to $d = 2$ and add the disclaimer that many other people's work was important in reaching the following conclusions. If you want the whole story and to have it told correctly, you should buy a copy of Grimmett's (1999) book.

Soon after Kesten's breakthrough, it was shown that $P_p(|C_0| \geq n)$ decayed exponentially fast for $p < p_c$ while for $p > p_c$ large finite clusters were very unlikely:

$$P_p(n \leq |C_0| < \infty) \leq C \exp(-\gamma n^{(d-1)/d}) \qquad \text{for } p > p_c.$$

Here one cannot do better than the power $n^{(d-1)/d}$ since a cube of vacant edges of radius r has probability $\exp(-cr^{d-1})$ and cuts off a volume of r^d. For a more recent look at estimates for the probability of a large cluster in supercritical percolation, see [105].

A second important consequence of Kesten's work is that it was possible for the first time to prove results about the behavior of various quantities near p_c. Taking $f(p) \approx |p - p_c|^\alpha$ to mean that $\log f(p)/\log|p - p_c| \to \alpha$, physicists tell us that near p_c we have

$$\theta(p) = P_p(|C_0| = \infty) \approx (p - p_c)^\beta \quad \text{as } p \downarrow p_c,$$
$$\chi(p) = E_p(|C_0|; |C_0| < \infty) \approx |p - p_c|^{-\gamma} \quad \text{as } p \to p_c$$

while if we use P_{cr} to denote P_p with $p = p_c$, then we can define three more critical exponents

$$P_{cr}(|C_0| \geq n) \approx n^{-1/\delta}$$
$$P_{cr}(\text{radius}(C_0) \geq n) \approx n^{-1/\delta_r}$$
$$P_{cr}((n, 0) \in C_0) \approx n^{-(d-2+\eta)}.$$

Kesten was the first [71] to prove bounds which show that in two dimensions if β, γ, and δ exist, then they are positive and finite. These insights were deepened in [81] when he gave a rigorous definition in two dimensions of physicists' "incipient infinite cluster at criticality" (where the probability of an infinite cluster is 0) by conditioning on the event that the origin is connected to the boundary of the box of radius n and letting $n \to \infty$.

The incipient infinite cluster is a fractal of dimension $(2-\eta)/(1-1/\delta)$. Simulations show that the cluster consists of many dangling ends and very little backbone, i.e., the part that would carry electricity if the origin was electrified and the boundary of the box was grounded. In order to prove a result that captures this mental picture of the structure of the cluster, Kesten considered in [82] and [83] random walk on various random graphs. In the case of a graph that is the family tree of a critical branching process conditioned on non-extinction, Kesten was able to show that the normalized height of the walker at time n, $n^{-1/3}h(X_n)$, converge to a limit. His results were less complete for walks on the "incipient infinite cluster" but he was able to establish subdiffusive behavior, i.e., show that $|X_n|/n^{0.5-\epsilon}$ was tight for some $\epsilon > 0$. For recent related work on the geometry of critical percolation clusters, in particular results about lowest crossing, see Kesten and Zhang [123].

The incipient infinite cluster is not only a mathematical curiosity but also a useful technical device. It allowed Kesten to show [89] that the three exponents that we introduced for P_{cr} were simply related in $d = 2$ (assuming they exist):

$$\eta = 2/\delta_r, \qquad \delta = 2\delta_r - 1. \tag{1.1.4}$$

The last two equalities are two of many scaling relationships that relate the behavior of various quantities at and near p_c. However, before we can state more of these relations we need to introduce a quantity that is more subtle but equally important as those introduced above.

Taking the simplest of several possible definitions we can define the *correlation length* by

$$\xi(p) = \left(\frac{1}{\chi(p)} \sum_y |y|^2 P_p(y \in C_0, |C_0| < \infty) \right)^{1/2}$$

and the corresponding exponent by $\xi(p) \approx |p - p_c|^{-\nu}$ as $p \to p_c$. (See [97] for a definition in terms of the exponential decay of various connection probabilities.) Intuitively, the correlation length gives the radius of a "typical" finite cluster. Using the definition above, Kesten was able to show [90] that the critical exponents of two dimensional percolation satisfy

$$\beta = \frac{2\nu}{\delta + 1} \qquad \gamma = 2\nu \cdot \frac{\delta - 1}{\delta + 1}. \tag{1.1.5}$$

These equalities, which can be guessed by back-of-the-envelope calculations, were widely accepted by physicists, but it required a fair amount of ingenuity for Kesten to prove them in the two dimensional case.

Another of the "obvious" facts about percolation that needed mathematical proof was the fact that the infinite cluster, when it existed, was unique. A first step in this direction was taken by Newman and Schulman (1981) who showed that, with probability one, there were 0, 1, or ∞ infinite clusters. The last possibility occurs for trivial reasons for percolation on trees. Not many people believed that infinitely many clusters was a reasonable possibility on $\mathbf{Z}^d$, but it took another half decade before Aizenman, Kesten, and Newman (see [88] and [91]) could prove this by relating uniqueness to the differentiability of the free energy, $\Delta(p)$. Though the proof was slow to be found, it did not last long as the best argument around. In

(1988) Burton and Keane discovered a very beautiful geometric proof that worked for a number of dependent models as well.

There are many reasons for being interested in percolation. When generalized to the oriented case and then to "continuous time," these results have led to a wealth of information about the contact process, one of the most basic interacting particle systems. See Durrett (1984) and Bezuidenhout and Grimmett (1990), (1991). On a different level, comparison with oriented percolation can be used to prove the existence of interesting behavior for particle systems with long range or fast stirring. See Durrett (1995a).

Physicists view percolation as a prototypical example of a system with phase transitions. In the case of the Potts model (a multicolor version of the Ising model) the connection is more than an analogy. Suppose we use the independent percolation measure P_p with $0 \le p \le 1$ to define a new measure for $q > 0$ by

$$dQ_{q,p}/dP_p = q^{C(\omega,\Lambda)} \cdot \frac{1}{Z(q,p,\Lambda)} \tag{1.1.6}$$

where $C(\omega, \Lambda)$ is the number of connected components in the box Λ (for some specified boundary conditions) and $Z(q, p, \Lambda)$ is the normalizing constant to make the $Q_{q,p}$ a probability measure. Then the infinite volume limit $Q_{q,p}$ gives the distribution of the q-state Potts model, and taking $q = 2$ we have the Ising model. This great idea came from Fortuin and Kastelyn (1972). For more on this see Grimmett (1995).

This connection was used by Aizenman, J. Chayes, L. Chayes, and Newman (1988) to prove the discontinuity of magnetization in the one dimensional $1/|x-y|^2$ Ising and Potts models by using earlier results of Aizenman and Newman (1986) for the analogous percolation process. Later [99, 100, 102], and [103], in joint work with various subsets of {Bricmont, Lebowitz, Schonmann}, Kesten used these ideas to study the asymptotic behavior of Ising, Potts, and Heisenberg models as the dimension gets large. More recently in [120] with Bezuidenhout and Grimmett, Kesten used this connection to show that the critical value β_c of the ferromagnetic Potts model is a *strictly* decreasing function of the strengths of the interactions in the process.

Turning away from critical values, our next topic is critical exponents, specifically "mean field bounds" on them. To explain this term, we note that percolation on a tree in which each node has degree $k + 1$ is essentially a Galton-Watson process in which each individual has a binomial(k, p) number of children. Calculations for branching processes (exercise for the reader) show that the critical value $p_c = 1/k$ and critical exponents $\beta = 1$, $\gamma = 1$, $\delta_r = 1$, and $\delta = 2$.

There are a number of results which show that the mean field values in general provide bounds for those of ordinary percolation. Aizenman and Newman (1984) showed that $\gamma \ge 1$. J.T. Chayes and L. Chayes (1986) showed $\beta \le 1$. Aizenman and Barsky (1987) showed $\delta \ge 2$. Going further in this direction, there are results which show that values of critical exponents do not take their mean field values in low dimensions. Reversing historical order we note that for $d = 2$, Kesten and Zhang [92] showed $\beta < 1$, while Kesten and van den Berg [79] showed that $\delta_r \geq 2 > 1$. The latter result is based on their famous inequality that the probability for two increasing events to occur disjointly is smaller than the product of their probabilities. For an up to date account of this inequality and its remarkable generalization to arbitrary events, see the article by Borgs, Chayes, and Randall in this volume.

Somewhat more surprising than the strict inequalities in low dimensions is the statement that above the critical dimension ($d_c = 6$ for percolation) all critical exponents take their mean field values. For a long time this statement was a claim that physicists made and mathematicians couldn't prove. The first steps toward a mathematical proof were taken by Aizenman and Newman (1984) who showed that if

$$\nabla \equiv \sum_{x,y} P_{cr}(0 \to x) P_{cr}(x \to y) P_{cr}(y \to 0) < \infty \qquad (1.1.7)$$

then $\gamma = 1$. Barsky and Aizenman (1988) showed that if the "triangle condition" was satisfied, then we also have $\beta = 1$ and $\delta = 2$. The final step was taken by Hara and Slade (1989, 1990) who showed that the triangle condition held (i) for the nearest neighbor case in $d \geq d_o$ (where $d_o \leq 19$, see Hara and Slade (1994)) or (ii) in $d > 6$ for a sufficiently spread out model. To be precise, they generalized the percolation model so that connections from x to y have probability $pL^{-d}g((y - x)/L)$, where g is a nice function, and showed that $\nabla < \infty$ if $L \geq L_o(d)$. Since there is no reason to believe that percolation with range 1 is different from range 10 or 100, (ii) gives a convincing demonstration that the critical dimension is ≤ 6.

Hara and Slade proved their results with the "lace expansion" which has turned out to be a powerful technique for understanding phase transitions in other systems. Recent applications of this method to lattice trees and the incipient infinite cluster are discussed in Slade's article in this volume. The limit in Slade's article involves a functional of super-Brownian motion, a process that is the subject of articles by LeGall and Cox, Durrett, and Perkins.

Results on percolation have continued to this day to be an important part of Kesten's work. Inspired by a seminar talk Larry Shepp gave at Cornell, Kesten and I solved a problem Shepp posed about long range percolation in one dimension, see [104]. In [119] he studied with Grimmett and Zhang random walk on the infinite cluster of bond percolation on $\mathbf{Z}^d$, showing that in the supercritical regime when $d \geq 3$ this random walk is a.s. transient. This conclusion was proved by considering the infinite percolation cluster as a random electrical network in which each open edge has unit resistance and showing that the effective resistance between a nominated point and points at infinity is almost surely finite.

Recently, Kesten has with Benjamini [130] and with Sidoravicius and Zhang [149], proved some fascinating results about the question: when can one with positive probability see every infinite word of 0's and 1's from a given site in a lattice of independent $\{0, 1\}$ valued random variables? To be precise, a word is a binary sequence $(v_1, v_2, \ldots) \in \Xi = \{0, 1\}^N$ where $N = \{1, 2, 3, \ldots\}$. We say that the word is seen from x if there is a self-avoiding path starting from a neighbor of x along which we see the word. (We start at a neighbor for the trivial reason that there can be only one value at a given site.) Let $S(v)$ be the words seen from a given vertex v and $S_\infty = \cup_v S(v)$ be the words seen from some vertex.

To relate this to our previous discussion, note that the classic question of percolation can be phrased as: "Is $(1, 1, 1, \ldots) \in S(v)$?" Another variant that has been investigated is AB percolation (see Wierman and Appel (1997) and references therein). In the current setting the question may be phrased as: "Is $(1, 0, 1, 0, \ldots) \in S(v)$?" Benjamini and Kesten [130] studied the case in which 0's and 1's each had probability 1/2 in the original product measure. They showed that in $d \geq 10$, $P(S_\infty = \Xi) = 1$ while for $d \geq 40$, $P(S(v) = \Xi \text{ for some } v) = 1$.

The dimensions 10 and 40 came from considering analogous questions for oriented percolation and hence are not sharp. To approach the question from the other end, consider the triangular lattice in two dimensions. In this case, due to the lack of percolation at $p = 1/2$, $(1, 1, 1, \ldots)$ cannot be seen, but Wierman and Appel (1987) have shown $(1, 0, 1, 0, \ldots)$ can be seen. To ask how many words can be seen, we can take a comprehensive look by introducing product measures with density $0 < \beta < 1$, ν_β on Ξ and let $\rho(\xi)$ be the probability the word ξ is seen from some starting point. In [149], Kesten, Sidoravicius, and Zhang showed that $\rho(\xi) = 1$ for ν_β almost every word ξ.

Finally, while I have been writing this article, Kesten has done his best to create new results faster than I can digest what he has done. In [151] he and Zhonggen Su investigated ρ-percolation. Letting $X(e) = 1$ if the bond is open and 0 if it is closed, they asked if there is an infinite oriented path $v_0 = 0, v_1, v_2, \ldots$ starting at the origin so that $\liminf_{n\to\infty}(1/n)\sum_{i=1}^{n} X(v_{i-1}, v_i) \geq \rho$. Defining the critical value in the obvious way, they considered $D_1 = \lim_{d\to\infty} d^{1/\rho} p_c(d)$ and an analogous limit, D_2, for site percolation showing that $D_1 < D_2$ and that neither of these values is the equal to the corresponding limit for regular d-ary trees.

1.2 First Passage Percolation

Again our subject originates in England, but this time more than a half decade later in the work of Hammersley and Welsh (1965). To formulate the model in d dimensions, we again imagine that the edges connecting neighboring sites in $\mathbf{Z}^d$ are channels. However, we now assume that the fluid can flow through any channel but will take an amount of time t_e to flow through edge e, where the $t_e \in [0, \infty]$ are independent and identically distributed random times.

With this set-up it is natural to ask: at what time $\tau(x, y)$ will fluid first appear at y if we turn on a source at x at time 0? If we let $e_1 = (1, 0, \ldots, 0)$, then the passage times $a_{m,n} = \tau(me_1, ne_1)$ do not have independent increments, but by exploiting an obvious subadditivity property $a_{0,m} + a_{m,n} \geq a_{0,n}$ one can conclude (see Kingman (1968) or Chapter 5 of Smythe and Wierman (1978)) that if $Et_e < \infty$

then

$$a_{0,n}/n \to \mu \quad \text{a.s. where } \mu = \inf_{m \geq 1} Ea_{0,m}/m. \tag{1.2.1}$$

Like many other people, Harry Kesten was attracted to the subject by Smythe and Wierman's (1978) monograph. Not surprisingly, his first result in this direction in [63] was closely related to his work on percolation. He showed that the time constant μ is 0 if and only if the atom at 0 in the passage time distribution is $\leq p_T$, the critical value for infinite mean cluster size. To put this result in context, we should mention that this was proved before it was known that $p_c = p_T$.

Having mentioned the possibility of $\mu = 0$, we will now define it out of existence by supposing for the rest of our discussion that the passage time distribution has $F(0) = 0$. There is nothing special about the direction e_1 in the limit theorem quoted above. Taking inspiration from Richardson (1973), one can define a propagation speed for each direction and then patch them together to get a shape theorem for the *wet region*

$$W_t = \{x \in \mathbf{Z}^d : \tau(0, x) \leq t\}.$$

To do this let $Q = [-1/2, 1/2]^d$ be the cube of side 1 and convert the wet region at time t into a solid blob by letting $\bar{W}_t$ be the union of $x + Q$ over all $x \in W_t$. Cox and Durrett (1980) showed that for any distribution F with $\lim_{x\to\infty} F(x) = 1$ (e.g, we do not have to assume the existence of a mean) there is a limiting convex set G so that for any $\epsilon > 0$

$$P(\bar{W}_t \subset (1+\epsilon)tG, \ |\bar{W}_t/t - G| \leq \epsilon) \to 1. \tag{1.2.2}$$

Here $|A|$ denotes the Lebesgue measure of A.

An easy consequence of this result is that if we let $b(0, n) = \min\{\tau(0, x) : x_1 = n\}$ be the *point to hyperplane passage time*, then for any distribution F we have

$$b_{0,n}/n \to \mu \quad \text{a.s. where } \mu = \inf_{m \geq 1} Ea_{0,m}/m. \tag{1.2.3}$$

In contrast, some moment condition is needed to have almost sure convergence of $a_{0,n}/n$ in (1.2.1). For, otherwise, the minimum of the $2d$ bonds ending at ne_1 may be $\geq \epsilon n$ infinitely often, spoiling the convergence. For $a_{0,n}/n$, Cox and Durrett (1980) showed that the necessary condition implicit in the previous sentence is sufficient for convergence to μ almost surely. The stubborn points are responsible for little holes in the limiting shape which force the complicated formulation. I would like to thank Harry Kesten for explaining this aspect of Cox and Durrett's work to me.

The time constant $\mu = \inf_{m\geq 1} Ea_{0,m}/m$ is a mysterious object. One can, with considerable pain, get upper bounds on μ by estimating $Ea_{0,n}/n$ for $n = 1$ or 2. However, to my knowledge it cannot be computed exactly in any case in which $\mu > \inf\{x : F(x) > 0\}$. Cox (1980) was the first to investigate continuity properties of μ as a function of the underlying distribution F. He proved that if the F_n were dominated by a single distribution G with finite mean and $F_n \Rightarrow F$ in the sense of weak convergence, then $\mu(F_n) \to \mu(F)$. As the "no moment" result in (1.2.3) might suggest, the domination condition is not needed. It was removed by Cox and Kesten [70] who showed that if $F_n \Rightarrow F$, then $\mu(F_n) \to \mu(F)$.

Eden (1961) was interested in the Markovian case of first passage percolation in which each edge had a mean one exponential distribution. Early simulations, and a little wishful thinking, suggested that the limit shape might be a ball. However in the mid 1980's a super-computer solved the problem by showing there was roughly a 2% difference between the speeds along the axis and on the line at 45 degrees. (See Zabolitsky and Stauffer (1986a,b).) In words, the fact that the L^1 distance to $(n/\sqrt{2}, n/\sqrt{2})$ is $\sqrt{2}n$ rather than n to the point $(n, 0)$ is almost exactly compensated by the fact that there are many more paths of minimum length to the first point. Kesten showed in Section 8 of [78] that in high dimensions the balance between these forces breaks down. If we suppose that the underlying distribution F has density function $F'(0) = 1$, then the time constant is of order $(\log d)/d$ while the passage time in the direction $(1, 1, \dots, 1)/\sqrt{d}$ is of order $1/d$.

There is only one special case in which we have some rigorous concrete information about the limiting shape G in (1.2.2). Consider two dimensions for simplicity and suppose that $P(t_e \geq 1) = 1$ and $P(t_e = 1) = p$. Since the fluid can move at most one unit per time, the limiting set must be contained in the diamond $\{(x, y) : |x|+|y| \leq 1\}$. Durrett and Liggett (1981) showed that if p was larger than the critical value for oriented percolation in two dimensions, then the boundary of the limiting shape contained an interval in $x + y = 1$.

Closely related to the topic of first passage percolation is the notion of random resistor networks. To formulate the model in two dimensions, we imagine that the edges connecting adjacent sites in the square lattice are resistors with random resistances r_e that are independent and identically distributed random values $\in [0, \infty]$. Grimmett and Kesten [74] investigated the bulk properties of random resistor networks consisting of $n \times n$ chunks of the square lattice, with a special interest in the case in which the values 1 and ∞ had probabilities p and $1-p$. They also looked at the flow through networks where edges have random capacities. This study involved a look at large deviations probabilities for the various passage times. In a second study [75] they examined properties of random electrical networks on complete graphs. The reader should not be surprised to hear that the results are more precise and detailed in this context.

Kesten's work on percolation and first passage percolation earned him his second invitation for a 45 minute lecture at the International Congress of Mathematicians, which met in Warsaw in 1983. The odd numbered year is not a typo. The Congress was delayed for a year due to political unrest in Poland. Like many mathematicians, Kesten showed his solidarity with Solidarity by not going to the Congress.

In 1984 Kesten lectured (with René Carmona and John Walsh) at École d'Été de Probabilités de Saint Flour XIV. Kesten's lecture notes cover many of the topics we have referred to above and in many cases present new refinements. We have already mentioned the asymptotics for large dimensions that were given in Section 8 of his notes. The work of Grimmett and Kesten [74] is taken further in Section 5 by giving large deviations results and rates of convergence of $Ea_{0,n}/n$ to the passage time, and in Section 7 with a look at convergence rates in the case $\mu = 0$ in $d = 2$. For more recent results on the last topic, see Kesten and Zhang [142].

The rate of convergence to the time constant given in (5.16) of the St. Flour notes was crude: $O((\log n)^{-1/(9d+3)})$ but it took several years before Kesten [117] and Alexander (1993) could improve that bound to the very respectable $O(n^{-1/2} \log n)$. Kesten showed in [117] that the fluctuations $a_{0,n} - Ea_{0,n}$ are at most diffusive, i.e., the variance of $a_{0,n}$ is $\leq Cn$. Novice readers might expect to hear next of a central limit theorem being proved. However, physicists tell us (see Kardar, Parisi, and Zhang (1986), Zabolitsky and Stauffer (1986a,b), and Krug and Spohn (1991)) that in two dimensions the standard deviation of the first passage time $t(0, (n, 0))$ is of order $n^{1/3}$.

The fluctuations in the passage times to ne_1 or, more geometrically, of the boundary of the wet region at time n, $\bar{W}_n$, can be used to define a critical exponent, χ, by declaring that they are $O(n^{\chi})$. In this new notation, Kesten's result is that $\chi \leq 1/2$, while physicists claim that $\chi = 1/3$. Lower bounds on the fluctuations have turned out to be more difficult. Pemantle and Peres (1994) and Newman and Piza (1995) have shown that, in $d = 2$, fluctuations diverge at least logarithmically fast. This result can be improved if one is willing to introduce hypotheses that seem reasonable, but that cannot at the moment be proved. Wehr and Aizenman (1990) studied an exponent ξ, defined so that the point the wet region first touches the hyperplane $x_1 = n$ is $O(n^{\xi})$, and proved that

$$\chi \geq \frac{1 - (d-1)\xi}{2}. \tag{1.2.4}$$

Combining this with Newman and Piza's (1995) result $\xi \leq 3/4$ in $d = 2$ gives $\chi \geq 1/8$. Weaker versions of the last conclusion can be proved without invoking any unverified hypotheses.

Seeking to understand the spread of first passage percolation, Newman (1995) introduced a graph that consists of the union of the time minimizing paths from one fixed point, say the origin, to all of the other points. It is easy to see that this graph must be a spanning tree, and that the spanning tree must have at least one infinite path, which Newman called a one sided geodesic. Proving that one sided geodesics exist going in all directions, or the more mysterious claim by physicists that two sided geodesics do not exist, has proved to be difficult. See Licea and Newman (1996). In this volume, Howard and Newman report on recent progress for models that take place on $\mathbf{R}^d$. Here rotational invariance can be used to great advantage once one pays the price of generalizing the lattice results to the new setting.

The exact solution for the critical value of two dimensional percolation rests on a duality between planar graphs. If one considers bond percolation in the three dimensions, then a natural dual two-dimensional object is a family of two dimensional plaquettes, i.e., the squares with side 1 perpendicular to the mid-point of the segments from x to $x + z$ where z is one of the six nearest neighbors of the origin: $(1, 0, 0)$, $(-1, 0, 0)$, $(0, 1, 0)$, $(0, -1, 0)$, $(0, 0, 1)$, and $(0, 0, -1)$. Aizenman, Chayes, Chayes, Fröhlich, and Russo (1983) showed that if we make plaquettes occupied or vacant with probabilities p and $1 - p$ and consider the event that there is a surface of occupied plaquettes with boundary exactly equal to a given rectangle, then the probability is of order exp(−area) if $p < p_c$, while it is of order exp(−surface) when $p > p_c$.

Inspired by this, and using analogies with the max-flow/min-cut theorem, Kesten considered in [87] a first passage percolation theory for random surfaces, i.e., the minimal cost surface that can be constructed with boundary equal to a given rectangle of side n. In three dimensions the asymptotic cost of such a loop is proportional to the area, and dividing by n^2 leads to a limit. This may all sound very straightforward, but it definitely is not since the topology of surfaces in $\mathbf{R}^3$ rears its ugly head. Kesten's paper was an important first step, but much remains to be done. A proper understanding of the issue involved would probably help us sort out the asymptotic behavior of the contact process on $[0, L]^d$. Specifically, the problem for the contact process is to show that if τ_L is the time the process dies out (i.e., reaches all sites vacant) starting from all sites occupied, then

$$(1/L^2) \log \tau_L \to \gamma \tag{1.2.5}$$

in probability as $L \to \infty$. This is known to be true in $d = 1$, see Durrett and Schonmann (1988); but only partial results exist in $d > 1$, see Mountford (1993).

In an opposite direction from the concept of surfaces with minimal weights is the notion of greedy lattice animals. To set up the problem, suppose we have i.i.d. positive random variables for the sites in the d-dimensional integer lattice, $\{X_v : v \in \mathbf{Z}^d\}$. Let M_n be the largest sum that we can get from a self-avoiding path of length n containing the origin, and let N_n be the largest sum for an animal (connected subset) of size n containing 0. In joint work with Cox, Gandolfi, and Griffin [121] and with Gandolfi [122], Kesten showed that if $EX_i^{d+a} < \infty$ for some $a > 0$, then $M_n/n \to \mu$ and $N_n/n \to \nu$ almost surely.

In an opposite direction from the notion of greedy lattice animals is that of minimal spanning trees. Let $X_1, X_2, \ldots$ be independent and identically distributed with common distribution μ that has support in $[0, 1]^d$, and choose a spanning tree T to minimize $M_\alpha = \sum_{e \in T} |e|^\alpha$ where $|e|$ is the length of the edge. Steele (1988) has shown that if $0 < \alpha < d$, then the minimum length satisfies

$$n^{-(d-\alpha)/d} M_\alpha \to c(\alpha, d) \int f(x)^{(d-\alpha)/d}\, dx \quad \text{a.s.} \tag{1.2.6}$$

where f is the density of the absolutely continuous part of μ and $C(\alpha, d)$ is a constant that only depends on α and d.

Aldous and Steele (1992) complemented this result by showing that for the uniform distribution when $\alpha = d$, $M_d \to c(d, d)$ in L^2. The central limit theorem had to wait for a while until Alexander (1996) and Kesten and Lee [137] independently showed that if μ is the uniform distribution, then for any $\alpha > 0$

$$n^{-(d-2\alpha)/2d} (M_\alpha - EM_\alpha) \Rightarrow \text{normal}(0, \sigma_{\alpha,d}^2) \tag{1.2.7}$$

where again $\sigma_{\alpha,d}^2$ is a constant that only depends on α and d.

Reversing direction yet again, we will motivate the last two topics in this section, by noting that (i) Eden's growth model can be thought of as a continuous time process in which vacant sites become occupied at a rate equal to the number of occupied neighbors, and (ii) if we assign the passage times to the sites instead and look at the embedded discrete time chain, then Eden's model becomes a process in which at each step a randomly chosen vacant site on the boundary of the wet region becomes occupied.

Taking (ii) first, we consider a new, more complicated model, called diffusion limited aggregation or DLA, in which boundary sites are added according to a non-uniform rule. Let $A_0 = \{0\}$, i.e., just the origin. Having defined A_n for $n \geq 0$, A_{n+1} is formed from A_n by releasing a particle at ∞ and letting it perform a nearest neighbor symmetric random walk on $\mathbf{Z}^d$ until it reaches a site on the boundary of A_n. Simulations of this process produce starfish like creatures. The intuition is easy to see: once arms with narrow valleys between them form, random walks are more likely to attach near the tips rather than traversing the fjords to get stuck in the interior.

It is a very difficult unsolved problem to show that arms form and the diameter of DLA grows at rate $n^{0.5+\epsilon}$. Kesten proved a result in the opposite direction in [93] showing that the arms of DLA can't grow any faster than $Cn^{2/3}$. In hindsight the answer is easy to see: the enhancement of adding at the tip is maximized if the configuration is always an interval (or a plus sign), and in this case the growth rate for the radius is 2/3. The proof of this result involves interesting estimates for hitting probabilities of random walks on $\mathbf{Z}^d$ [94] and has led to relationships between solutions to discrete and continuous Dirichlet problems [112]. Physicists, in their quest for new and exciting pictures generalized the original model to include versions where the probability of attachment is a power η of the "harmonic measure," i.e., the hitting distribution for the random walk. Kesten [113] kept up with them as best as he could, proving results to explain their simulations.

Returning now to (i), Kesten and Schonmann [133] considered a variant of Eden's growth model in which each site on $\mathbf{Z}^d$ becomes occupied at rate 1 if the site has at least θ occupied neighbors, at rate ϵ if at least one but $< \theta$ occupied neighbors, and at rate 0 if it has no occupied neighbor. In the case $\theta = 2$ they were able to show that the asymptotic growth rate for the model was $O(\epsilon^{1/d})$, i.e., was bounded above and below by constant multiples of this quantity, and that the limiting shape after rescaling is a cube as $\epsilon \to 0$. The model with $\theta = 3$ is not interesting in two dimensions, but in $d = 3$ presents an intriguing open problem that seems to have connections to so-called bootstrap percolation. See Aizenman and Lebowitz (1988).

1.3 Random Walks

Having taken two trips through the most recent twenty years of Kesten's work, we now go back to a time when symmetric random walk transition probabilities were better known as Toeplitz matrices: [11, 12, 18]. The natural place to start is with

Kesten's Ph.D. thesis in which he considered symmetric random walks on groups. Let G be a countable group and let $p = (p_x)_{x \in G}$ be a symmetric probability distribution whose support generates G. Consider the random walk on G in which every step corresponds to right multiplication by x with probability p_x. Kesten's thesis explored connections between the spectrum of the transition probability (as an operator on $L^2(G)$) and the structure of the group G. The operator is self-adjoint (since the random walk is symmetric) and has norm ≤ 1, so its spectrum is a subset of $[-1, 1]$. Kesten showed that the spectral radius $\lambda(G, p)$ is the maximal value in the spectrum and

$$\lambda(G, p) = \lim_{n \to \infty} a_{2n}^{1/2n}$$

where a_{2n} is the probability that the walk is back at the origin after $2n$ steps. Thus $\lambda(G, p) < 1$ if and only if $a_{2n} \to 0$ exponentially fast.

Kesten was especially interested in determining when $\lambda(G, p) = 1$. He showed that this is a property of G alone and does not depend on the choice of p. He studied what happened for finite direct products and proved comparison results between groups and their normal subgroups and quotients. Using that machinery he showed that if G is an Abelian group, then $\lambda(G, p) = 1$, while if $\lambda(G, p) = 1$, then G has no free subgroups on more than one generator. He calculated $\lambda(G, p)$ explicitly if G is free and p assigns equal probability to the generators and their inverses. Later in [5] he showed that $\lambda(G, p) = 1$ if and only if G is amenable. This famous result is now known as Kesten's criterion for amenability.

Kesten returned to random walks on groups in [32] where he considered, among other things, the question of recurrence. Kesten's results and questions inspired a generation of workers, culminating in Varopoulos's beautiful solution to "Kesten's conjecture": Simple random walk on a finitely generated group G is recurrent if and only if G is virtually Abelian of rank ≤ 2.

From the first few entries in his publication list, you can see that even as a young man Kesten was already hard at work solving other people's problems. An interesting thread in his early work is what one might call ergodic number theory. Kac and Kesten [3] considered the transformation $Tx = 1/x - [1/x]$ of the unit interval, where $[1/x]$ is the integer part of $1/x$ and $[1/T^{n-1}x]$ gives the digits in the continued fraction representation of x. Using Lévy's result that T is rapidly mixing, they were able to show that the number of times a specified digit occurs among the first n digits in a continued fraction representation is asymptotically normally distributed. This was published in the Bulletin of the American Mathematical Society, although they later learned that the result was due to Doeblin (1940).

A second set of results in this direction concern uniform distribution mod 1. Let $f(\xi)$ be the indicator function of the interval $[0, t]$ extended to be periodic with period 1, that is, $f(\xi + 1) = f(\xi)$. In [8] and [13] Kesten showed that if X and Y are independent and uniform on $[0, 1]$, then

$$(\log n)^{-1} \sum_{k=1}^{n} \{f(Y + kX) - t\} \tag{1.3.1}$$

has a limiting Cauchy distribution. This contrasts sharply with the behavior Kac (1946) observed for lacunary series, where $n^{-1/2}\sum_{k=1}^{n}\{f(2^k X)-t\}$ converges to a normal distribution. Fine (1954) sharpened Kac's result by proving convergence of finite dimensional distributions for the process indexed by t, while Ciesielski and Kesten [15] established tightness to complete the proof of weak convergence to a limiting Gaussian process. Related results and refinements were proved by Kesten in [14, 24, 29], and [30].

Like a good mystery story, parts of Kesten's early work foreshadow later developments. Inspired by work of Bellman (1954) for i.i.d. sequences, Furstenberg and Kesten [19] considered products ${}^nY^1 = X^n X^{n-1}\cdots X^1$ where the X^i are an ergodic stationary sequence of $k\times k$ matrices, and showed that if $E(\log^+\|X^1\|)<\infty$, then with probability one

$$\lim_{n\to\infty} n^{-1}\log\|{}^nY^1\| = \lim_{n\to\infty} n^{-1}E\log\|{}^nY^1\|. \tag{1.3.2}$$

Nowadays this is a textbook application of Kingman's (1968) subadditive ergodic theorem. See e.g., pages 367–369 of Durrett (1995b). While the above law of large numbers for the norm of the matrix, $\|{}^nY^1\|$ and related results for the entries ${}^nY^1_{ij}$ are widely known and definitive, the corresponding central limit question has not been much investigated (see [9] and Ishitani (1977)) and still has room for improvement. For more recent work on products of random matrices see Kesten and Spitzer [76], Cohen and Newman (1984), and the collection of papers from a 1984 AMS Summer Research Conference edited by Cohen, Kesten, and Newman (1984).

A second harbinger of the future, again related to subadditivity, is Kesten's work on self-avoiding walks in [22] and [23]. Let χ_n be the number of self-avoiding walks on the integer lattice in d dimensions that start at the origin. The obvious inequality $\chi_m\chi_n\le\chi_{m+n}$ leads one easily to the conclusion that

$$(\chi_n)^{1/n}\to\beta_d\equiv\inf_{m\ge1}(\chi_m)^{1/m}. \tag{1.3.3}$$

It has long been conjectured (see Hammersley (1961) and references therein) that the ratio $\chi_{n+1}/\chi_n\to\beta_d$. Kesten [22] proved more and less than this when he showed that

$$|\chi_{n+2}/\chi_n-\beta_d^2|\le Cn^{-1/3}. \tag{1.3.4}$$

The result in (1.3.3) clearly allows one to compute upper bounds on β_d by calculating χ_m for small values of m. Results about the limit β_d are much harder to come by. Kesten [23] attacked this question by considering $\chi_{n,2r}(d)$ = the number of n-step walks on the integer lattice in d- dimensions with no loops of $2r$ steps or less. He showed that $\beta_{d,2r}=\lim_{n\to\infty}(\chi_{n,2r}(d))^{1/n}$ existed and satisfied $\beta_{d,2r}-\beta_d=O(d^{-r})$ as $d\to\infty$. Taking $r=2$ leads to an asymptotic expansion

$$\beta_d=2d-1-\frac{1}{2d}+O(1/d^2). \tag{1.3.5}$$

In the three and a half decades since Kesten's paper there has been an explosion of results on self-avoiding random walks and related processes. See the books by

Madras and Slade (1993) and Lawler (1991), and the paper by Lawler on "loop-erased walk" in this volume.

Returning to ordinary random walks on the integer lattice, let $p^k(x, y)$ be the probability of going from x to y in k steps, let T be the time of the first visit to the origin, 0, and let $r_n = P_0(T > n)$. Kesten, Ornstein, and Spitzer [17] proved that for all $x \neq 0$,

$$\lim_{n\to\infty} \frac{P_x(T > n)}{P_0(T > n)} = a(x), \tag{1.3.6}$$

where $a(x) = \sum_{k=0}^{\infty} p^k(0, 0) - p^k(x, 0)$, which exists by earlier work of Spitzer (1962). Note that the last result holds for ANY random walk. This was the first of many ratio limit theorems for arbitrary random walks. See Kesten and Spitzer [20], and Kesten [21, 38]. These three papers cover more than 100 pages in *Journal d'Analyse Mathématique*, so it would be difficult to even sketch their contents here.

Inspired by the pioneering work of Hunt (1957–1958) developing a potential theory for transient Markov process, the 60's were the golden age of potential theory of random walks. Using the notation of the previous paragraph, this can be defined as the study of the potential kernels $\sum_{k=0}^{\infty} p^k(0, x)$ and $\sum_{k=0}^{\infty} p^k(0, 0) - p^k(x, 0)$, the former appropriate for the transient and the latter for the recurrent case. Spitzer's beautiful (1964) book contained definitive results for random walks on the d-dimensional integer lattice, which Kesten and Spitzer [26] generalized to countably infinite Abelian groups. Ornstein (1969) and Port and Stone (1969) generalized this work to R^d and to locally compact Abelian groups.

The work of Kesten and Spitzer [26] referred to in the previous paragraph led to Kesten's paper on "The Martin boundary of recurrent random walks on countable groups" which appeared in the Proceedings of the 5th Berkeley Symposium. The breadth and depth of talent at that meeting can be illustrated by noting that Part II of Volume II featured papers by Blumenthal and Getoor, Breiman, Dynkin, Kakutani, Karlin and McGregor, Kesten, Kendall, Kunita and Watanabe, Lamperti, Neveu, Ornstein, Ray, Rosenblatt, Smith, and Spitzer.

Kesten gave a 45 minute talk at the 1970 International Congress in Nice [39] on "Hitting of sets by processes with stationary independent increments." The highlight of that talk was his result announced earlier in the Bulletin of the AMS [36] and presented in detail in a 129 page volume of the Memoirs of the AMS [37] giving necessary and sufficient conditions for processes with independent increments to hit points with positive probability. As Kesten [36] explains in his announcement (see that paper for precise references) he was motivated by earlier work of Lévy, Erdös, Kac, and Port, who resolved the question for symmetric stable processes, and by a convolution equation of Chung that P.A. Meyer had shown was related to the probabilities of hitting points. From the list of people who had worked on the problem and the fact that Neveu and McKean had already published false solutions of Chung's problem, you can see that its solution was an impressive achievement. With his characteristic modesty, Kesten told me when I was quizzing him about some of the details of his early work, that "I was very happy when I was able to solve that problem." Kesten did other work in this general area concerning "positivity intervals for stable processes" [19], finding results that generalized the arcsine law, and on "Lévy processes with a nowhere dense range" which considered related questions concerning the complement of the range.

Continuing to track Kesten's career through his prestigious lectures, our next stop is his 1971 Rietz Lecture given at the annual meeting of the IMS in Fort Collins Colorado, September 20–23, 1971. As the associated paper [45] indicates, his intent was to survey generalizations or analogues of classical limit theorems which do not make a priori moment or smoothness assumptions on the underlying distribution, e.g., results that hold for any random walk. Three topics were discussed: (i) ratio limit theorems, which we have touched on above; (ii) a concentration function inequality that gives an upper bound on the probability a random walk lies in an interval of length L, and (iii) the set of accumulation points of normalized random walk, i.e, given a normalizing sequence γ_n,

$$A(F, \gamma_n) = \cap_{m=1}^{\infty} \overline{\{S_n/\gamma_n : n \geq m\}} \tag{1.3.7}$$

where the bar denotes closure, and F is the distribution function for a single step.

The main result in category (ii) is the one proved in [35]. Its name is a mouthful: "A sharper form of the Doeblin-Lévy-Kolmogorov-Rogozin inequality" but I have fond memories of it, since it was very useful in Bramson, Durrett, and Swindle (1989) by providing uniform upper bounds for the sequence of random walks we considered. Restricting our attention to the simplest case of one dimensional random walks, the results in (iii) start from the simple observation that if F has finite mean μ, then the law of large numbers implies $A(F, n) = \{\mu\}$. The single points $+\infty$ and $-\infty$ are obviously possible limits. Kesten showed in [34] that if $A(F, n)$ contains at least two points, then it must contain $+\infty$ and $-\infty$. Conversely, any closed set A of $[-\infty, \infty]$ that contains $+\infty$ and $-\infty$ is $A(F, n)$ for some distribution F.

Turning to smaller normalizations, we note that if F has mean 0 and finite variance σ^2, then Strassen's (1964) version of the law of the iterated logarithm implies

$$A(F, (2n \log\log n)^{1/2}) = [-\sigma, \sigma]. \tag{1.3.8}$$

In [41] Kesten solved his own open problem by showing that (assuming F is not a point mass at 0) (a) if $\alpha < 1/2$ and $A(F, n^{-\alpha})$ has a finite limit point, then it contains all real numbers, while (b) if $A(F, n^{-1/2})$ has a finite limit point, then it contains a half line of values $(-\infty, b]$ or $[b, \infty)$. It is natural to conjecture, as Kesten did in [41], that in case (b) $A(F, n^{-1/2}) = [-\infty, \infty]$, but this seems to be an open problem.

Erickson and Kesten [49] introduced the notion of strong limit points of random walks as the set of values $B(F, n^{-\alpha})$ so that $n_k^{-\alpha} S_{n_k} \to b$ for some deterministic sequence n_k. If $\alpha \leq 1/2$ and F is not a point mass at 0, $B(F, n^{-\alpha}) = \emptyset$. Depending upon F and the value of α, $B(F, n^{-\alpha})$ may be $\emptyset$, $\{\infty\}$, $\{-\infty\}$, $[0, \infty]$, $[-\infty, 0]$, or $\mathbf{R}$. Four theorems and five examples in [49] carefully described the possible behaviors, though in words we have heard more than once in Kesten's work "we do not give the proof of Theorem 4 since it is rather lengthy." For more on strong limit points, see Kesten and Maller [144].

Our next topic, Random Walks in Random Environments, are not random walks at all. In the formulation of Solomon's Ph.D. thesis (1975) they are discrete time birth and death chains X_m on the integers in which $x \to x+1$ has probability α_x and $x \to x-1$ has probability $1-\alpha_x$, where the environment α_x is a sequence of independent and identically distributed random variables, and we suppose for simplicity here that $0 < \epsilon \le \alpha_x \le 1-\epsilon < 1$. Let $\sigma = (1-\alpha_0)/\alpha_0$. It is a straightforward exercise in the theory of birth and death chains to show that X_n is recurrent if and only if $E \ln \sigma = 0$. However, this simple problem and some related questions about branching processes in random environments inspired Kesten [48] to produce some very nice results on "random difference equations."

Returning to the original problem, things become very interesting when one considers limit theorems. Solomon (1975) showed that

$$\text{If } E\sigma < 1 \text{ then } \lim_{n\to\infty} X_n/n = (1-E\sigma)/(1+E\sigma).$$

$$\text{If } E(\sigma^{-1}) < 1 \text{ then } \lim_{n\to\infty} X_n/n = -(1-E(\sigma^{-1})/(1+E(\sigma^{-1}).$$

$$\text{If } (E\sigma)^{-1} \le 1 \le E(\sigma^{-1}) \text{ then } \lim_{n\to\infty} X_n/n = 0.$$

Kesten, Kozlov and Spitzer [53] probed the middle ground where $E \ln \sigma < 0$ but $E\sigma \ge 1$. In this case if one defines κ by $E\sigma^\kappa = 1$, then

$$\lim_{t\to\infty} P(t^{-\kappa} X_t \le x) = 1 - L_\kappa(x^{-1/\kappa}), \tag{1.3.9}$$

where L_κ is the stable law with index κ. Other non-normal limit theorems were found for $1 \le \kappa \le 2$ in [53] and generalized by Kesten and Kawazu [77]. Ritter (1976) proved some results about the critical case $E \ln \sigma = 0$ in his thesis, but a complete solution had to wait until Sinai (1982) showed that $(\log n)^{-2} X_n$ converges in distribution to a nondegenerate limit defined in terms of a functional of a Brownian motion associated with the environment. The distribution of the limit was later calculated by Kesten [84].

If one drops the assumption of nearest neighbor jumps in $d = 1$, the problem becomes technically more difficult. Key's (1984) thesis gives results for the finite range model in $d = 1$. The answers are not as explicit as in the nearest neighbor case since they are most naturally framed in terms of Lyapunov exponents of random matrices, which typically cannot be computed explicitly. While the finite range case in $d = 1$ is hard, the nearest neighbor model in $d > 1$ proved to be almost impossible. Some remarkably clever arguments were used by Kalikow in his (1981) thesis to prove transience of some lopsided models in $d = 2$. However, little was known in $d \ge 2$ until Bricmont and Kupiainen (1991), (1992) used rigorous renormalization group methods to show that the critical dimension was 2, i.e., one has central limit theorem behavior in $d > 2$. For an overview of this and related work see the text of Kupiainen's (1990) talk at the International Congress of Mathematicians in Kyoto.

Kesten's two papers with Papanicolaou, [60] and [64], studied a different type of motion in a random environment. The first studied turbulent diffusion, that is solutions to $dx(t)/dt = V(x(t))$ with $V(x) = v + \epsilon F(x)$ where $v \neq 0$, ϵ is small, and F is a mean zero stationary random vector field. When F satisfied suitable hypotheses they were able to show that as $\epsilon \to 0$, $x^\epsilon(t) = x(t/\epsilon^2) - vt/\epsilon^2$ converged to a diffusion process with constant coefficients that came from averaging F. The second paper, [64], used similar methods to study stochastic acceleration $d^2x(t)/dt^2 = \epsilon F(x(t))$.

Readers who recall that Kesten's result "$p_c = 1/2$" is in [67] realize that we have now almost reached the beginning of Kesten's work on percolation and first passage percolation. He still wrote beautiful papers on questions about random walks: [58] proves a conjecture of Erickson to the effect that any genuinely d-dimensional random walk S_n in $d \geq 3$ goes to infinity at least as fast as simple random walk. However, increasingly his work on random walk was motivated by ideas from physics. An example is his work with Spitzer [62] on random walk in random scenery. They studied the limiting behavior of sums of the form $W_n = \sum_{k=1}^n \xi(S_k)$ where S_k is a random walk on the integers and the $\xi(x)$ are i.i.d. and independent of S_k. They found that $n^{-3/4} W_{nt}$ converged weakly to a limit Δ_t that had stationary increments and was self-similar, i.e., Δ_{ct} has the same distribution as $c^{3/4}\Delta_t$.

Physics was not the only science to provide Kesten with problems. He had been for some time interested in models for population growth. See [40, 42, 46], and [55]. In [65] and [66] he studied the number of alleles in the stepwise mutation model. Simulations of Ohta and Kimura had suggested that the number of different alleles $\Lambda(N)$ found in a population of size N remained bounded in distribution as $N \to \infty$, but Kesten showed that it went to infinity very slowly. To state his result, we begin with the rapidly increasing sequence defined by $\gamma_0 = 0$ and $\gamma_{k+1} = \exp(\gamma_k)$ for $k \geq 0$. This begins

$$\gamma_1 = e, \quad \gamma_2 = 15.15, \quad \gamma_3 = 3,814,279, \quad \gamma_4 > 10^{1,656,620}$$

so the inverse function $\lambda(n) = \max\{k : \gamma_k \leq n\}$ grows very slowly. Kesten's result says (in the symmetric nearest neighbor case) that

$$P(|\Lambda(N) - \lambda(N)| > \log \lambda(N)) \to 0$$

as $N \to \infty$. A second contact with biology can be seen in his work with Ogura [69] giving recurrence properties of Lotka-Volterra models with random fluctuations.

From biology we move next to the study of river networks. The problem studied in [107] came from a sabbatical visit to Cornell by Ed Waymire. The problem may be formulated in a purely mathematical way as follows. Consider the family tree of a branching process starting from a single progenitor and conditioned to have ν edges (total progeny). To each edge e we associate a weight $W(e)$ which we think of as the length of the edge. Interest then focuses on the height of the tree, i.e., the maximum sum of weights that can be achieved by a self- avoiding path starting at the progenitor. Kesten refined the results in [126, 134, 135]. Some of this later work was inspired by connections with Aldous' (1993) continuum random tree and LeGall's (1991) random snake construction of super-processes.

Our next paper [110] involves a sabbatical visit to Cornell by Greg Lawler and a letter from Spataru to Spitzer. Spataru's question, after a little rewriting, asks: Suppose that a casino offers p fair games. Can we make money playing the games (a) according to a fixed schedule or (b) using a strategy that depends on our wealth? Somewhat surprisingly the answer to (a) is yes. Let $\alpha < 1/2$. If we choose p large enough and construct the fair games carefully then our fortune at time n will have $\liminf S_n/n^\alpha = \infty$. In the other direction if the fair games all have finite variance then the answer to (b) is No. A more detailed study of these questions was carried out in [112].

The phenomena in the last paragraph come from the fact that the behavior of the first n steps of a random walk is dictated primarily by the part of the distribution F between $F^{-1}(1/n)$ and $F^{-1}(1-1/n)$ and this truncated distribution may have a mean different from 0. This trimming of the distribution may be used for good rather than evil. In the 70's and 80's statisticians realized that the removal of outliers from a random sample led to robust estimators with reduced variability, and probabilists realized that trimming could produce central limit behavior from distributions with even heavy tails.

Given Kesten's expertise with random walks, it was natural for him to get involved in this area, where much of his work has been done in collaboration with Ross Maller. The eight papers cited in this paragraph total almost 300 journal pages, so we will just mention some random results to arouse the reader's interest. [114] and [115] concern conditions which guarantee that a sum of independent random variables is much larger than the largest summand. These results are of interest in relation to the law of large numbers, central limit behavior, and law of the iterated logarithm. In [115] a necessary and sufficient condition for $P(S_n \geq 0) \to 1$, answering a simple sounding problem first mentioned by Révész. [125] and [131] show that deleting a fixed finite number of terms cannot affect asymptotic normality of normed sums, that is, the trimmed sum has a limit if and only if the untrimmed one does. In the other direction, [128] shows that a fixed trimming can have a significant effect. The original random walk may be recurrent while the trimmed one is transient. [141] studies questions from the renewal theory for random walks with $S_n \to \infty$. [145] and [146] concern random walks crossing curved, e.g., power law, boundaries.

A somewhat less technical problem, at least in its formulation, is the question: can you distinguish sceneries by observing them along a random walk path? Given is a sequence of random variables $\xi_x, x \in \mathbf{Z}$ taking values in a finite alphabet and a symmetric nearest neighbor random walk S_n on the integers starting at 0. A robot walks according to S_n and calls out the symbols she sees: $\xi(S_0), \xi(S_1), \ldots$. The question is: can we reconstruct the underlying scenery from this information? Partial results can be found in [138, 139], and [147]. However, it was Harry's student Henry Matzinger who was finally able to answer the question in the affirmative in his Ph.D. thesis. For three or more symbols this involves the pretty idea of using the observed sequence to define a self-intersecting path on a tree in order to disentangle the underlying sequence. We leave it to the reader to fill in the remaining details and to contemplate the more complicated case of an alphabet with two symbols.

In the other direction, if you can't do the case in which the ξ_x are i.i.d. uniform on (0,1), you should not operate a motorized vehicle.

The last stop on our random walk through the theory of random walks is Kesten's recent work [150] with van den Berg on the asymptotic density in a coalescing random walk model. Particles perform continuous time random walks on $\mathbf{Z}^d$ but interact only when a particle jumps onto a site at which there are j particles present, in which case the jumping particle is removed with probability p_j. If we start with at most one particle per site and have $p_1 = 1$, this is classical coalescing random walk. In this case asymptotics for the density are known from work of Sawyer (1979) and Bramson and Griffeath (1980). Kesten and van den Berg show (under some natural assumptions) that in $d \geq 6$ the density of particles $u(t) \sim C(d)/t$. The last sentence should be correct with 6 replaced by 3. For the reader who wants to study this question I have some good news and some bad news: it would even be interesting to extend the methods of [151] to prove the Bramson-Griffeath-Sawyer result in $d \geq 3$.

1.4 Denouement

At this point we have exhausted the author, but not Kesten's publication list. I would like to thank Maury Bramson, Ken Brown, Geoff Grimmett, Harry Kesten, Ross Maller, Chuck Newman, Yuval Peres, and Gordon Slade for reading various drafts and making numerous corrections. They are, of course, responsible for all errors that remain, even in the parts that they never read. I would like to express my appreciation to Harry Kesten, not only for the lessons he gave me in connection with rewriting this paper, but also for his insights and his friendship for the twenty years I have known him. He is not only a brilliant mathematician, but also one of the nicest people you could hope to meet.

REFERENCES

Aizenman, M. and Barsky, D.J. (1987) Sharpness of the phase transition in percolation models. *Comm. Math. Phys.* **86**, 1–48

Aizenman, M., Chayes, J.T., Chayes, L., Föhlich, J., and Russo, L. (1983) On a sharp transition from area law to perimeter law in a system of random surfaces. *Comm. Math. Phys.* **92**, 19–69

Aizenman, M., Chayes, J.T., Chayes, L., and Newman, C.M. (1988) Discontinuity of the magnetization in the one dimensional $1/|x-y|^2$ Ising and Potts models. *J. Stat. Phys.* **50**, 1–40

Aizenman, M., and Lebowitz, J. (1988) Metastability effects in bootstrap percolation. *J. Stat. Phys.* **21**, 3801-3813

Aizenman, M. and Newman, C.M. (1984) Tree graph inequalities and critical behavior in percolation models. *J. Stat. Phys.* **36**, 107–143

Aizenman, M. and Newman, C.M. (1986) Discontinuity of the percolation density in one dimensional $1/|x-y|^2$ percolation models. *Comm. Math. Phys.* **107**, 611–647

Aldous, D. (1993) The continuum random tree III. *Ann. Prob.* **21**, 248–289

Aldous, D. and Steele, J.M. (1992) Asymptotics for Euclidean minimal spanning trees on random points. *Prob. Th. Rel. Fields.* **92**, 247–258

Alexander, K.S. (1993) A note on some rates of convergence for first passage percolation *Ann. Appl. Prob.* **3**, 81–90

Alexander, K.S. (1996) The RSW theorem for continuum percolation and the CLT for Euclidean minimal spanning trees. *Ann. Appl. Prob.* **3**, 1033–1046

Barsky, D.J. and Aizenman, M. (1988) Percolation critical exponents under the triangle condition. *Preprint.*

Bezuidenhout, C. and Grimmett, G. (1990) The critical contact process dies out. *Ann. Prob.* **18**, 1462–1482

Bezuidenhout, C. and Grimmett, G. (1991) Exponential decay for subcritical contact and percolation processes. *Ann. Prob.* **19**, 984–1009

Bramson, M., Durrett, R., and Swindle, G. (1989) Statistical mechanics of crabgrass. *Ann. Prob.* **17**, 444-481

Bramson, M. and Griffeath, D. (1980) Asymptotics for interacting particle systems on $\mathbf{Z}^d$. *Z. fur Wahr.* **53**, 183–196

Bricmont, J. and Kupiainen, A. (1991a) Renormalization-group for diffusion in a random medium. *Phys. Rev. Letters* **66**, 1689-1692

Bricmont, J. and Kupiainen, A. (1991b) Random walks in asymmetric environments. *Comm. Math. Phys.* **142**, 345–420

Broadbent, S.R. and Hammersley, J.M. (1957) Percolation processes. *Proc. Camb. Phil. Soc.* **53**, 629–645

Burton, R.M., and Keane, M. (1988) Density and uniqueness in percolation. *Comm. Math. Phys.* **121**, 501–505

Cohen, J.E., Kesten, H., and Newman, C.M. (1984) *Random Matrices and Their Applications.* Contemporary Mathematics, Vol. 50, American Mathematical Society, Providence, RI

Cohen, J.E., and Newman, C.M. (1984) The stability of large random matrices and their products. *Ann. Prob.* **12**, 283–310

Cox, J.T. and Durrett, R. (1981) Some limit theorems for percolation processes with necessary and sufficient conditions. *Ann. Prob.* **9**, 583–603

Doeblin, W. (1940) Remarques sur la theorie métrique des fractions continues. *Compositio Math.* **7**, 353–371

Durrett, R. (1984) Oriented percolation in two dimensions. *Ann. Prob.* **12**, 999–1040

Durrett, R. (1995a) Ten lectures on particle systems. Pages 97-201 in Lecture Notes in Mathematics 1608, Springer-Verlag, New York

Durrett, R. (1995b) *Probability: Theory and Examples.* 2nd Edition. Duxbury Press, Belmont, CA

Durrett, R. and Liggett, T.M. (1981) The shape of the limit set in Richardson's growth model. *Ann. Prob.* **9**, 186–193

Durrett, R., and Schonmann, R.H. (1988) The contact process on a finite set, II. *Ann. Prob.* **16**, 1570–1583

Eden, M. (1961) A two dimensional growth process. Pages 223–239 in Vol. IV of the *Proceedings of the 4th Berkeley Symposium.* U. of California Press, San Francisco

Fine, N.J. (1954) On the asymptotic distribution of certain sums. *Proc. AMS.* **5**, 243–252

Fortuin, C.M. and Kastelyn, P.W. (1972) On the random cluster model, I. Introduction and relation to other models. *Physica* **57**, 536–564

Grimmett, G. (1995) The stochastic random-cluster process and uniqueness of random-cluster measures. *Ann. Probab.* **24**, 1461–1510

Grimmett, G. (1999) *Percolation.* Second edition. Springer-Verlag, New York

Hammersley, J.M. (1959) Bornes supérieures de la probabilité critique dans un process de filtration. Pages 17–37 in *Le calcul de probabilités et ses applications.* CNRS, Paris

Hammersley, J.M. (1961) The number of polygons on a lattice. *Proc. Camb. Phil. Soc.* **57**, 516–523

Hammersley, J.M. and Welsh, D.J.A. (1965) First passage percolation, sub- additive processes, stochastic networks, and generalized renewal theory. In *Bernoulli, Bayes, and Laplace.* Edited by J. Neyman and L. LeCam. Springer-Verlag, Berlin

Hara, T., and Slade, G. (1989) The triangle condition in percolation. *Bull AMS.* **21**, 269–273

Hara, T., and Slade, G. (1990) Mean field critical behaviour for percolation in high dimensions. *Comm. Math. Phys.* **128**, 333-391

Hara, T., and Slade, G. (1994) Mean field behaviour and the lace expansion. In *Probability and Phase Transition* edited by G. Grimmett, Kluwer, Dordrecht

Harris, T.E. (1960) A lower bound for the critical probability in a certain percolation process. *Proc. Camb. Phil. Soc.* **56**, 13–20

Hunt, G.A. (1957–1958) Markoff processes and potential, I–III. *Illinois J. Math.* **1**, 44–93, 316–319; **2**, 151–213

Ishitani, H. (1977) A central limit theorem for the subadditive process and its application to products of random matrices. *RIMS, Kyoto* **12**, 565–575

Kac, M. (1946) On the distribution of values of sums of the type $\sum f(2^k t)$. *Annals of Math.* **47**, 33–49

Kalikow, S.A. (1981) Generalized random walk in a random environment. *Ann. Prob.* **9**, 753–768

Kardar, M., Parisi, G., and Zhang, Y.C. (1986) Dynamic scaling of growing interfaces. *Phys. Rev. Lett.* **56**, 889–892

Key, E. (1984) Recurrence and transience for random walk in random environment. *Ann. Prob.* **12**, 529–560

Kingman, J.F.C. (1968) The ergodic theory of subadditive processes. *J. Roy. Stat. Soc. B* **30**, 499–510

Kingman, J.F.C. (1968) The ergodic theory of subadditive stochastic processes. *J. Roy. Stat. Soc. B*, **30**, 499–510

Krug, J. and Spohn, H. (1991) Kinetic roughening of growing surfaces. Pages 479–582 in *Solids Far from Equilibrium: Growth, Morphology, and Defects.* Cambridge U. Press, Cambridge, UK

Kupiainen, A. (1990) Renormalization group and random systems. Pages 1363-1372 in *Proceedings of the International Congress of Math, Kyoto.* (1990), Springer-Verlag, New York

Lawler, G. (1991) *Intersections of Random Walks.* Birkhauser, Boston

LeGall, J.F. (1991) Brownian excursions, trees, and measure valued branching processes. *Ann. Prob.* **19**, 1399–1439

Licea, C. and Newman, C.M. (1996) Geodesics in two-dimensional first passage percolation. *Ann. Prob.* **24**, 399–410

Madras, N., and Slade, G. (1993) *Self-Avoiding Random Walk.* Birkhauser, Boston

Mountford, T. S. (1993) A metastable result for the finite multi-dimensional contact process. *Canad. Math. Bull.* **36**, 216–222

Newman, C.M. (1995) A surface view of first-passage percolation. Pages 1017–1023 in *Proceedings of the International Congress of Mathematicians, Berkeley.* Birkhäuser, Basel

Newman, C.M. and Piza, M.S.T. (1995) Divergence of shape fluctuations in two dimensions. *Ann. Probab.* **23**, 977-1005

Newman, C.M., and Schulman, L.S. (1981) Infinite clusters in percolation models. *J. Stat. Phys.* **26**, 613–628

Pemantle, R. and Peres, Y. (1994) Planar first passage percolation times are not tight. Pages 261–264 in *Probability and Phase Transition*, edited by G. Grimmett. Kluwer, Dordrecht.

Port, S. and Stone, C.J. (1969) Potential theory of random walks on Abelian groups. *Acta Math.* **122**, 19–114

Richardson, D. (1973) Random growth in a tessellation. *Proc. Camb. Phil. Soc.* **74**, 515–528

Ritter, G.A. (1976) Random walk in a random environment, critical case. *Ph.D. Dissertation, Cornell U.*

Russo, L. (1978) A note on percolation *Z. fur Wahr.* **43**, 39–48

Sawyer, S. (1979) A limit theorem for patch sizes in a selectively neutral immigration model. *J. Appl. Prob.* **16**, 482–495

Seymour, P.D., and Welsh, D.J.A. (1978) Percolation probabilities on the square lattice. *Ann. Discrete Math.* **3**, 227–245

Smythe, R.T. and Wierman, J.C. (1978) *First passage percolation on the square lattice.* Lecture Notes in Mathematics 671, Springer-Verlag, Berlin

Sinai, Y.G. (1982) The limiting behavior of a one-dimensional random walk in a random medium. *Th. Prob. Appl.* **27**, 256–268

Solomon, F. (1975) Random walks in a random environment. *Ann. Prob.* **3**, 1–31

Spitzer, F. (1962) Hitting probabilities. *J. Math. Mech.* **11**, 593–614

Spitzer, F. (1964) *Principles of Random Walk.* Van Nostrand, New York

Strassen, V. (1964) An invariance principle for the law of the iterated logarithm. *Z. fur Wahr.* **3**, 211–226

Sykes, M.F. and Essam, J.W. (1964) Exact critical percolation probabilities for site and bond percolation in two dimensions. *J. Math. Phys.* **5**, 1117–1127

Wehr, J. and Aizenman, M. (1990) Fluctuations of extensive functions of quenched random couplings. *J. Stat. Phys.* **60**, 287–306

Zabolitsky, J.G. and Stauffer, D. (1986a) Simulation of large Eden clusters. *Phys. Rev. A* **34**, 1523–1530

Zabolitsky, J.G. and Stauffer, D. (1986b) Dynamic scaling of Eden cluster surfaces. *Phys. Rev. Letters* **57**, 1809

Rick Durrett
Department of Mathematics
Cornell University
Ithaca, NY 14853
rtd1@cornell.edu

Publications of Harry Kesten

[1] (with J. Th. Runnenberg) Priority in waiting line problems. *Proc. Koninklijke Nederlandse Akad. van Wetenschappen, Ser. A.* **60** (1957), 312–336

[2] Accelerated stochastic approximation. *Ann. Math. Stat.* **29** (1958), 41–59

[3] (with M. Kac) On rapidly mixing transformations and an application to continued fractions. *Bulletin AMS.* **64** (1958), 283–287

[4] Symmetric random walks on groups. *Trans. AMS* **92** (1959), 336–354

[5] Full Banach mean values on countable groups. *Math. Scand.* **7** (1959), 146–156

[6] (with N. Morse) A property of the multinomial distribution. *Ann. Math. Stat.* **30** (1959), 120–127

[7] On a series of cosecants II. *Proc. Koninklijke Nederlandse Akad. van Wetenschappen, Ser. A.* **62** (1959), 110–119

[8] Uniform distributions mod 1. *Annals of Math.* **71** (1960), 445–471

[9] (with H. Furstenberg) Products of random matrices. *Ann. Math. Stat.* **31** (1960), 457–469

[10] Some remarks on the capacity of compound channels in the semi-continuous case. *Information and Control.* **4** (1961), 169–184

[11] On a theorem of Spitzer and Stone and random walks with absorbing barriers. *Illinois J. Math.* **5** (1961), 246–262

[12] Random walks with absorbing barriers and Toeplitz forms. *Illinois J. Math.* **5** (1961), 267–290

[13] Uniform distributions mod 1, II. *Acta Arithmetica.* **7** (1962), 355–380

[14] Some probabilistic theorems on diophantine approximation. *Trans. AMS.* **103** (1962), 189–217

[15] (with Z. Ciesielski) A limit theorem for the fractional parts of the sequence $2^k t$. *Proc. AMS.* **13** (1962), 596–600

[16] Occupation times for Markov and semi-Markov chains. *Trans. AMS.* **103** (1962), 82–112

[17] (with D. Ornstein and F. Spitzer) A general property of random walk. *Bulletin AMS* **68** (1962), 526–528

[18] On extreme eigenvalues of translation kernels and Toeplitz matrices. *J. d'Analyse Math.* **10** (1962/63), 117–138

[19] Positivity intervals of stable processes. *J. Math. Mech.* **12** (1963), 391–410

[20] (with F. Spitzer) Ratio theorems for random walks, I. *J. d'Analyse Math.* **11** (1963), 285–322

[21] Ratio theorems for random walks, II. *J. d'Analyse Math.* **11** (1963), 323–379

[22] On the number of self-avoiding walks. *J. Math. Phys.* **4** (1963), 960–969

[23] On the number of self-avoiding walks, II. *J. Math. Phys.* **5** (1964), 1128–1137

[24] The discrepancy of random sequences $\{kx\}$. *Acta Arithmetica* **10** (1964), 183–213

[25] An iterated logarithm law for local time. *Duke Math. J.* **32** (1965), 447–456

[26] (with F. Spitzer) Random walks on countably infinite Abelian groups. *Acta Mathematica* **114** (1965), 237–265

[27] (with B.P. Stigum) A limit theorem for multidimensional Galton-Watson processes. *Ann. Math. Stat.* **37** (1966), 1211–1223

[28] (with B.P. Stigum) Additional limit theorems for indecomposable multidimensional Galton-Watson processes. *Ann. Math. Stat.* **37** (1966), 1463–1481

[29] (with V.T. Sós) On two problems of Erdös, Szüs, and Turán concerning diophantine approximations. *Acta Arithmetica* **12** (1966), 183–192

[30] On a conjecutre of Erdös and Szüs related to uniform distribution mod 1. *Acta Arithmetica* **12** (1966), 193–212

[31] (with P. Ney and F. Spitzer) The Galton-Watson process with mean one and finite variance. *Teorija Verojatnostei i ee Primenenija* **11** (1966), 579–611

[32] The Martin boundary of recurrent random walks on countable groups. *Proc. 5th Berkeley Symposium on Math. Stat. and Prob.*, Vol. II, Part 2 (1967), 51–74

[33] (with B.P. Stigum) Limit theorems for decomposable multidimensional Galton-Watson processes. *J. Math. Anal. Appl.* **7** (1967), 309–338

[34] A Tauberian theorem for random walk. *Israel J. Math.* **6** (1968), 279–294

[35] A sharper form of the Doeblin-Lévy-Kolmogorov-Rogozin inequality for concentration functions. *Math. Scand.* **25** (1969), 133–144

[36] A convolution equation and hitting probabilities of single points for processes with stationary independent increments. *Bulletin AMS.* **75** (1969), 573–578

[37] Hitting probabilities of single points for processes with stationary independent increments. *Memoirs of the AMS,* No. 93, (1969)

[38] A ratio limit theorem for symmetric random walk. *J. d'Analyse Math.* **23** (1970), 199–213

[39] Hitting of sets by processes with stationary independent increments. *Proc. Int. Cong. Math., Nice* (1970) Gauthier-Villar, Paris

[40] Quadratic transformations: A model for population growth. *Adv. Appl. Prob.* 2 (1970), 1–82 and 179–228

[41] The limit points of a normalized random walk. *Ann. Math. Stat.* **41** (1970), 1173–1205

[42] Some nonlinear stochastic growth models. *Bulletin AMS.* **77** (1971), 492–551

[43] (with R.K. Getoor) Continuity of local times for Markov processes. *Compositio Math.* **24** (1972), 277-303

[44] (with B.P. Stigum) Balanced growth under uncertainty in decomposable models. Pages 339–381 in *Essays on Economic Behavior under Uncertainty.* Edited by M.S. Balch, D.L. McFadden, and S.Y. Wu. (1974) American Elsevier, New York

[45] Sums of independent random variables - without moment conditions. *Ann. Math. Stat.* **43** (1972), 701–732

[46] Limit theorems for stochastic growth models. *Adv. Appl. Prob.* **4** (1972), 193–232 and 393–428

[47] Renewal theory for functionals of a Markov chain with general state space. *Ann. Prob.* **2** (1974), 355–386

[48] Random difference equations and renewal theory for products of random matrices. *Acta Mathematica.* **131** (1973), 208–248

[49] (with K.B. Erickson) Strong and weak points of a normalized random walk. *Ann. Prob.* **2** (1974), 553–579

[50] (with F. Spitzer) Controlled Markov chains. *Ann. Prob.* **3** (1975), 32–40

[51] Sums of stationary sequences cannot grow slower than linearly. *Proc. AMS.* **49** (1975), 205–211

[52] Lévy processes with nowhere dense range. *Indiana Math. J.* **25** (1976), 45–64

[53] (with M.V. Kozlov and F. Spitzer) A limit law for random walk in a random in a random environment. *Comp. Math.* **30** (1975), 145–168

[54] Existence and uniqueness of countable one dimensional Markov random fields. *Ann. Prob.* **4** (1976), 557–569

[55] Recurrence criteria for multi-dimensional Markov chains and multi-dimensional linear birth and death processes. *Adv. Appl. Prob.* **8** (1976), 58–87

[56] (with G.L. O'Brien) Examples of mixing sequences. *Duke Math. J.* **43** (1976) 405–415

[57] A renewal theorem for random walk in random environment. Pages 67–77 in *Proc. Symp. Pure Math.* **31**, (1977) American Math. Society, Providence, RI

[58] Erickson's conjecture on the rate of escape of d-dimensional random walk. *Trans. AMS* **240** (1978), 65–113

[59] The speed of convergence of a martingale. *Israel Math. J.* **32** (1979), 83–96

[60] (with G. Papanicolaou) A limit theorem for turbulent diffusion. *Comm. Math. Phys.* **65** (1979), 97–128

[61] Positive harmonic functions. *Proc. Symp. Pure Math.* **35**, part I, (1979) American Math. Society, Providence, RI

[62] (with F. Spitzer) A limit theorem related to a new class of self-similar processes. *Z. fur Wahr.* **50** (1979), 5–25

[63] On the time constant and path length of first-passage percolation. *Adv. Appl. Prob.* **12** (1980), 848–863

[64] (with G. Papanicolaou) A limit theorem for stochastic acceleration. *Comm. Math. Phys.* **78** (1980), 19–63

[65] The number of alleles in electrophoretic experiments. *Theor. Pop. Biol.* **18** (1980), 290–294

[66] The number of distinguishable alleles according to the Ohta-Kimura model of neutral mutation. *J. Math. Biol.* **10**, 167–187

[67] The critical probability for bond percolation on the square lattice equals 1/2. *Comm. Math. Phys.* **74** (1980), 41–59

[68] Random processes in random enviornments. Pages 82–92 in the proceedings of *Models of biological growth and spread.* Springer Lecture Notes in Biomathematics, Vol. 38, Springer-Verlag, NY

[69] (with Y. Ogura) Recurrence properties of Lotka-Volterra models with random fluctuations. *J. Math. Soc. Japan.* **32** (1981), 335–366

[70] (with J.T. Cox) On the continuity of the time constant of first-passage percolation. *J. Appl. Prob.* **18** (1981), 809–819

[71] Analyticity properties and power estimates of functions in percolation theory. *J. Stat. Phys.* **25** (1981), 717–756

[72] *Percolation Theory for Mathematicians.* (1982) Birkhäuser, Boston

[73] Percolation theory and resistance of random elctrical networks. *Proc. Int. Cong. Math., Warsaw*, (1983), Vol. 2, 1081–1088

[74] (with G. Grimmett) First-passage percolation, network flows and electrical resistances. *Z. fur Wahr.* **66** (1984), 335–366

[75] (with G. Grimmett) Random electrical networks and complete graphs. *J. London Math. Soc.*, Series 2. **30** (1984), 171–192

[76] (with F. Spitzer) Convergence in distribution of products of random matrices. *Z. fur Wahr.* **67** (1984), 363–386

[77] (with K. Kawazu) On birth and death processes in symmetric random environments. *J. Stat. Phys.* **37** (1984), 561–576

[78] Aspects of first passage percolation. Pages 126–264 in *École d'Été de Probabilités de Saint Flour XIV*, (1984), Lecture Notes in Mathematics, 1180, Springer-Verlag, New York

[79] (with J. van den Berg) Inequalities with application to percolation and reliability. *J. Appl. Prob.* **22** (1985), 556–569

[80] First-passage percolation and a higher generalization. Pages 235–251 in *Particle Systems, Random Media and Large Deviaitons.* Contemporary Math., Vol. 41, American Mathematical Society, Providence, RI

[81] The incipient infinite cluster in two-dimensional percolation. *Prob. Th. Rel. Fields.* **73** (1986), 369–394

[82] Subdiffusive behavior of a random walk on a random cluster. *Ann. Inst. H. Poincaré* **22** (1986), 425–487

[83] Random walks on random clusters at criticality. Pages 211–218 in *Probabilistic Mathods in Math. Physics,* edited by K. Itô and N. Ikeda. (1987) Kinokunyia Co., Ltd.

[84] The limit distribution of Sinai's random walk in random environment. *Physica A.* **138** (1986), 299–309

[85] Comment on "Evidence for Ising-type critical phenomena in two-dimensional percolation." *Phys. Rev. Lett.* **56** (1986), 1210

[86] The influence of Mark Kac on probability. *Ann. Prob.* **14** (1986), 1103–1128. Correction in *Ann. Prob.* **15** (1987), 1228

[87] Surfaces with minimal random weights and maximal flows: a higher dimensional generalization of first passage percolation. *Illinois. J. Math.* **31** (1987), 99–166

[88] (with M. Aizenman and C.M. Newman) Uniqueness of the infinite cluster and related results in percolation. Pages 13–20 in *Percolation Theory and Ergodic Theory of Infinite Particle Systems.* IMA Volumes in Math. Appl., Vol. 8, (1987), Springer-Verlag, New York

[89] A scaling relation at criticality for 2D-percolation. Pages 203–212 in *Percolation Theory and Ergodic Theory of Infinite Particle Systems.* IMA Volumes in Math. Appl., Vol. 8, (1987), Springer-Verlag, New York

[90] Scaling relations for 2D-percolation. *Comm. Math. Phys.* **109** (1987), 109–156

[91] (with M. Aizenman and C.M. Newman) Uniqueness of the infinite cluster and continuity of connectivity functions for short and long range percolation. *Comm. Math. Phys.* **111** (1987), 505–531

[92] (with Y. Zhang) Strict inequalities for some critical exponents in 2D-percolation. *J. Stat. Phys.* **46** (1987), 1031–1055

[93] How long are the arms in DLA? *J. Phys. A.* **20** (1987), L29–L33

[94] Hitting probabilities of random walks on $\mathbf{Z}^d$. *Stoch. Proc. and Appl.* **25** (1987), 165–184

[95] Percolation theory and first-passage percolation. *Ann. Prob.* **15** (1987), 1231–1271

[96] (with J.T.Bendler, M.Dishon and G.H.Weiss) A first passage time problem for random walk occupancy. *J. Stat. Phys.* **50** (1988), 1069–1087

[97] (with J.T.Chayes, L.Chayes, G.Grimmett, R.Schonmann) The correlation length for the high density phase of Bernoulli percolation. *Ann. Prob.* **17** (1989) 1277–1302

[98] Recent results in rigorous percolation theory. Pages 217–232 in *Colloque Paul Lévy sur les Processus Stochastiques*, Astérisque, Paris vol. **157/158**, (1988)

[99] with J.Bricmont, J.Lebowitz and R.Schonmann) A note on the large dimensional Ising model. *Comm. Math. Phys.* **122** (1989), 597–607

[100] (with R.Schonmann) Behavior in large dimensions of the Potts and Heisenberg model. *Rev. in Math. Phys.* **1** (1990), 147–182

[101] Correlation length and critical probabilities of slabs for percolation, unpublished preprint, 1988.
[102] Asymptotics in high dimensions for percolation. Pages 219–240 in *Disorder in Physical Systems*, (1990) edited by G. Grimmett and D.J.A. Welsh, Oxford Univ. Press, Oxford, UK
[103] Asymptotics in high dimensions for the Fortuin-Kasteleyn random cluster model. Pages 57-85 in *Spatial Stochastic Processes*, (1992), edited by K.S. Alexander and J.C. Watkins, Birkhäuser, Boston
[104] (with R. Durrett) The critical parameter for connectedness of some random graphs. Pages 161–176 in *A Tribute to Paul Erdös*, (1990) edited by A. Baker, B. Bollobas and A. Hajnal, Cambridge Univ. Press, Cambridge, UK
[105] (with Y. Zhang) The probability of a large finite cluster in supercritical Bernoulli percolation. *Ann. Prob.* **18** (1990), 537–555
[106] Supercritical branching processes with countably many types and the size of random Cantor sets. Pages 103-121 in *Probability, Statistics and Mathematics, Papers in honor of Samuel Karlin*, (1989) edited by T.E. Anderson and K. Athreya, Academic Press, Orlando, FL
[107] (with R. Durrett and E. Waymire) On weighted heights of random trees, *J. Theoret. Prob.* **4** (1991), 223-237
[108] Connectivity of certain graphs on halfspaces, quarter spaces,..., Pages 91-104 in *Probability Theory*, 1992 edited by L.H.Y. Chen, K.P. Choi, K. Hu and J-H. Lou, Walter de Gruyter & Co, Berlin
[109] Upper bounds for the growth rate of DLA, *Physica A* **168** (1990), 529–535
[110] (with R. Durrett and G. Lawler) Making money from fair games. Pages 255–267 in *Random walk, Brownian motion and interacting particle systems* (1991), edited by R. Durrett and H. Kesten, Birkhäuser, Boston
[111] (with G. Lawler) A necessary condition for making money from fair games. *Ann. Prob.* **20** (1992), 855-882.
[112] Relations between solutions to a discrete and continuous Dirichlet problem. Pages 309-321 in *Random walk, Brownian motion and interacting particle systems* (1991), edited by R. Durrett and H. Kesten, Birkhäuser, Boston
[113] Some caricatures of multiple contact diffusion-limited aggregation and the eta-model. Pages 179-227 in *Stochastic Analysis* (1991), edited by M.T. Barlow and N.H. Bingham, Cambridge Univ. Press, Cambridge, UK
[114] (with R. Maller) Ratios of trimmed sums and order statistics. *Ann. Prob.* **20** (1992), 1805–1842
[115] (with R. Maller) Infinite limits and infinite limit points of random walks and trimmed sums. *Ann. Prob.* **22** (1994), 1473–1513
[116] (with J. van den Berg) Inequalities for the time constant in first-passage percolation. *Ann. Appl. Prob.* **3** (1993), 56–80
[117] On the speed of convergence in first-passage percolation, *Ann. Appl. Prob.* **3** (1993), 296–338
[118] An absorption problem for several Brownian motions. Pages 59–72 in *Seminar on Stochastic Processes 1991*, Birkhäuser, Boston
[119] (with G. R. Grimmett and Y. Zhang) Random walk on the infinite cluster of the percolation model. *Prob. Th. Rel. Fields* **96** (1993), 33–44
[120] (with C. E. Bezuidenhout and G. R. Grimmett) Strict inequality for critical values of Potts models and random cluster processes, *Comm. Math. Phys.* **158** (1993), 1–16.

[121] (with J. T. Cox, A. Gandolfi and P. Griffin) Greedy lattice animals I: Upper bounds. *Ann. Appl. Prob.* **3** (1993), 1151–1169
[122] (with A. Gandolfi) Greedy lattice animals II: Linear growth. *Ann. Appl. Prob.* **4** (1994), 76–107
[123] (with Y. Zhang) The tortuosity of occupied crossings of a box in critical percolation. *J. Stat. Phys.* **70** (1993), 599–611
[124] First- and last-passage percolation. *Proceedings of a Statistics Conference in Rio de Janeiro, Brasil,* July 1992
[125] Convergence in distribution of lightly trimmed and untrimmed sums are equivalent. *Math. Proc. Camb. Phil. Soc.* **113** (1993), 615–638
[126] A limit theorem for weighted branching process trees. Pages 153–166 in *The Dynkin Festschrift. Markov Processes and their Applications*, (1994), edited by M. I. Freidlin, Birkhäuser, Boston
[127] Frank Spitzer's work on random walk and Brownian motion. *Ann. Prob.* **21** (1993), 593–607
[128] (with R. A. Maller) The effect of trimming on the law of large numbers and generalized laws of the iterated logarithm. *Proc. London Math. Soc.* **71** (1995), 441–480
[129] (with P. A. Ferrari, S. Martinez and P. Picco) Existence of quasi stationary distributions. A renewal dynamical approach, *Ann. Prob.* **23** (1995), 501–521
[130] (with I. Benjamini) Percolation of arbitrary words in $\{0, 1\}^{\mathbf{N}}$, *Ann. Prob.* **23** (1995), 1024–1060
[131] (with R. A. Maller) Random deletion does not affect asymptotic normality or quadratic negligibility. *J. Multivariate Anal.* **63** (1997), 136–179
[132] A ratio limit theorem for (sub)Markov chains on $\{1, 2, \ldots\}$ with bounded jumps. *Adv. Appl. Prob.* **27** (1995), 652–691
[133] (with R. Schonmann) On some growth models with a small parameter. *Prob. Th. Rel. Fields.* **101** (1995), 435–468
[134] (with Boris Pittel) A local limit theorem for the number of nodes, the height and the number of final leaves in critical branching process trees. *Random Structures & Algorithms* **8** (1996), 243–299
[135] Branching random walk with a critical branching part. *J. Theoret. Prob.* **8** (1995), 921–962
[136] (with P. Ferrari and S. Martinez) R-positivity, quasi-stationary distributions and ratio limit theorems for a class of probabilistic automata. *Ann. Appl. Prob.* **6** (1996), 577–616
[137] (with Sungchul Lee) The central limit theorem for weighted minimal spanning trees on random points. *Ann. Appl. Prob.* **6** (1996), 495–527
[138] (with I. Benjamini) Distinguishing sceneries by observing the scenery along a random walk path. *J. d'Analyse Math.* **69** (1996), 97–135
[139] Detecting a single defect in a scenery by observing the scenery along a random walk path. Pages 117-183 in *Itô's Stochastic Calculus and Probability Theory*, (1996), edited by N. Ikeda, S. Watanabe, M. Fukushima and H. Kunita, Springer-Verlag, New York
[140] On the non-convexity of the time constant in first-passage percolation, *Electronic Comm. Probab.* **1** (1996) 1–6; `http://www.math.washington.edu/~ejpecp`
[141] (with R. A. Maller) Two renewal theorems for general random walks tending to infinity. *Prob. Th. Rel. Fields* **106** (1996), 1–38

[142] (with Y. Zhang) A central limit theorem for "critical" first-passage percolation in two dimensions. *Prob. Th. Rel. Fields* **107** (1997), 137–160

[143] A universal form of the Chung-type law of the iterated logarithm. *Ann. Prob.* **25** (1997), 1588–1620

[144] (with R. A. Maller) Divergence of a random walk through deterministic and random subsequences. *J. Theoret. Prob.* **10** (1997) 395–427

[145] (with R. A. Maller) Random walks crossing high level curved boundaries. *J. Theoret. Prob.* **11** (1998), 1019–1074

[146] (with R. A. Maller) Random walks crossing power law boundaries. *Studia Sci. Math. Hungarica.* **34** (1998), 219–252

[147] Distinguishing and reconstructing sceneries from observations along random walk paths, in DIMACS series in Discrete Math. and Computer Science

[148] (with C. Borgs, J. T. Chayes and J. Spencer) Uniform boundedness of critical crossing probabilities implies hyperscaling, in preparation

[149] (with V. Sidoravicius and Y. Zhang) Almost all words are seen in critical site percolation on the triangular lattice. *Electronic J. Prob.** **3** (1998), paper 10

[150] (with J. van den Berg) Asymptotic density in a coalescing random walk model. *Preprint*

[151] (with Z.G. Su) Asymptotic behavior of the critical probability of for ρ-percolation in high dimension. *Preprint*

Ph.D. Students of Harry Kesten

E. Granier[1], 1962

T.A. Ryan, 1968

B. Belkin, 1968

D.L. Tanny, 1975

G. Ritter, 1976

S. Kalikow, 1977

K. Ichihara, 1977

M. Bramson, 1977

J. Hutton, 1979

E. Key, 1983

J.H. Lou, 1983

R. Roy, 1987

Y. Zhang, 1990

S. Lee, 1994

H. Matzinger, 1998

D. Stephenson, 1999

---

[1]joint supervision with A. Dvoretsky

2

Lattice Trees, Percolation and Super-Brownian Motion

Gordon Slade

ABSTRACT This paper surveys the results of recent collaborations with Eric Derbez and with Takashi Hara, which show that integrated super-Brownian excursion (ISE) arises as the scaling limit of both lattice trees and the incipient infinite percolation cluster, in high dimensions. A potential extension to oriented percolation is also mentioned.

Keywords: Lattice trees, percolation, super-Brownian motion, integrated super-Brownian excursion, scaling limit.

AMS Subject Classifications: 60J80, 60K35, 82B41, 82B43.

2.1 Introduction

This paper concerns lattice trees and percolation on $\mathbb{Z}^d$. From the point of view of statistical mechanics, one of the fundamental problems in the study of these models is the construction and analysis of the scaling limit, in which the lattice spacing goes to zero. Control of the scaling limit is closely related to control of the model's critical exponents. General features of the scaling limit are beginning to emerge [1, 2], but much work remains to be done. In particular, there is still no proof of the existence of a single critical exponent for either model in low dimensions.

However, in high dimensions, there has been recent progress for both models. This progress has relied on the fact that the scaling limits in high dimensions turn out to involve integrated super-Brownian excursion (ISE), a close relative of super-Brownian motion (SBM). SBM is a fundamental example of a measure-valued process, a class of objects that has been intensively studied in the probability literature [9, 10, 23].

2.2 SBM and ISE

We will not give a precise mathematical definition of super-Brownian motion here. Our goal in this section is to introduce the key functions associated with ISE that will appear in the results for lattice trees and percolation. A more detailed description of SBM can be found in the article by Cox, Durrett and Perkins in this volume [8], which describes how SBM arises as the scaling limit also for the voter model and the contact process.

SBM can be constructed as an appropriate scaling limit of a critical branching random walk on $\mathbb{Z}^d$, originating from a single initial particle, in the limit as the lattice spacing is shrunk to zero. Such a construction is described in [8], and defines SBM as a remarkable Markov process in which the state at any particular time is a random finite measure on $\mathbb{R}^d$ representing the mass density of particles present at that time. The process dies out in finite time. The entire family tree of SBM is a random finite measure on $\mathbb{R}^d$ and is referred to as the historical process. For dimensions $d \geq 4$, it is almost surely supported on a subset of $\mathbb{R}^d$ having Hausdorff dimension 4 [9, 25].

The mean measure of SBM at time t, which represents the mass density at time t averaged over all family trees, is a deterministic measure that is absolutely continuous with respect to Lebesgue measure in all dimensions $d \geq 1$. In fact, it is the probability measure on $\mathbb{R}^d$ with density

$$p_t(x) = \frac{1}{(2\pi t)^{d/2}} e^{-x^2/2t}. \tag{2.2.1}$$

This density is the transition density for Brownian motion in $\mathbb{R}^d$ to travel from 0 to x in time t.

ISE is the random measure on $\mathbb{R}^d$ obtained by conditioning the historical process to be a probability measure on $\mathbb{R}^d$. Alternately, it can be constructed from critical branching random walk on $n^{-1/4}\mathbb{Z}^d$, starting from a single particle and conditional on a fixed size n for the total size of the initial particle's family tree up to extinction, in the limit $n \to \infty$. The law of ISE is a probability measure μ_{ISE} on the space $M_1(\mathbb{R}^d)$ consisting of probability measures on $\mathbb{R}^d$ and equipped with the topology of weak convergence. The mean of μ_{ISE} is a deterministic probability measure on $\mathbb{R}^d$, which corresponds to averaging over all family trees resulting from the initial particle, under the basic unit mass condition required by ISE. Define

$$a^{(2)}(x,t) = te^{-t^2/2} p_t(x). \tag{2.2.2}$$

The mean ISE measure is absolutely continuous with respect to Lebesgue measure in all dimensions $d \geq 1$. Its density with respect to Lebesgue measure is the function

$$A^{(2)}(x) = \int_0^\infty a^{(2)}(x,t)dt = \int_0^\infty te^{-t^2/2} p_t(x)dt. \tag{2.2.3}$$

The functions (2.2.2) and (2.2.3) are ISE two-point functions. A discussion of higher point functions requires the notion of *shape*, which is defined as follows. We start with an *m-skeleton*, which is a tree having m unlabelled external vertices of degree 1 and $m-2$ unlabelled internal vertices of degree 3, and no other vertices. An *m-shape* is a tree having m labelled external vertices of degree 1 and $m-2$ unlabelled internal vertices of degree 3, and no other vertices, *i.e.*, an m-shape is a labelling of an m-skeleton's external vertices by the labels $0, 1, \ldots, m-1$. When

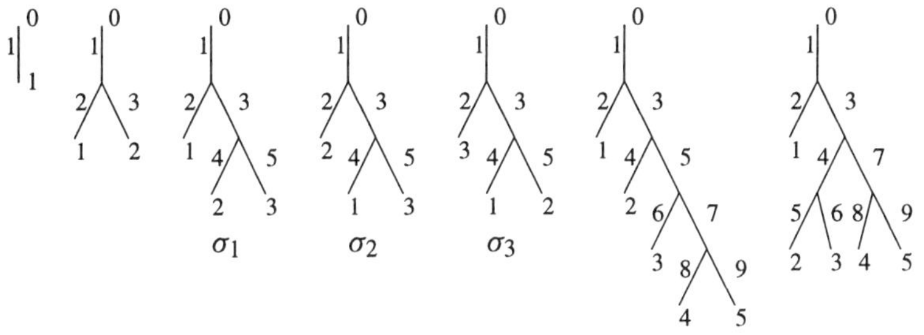

FIGURE 2.1. The shapes for $m = 2, 3, 4$, and examples of the $7!! = 7 \cdot 5 \cdot 3 = 105$ shapes for $m = 6$. The shapes' edge labellings are arbitrary but fixed.

m is clear from the context, we will refer to an m-shape simply as a shape. For notational convenience, we associate to each m-shape an arbitrary labelling of its $2m-3$ edges, with labels $1, \ldots, 2m-3$. This arbitrary choice of edge labelling is fixed once and for all. Thus an m-shape σ is a labelling of an m-skeleton's external vertices together with a corresponding specification of edge labels. Let Σ_m denote the set of m-shapes. There is a unique shape for $m = 2$ and $m = 3$, and $(2m-5)!!$ distinct shapes for $m \geq 4$ (see [14, (5.96)] for a proof). In this notation, $(-1)!! = 1$ and $(2j+1)!! = (2j+1)(2j-1)!!$ for $j \geq 0$.

Let $m \geq 2$. Given a shape $\sigma \in \Sigma_m$, we associate to edge j (oriented away from vertex 0) a nonnegative real number t_j and a vector y_j in $\mathbb{R}^d$. Writing $\vec{y} = (y_1, \ldots, y_{2m-3})$ and $\vec{t} = (t_1, \ldots, t_{2m-3})$, we define

$$a^{(m)}(\sigma; \vec{y}, \vec{t}) = \left(\sum_{i=1}^{2m-3} t_i \right) e^{-(\sum_{i=1}^{2m-3} t_i)^2/2} \prod_{i=1}^{2m-3} p_{t_i}(y_i) \tag{2.2.4}$$

and

$$A^{(m)}(\sigma; \vec{y}) = \int_0^\infty dt_1 \cdots \int_0^\infty dt_{2m-3}\, a^{(m)}(\sigma; \vec{y}, \vec{t}). \tag{2.2.5}$$

Then $\int_{\mathbb{R}^{d(2m-3)}} A^{(m)}(\sigma; \vec{y}) d\vec{y} = 1/(2m-5)!!$, so the sum of this integral over shapes $\sigma \in \Sigma_m$ is equal to 1. Let $\vec{k} \cdot \vec{y} = \sum_{j=1}^{2m-3} k_j \cdot y_j$, with each $k_j \in \mathbb{R}^d$. The Fourier integral transform $\hat{A}^{(m)}(\sigma; \vec{k}) = \int_{\mathbb{R}^{d(2m-3)}} A^{(m)}(\sigma; \vec{y}) e^{i\vec{k}\cdot\vec{y}} d\vec{y}$ is given by

$$\hat{A}^{(m)}(\sigma; \vec{k}) = \int_0^\infty dt_1 \cdots \int_0^\infty dt_{2m-3}\, \hat{a}^{(m)}(\sigma; \vec{k}, \vec{t}), \tag{2.2.6}$$

with

$$\hat{a}^{(m)}(\sigma; \vec{k}, \vec{t}) = \left(\sum_{i=1}^{2m-3} t_i \right) e^{-(\sum_{i=1}^{2m-3} t_i)^2/2} \prod_{i=1}^{2m-3} e^{-k_i^2 t_i/2}. \tag{2.2.7}$$

The lth moment measure $M^{(l)}$ for ISE can be written in terms of $A^{(l+1)}$, for $l \geq 1$. This is a deterministic measure which is absolutely continuous with respect to Lebesgue measure on $\mathbb{R}^{dl}$. The first moment measure $M^{(1)}$ has density $A^{(2)}(x)$.

The second moment measure $M^{(2)}$ has density $\int A^{(3)}(y, x_1 - y, x_2 - y)d^d y$. In general, the density of $M^{(l)}$ at $x_1, \ldots, x_l$, for $l \geq 3$, is given by integrating $A^{(l+1)}(\sigma; \vec{y})$ over $\mathbb{R}^{d(l-1)}$ and then summing over the $(2l-3)!!$ shapes σ. Here $\vec{y}$ consists of integration variables y_j corresponding to the edges j on paths from vertex 0 to vertices of degree 3 in σ, and the other y_a are fixed by the requirement that each external vertex x_i is given by the sum of the y_e over the edges e connecting vertices 0 and i in σ. Thus, the integration corresponds to integrating over the $l-1$ internal vertices, with the $l+1$ external vertices fixed at $0, x_1, \ldots, x_l$. For example, the contribution to the density of $M^{(3)}$ due to σ_1 of Figure 2.1 is $\int A^{(4)}(\sigma_1; y_1, x_1 - y_1, y_3, x_2 - y_1 - y_3, x_3 - y_1 - y_3)d^d y_1 d^d y_3$.

ISE and the functions (2.2.4) and (2.2.5) are further discussed in [5] (see also [4, 11, 24]). A construction of ISE as the scaling limit of branching random walk conditioned on the total size of the family tree, including a derivation of these functions, is given in [6].

2.3 Generating Functions

For our applications to lattice trees and percolation, it will be essential to understand that the Fourier integral transforms of $a^{(m)}$ and $A^{(m)}$, $m = 2, 3, 4, \ldots$, occur in the asymptotic behaviour of certain generating function coefficients. This connection between ISE and generating functions was pointed out in [11].

The relevant generating functions are defined as follows. For $k \in \mathbb{R}^d$, define $C^{(2)}_{z,\zeta}(k)$ and $c^{(2)}_{n,s}(k)$ by

$$C^{(2)}_{z,\zeta}(k) = \frac{2}{k^2 + 2^{3/2}\sqrt{1-z} + 2(1-\zeta)} = \sum_{s,n=0}^{\infty} c^{(2)}_{n,s}(k) z^n \zeta^s, \quad |\zeta|, |z| < 1, \tag{2.3.1}$$

where the square root has branch cut $[1, \infty)$ and is positive for real $z < 1$. For $m \geq 2$, given a shape $\sigma \in \Sigma_m$, to edge j we associate $k_j \in \mathbb{R}^d$ and $\zeta_j \in \mathbb{C}$, with $|\zeta_j| < 1$. We write $\vec{k} = (k_1, \ldots, k_{2m-3})$ and $\vec{\zeta} = (\zeta_1, \ldots, \zeta_{2m-3})$, and define

$$C^{(m)}_{z,\vec{\zeta}}(\sigma; \vec{k}) = \prod_{j=1}^{2m-3} C^{(2)}_{z,\zeta_j}(k_j) = \sum_{s_1,\ldots,s_{2m-3}=0}^{\infty} \sum_{n=0}^{\infty} c^{(m)}_{n,\vec{s}}(\sigma; \vec{k}) z^n \prod_{j=1}^{2m-3} \zeta_j^{s_j}. \tag{2.3.2}$$

We write $b^{(m)}_{\vec{s}}(\sigma; \vec{k}) = \sum_{n=0}^{\infty} c^{(m)}_{n,\vec{s}}(\sigma; \vec{k})$ for the coefficient of $\prod_{j=1}^{2m-3} \zeta_j^{s_j}$ in $C^{(m)}_{1,\vec{\zeta}}(\sigma; \vec{k})$. Writing $\vec{1} = (1, \ldots, 1)$, we denote the coefficient of z^n in $C^{(m)}_{z,\vec{1}}(\sigma; \vec{k})$ by $c^{(m)}_n(\sigma; \vec{k}) = \sum_{s_1,\ldots,s_{2m-3}=0}^{\infty} c^{(m)}_{n,\vec{s}}(\sigma; \vec{k})$.

The coefficients $b^{(m)}_{\vec{s}}(\sigma; \vec{k})$ are easily identified from the fact that $C^{(2)}_{1,\zeta}(k)$ is the sum of a geometric series in ζ, namely

$$C^{(2)}_{1,\zeta}(k) = \frac{2}{k^2 + 2(1-\zeta)} = \sum_{s=0}^{\infty} \frac{1}{(1 + k^2/2)^{s+1}} \zeta^s. \tag{2.3.3}$$

Therefore $b_{\vec{s}}^{(m)}(\sigma;\vec{k}) = \prod_{j=1}^{2m-3}(1+k_j^2/2)^{-(s_j+1)}$. For $t_j \in [0,\infty)$, the Fourier transform of the Brownian transition density (2.2.1) then emerges as the $m=2$ case of the limit

$$\lim_{n\to\infty} b_{\lfloor \vec{t}n \rfloor}^{(m)}(\sigma;\vec{k}n^{-1/2}) = \prod_{j=1}^{2m-3} e^{-k_j^2 t_j/2}. \tag{2.3.4}$$

Here $\lfloor \vec{t}n \rfloor$ denotes the vector with components $\lfloor t_j n \rfloor$.

For the ISE m-point function (2.2.5), we consider the generating function $C_{z,\vec{1}}^{(m)}(\sigma;\vec{k}) = \prod_{j=1}^{2m-3} 2(k_j^2 + 2^{3/2}\sqrt{1-z})^{-1}$. By Cauchy's theorem,

$$c_n^{(m)}(\sigma;\vec{k}) = \frac{1}{2\pi i}\oint_{\Gamma} C_{z,\vec{1}}^{(m)}(\sigma;\vec{k})\frac{dz}{z^{n+1}}, \tag{2.3.5}$$

where Γ is a circle centred at the origin with radius less than 1. By deforming the contour to the branch cut $[1,\infty)$ of the square root, it can be shown that for any $m \geq 2$, $k \in \mathbb{R}^d$,

$$c_n^{(m)}(\sigma;\vec{k}n^{-1/4}) \sim \frac{1}{\sqrt{2\pi}} n^{m-5/2} \hat{A}^{(m)}(\sigma;\vec{k}) \tag{2.3.6}$$

as $n \to \infty$. Here $f(n) \sim g(n)$ denotes $\lim_{n\to\infty} f(n)/g(n) = 1$. A proof of (2.3.6) is given in the proof of [12, Theorem 1.1].

The functions $\hat{a}^{(m)}(\sigma;\vec{k},\vec{t})$ arise from an appropriate joint limit of the coefficients $c_{n,\vec{s}}^{(m)}(\sigma;\vec{k})$. Namely, for $m \geq 2$,

$$c_{n,\lfloor \vec{t}n^{1/2} \rfloor}^{(m)}(\sigma;\vec{k}n^{-1/4}) \sim \frac{1}{\sqrt{2\pi}}\frac{1}{n}\hat{a}^{(m)}(\sigma;\vec{k},\vec{t}) \tag{2.3.7}$$

as $n \to \infty$. A proof is given in the proof of [12, Theorem 1.2].

One might wonder at this point what any of this has to do with lattice trees or percolation. The connection is that some of these models' key thermodynamic functions have the form of the above generating functions in high dimensions, and this links them to ISE.

2.4 Lattice Trees

A lattice tree in $\mathbb{Z}^d$ is a finite connected set of lattice bonds containing no cycles. For the nearest-neighbour model, the bonds are nearest-neighbour bonds $\{x, y\}$, $x, y \in \mathbb{Z}^d$, $\|x-y\|_1 = 1$. We will also consider "spread-out" lattice trees constructed from bonds $\{x, y\}$ with $0 < \|x-y\|_\infty \leq L$. The parameter L will later be taken to be large but finite. We associate the uniform probability measure to the set of all n-bond lattice trees which contain the origin.

In this section, we will describe results showing that in high dimensions the scaling limit of lattice trees of size n, with space scaled by a multiple of $n^{-1/4}$, is ISE. Thus lattice trees in high dimensions behave like branching random walk.

We define the one-point function $t_n^{(1)}$ to be the number of n-bond lattice trees containing the origin, with $t_0^{(1)} = 1$. By a subadditivity argument, there is a positive constant z_c (depending on d, and on L for the spread-out model) such that $\lim_{n\to\infty}[t_n^{(1)}]^{1/n} = z_c^{-1}$.

Next, we would like to define the higher-point functions $t_n^{(m)}(\sigma; \vec{y}, \vec{s})$, for $m \geq 2$. These functions count lattice trees with a certain property. To describe this, we need some definitions. Let $\sigma \in \Sigma_m$, and associated to each edge j in σ, let $y_j \in \mathbb{Z}^d$ and let s_j be a nonnegative integer ($j = 1, \ldots, 2m-3$). First, we introduce the notion of backbone. Given a lattice tree T containing the sites $0, x_1, \ldots, x_{m-1}$, we define the *backbone* B of $(T; 0, x_1, \ldots, x_{m-1})$ to be the subtree of T spanning $0, x_1, \ldots, x_{m-1}$. There is an induced labelling of the external vertices of the backbone, in which vertex x_l is labelled l. Ignoring vertices of degree 2 in B, this backbone is equivalent to a shape σ_B or to its modification by contraction of one or more edges to a point. (In the latter case, as we will discuss further in Appendix 2.A, the choice of σ_B may not be unique.) Next, we need a notion of compatibility. Restoring vertices of degree 2 in B, let b_j denote the length of the backbone path corresponding to edge j of σ_B, with $b_j = 0$ for any contracted edge. We say that $(T; 0, x_1, \ldots, x_{m-1})$ is *compatible* with $(\sigma; \vec{y}, \vec{s})$ if σ_B can be chosen (when not uniquely determined) such that $\sigma_B = \sigma$, if $b_j = s_j$ for all edges j of σ, and if the backbone path corresponding to j undergoes the displacement y_j for all edges j of σ.

Then we define $t_n^{(m)}(\sigma; \vec{y}, \vec{s})$ to be the number of n-bond lattice trees T, containing the origin, for which there are sites $x_1, \ldots, x_{m-1} \in T$ such that $(T; 0, x_1, \ldots, x_{m-1})$ is compatible with $(\sigma; \vec{y}, \vec{s})$. See Figure 2.2. We also define

$$t_n^{(m)}(\sigma; \vec{y}) = \sum_{\vec{s}} t_n^{(m)}(\sigma; \vec{y}, \vec{s}), \tag{2.4.1}$$

where the sum over $\vec{s}$ denotes a sum over the nonnegative integers s_j. We will make use of Fourier transforms with respect to the $\vec{y}$ variables, for example,

$$\hat{t}_n^{(m)}(\sigma; \vec{k}) = \sum_{\vec{y}} t_n^{(m)}(\sigma; \vec{y}) e^{i\vec{k}\cdot\vec{y}}, \quad k_j \in [-\pi, \pi]^d. \tag{2.4.2}$$

For $m = 2, 3$ there is only one shape and we will sometimes omit it from the notation.

Define

$$G_{z,\vec{\zeta}}^{(m)}(\sigma; \vec{y}) = \sum_{n=0}^{\infty} \sum_{\vec{s}} t_n^{(m)}(\sigma; \vec{y}, \vec{s}) z^n \prod_{j=1}^{2m-3} \zeta_j^{s_j}. \tag{2.4.3}$$

The sum over $\vec{y} \in \mathbb{R}^{d(2m-3)}$ of (2.4.3) is finite for $|z| < z_c$ and $|\zeta_j| \leq 1$, for all m.

In terms of critical exponents, the Fourier transform of the two-point function $G_{z,1}^{(2)}(y)$ is believed to behave asymptotically as

$$\hat{G}_{z_c,1}^{(2)}(k) \sim \frac{c_1}{k^{2-\eta}} \text{ as } k \to 0, \quad \hat{G}_{z,1}^{(2)}(0) \sim \frac{c_2}{(1 - z/z_c)^{\gamma}} \text{ as } z \to z_c, \tag{2.4.4}$$

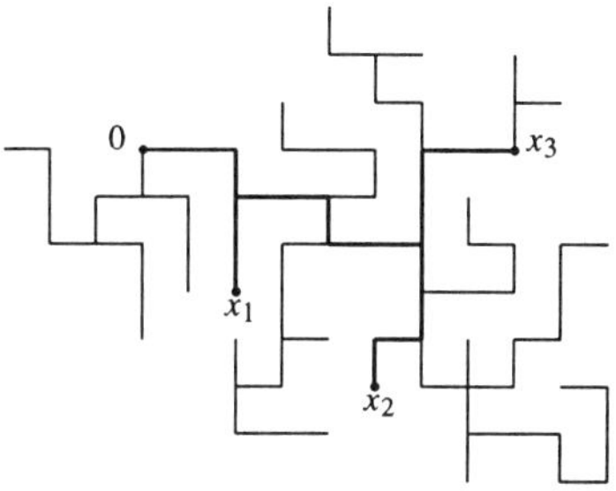

FIGURE 2.2. A 2-dimensional lattice tree contributing to $t_{78}^{(4)}(\sigma_1; \vec{y}, \vec{s})$, with σ_1 depicted in Figure 2.1, $\vec{y} = ((2,-1),(0,-2),(4,-1),(-1,-3),(2,2))$, $\vec{s} = (3,2,5,4,4)$.

with the mean-field values $\eta = 0$ and $\gamma = \frac{1}{2}$ for $d > 8$. For $d > 8$, the simplest combination of the two asymptotic relations in (2.4.4) that could be hoped for is

$$\hat{G}_{z,1}^{(2)}(k) = \frac{C_1}{D_1^2 k^2 + 2^{3/2}(1 - z/z_c)^{1/2}} + \text{error}, \tag{2.4.5}$$

where C_1 and D_1 are positive constants depending on d and L. The error term is meant to be of lower order than the main term, in some suitable sense, as $k \to 0$ and $z \to z_c$.

An optimist expecting to find ISE and familiar with (2.3.1) and (2.3.7) could also hope that, for $d > 8$,

$$\hat{G}_{z,\zeta}^{(2)}(k) = \frac{C_1}{D_1^2 k^2 + 2^{3/2}(1 - z/z_c)^{1/2} + 2T_1(1-\zeta)} + \text{error}, \tag{2.4.6}$$

and that there is an approximate independence of the form

$$\hat{G}_{z,\vec{\zeta}}^{(m)}(\sigma; \vec{k}) = v_1^{m-2} \prod_{j=1}^{2m-3} \hat{G}_{z,\zeta_j}^{(2)}(k_j) + \text{error}. \tag{2.4.7}$$

Here v_1 is a finite positive constant which translates the self-avoidance interactions of lattice trees into a renormalized vertex factor. For the nearest-neighbour model with d sufficiently large, and for spread-out models for $d > 8$ with L sufficiently large, relations of the form (2.4.6) and (2.4.7) are proved in [12], for all $m \geq 2$ if $\vec{\zeta} = \vec{1}$ and for $m = 2, 3$ for general $\vec{\zeta}$. The results given below arise as a consequence.

We define

$$p_n^{(m)}(\sigma; \vec{y}) = \frac{t_n^{(m)}(\sigma; \vec{y})}{\sum_{\sigma \in \Sigma_m} \hat{t}_n^{(m)}(\sigma; \vec{0})}, \tag{2.4.8}$$

which is a probability measure on $\Sigma_m \times \mathbb{Z}^{d(2m-3)}$. The following theorem, whose proof extends the methods of [17, 18], shows that (2.4.8) has the corresponding ISE density as its scaling limit in high dimensions. In its statement, the scaling of $\vec{k}$ by $D_1^{-1} n^{-1/4}$ corresponds to scaling down the lattice spacing by $D_1^{-1} n^{-1/4}$.

Theorem 2.4.1 ([11, 12]) *Let $m \geq 2$ and $k_j \in \mathbb{R}^d$ $(j = 1, \ldots, 2m-3)$. For nearest-neighbour lattice trees in sufficiently high dimensions $d \geq d_0$, and for spread-out lattice trees with $d > 8$ and L sufficiently large depending on d, there are constants c_1, D_1 depending on d and L, such that*

$$\hat{t}_n^{(m)}(\sigma; \vec{k} D_1^{-1} n^{-1/4}) \sim c_1 n^{m-5/2} z_c^{-n} \hat{A}^{(m)}(\sigma; \vec{k}) \quad (n \to \infty). \tag{2.4.9}$$

In particular,

$$\lim_{n \to \infty} \hat{p}_n^{(m)}(\sigma; \vec{k} D_1^{-1} n^{-1/4}) = \hat{A}^{(m)}(\sigma; \vec{k}).$$

It is a corollary of Theorem 2.4.1 that high-dimensional lattice trees converge weakly to ISE, as we now explain. Given an n-bond lattice tree T containing the origin, we define μ_n^T to be the probability measure on $\mathbb{R}^d$ which assigns mass $(n+1)^{-1}$ to the each of the $n+1$ points $x D_1^{-1} n^{-1/4}$, for $x \in T$. Let $M_1(\mathbb{R}^d)$ be the space of probability measures on $\mathbb{R}^d$. We then define a probability measure μ_n on $M_1(\mathbb{R}^d)$, supported on the μ_n^T, by $\mu_n(\mu_n^T) = (t_n^{(1)})^{-1}$ for each n-bond T containing 0. In this way, n-bond lattice trees induce a random probability measure on $\mathbb{R}^d$. Let $\dot{\mathbb{R}}^d$ denote the one-point compactification of $\mathbb{R}^d$, and let $M_1(\dot{\mathbb{R}}^d)$ denote the compact set of probability measures on $\dot{\mathbb{R}}^d$, under the topology of weak convergence. We regard $M_1(\mathbb{R}^d)$ as embedded in $M_1(\dot{\mathbb{R}}^d)$.

Corollary 2.4.1 *For nearest-neighbour lattice trees in sufficiently high dimensions $d \geq d_0$, and for spread-out lattice trees with $d > 8$ and L sufficiently large, μ_n converges weakly to μ_{ISE}, as measures on $M_1(\dot{\mathbb{R}}^d)$.*

The weak convergence in Corollary 2.4.1 is the assertion that for any continuous function F on $M_1(\dot{\mathbb{R}}^d)$,

$$\lim_{n \to \infty} \int_{M_1(\dot{\mathbb{R}}^d)} F(\nu) d\mu_n(\nu) = \int_{M_1(\dot{\mathbb{R}}^d)} F(\nu) d\mu_{\mathrm{ISE}}(\nu). \tag{2.4.10}$$

The argument leading from Theorem 2.4.1 to Corollary 2.4.1 is presented in Appendix 2.A.

For a more refined statement than Theorem 2.4.1, we define

$$p_n^{(m)}(\sigma; \vec{y}, \vec{s}) = \frac{t_n^{(m)}(\sigma; \vec{y}, \vec{s})}{\sum_{\sigma \in \Sigma_m} \hat{t}_n^{(m)}(\sigma; \vec{0})}, \tag{2.4.11}$$

which is a probability measure on $\Sigma_m \times \mathbb{Z}^{d(2m-3)} \times \mathbb{Z}_+^{2m-3}$.

Theorem 2.4.2 ([11, 12]) *Let $m = 2$ or $m = 3$, $k_j \in \mathbb{R}^d$, and $t_j \in (0, \infty)$ $(j = 1, \ldots, 2m-3)$. For nearest-neighbour lattice trees in sufficiently high dimensions $d \geq d_0$, and for spread-out lattice trees with $d > 8$ and L sufficiently large depending on d, there is a constant T_1 depending on d and L, such that*

$$\hat{t}_n^{(m)}(\sigma; \vec{k} D_1^{-1} n^{-1/4}, \lfloor \vec{t}\, T_1 n^{1/2} \rfloor) \sim c_1 T_1^{-(2m-3)} n^{-1} z_c^{-n} \hat{a}^{(m)}(\sigma; \vec{k}, \vec{t}) \quad (n \to \infty).$$

In particular,

$$\lim_{n \to \infty} (T_1 n^{1/2})^{2m-3}\, \hat{p}_n^{(m)}(\sigma; \vec{k} D_1^{-1} n^{-1/4}, \lfloor \vec{t}\, T_1 n^{1/2} \rfloor) = \hat{a}^{(m)}(\sigma; \vec{k}, \vec{t}). \tag{2.4.12}$$

We believe that Theorem 2.4.2 holds for all $m \geq 2$, but technical difficulties arise for $m \geq 4$ and the theorem has been proved only for $m = 2$ and $m = 3$. Theorem 2.4.2 indicates that, at least for $m = 2$ and $m = 3$, skeleton paths with length of order $n^{1/2}$ are typical. This is Brownian scaling, since distance is scaled as $n^{1/4}$. The statement of Theorem 2.4.2 for $m = 3$ in [11, 12] incorrectly included the case where $t_j = 0$ for one or two values of j, for which different constants occur, in fact, in the asymptotic formula for $\hat{t}_n^{(3)}(\sigma; \vec{k}D_1^{-1}n^{-1/4}, \lfloor \vec{t}\,T_1 n^{1/2} \rfloor)$.

We expect that the above results for lattice trees should apply also to lattice animals for $d > 8$, yielding ISE for their scaling limit for $d > 8$. This would be consistent with the general belief that lattice trees and lattice animals have the same scaling properties in all dimensions.

2.5 Percolation

Consider independent Bernoulli bond percolation on $\mathbb{Z}^d$, either nearest-neighbour or spread-out, with p fixed and equal to its critical value p_c [14]. Bonds are pairs $\{x, y\}$ of sites in $\mathbb{Z}^d$, with $\|x - y\|_1 = 1$ for the nearest-neighbour model and $0 < \|x - y\|_\infty \leq L$ for the spread-out model. Let $C(0)$ denote the random set of sites connected to 0, let $|C(0)|$ denote the cardinality of $C(0)$, and let

$$\tau^{(2)}(x; n) = P_{p_c}(C(0) \ni x, |C(0)| = n) \qquad (2.5.1)$$

denote the probability at the critical point that the origin is connected to x via a cluster containing n sites. We define a generating function

$$\tau_z^{(2)}(x) = \sum_{n=1}^{\infty} \tau^{(2)}(x; n) z^n, \qquad (2.5.2)$$

which converges absolutely if $|z| \leq 1$. Assuming no infinite cluster at p_c, $\tau_1^{(2)}(x)$ is the probability that 0 is connected to x.

The conventional definitions [14, Section 7.1] of the critical exponents η and δ suggest that

$$\hat{\tau}_1^{(2)}(k) \sim \frac{c_1}{k^{2-\eta}} \text{ as } k \to 0, \quad \hat{\tau}_z^{(2)}(0) \sim \frac{c_2}{(1-z)^{1-1/\delta}} \text{ as } z \to 1, \qquad (2.5.3)$$

but there is still no proof of existence of these exponents except in high dimensions. Assuming the mean-field values $\eta = 0$ and $\delta = 2$ above six dimensions, the simplest combination of the above asymptotic relations for $d > 6$ would be

$$\hat{\tau}_z^{(2)}(k) = \frac{C_2}{D_2^2 k^2 + 2^{3/2}(1-z)^{1/2}} + \text{error}, \qquad (2.5.4)$$

for some constants C_2, D_2. This is analogous to (2.4.5). The following theorem shows that this behaviour is what does occur for sufficiently spread-out percolation above six dimensions.

Theorem 2.5.1 ([15, 20]) *Let $k \in [-\pi,\pi]^d$, $z \in [0,1)$. For spread-out percolation with $d > 6$ and L sufficiently large, there are functions $\epsilon_1(z)$ and $\epsilon_2(k)$ with $\lim_{z\to 1}\epsilon_1(z) = \lim_{k\to 0}\epsilon_2(k) = 0$, and constants C_2 and D_2 depending on d and L, such that*

$$\hat{\tau}_z^{(2)}(k) = \frac{C_2}{D_2^2 k^2 + 2^{3/2}(1-z)^{1/2}}\,[1+\epsilon(z,k)] \tag{2.5.5}$$

with $|\epsilon(z,k)| \le \epsilon_1(z) + \epsilon_2(k)$.

In view of (2.3.6), Theorem 2.5.1 is highly suggestive that ISE occurs as a scaling limit for percolation, but the control of the error term in (2.5.5) is too weak to obtain bounds on $\hat{\tau}^{(2)}(kD_2^{-1}n^{-1/4};n)$ via contour integration. However, for the nearest-neighbour model in sufficiently high dimensions, better control of the error terms has been obtained, for *complex* z with $|z| < 1$, leading to the following theorem. The theorem also gives a result for the three-point function

$$\tau^{(3)}(x,y;n) = P_{p_c}(x,y \in C(0), |C(0)| = n), \tag{2.5.6}$$

in terms of its Fourier transform

$$\hat{\tau}^{(3)}(k,l;n) = \sum_{x,y\in\mathbb{Z}^d} \tau^{(3)}(x,y;n)e^{ik\cdot x + il\cdot y}. \tag{2.5.7}$$

Theorem 2.5.2 ([16, 20]) *Fix $k,l \in \mathbb{R}^d$ and any $\epsilon \in (0,\frac{1}{2})$. There is a d_0 such that for nearest-neighbour percolation with $d \ge d_0$, there are constants C_2, D_2 (depending on d) such that as $n \to \infty$*

$$\hat{\tau}^{(2)}(kD_2^{-1}n^{-1/4};n) = \frac{C_2}{\sqrt{8\pi n}}\hat{A}^{(2)}(k)[1+O(n^{-\epsilon})], \tag{2.5.8}$$

$$\hat{\tau}^{(3)}(kD_2^{-1}n^{-1/4}, lD_2^{-1}n^{-1/4};n) = \frac{C_2}{\sqrt{8\pi}}n^{1/2}\hat{A}^{(3)}(k+l,k,l)[1+O(n^{-\epsilon})]. \tag{2.5.9}$$

It follows from (2.5.8) that

$$P_{p_c}(|C(0)| = n) = n^{-1}\hat{\tau}^{(2)}(0;n) = C_2(8\pi)^{-1/2}n^{-3/2}[1+O(n^{-\epsilon})]. \tag{2.5.10}$$

This shows that the critical exponent δ, defined by $P_{p_c}(|C(0)| = n) \approx n^{-1-1/\delta}$, is given by $\delta = 2$ in high dimensions.

The variables in (2.5.9) are arranged schematically as:

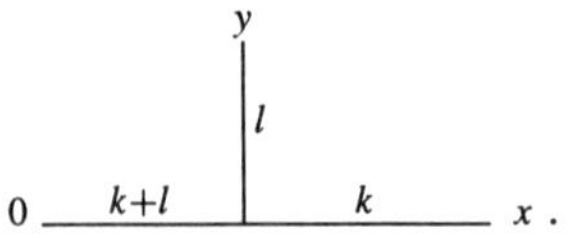

To obtain (2.5.9), we work with the generating function

$$\hat{\tau}_z^{(3)}(k,l) = \sum_{n=1}^{\infty} \tau^{(3)}(k,l;n)z^n, \tag{2.5.11}$$

and prove that there is a positive constant v_2 such that

$$\hat{\tau}_z^{(3)}(k,l) = v_2 \hat{\tau}_z^{(2)}(k+l)\hat{\tau}_z^{(2)}(k)\hat{\tau}_z^{(2)}(l) + \text{error}. \tag{2.5.12}$$

An asymptotic relation in the spirit of (2.5.12), with $k = l = 0$, was conjectured for $d > 6$ already in [3].

We expect that Theorem 2.5.2 should extend to general m-point functions, for all $m \geq 2$, but this has not been proven. This is essentially the conjecture of [20] that the scaling limit of the incipient infinite cluster is ISE for $d > 6$. We now discuss this conjecture in more detail.

Given a site lattice animal S containing n sites, one of which is the origin, define the probability measure $\nu_n^S \in M_1(\mathbb{R}^d)$ to assign mass n^{-1} to $xD_2^{-1}n^{-1/4}$, for each $x \in S$. We define ν_n to be the probability measure on $M_1(\mathbb{R}^d)$ which assigns probability $P_{p_c}(C(0) = S \mid |C(0)| = n)$ to ν_n^S, for each S as above. We regard the limit of ν_n, as $n \to \infty$, as the scaling limit of the incipient infinite cluster. This is related to one of Kesten's definitions of the incipient infinite cluster [22], but here we are taking the lattice spacing to zero as $n \to \infty$. The conjecture of [20] is that, as in Corollary 2.4.1 above, ν_n converges weakly to μ_{ISE} for $d > 6$.

The conjecture is supported by Theorem 2.5.2. In fact, the characteristic functions $\hat{N}_n^{(1)}(k)$ and $\hat{N}_n^{(2)}(k,l)$ of the first and second moment measures $N_n^{(1)}$ and $N_n^{(2)}$ of ν_n are given by

$$\hat{N}_n^{(1)}(k) = \frac{\hat{\tau}^{(2)}(kD_2^{-1}n^{-1/4}; n)}{\hat{\tau}^{(2)}(0; n)}, \tag{2.5.13}$$

$$\hat{N}_n^{(2)}(k,l) = \frac{\hat{\tau}^{(3)}(kD_2^{-1}n^{-1/4}, lD_2^{-1}n^{-1/4}; n)}{\hat{\tau}^{(3)}(0,0; n)}, \tag{2.5.14}$$

and in high dimensions these converge respectively to the characteristic functions $\hat{A}^{(2)}(k)$ and $\hat{A}^{(3)}(k+l, k, l)$ of the corresponding ISE moments, by Theorem 2.5.2.

2.6 Oriented Percolation

Consider independent oriented percolation on $\mathbb{Z}^d \times \mathbb{Z}_+$. Bonds are directed and are of the form $((x,n),(y,n+1))$, with $x, y \in \mathbb{Z}^d$ obeying $\|x-y\|_1 = 1$ for the nearest-neighbour model and obeying $0 < \|x-y\|_\infty \leq L$ for the spread-out model. Bonds are occupied with probability p. We write $(x,m) \to (y,n)$ if there is an oriented path from (x,m) to (y,n) consisting of occupied bonds, and define $C(x,m) = \{(y,n) : (x,m) \to (y,n)\}$. Let

$$\sigma^{(2)}((x,n); N) = P_{p_c}(C(0,0) \ni (x,n), |C(0,0)| = N) \tag{2.6.1}$$

denote the probability at the oriented percolation critical point that $(0,0)$ is connected to (x,n) via a cluster containing N sites. We denote the Fourier transform with respect to x by

$$\hat{\sigma}^{(2)}((k,n); N) = \sum_{x \in \mathbb{Z}^d} \sigma^{(2)}((x,n); N) e^{ik \cdot x}, \quad k \in [-\pi, \pi]^d, \tag{2.6.2}$$

and define

$$\hat{\sigma}^{(2)}_{z,\zeta}(k) = \sum_{N=1}^{\infty}\sum_{n=0}^{\infty} \hat{\sigma}^{(2)}((k,n);N) z^N \zeta^n, \quad |z|, |\zeta| < 1. \tag{2.6.3}$$

The symmetry under $x \to -x$ is responsible for the absence of a term linear in k in the denominators of (2.4.5) and (2.5.5). This symmetry applies also for oriented percolation, but there is no such symmetry for the time variable n and a term linear in $(1-\zeta)$ should appear. Thus we expect that above the upper critical dimension, i.e., for $d+1>5$,

$$\hat{\sigma}^{(2)}_{z,\zeta}(k) = \frac{C_3}{D_3^2 k^2 + 2^{3/2}\sqrt{1-z} + 2T_3(1-\zeta)} + \text{error} \tag{2.6.4}$$

as $(k,z,\zeta) \to (0,1,1)$. An upper bound for (2.6.3) of the form $(k^2 + |1-\zeta|)^{-1}$ was obtained for $z=1$ in [27], for the nearest-neighbour model in sufficiently high dimensions and for sufficiently spread-out models when $d+1>5$. This is consistent with (2.6.4).

Apart from constants, the form of (2.6.4) is identical to the generating function $C^{(2)}_{z,\zeta}(k)$ defined in (2.3.1). As in (2.3.7), if (2.6.4) accurately captures the behaviour of the two-point function, as $N \to \infty$ we would have

$$\hat{\sigma}^{(2)}((kD_3^{-1}N^{-1/4}, \lfloor tT_3 N^{1/2} \rfloor); N) \sim C_3 T_3^{-1} \frac{1}{\sqrt{8\pi}N} \hat{a}^{(2)}(k,t). \tag{2.6.5}$$

This suggests ISE as the scaling limit, when time and space are scaled respectively by $N^{-1/2}$ and $N^{-1/4}$. The ISE time variable corresponds simply to the direction of orientation.

Consider now the limit in which the cluster size N is summed over rather than fixed, with $n \to \infty$ and space scaled by $n^{-1/2}$. Summing over N removes any conditioning on the cluster size, so SBM becomes relevant as the scaling limit, rather than ISE. According to the above picture, we can expect that

$$\hat{\sigma}^{(2)}_{1,\zeta}(k) = \frac{C_3}{D_3^2 k^2 + 2T_3(1-\zeta)} + \text{error}. \tag{2.6.6}$$

As in (2.3.4), with sufficient control on the error (2.6.6) implies

$$2C_3^{-1}T_3 \lim_{n\to\infty} \sum_{N=1}^{\infty} \hat{\sigma}^{(2)}((kT_3^{1/2}D_3^{-1}n^{-1/2}, \lfloor tn \rfloor); N) = e^{-k^2t/2}. \tag{2.6.7}$$

In fact, (2.6.6)–(2.6.7) were proven in [28] for the nearest-neighbour model in sufficiently high dimensions and for sufficiently spread-out models when $d+1>5$. Work is in progress with Derbez and van der Hofstad to prove a corresponding result for higher-order connectivity functions, to obtain a stronger statement of convergence to SBM. This work in progress is based on the inductive method of [21], which bypasses the use of generating functions and the difficulties associated with their inversion.

The above picture relating SBM and oriented percolation can be contrasted with the results of [13] (see also [8]). In [13], it is shown that SBM arises as the scaling limit of a critical contact process for $d \geq 2$. The scaling limit of [13] is for the infinitely spread-out contact process, in the limit $L \to \infty$ (sometimes called the Kac limit). This is a mean-field limit, for which the non-gaussian behaviour expected below $d+1=5$ when L is finite is no longer relevant.

2.7 The Lace Expansion

The method of proof of the above results is based on the lace expansion, which was first introduced in [7] in the context of self-avoiding walks. Reviews of work on the lace expansion prior to the work described in this paper can be found in [19, 26]. The extensions required to prove the results of Sections 2.4 and 2.5 make use of a double lace expansion and it is beyond the scope of this paper to indicate any details. Details can be found in [12, 15, 16].

Appendix

2.A Proof of Corollary 2.4.1

In this appendix, we show how Corollary 2.4.1 follows from Theorem 2.4.1. The corollary follows in a straightforward way via [10, Lemma 2.4.1(b)], which asserts that weak convergence of moment measures implies weak convergence of random probability measures (on a compact set). However, there is one subtlety. This point was overlooked in [11, 12], and we take this opportunity to clarify it.

For $l \geq 1$, let $s_n^{(l+1)}(x_1, \ldots, x_l)$ denote the number of n-bond lattice trees containing the lattice sites $0, x_1, \ldots, x_l$. To abbreviate the notation, we will write $\tilde{x} = (x_1, \ldots, x_l)$. The lth moment measure $M_n^{(l)}$ of μ_n is the deterministic probability measure on $\mathbb{R}^{dl}$ which places mass

$$r_n^{(l+1)}(\tilde{x}) = \frac{1}{(n+1)^l} \frac{1}{t_n^{(1)}} s_n^{(l+1)}(\tilde{x}) \tag{2.A.1}$$

at $\tilde{x} D_2^{-1} n^{-1/4}$, for $\tilde{x} \in \mathbb{Z}^{dl}$. The characteristic function $\hat{M}_n^{(l)}(k)$ of $M_n^{(l)}$ is given by

$$\hat{M}_n^{(l)}(\tilde{k}) = \hat{r}_n^{(l+1)}(\tilde{k} D_2^{-1} n^{-1/4}), \tag{2.A.2}$$

where, writing $\tilde{k} = (k_1, \ldots, k_l)$ and $\tilde{k} \cdot \tilde{x} = k_1 \cdot x_1 + \cdots + k_l \cdot x_l$,

$$\hat{r}_n^{(l+1)}(\tilde{k}) = \sum_{\tilde{x}} r_n^{(l+1)}(\tilde{x}) e^{i\tilde{k}\cdot\tilde{x}}. \tag{2.A.3}$$

Since $\hat{s}_n^{(l+1)}(\tilde{0}) = (n+1)^l t_n^{(1)}$, we have

$$\hat{M}_n^{(l)}(\tilde{k}) = \frac{\hat{s}_n^{(l+1)}(\tilde{k}D_1^{-1}n^{-1/4})}{\hat{s}_n^{(l+1)}(\tilde{0})}. \tag{2.A.4}$$

To prove convergence of the moment measures of μ_n to those of ISE, it suffices to show that, for each $l \geq 1$, $\hat{M}_n^{(l)}(k)$ converges to the characteristic function $\hat{M}^{(l)}(k)$ of the corresponding ISE moment measure described under (2.2.7). For $l = 1$, this is an immediate consequence of (2.A.4) and Theorem 2.4.1, since $\hat{s}_n^{(2)}(k) = \hat{t}_n^{(2)}(k)$ and $\hat{M}^{(1)}(k) = \hat{A}^{(2)}(k)$. Similarly, for $l = 2$, there is a unique shape and $\hat{s}_n^{(3)}(k_1, k_2) = \hat{t}_n^{(3)}(k_1+k_2, k_1, k_2)$. Since $\hat{M}^{(2)}(k_1, k_2) = \int A^{(3)}(y, x_1 - y, x_2 - y)e^{ik_1 \cdot x_1}e^{ik_2 \cdot x_2}d^d y d^d x_1 d^d x_2 = \hat{A}^{(3)}(k_1 + k_2, k_1, k_2)$, convergence of the second moments follows directly from Theorem 2.4.1.

The convergence of the third and higher moments follows similarly, apart from one detail. For $l \geq 3$, there is more than one shape, and

$$\hat{M}^{(l)}(\tilde{k}) = \sum_{\sigma \in \Sigma_{l+1}} \hat{A}^{(l+1)}(\sigma; \vec{k}) \tag{2.A.5}$$

with each of the $2l - 1$ components of $\vec{k}$ given by a specific linear combination (depending on σ) of the l components of $\tilde{k}$. For example, for $l = 3$ and the shape σ_1 of Figure 2.1, $(\sigma_1; \vec{k}) = (\sigma_1; k_1 + k_2 + k_3, k_1, k_2 + k_3, k_2, k_3)$. If it were the case that $\hat{s}_n^{(l+1)}(\tilde{k})$ were equal to $\sum_{\sigma \in \Sigma_{l+1}} \hat{t}_n^{(l+1)}(\sigma; \vec{k})$, convergence of all moments would be immediate since Theorem 2.4.1 implies that

$$\lim_{n\to\infty} \frac{\sum_{\sigma \in \Sigma_{l+1}} \hat{t}_n^{(l+1)}(\sigma; \vec{k}D_1^{-1}n^{-1/4})}{\sum_{\sigma \in \Sigma_{l+1}} \hat{t}_n^{(l+1)}(\sigma; \vec{0})} = \sum_{\sigma \in \Sigma_{l+1}} \hat{A}^{(l+1)}(\sigma; \vec{k}). \tag{2.A.6}$$

But $\hat{s}_n^{(l+1)}(\tilde{k})$ is not equal to $\sum_{\sigma \in \Sigma_{l+1}} \hat{t}_n^{(l+1)}(\sigma; \vec{k})$, because it is not the case that $s_n^{(l+1)}(\tilde{x})$ is equal to the sum of $t_n^{(l+1)}(\sigma; \vec{y})$ over all $(\sigma; \vec{y})$ that are consistent with $\tilde{x}$ in the sense that the x_i are given by the sum of the y_j as prescribed by the shape σ. The discrepancy arises from degenerate lattice tree configurations, containing sites $x_1, \ldots, x_l$, which can correspond to more than one choice of $(\sigma; \vec{y})$. These configurations can only occur when $l \geq 3$ and at least one y_j is zero.

For example, there is a unique 1-bond lattice tree containing 0 and the site $e_1 = (1, 0, \ldots, 0)$, and hence $s_1^{(4)}(0, 0, e_1) = 1$. However, this lattice tree containing the sites $x_1 = x_2 = 0$, $x_3 = e_1$ contributes to each of $t_1^{(4)}(\sigma_1; 0, 0, 0, 0, e_1)$, $t_1^{(4)}(\sigma_2; 0, 0, 0, 0, e_1)$ and $t_1^{(4)}(\sigma_3; 0, e_1, 0, 0, 0)$. See Figure 2.1. Thus it is not the case, in general, that $s_n^{(l+1)}(\tilde{x})$ is given by the sum of $t_n^{(l+1)}(\sigma; \vec{y})$ over all corresponding $(\sigma; \vec{y})$. The assertion of [11, (3.4)] and [12, (1.11)] that $\sum_{\sigma \in \Sigma_{l+1}} \hat{t}_n^{(l+1)}(\sigma; \vec{0})$ equals $(n+1)^l t_n^{(1)}$ implicitly assumed uniqueness of $(\sigma; \vec{y})$ and is incorrect for $l \geq 3$. This false assertion was not needed in [11, 12], as it can be replaced by [12, (2.14)-(2.15)] with $\vec{k} = \vec{0}$ (i.e., (2.4.9) above) and [12, (1.12)] to conclude that

$$\sum_{\sigma\in\Sigma_{l+1}} \hat{t}_n^{(l+1)}(\sigma;\vec{0}) \sim c_1 n^{l-3/2} z_c^{-n} \sim (n+1)^l t_n^{(1)}, \tag{2.A.7}$$

which is sufficient for [11, 12]. The degenerate cases appear in error terms to (2.A.7) and do not affect the leading behaviour.

In view of (2.A.4)–(2.A.6), to prove convergence of the lth moments, for $l \geq 3$, it suffices to show that

$$\left| \hat{s}_n^{(l+1)}(\tilde{k}) - \sum_{\sigma\in\Sigma_{l+1}} \hat{t}_n^{(l+1)}(\sigma;\vec{k}) \right| \leq O(n^{l-2} z_c^{-n}). \tag{2.A.8}$$

This difference then constitutes an error term, down by $n^{-1/2}$ compared to $\hat{s}_n^{(l+1)}(\tilde{k})$, by Theorem 2.4.1. The remainder of the proof is devoted to obtaining (2.A.8).

Let $l \geq 3$, and recall the definition of compatibility above (2.4.1). If the backbone of $(T; 0, x_1, \ldots, x_l)$ comprises $2l-1$ *nontrivial* paths (each having length greater than zero), then $\tilde{x}$ induces a labelling of the external vertices of an $(l+1)$-skeleton and there is therefore a unique compatible $(\sigma; \vec{y}, \vec{s})$. Whether or not the backbone comprises $2l-1$ nontrivial paths, given $(\sigma; \vec{y}, \vec{s})$ compatible with the backbone, the $2l-1$ backbone displacements $\vec{y}$ and their lengths $\vec{s}$ (possibly zero) are uniquely determined by σ and $(T; 0, x_1, \ldots, x_l)$. Nonuniqueness of $(\sigma; \vec{y}, \vec{s})$ thus requires at least one of the backbone paths to be trivial, and, in such a degenerate case, the maximum possible number of compatible choices for $(\sigma; \vec{y}, \vec{s})$ is the number of shapes, which is $(2l-3)!!$. Let $u_n^{(l+1)}(\tilde{x})$ denote the number of n-bond lattice trees for which each of the $2l-1$ backbone paths is nontrivial, and let $e_n^{(l+1)}(\tilde{x})$ denote the number of n-bond lattice trees for which at least one backbone path has a zero displacement. Then $s_n^{(l+1)}(\tilde{x}) = u_n^{(l+1)}(\tilde{x}) + e_n^{(l+1)}(\tilde{x})$, and, for $l \geq 3$,

$$\left| \hat{s}_n^{(l+1)}(\tilde{k}) - \sum_{\sigma\in\Sigma_{l+1}} \hat{t}_n^{(l+1)}(\sigma;\vec{k}) \right| \leq [(2l-3)!! - 1]\hat{e}_n^{(l+1)}(\tilde{0}). \tag{2.A.9}$$

It suffices to argue that the right side of (2.A.9) is at most $O(n^{l-2} z_c^{-n})$. For this, we introduce the generating function $E^{(l+1)}(z) = \sum_n \hat{e}_n^{(l+1)}(\tilde{0}) z^n$. Let $\chi(z) = \sum_x G_z^{(2)}(x)$. It can be shown using standard bounds that $|E^{(l+1)}(z)| \leq O(\chi(|z|)^{2l-2})$, where the power $2l-2$ arises because at least one of the $2l-1$ backbone paths is trivial. Using the methods of [12], this can be refined to $|E^{(l+1)}(z)| \leq O(|\chi(z)|^3 \chi(|z|)^{2l-5})$, uniform in $|z| < z_c$. It follows from [18, (1.12)] that $|E^{(l+1)}(z)| \leq O(|1 - z/z_c|^{-3/2}(1 - |z|/z_c)^{-l+5/2})$. Then [12, Lemma 3.2(i)] implies the desired bound $\hat{e}_n^{(l+1)}(\tilde{0}) \leq O(n^{l-2} z_c^{-n})$. □

Acknowledgments: This work was supported in part by NSERC. It is a pleasure to thank Eric Derbez and Takashi Hara for the enjoyable collaborations that led to the results described in this paper, and Christian Borgs, Jennifer Chayes, Remco van der Hofstad and Ed Perkins for valuable conversations. This paper was written primarily during a visit to Microsoft Research.

REFERENCES

[1] M. Aizenman. Scaling limit for the incipient spanning clusters. In K.M. Golden, G.R. Grimmett, R.D. James, G.W. Milton, and P.N. Sen, editors, *Mathematics of Materials: Percolation and Composites*. Springer-Verlag, New York, (1997).

[2] M. Aizenman and A. Burchard. Hölder regularity and dimension bounds for random curves. *Duke Math. J.* To appear.

[3] M. Aizenman and C.M. Newman. Tree graph inequalities and critical behavior in percolation models. *J. Stat. Phys.*, **36**:107–143, (1984).

[4] D. Aldous. The continuum random tree III. *Ann. Probab.*, **21**:248–289, (1993).

[5] D. Aldous. Tree-based models for random distribution of mass. *J. Stat. Phys.*, **73**:625–641, (1993).

[6] C. Borgs, J.T. Chayes, R. van der Hofstad, and G. Slade. Mean-field lattice trees. *Ann. Combinat.* To appear.

[7] D.C. Brydges and T. Spencer. Self-avoiding walk in 5 or more dimensions. *Commun. Math. Phys.*, **97**:125–148, (1985).

[8] T. Cox, R. Durrett, and E.A. Perkins. Rescaled particle systems converging to super-Brownian motion. In this volume.

[9] D. Dawson and E. Perkins. Measure-valued processes and renormalization of branching particle systems. In R. Carmona and B. Rozovskii, editors, *Stochastic Partial Differential Equations: Six Perspectives*. AMS Math. Surveys and Monographs, (1998).

[10] D.A. Dawson. Measure-valued Markov processes. In *Ecole d'Eté de Probabilités de Saint-Flour 1991. Lecture Notes in Mathematics #1541*, Springer, Berlin, (1993).

[11] E. Derbez and G. Slade. Lattice trees and super-Brownian motion. *Canad. Math. Bull.*, **40**:19–38, (1997).

[12] E. Derbez and G. Slade. The scaling limit of lattice trees in high dimensions. *Commun. Math. Phys.*, **193**:69–104, (1998).

[13] R. Durrett and E.A. Perkins. Rescaled contact processes converge to super-Brownian motion for $d \geq 2$. To appear in *Probab. Th. Rel. Fields*.

[14] G. Grimmett. *Percolation*. Springer, Berlin, (1989).

[15] T. Hara and G. Slade. The scaling limit of the incipient infinite cluster in high-dimensional percolation. I. Critical exponents. Preprint.

[16] T. Hara and G. Slade. The scaling limit of the incipient infinite cluster in high-dimensional percolation. II. Integrated super-Brownian excursion. Preprint.

[17] T. Hara and G. Slade. On the upper critical dimension of lattice trees and lattice animals. *J. Stat. Phys.*, **59**:1469–1510, (1990).

[18] T. Hara and G. Slade. The number and size of branched polymers in high dimensions. *J. Stat. Phys.*, **67**:1009–1038, (1992).

[19] T. Hara and G. Slade. Mean-field behaviour and the lace expansion. In G. Grimmett, editor, *Probability and Phase Transition*, Kluwer, Dordrecht, (1994).

[20] T. Hara and G. Slade. The incipient infinite cluster in high-dimensional percolation. *Electron. Res. Announc. Amer. Math. Soc.*, **4**:48–55, (1998). `http:\\` www.ams.org\era\.

[21] R. van der Hofstad, F. den Hollander, and G. Slade. A new inductive approach to the lace expansion for self-avoiding walks. *Probab. Th. Rel. Fields*, **111**:253–286, (1998).
[22] H. Kesten. The incipient infinite cluster in two-dimensional percolation. *Probab. Th. Rel. Fields*, **73**:369–394, (1986).
[23] J.-F. Le Gall. Branching processes, random trees and superprocesses. In *Proceedings of the International Congress of Mathematicians, Berlin, 1998*, volume III, pages 279–289, (1998). *Documenta Mathematica*, Extra Volume ICM 1998.
[24] J.-F. Le Gall. The uniform random tree in a Brownian excursion. *Probab. Th. Rel. Fields*, **96**:369–383, (1993).
[25] J.-F. Le Gall. The Hausdorff measure of the range of super-Brownian motion. In this volume.
[26] N. Madras and G. Slade. *The Self-Avoiding Walk*. Birkhäuser, Boston, (1993).
[27] B.G. Nguyen and W-S. Yang. Triangle condition for oriented percolation in high dimensions. *Ann. Probab.*, **21**:1809–1844, (1993).
[28] B.G. Nguyen and W-S. Yang. Gaussian limit for critical oriented percolation in high dimensions. *J. Stat. Phys.*, **78**:841–876, (1995).

Department of Mathematics and Statistics
McMaster University
Hamilton, ON, Canada L8S 4K1

Present address:
Department of Mathematics
University of British Columbia
Vancouver, BC, Canada V6T 1Z2
slade@math.ubc.ca

3

Percolation in $\infty + 1$ Dimensions at the Uniqueness Threshold

Roberto H. Schonmann

ABSTRACT For independent density p site percolation on the (transitive non-amenable) graph $\mathbb{T}_b \times \mathbb{Z}$, where $\mathbb{T}_b$ is a homogeneous tree of degree $b + 1$, and b is supposed to be large, it is shown that for $p = p_u = \inf\{p\colon \text{a.s. there is a unique infinite cluster}\}$ there are a.s. infinitely many infinite clusters. This contrasts with a recent result of Benjamini and Schramm, according to whom for transitive non-amenable planar graphs there is a.s. a unique infinite cluster at p_u.

Keywords: Percolation, transitive graphs, Cayley graphs, number of infinite clusters, critical point.

AMS Subject Classification: 60K35.

3.1 Introduction and Results

Percolation theory has for a long time been a central area of research in both Probability and Mathematical Physics. Two complementary reasons concur for this. In the words of Kesten in the preface to [Kes2]: "... it is a source of fascinating problems of the best kind a mathematician can wish for: problems which are easy to state with a minimum of preparation, but whose solutions are (apparently) difficult and require new methods. At the same time many of the problems are of interest to or proposed by statistical physics and not dreamed up merely to demonstrate ingenuity."

Until recent years most of the interest in percolation concerned the case of the d-dimensional cubic lattice $\mathbb{Z}^d$ (and other graphs adapted to the geometry of Euclidean space $\mathbb{R}^d$ in some periodic fashion; see [Kes2] and [Gri]). This state of affairs has been changing with the perception that, at least from the mathematical viewpoint, the proper setting for studying percolation problems is a more general graph. [Lyo2] studied percolation on trees and found a sharp relation between the value of the critical point and the branching number of the tree. The paper [GN] showed that phenomena which are excluded on $\mathbb{Z}^d$ can occur on other (decent) graphs, e.g., we can have the presence of a regime with infinitely many infinite clusters, and another one with a single infinite cluster. More recently a number of papers have appeared presenting a series of results and raising a number of natural and interesting questions concerning percolation on very general graphs. This includes the papers [BS1, Lal1, BB, BLPS1, BLPS2, HP, Sch2, BS2, HPS]. Moreover the study of percolation on graphs is part of a larger trend of studying processes which also include interacting particle systems and statistical mechanics models on such general graphs. In this connection the reader is invited to consult, for instance, [Lyo1, NW, Pem, SeS, Wu3, SaS1, SaS3, ST, JS, BHW], and references therein. Good reasons exist for the study of such processes, even if one's ultimate interest lies in the processes on $\mathbb{Z}^d$. The study of processes on more general graphs has given rise to new techniques and has helped in elucidating which mathematical features of $\mathbb{Z}^d$ are crucial for the behavior of a process there to be what it is. This study may also provide new insights on what to expect on $\mathbb{Z}^d$.

Before we turn to our main topic, we will review some of the relevant concepts and results pertaining to percolation on an infinite connected graph of bounded degree $G = (V, E)$. Here V is the (necessarily countable) set of vertices (also called sites) of G, and E is the set of edges (also called bonds) of G. Sites will be said to be neighbors if they belong to a common edge. A chain is a sequence $x_0, x_1, \dots, x_n$, of distinct sites in which for each i, x_i is neighbor to x_{i+1}; the length of this chain is n, and x_0 and x_n are its end-points. The distance $d(y, z)$ between sites y and z is the minimal length of the chains which have y and z as end-points. The distance $d(y, S)$ between a site y and a set of sites S is the minimal distance between y and any site in S. The ball of center $x \in V$ and radius n is the set

$$B(x, n) = \{y \in V: d(x, y) \le n\}.$$

The outer boundary of a set $S \subset V$ is the set

$$\partial S = \{y \notin S: d(y, S) = 1\}.$$

Given a pair of graphs $G_1 = (V_1, E_1)$ and $G_2 = (V_2, E_2)$, we will let $G_1 \times G_2$ denote the graph which has as vertices the elements of the Cartesian product $V_1 \times V_2$ and an edge connects (x, u) to (y, v) iff either $x = y$ and $\{u, v\} \in E_2$ or else $\{x, y\} \in E_1$ and $u = v$.

An isomorphism between two graphs, $G_1 = (V_1, E_1)$ and $G_2 = (V_2, E_2)$, is a one-to-one mapping from V_1 onto V_2 which preserves the graph structure, i.e., such that the set of edges of G_2 can be obtained as the set of pairs of images of vertices of G_1 which form edges. An isomorphism between a graph G and itself is called an automorphism of G.

A graph is said to be transitive (or homogeneous) if for each pair x and y of its vertices there is an automorphism of the graph which maps x into y. Intuitively speaking, in a transitive graph all the vertices play exactly the same role.

A graph is said to be quasi-transitive if there is a finite set of vertices, V_0, with the property that each vertex of the graph can be mapped into one of the vertices of V_0 by an automorphism. Intuitively speaking, in a quasi-transitive graph there is a finite number of types of vertices and all the vertices of the same type play exactly the same role.

A graph is said to be amenable if there is a sequence of finite sets $S_n \subset V$, such that $\lim_{n\to\infty} |\partial S_n|/|S_n| = 0$.

An important class of transitive graphs is that of the Cayley graphs of finitely generated groups. Suppose that V is such a group and that g is a finite symmetric set of generators for it. Then the (right) Cayley graph of V for this set of generators is the graph $G = (V, E)$ which has $E = \{\{x, y\} : x, y \in V,\ y = xz \text{ for some } z \in g\}$. Cayley graphs are easily seen to be transitive. A number of important facts conjectured to hold for percolation on the full class of transitive graphs has been proved for Cayley graphs. This is so because for Cayley graphs a very powerful technique known as "mass transport technique" is available (see [BLPS1, BLPS2, HP]). This basic tool is also available for a larger class of graphs, called unimodular quasi-transitive graphs, but this class of graphs is known not to contain all the transitive graphs (see definition and discussion in [BLPS1]).

We will consider independent site percolation on graphs. Independent bond percolation is a particular case thereof, since the edges of a graph can be considered as the sites of another graph. (Note that in this procedure, if the original graph is quasi-transitive, then so is also the new graph.) Declare each site of a graph $G = (V, E)$ to be independently occupied with probability p and to be vacant otherwise. The corresponding probability measure will be denoted by $\mathbb{P}_p = \mathbb{P}$, and the corresponding expectation will be denoted by $\mathbb{E}_p = \mathbb{E}$. We will use the notation $A \longleftrightarrow B$ to denote the event that there is a fully occupied chain with one endpoint in $A \subset V$ and the other one in $B \subset V$. In this case we say that A is connected to B. The notation $A \overset{C}{\longleftrightarrow} B$ will denote the event that the mentioned chain is contained in the set C. The notation $A \longrightarrow \infty$ will denote the event that some site of the set $A \subset V$ is connected by occupied chains to infinitely many sites. As usual, we may replace singleton sets with their unique element, in order to simplify the notation. Also to simplify the notation, we may confuse a graph with its set of vertices.

The connected components of the graph which has as vertices the occupied sites, and as edges the edges in E which connect any two of these vertices, are called clusters. The random variable N will denote the number of infinite clusters. The following result was proved first in [HP] in the case of unimodular quasi-transitive graphs and then in [Sch2] with the assumption of unimodularity removed.

Theorem 3.1.1 *Suppose that G is a quasi-transitive graph. Then there are two critical points $0 < p_c \le p_u \le 1$ such that*

- *if $p < p_c$, then $\mathbb{P}_p(N = 0) = 1$,*
- *if $p_c < p < p_u$, then $\mathbb{P}_p(N = \infty) = 1$,*
- *if $p_u < p$, then $\mathbb{P}_p(N = 1) = 1$.*

The fact that for each value of p one of the three events in the alternatives above must have probability 1 is an older result from [NS], based on the 0-1 law for automorphism invariant events. The extra non-trivial information in the theorem is the statement that the uniqueness of the infinite cluster is a monotone property of p, i.e., once it holds for a certain value of p it will also hold for larger values of p. It is easy to see that without making any assumption on the graph this would be false.

There are, of course, cases in which p_c, p_u, or both are degenerate, i.e., take the value 1, or are identical to each other. The most remarkable result in this direction is that $p_c = p_u$ on $\mathbb{Z}^d$. This was first proved in [AKN], and later in a more elementary and also more general fashion in [BK]. The [BK] proof can actually be adapted to show the same result for all the amenable quasi-transitive graphs.

We will denote by $\mathbb{T}_b$ the homogeneous tree with degree $b+1$. The case $b = 1$ corresponds to the linear chain $\mathbb{Z}$, which has $p_c = p_u = 1$. When $b \geq 2$ we have from elementary branching process theory that $p_c = 1/b < 1 = p_u$.

The first example of a transitive graph for which it was shown that there is a non-degenerate interval of values of p on which $\mathbb{P}_p(N = \infty) = 1$ and another one on which $\mathbb{P}_p(N = 1) = 1$ – so that in the terminology above $p_c < p_u < 1$ – is the graph $\mathbb{T}_b \times \mathbb{Z}$, with b large enough. This was done in [GN] for bond percolation, but the same arguments apply also to site percolation. Note that the graph $\mathbb{T}_b \times \mathbb{Z}$ is not only transitive, but is also a Cayley graph. Further results of this nature appear in [BS1] and [Lal1].

It is very natural to ask what happens at the critical points p_c and p_u. A standard conjecture is that for all the quasi-transitive graphs for which $p_c < 1$, $P_{p_c}(N = 0) = 1$. This is known to be true for a class of planar graphs which includes bond and site percolation on $\mathbb{Z}^2$ (this was proved in one case in [Kes1] and generalized in [Rus]). And it is also known to hold for another class of graphs which includes bond and site percolation on $\mathbb{Z}^d$ with large d (this was proved in [HS] – in that paper the so called triangle condition was verified; a proof that this condition implies absence of percolation at p_c can be found in the union of the papers [AN] and [New] or in the paper [BA]). More recently it has been shown in [BLPS1] and [BLPS2] that, for bond or site percolation on any unimodular quasi-transitive non-amenable graph, this conjecture is also true.

Regarding now the behavior at p_u, [BS2] proved that, for planar non-amenable transitive graphs, $\mathbb{P}_{p_u}(N = 1) = 1$. Here we will prove that in contrast

Theorem 3.1.2 *For the graph $\mathbb{T}_b \times \mathbb{Z}$ with $b \geq 2$,*

$$\mathbb{P}_{p_u}(N = 1) = 0.$$

So if b is such that $p_c < p_u$, then

$$\mathbb{P}_{p_u}(N = \infty) = 1.$$

Therefore we learn that, for transitive graphs (or even Cayley graphs) with $p_c < p_u < 1$, both behaviors are possible at p_u: we can have $\mathbb{P}_{p_u}(N = 1) = 1$ or $\mathbb{P}_{p_u}(N = \infty) = 1$. It would be interesting to know for which graphs each alternative holds. The technique in this paper can be generalized to a small extent, so that if H is a quasi-transitive graph, then in the statement of Theorem 3.1.2, we can have the graph $\mathbb{T}_b \times H$ in place of the graph $\mathbb{T}_b \times \mathbb{Z}$. The proof can also be adapted to bond percolation on the same graphs. The main point of this paper is, nevertheless, to simply provide an example which shows that a certain type of behavior is possible. For this reason, and since the generalizations would require more involved notation, we restrict ourselves to only proving Theorem 3.1.2 as it is stated above. Note that [BS1, Question 5] is now answered by Theorem 3.1.1, the work of [BS2] and Theorem 3.1.2.

The condition that b be large for $p_c < p_u$ to happen can be improved. Using techniques from [Sta] and from the present paper one can show that $p_c < p_u$ for all $b \geq 2$, for bond or site percolation. This proof will be omitted from this paper since it is quite technical and moreover it is a very special result; it is conjectured that, for all the non-amenable quasi-transitive graphs, one should have $p_c < p_u$. Similarly, using also techniques from [Sch1], one can prove that the so called triangle condition holds below p_u when $b \geq 2$. This improves a result of [Wu1], and implies that several critical exponents (at p_c) take their mean-field value when $b \geq 2$. Again, the proof is omitted from this paper for the same reason explained above.

One of the ingredients of our proof of Theorem 3.1.2 will be a characterization of p_u in terms of connectivities between large balls. This result, from [Sch2], states the following.

Theorem 3.1.3 *For every quasi-transitive graph*

$$p_u = \inf \left\{ p\colon \lim_{n\to\infty} \inf_{x,y\in V} \mathbb{P}_p(B(x,n) \longleftrightarrow B(y,n)) = 1 \right\}. \tag{3.1.1}$$

Another ingredient of the proof will be the behavior of a function $\alpha(p)$, defined below, which is simply an inverse correlation length. To define it, we need next to present some standard arguments.

We will think of $\mathbb{Z}$ as a subgraph of $\mathbb{T}_b$, so that it will make sense to talk about the site $n \in \mathbb{T}_b$, and let $n \to \infty$.

From Harris' inequality we have $\mathbb{P}_p((0,0) \longleftrightarrow (m+n,0)) \geq \mathbb{P}_p((0,0) \longleftrightarrow (m,0))\mathbb{P}_p((0,0) \longleftrightarrow (n,0))$, and therefore

$$\lim_{n\to\infty} \left(\mathbb{P}_p((0,0) \longleftrightarrow (n,0))\right)^{1/n} = \sup_{n\geq 1}\{(\mathbb{P}_p((0,0) \longleftrightarrow (n,0)))^{1/n}\} = \alpha(p). \tag{3.1.2}$$

Note that

$$\mathbb{P}_p((0,0) \longleftrightarrow (n,0)) \leq (\alpha(p))^n, \tag{3.1.3}$$

for $n \geq 0$. Use now (3.1.2) to write

$$\alpha(p) = \sup_{n\geq 1} \sup_{k\geq 1} \left(\mathbb{P}_p\left((0,0) \overset{B((0,0),k)}{\longleftrightarrow} (n,0)\right)\right)^{1/n},$$

and observe that here we are taking the supremum of a continuous function of p. Therefore $\alpha(\cdot)$ is lower-semi-continuous and since it is clearly non-decreasing, it is also left-continuous.

The technically most involved part of the proof of Theorem 3.1.2 is the proof of the following proposition, which is deferred to Section 3.2.

Proposition 3.1.1 *If* $\alpha(p) > 1/\sqrt{b}$ *then*

$$\lim_{n\to\infty} \inf_{x,y\in V} \mathbb{P}_p(B(x,n) \longleftrightarrow B(y,n)) = 1. \tag{3.1.4}$$

From this proposition and (3.1.1) we learn that for $p < p_u$ we must have $\alpha(p) \le 1/\sqrt{b}$. By the left-continuity of $\alpha(\cdot)$ we obtain then

$$\alpha(p_u) \le 1/\sqrt{b}.$$

But when $P_p(N = 1) = 1$, we obtain from Harris' inequality and transitivity of the graph that, for all $x, y \in V$,

$$\mathbb{P}_p(x \longleftrightarrow y) \ge \mathbb{P}_p(x \longrightarrow \infty, y \longrightarrow \infty) \ge \mathbb{P}_p(x \longrightarrow \infty)\mathbb{P}_p(y \longrightarrow \infty) > 0,$$

uniformly in x, y, so that in particular $\alpha(p) = 1$. This concludes the proof of Theorem 3.1.2 based on Proposition 3.1.1.

The proof of Proposition 3.1.1 is the subject of Section 3.2; a heuristic reasoning that gives meaning to the value $1/\sqrt{b}$ in Proposition 3.1.1 will also be provided there. Section 3.3 will contain some remarks about the origin of the approach in this paper and its connection with the study of the contact process on homogeneous trees.

This section ends with the following insight, which is inspired by Theorem 3.1.2. As mentioned before, the standard conjecture about the behavior of independent percolation on transitive graphs at p_c, when $0 < p_c < 1$, is that one should have $\mathbb{P}_{p_c}(N = 0) = 1$. This conjecture seems to be motivated mostly by the fact that there is now a number of cases in which it has been proved (see above) and there is no known counterexample. In contrast we know now that we can have $p_c < p_u < 1$ with each one of the possible behaviors at p_u: $\mathbb{P}_{p_u}(N = 1) = 1$ and $\mathbb{P}_{p_u}(N = \infty) = 1$. Couldn't it then also be the case that at p_c we do not always have the same behavior, but that for some transitive graphs we have $0 < p_c < 1$ and $\mathbb{P}_{p_c}(N \neq 0) = 1$?

3.2 Proof of Proposition 3.1.1

We will use the following notation for subsets of $\mathbb{Z}$. For $j \le k$, $[j, k] = \{j, \dots, k\}$; for $j + 1 \le k$, $]j, k] = \{j + 1, \dots, k\}$, $[j, k[= \{j, \dots, k - 1\}$; for $j + 1 \le k - 1$, $]j, k[= \{j + 1, \dots, k - 1\}$.

Regarding some subgraphs of $\mathbb{T}_b$, we will use the following notation. The tree obtained by removing all the sites which can only reach the site 0 through a chain which contains the site -1 is denoted $\mathbb{T}_b^+$. Given a site $x \neq 0$, we denote by $\mathbb{T}_b(x)$ the tree obtained by removing the sites which can reach the site 0 through a chain which does not contain x. Note that each graph $\mathbb{T}_b(x)$ is isomorphic to $\mathbb{T}_b^+$, with x taking the role of 0.

Given a subgraph H of $\mathbb{T}_b$, the notation $S_H(n)$ will denote the set of sites which in $\mathbb{T}_b$ are at distance n from 0 and are also sites of H.

Similarly to the considerations related to the definition of $\alpha(p)$, we have also some further standard facts. For each k the following limit exists.

$$\lim_{n\to\infty}\left(\mathbb{P}\left((0,0) \overset{\mathbb{T}_b\times[-k,k]}{\longleftrightarrow} (n,0)\right)\right)^{1/n} = \sup_{n\geq 1}\left\{\left(\mathbb{P}\left((0,0) \overset{\mathbb{T}_b\times[-k,k]}{\longleftrightarrow} (n,0)\right)\right)^{1/n}\right\} = \alpha_k(p),$$

and the bound

$$\mathbb{P}\left((0,0) \overset{\mathbb{T}_b\times[-k,k]}{\longleftrightarrow} (n,0)\right) \leq (\alpha_k(p))^n, \tag{3.2.1}$$

holds for all $n \geq 0$.

The following lemma is also standard.

Lemma 3.2.1

$$\lim_{k\to\infty} \alpha_k(p) = \alpha(p).$$

PROOF. It is clear that for all k, $\alpha_k(p) \leq \alpha(p)$ and that $\alpha_k(p)$ increases with k to some limit bounded above by $\alpha(p)$. On the other hand, by (3.1.2) and (3.2.1), for any small $\epsilon > 0$ there is an n such that

$$(\alpha(p)-\epsilon))^n \leq \mathbb{P}((0,0) \longleftrightarrow (n,0)) = \lim_{k\to\infty} \mathbb{P}\left((0,0) \overset{\mathbb{T}_b\times[-k,k]}{\longleftrightarrow} (n,0)\right)$$
$$\leq \lim_{k\to\infty} (\alpha_k(p))^n = \left(\lim_{k\to\infty} \alpha_k(p)\right)^n.$$

Therefore $\lim_{k\to\infty}\alpha_k(p) \geq \alpha(p)-\epsilon$. Since ϵ is arbitrary, we have $\lim_{k\to\infty}\alpha_k(p) \geq \alpha(p)$, finishing the proof. □

For $x \in \mathbb{T}_b$, let now E_x be the event that all the sites in $\{x\} \times [0, s[$ are occupied, and define

$$Y_{n,s} = \mathbb{P}\left(E_0 \cap E_n \cap \left\{(0,0) \overset{\mathbb{T}_b^+\times[0,s[}{\longleftrightarrow} (n, s-1)\right\}\right),$$

Lemma 3.2.2 *There exists a sequence $(s(n))_{n\geq 1}$, $s(n) \geq 1$, such that*

$$\liminf_{n\to\infty} \left(Y_{n,s(n)}\right)^{1/n} \geq \alpha(p).$$

PROOF. By Harris' inequality, when s is even,

$$Y_{n,s} \geq p^{2s}\, \mathbb{P}\left((0, s/2) \overset{\mathbb{T}_b\times[0,s[}{\longleftrightarrow} (n, s/2)\right).$$

So that the lemma follows at once from Lemma 3.2.1 and the homogeneity of the graph $\mathbb{T}_b \times \mathbb{Z}$. □

The statement and the proof of the next lemma will be easier to understand with the following picture in mind. Suppose that for some $C_1, C_2 > 0$ the probability of connecting $(0, 0)$ to $(C_1 i, C_2 i)$ is of order $\geq \alpha^{C_1 i}$, when i is large. Then the probability of connecting $(0, 0)$ to $(0, 2C_2 i)$ by a path which goes through $(C_1 i, C_2 i)$ (or any other fixed site in $S_{\mathbb{T}_b}(C_1 i) \times \{C_2 i\}$) is of order $\geq \alpha^{2C_1 i}$. There are $\sim b^{C_1 i}$ vertices in $\mathbb{T}_b$ at distance $C_1 i$ from 0. So, summing over the sites of $S_{\mathbb{T}_b}(C_1 i) \times \{C_2 i\}$, we get an heuristic lower bound $(b\alpha^2)^{C_1 i}$ for the order of magnitude of the probability of $\{(0, 0) \longleftrightarrow (0, 2C_2 i)\}$. Of course, this reasoning only makes sense when $b\alpha^2 << 1$, but it also indicates what should happen when α approaches $1/\sqrt{b}$ and becomes larger than this value. Technically the structure of $\mathbb{T}_b$ allows us to find enough "independence" to show that the heuristic prediction, contained in the next lemma, is correct. (In the proof of this lemma, the role of C_1 will be played by the quantity nl and the role of C_2 will be played by the quantity sl.)

Lemma 3.2.3 *If $\alpha(p) > 1/\sqrt{b}$ then*

$$\inf_{z \in \mathbb{Z}} \mathbb{P}\left((0, 0) \overset{\mathbb{T}_b^+ \times \mathbb{Z}}{\longleftrightarrow} (0, z) \right) > 0. \tag{3.2.2}$$

Proof. Suppose that $\alpha(p) > 1/\sqrt{b}$. Thanks to Lemma 3.2.2, we can take n and $s \geq 1$ such that

$$\left(Y_{n,s}\right)^{1/n} = a > 1/\sqrt{b}. \tag{3.2.3}$$

We will show that if, for a proper choice of a positive integer l, we set

$$r_i = \mathbb{P}\left((0, 0) \overset{\mathbb{T}_b^+ \times [0, 2ils]}{\longleftrightarrow} (0, 2ils) \right), \qquad i = 0, 1, 2, \ldots$$

then

$$\inf_{i=0,1,2,\ldots} r_i > 0. \tag{3.2.4}$$

This clearly suffices for our purposes, since then, by Harris' inequality and symmetry,

$$\inf_{z \in \mathbb{Z}} \mathbb{P}\left((0, 0) \overset{\mathbb{T}_b^+ \times \mathbb{Z}}{\longleftrightarrow} (0, z) \right) \geq p^{2ls} \inf_{i=0,1,2,\ldots} r_i > 0.$$

For $j = 1, 2, \ldots$ set

$$\bar{Z}_j = \left| \left\{ x \in S_{\mathbb{T}_b^+}(jn) : (0, 0) \overset{\mathbb{T}_b^+ \times [0, js[}{\longleftrightarrow} (x, js - 1) \right\} \right|.$$

Clearly $\bar{Z}_j \geq Z_j$, where $(Z_j)_{j=0,1,2,\ldots}$ is a branching process with offspring distribution identical to the distribution of $\bar{Z}_1$. This distribution is concentrated on $\{0, 1, \ldots, b^n\}$ and has mean $\mu_n \geq b^n Y_{n,s} = (ba)^n$. Since the offspring distribution has a finite support and in particular a finite second moment, it follows from standard branching-process theory (see, e.g., [Dur2, Section 4.4, example 4.3, p. 254]) that for some random variable X with mean $\mathbb{E}(X) = 1$

$$\frac{Z_j}{(\mu_n)^j} \to X \qquad \text{a.s.} \qquad \text{as } j \to \infty.$$

In particular, there is $\epsilon > 0$ such that

$$\mathbb{P}\left(\bar{Z}_l \geq \frac{(ba)^{nl}}{2}\right) \geq \mathbb{P}\left(Z_l \geq \frac{(ba)^{nl}}{2}\right) \geq \mathbb{P}\left(Z_l \geq \frac{(\mu_n)^l}{2}\right) \geq \epsilon, \tag{3.2.5}$$

for all large enough l.

Choose now l large enough for (3.2.5) to hold, and also so that

$$\left(1 - \frac{\epsilon}{2}a^{nl}\right)^{\frac{(ba)^{nl}}{2}} \leq \frac{1}{2}. \tag{3.2.6}$$

This last requirement can be fulfilled because

$$\limsup_{l\to\infty} \left(1 - \frac{\epsilon}{2}a^{nl}\right)^{\frac{(ba)^{nl}}{2}} \leq \limsup_{l\to\infty} \exp\left(-\frac{\epsilon}{4}(ba^2)^{nl}\right) = 0,$$

since $ba^2 > 1$ by (3.2.3).

Next we will show inductively in i that

$$r_i \geq \frac{\epsilon}{2}a^{nl}, \tag{3.2.7}$$

verifying therefore the validity of (3.2.4).

For $i = 0$ inequality (3.2.7) is clearly true (since $r_0 = 1$ and $\epsilon, a \leq 1$), and we will show now that, if it is true for i, it is true for $i + 1$. From the spherical symmetry of $\mathbb{T}_b$ we have

$$\begin{aligned} r_{i+1} \geq{} & \mathbb{P}\left((0,0) \overset{\mathbb{T}_b^+ \times [0,(2i+1)ls]}{\longleftrightarrow} S_{\mathbb{T}_b^+}(nl) \times \{(2i+1)ls\}\right) \\ & \cdot \mathbb{P}\left((nl, (2i+1)ls + 1) \overset{\mathbb{T}_b^+ \times [(2i+1)ls+1, 2(i+1)ls]}{\longleftrightarrow} (0, 2(i+1)ls)\right). \end{aligned} \tag{3.2.8}$$

We will now look separately at each one of the factors in the right hand side of (3.2.8). First, noting that the graphs $\mathbb{T}_b(x) \times [ls, (2i+1)ls]$, $x \in S_{\mathbb{T}_b^+}(nl)$ are disjoint and isomorphic to $\mathbb{T}_b^+ \times [0, 2ils]$, we can use independence to write

$$\begin{aligned} & \mathbb{P}\left((0,0) \overset{\mathbb{T}_b^+ \times [0,(2i+1)ls]}{\longleftrightarrow} S_{\mathbb{T}_b^+}(nl) \times \{(2i+1)ls\}\right) \\ & \qquad \geq \mathbb{P}\left(\bar{Z}_l \geq \frac{(ba)^{nl}}{2}\right) \cdot \left(1 - (1 - r_i)^{\frac{(ba)^{nl}}{2}}\right). \end{aligned}$$

Using also symmetries of the graph $\mathbb{T}_b \times \mathbb{Z}$ and independence on disjoint subsets, we have for the second factor in the right hand side of (3.2.8),

$$\mathbb{P}\left((nl, (2i+1)ls+1) \overset{\mathbb{T}_b^+ \times [(2i+1)ls+1, 2(i+1)ls]}{\longleftrightarrow} (0, 2(i+1)ls)\right)$$

$$= \mathbb{P}\left((nl, 0) \overset{\mathbb{T}_b^+ \times [0, ls[}{\longleftrightarrow} (0, ls-1)\right) \geq (Y_{n,s})^l.$$

Combining the last three displays and using (3.2.5), the induction hypothesis (3.2.7), (3.2.6) and (3.2.3) we obtain

$$r_{i+1} \geq \mathbb{P}\left(\bar{Z}_l \geq \frac{(ba)^{nl}}{2}\right) \cdot \left(1 - (1 - r_i)^{\frac{(ba)^{nl}}{2}}\right) \cdot (Y_{n,s})^l \geq \frac{\epsilon}{2} a^{nl}.$$

This completes the proof of (3.2.7) and of Lemma 3.2.3. □

PROOF OF PROPOSITION 3.1.1. The infimum in (3.1.4) can be taken over $x = (0, 0)$, $y = (z, t)$, with $z \in \mathbb{T}_b^+$. We suppose below that x and y are of this form.

Note that Lemma 3.2.3 implies

$$\mathbb{P}\left((0, 0) \overset{\mathbb{T}_b^+ \times \mathbb{Z}}{\longleftrightarrow} (0, j) \text{ for infinitely many } j \geq 0\right) > 0. \tag{3.2.9}$$

Let $A_1(n)$ be the event that some $v \in S_{\mathbb{T}_b \setminus \mathbb{T}_b^+}(n)$ is connected inside of $\mathbb{T}_b(v) \times \mathbb{Z}$ to infinitely many sites in $\{v\} \times \mathbb{Z}_+$. Noting that the b^{n-1} graphs $\mathbb{T}_b(v) \times \mathbb{Z}$, $v \in S_{\mathbb{T}_b \setminus \mathbb{T}_b^+}(n)$ are disjoint and isomorphic to $\mathbb{T}_b^+ \times \mathbb{Z}$ we obtain from (3.2.9)

$$\lim_{n \to \infty} \mathbb{P}(A_1(n)) = 1. \tag{3.2.10}$$

In order to define the next event of interest, we let $|z|$ be the distance in the graph $\mathbb{T}_b$ between the root and the site z, and we set

$$\mathbb{T}_b(z, n) = \bigcup_{u \in S_{\mathbb{T}_b(z)}(|z|+n)} \mathbb{T}_b(u),$$

Note that this is a disjoint union of copies of $\mathbb{T}_b^+$. Now let $A_2(n, y)$ be the event that some $v \in S_{\mathbb{T}_b \setminus \mathbb{T}_b^+}(n)$ is connected inside of $(\mathbb{T}_b(z, n))^c \times \mathbb{Z}$ to all sites in $\partial \mathbb{T}_b(z, n) \times \{j\}$ for some $j \in \mathbb{Z}_+$. It is clear that for all n,

$$\mathbb{P}(A_2(n, y) \mid A_1(n)) = 1. \tag{3.2.11}$$

Since the graph $\mathbb{T}_b(z, n) \times \mathbb{Z}$ is a disjoint union of b^n copies of $\mathbb{T}_b^+ \times \mathbb{Z}$, we obtain, using Lemma 3.2.3 again and independence on disjoint subsets of the graph $\mathbb{T}_b \times \mathbb{Z}$, that

$$\lim_{n \to \infty} \mathbb{P}(B(x, n) \longleftrightarrow B(y, n) \mid A_1(n) \cap A_2(n, y)) = 1, \tag{3.2.12}$$

uniformly in $y = (z, t)$.

Combining (3.2.10), (3.2.11) and (3.2.12), we obtain (3.1.4). □

3.3 Relations with Work on the Contact Process

In this section we will explain how the work in the present paper is related to work on the contact process. The contact process is one of the fundamental examples of an interacting particle system and can be described as follows. At each time $t \geq 0$ each site of a graph $G = (V, E)$ can be healthy (state 0) or infected by a disease (state 1). Infected sites recover at rate 1, independently of anything else, while each healthy site is infected at a rate which is the product of a constant $\lambda > 0$ by the number of its neighboring sites which are infected. The constant λ is called the infection parameter and plays a role similar to that of p in percolation. In what follows we will generally suppose that the reader is familiar with this model, otherwise we refer him/her to [Lig1] or [Dur1] for introductions to it.

That close relations exist between percolation and the contact process is not a surprise, since the graphical construction of the contact process shows that this model is identical to a percolation process which is partially oriented and continuous. This percolation process takes place on $V \times \mathbb{R}_+$, which can be seen as the space-time setting for the model. Most of the relations between work on percolation and work on the contact process have been in the direction of percolation results and techniques leading to developments in our understanding of the contact process. For instance [DG] introduced a rescaling scheme for the one-dimensional contact process and obtained a series of exponential estimates for this model inspired by the rescaling scheme for percolation introduced in [Rus]. One of the most fundamental papers ever written about the contact process, the paper [BG], built on a dynamic rescaling technique devised for percolation in [BGN]. What is particularly interesting about the percolation–contact-process relation in the present paper is that here the relation is inverted, with contact process techniques feeding back to percolation theory.

The relatively long digression provided below seems to be worthwhile for two reasons. For readers who are interested in following the developments in percolation and contact process theory it will explain and emphasize some important connections. Readers who study the papers referred to below will certainly find the technical work in Section 3.2 above less mysterious. The other reason is to advertise once more this important point: the study of statistical mechanics models, percolation and interacting particle systems are interrelated in many ways and progress in each one of them may lead to progress in the others, sometimes in unexpected ways.

The paper [GN] seems to have been one of the sources of motivation for the paper [Pem], where the study of the contact process on homogeneous trees was started. The reason for this connection should be clear from the nature of the space-time setting for this contact processes. While [GN] showed that percolation on $\mathbb{T}_b \times \mathbb{Z}$ (with b large) has three phases, with $N = 0, \infty, 1$, respectively, [Pem] showed that the contact process on $\mathbb{T}_b$ (with b large) has three phases also. In the first one the process (actually the infection) dies out, i.e., if started from a finite set of infected sites, eventually the infection will disappear. In the second phase there is survival (of the infection) in the global sense, meaning that the system does not die out, but there is no local survival (of the infection) as defined below. In the third phase there is local survival (of the infection), meaning that if the process is started from a finite set of infected sites, then with positive probability every site will be infected at arbitrarily large times. The values of λ at the boundary between the first regime and the second is denoted λ_1, and the one between the second and third regimes is denoted λ_2.

It is worth pointing out that while percolation on $\mathbb{T}_b$ is basically a trivial model, since it amounts to a simple branching process, other processes like the Ising model on $\mathbb{T}_b$ have also an interesting theory (see [Geo, Chapter 12]).

[Pem] was followed by a large number of papers dedicated to the study of the contact process on $\mathbb{T}_b$ and some related processes. In chronological order we have: [MS, MSZ, DS, Wu2, Zha, Lig2, Sta, Lig3, LS, SaS2, Lal2], and [Sch1]. We will explain now how the work in some of these papers led to the work presented here. [Zha] proved the following two results: (a) for $\lambda > \lambda_2$ complete convergence holds (roughly speaking, if the process is started from any set of infected sites and it survives, then we have convergence in distribution to a unique non-trivial invariant probability distribution); (b) at $\lambda = \lambda_2$ the process does not survive locally. [Lig3] introduced an analogue of the quantity $\alpha(p)$ in the current paper (let's call it $\alpha_{\text{cp}}(\lambda)$); he obtained several of its properties and conjectured that for $\lambda \leq \lambda_2, \alpha_{\text{cp}}(\lambda) \leq 1/\sqrt{b}$. This conjecture was proved by [LS], who also stressed, as pointed out by Liggett, that this provides a simplified proof of the result (b) of [Zha]. In parallel to these developments, and motivated by the work on the contact process on $\mathbb{T}_b$, [SaS1] and [SaS3] studied the contact process on fairly arbitrary graphs. It was shown in the first of these papers that, for the contact process on a quasi-transitive graph, the threshold for complete convergence (with survival) can be characterized in a way which is analogous to the characterization of p_u in (3.1.1) (see [SaS1, Theorem 2, parts (b) and (i)]). A combination of this result from [SaS1] with ideas from [Zha] and [LS] led to the paper [SaS2], in which a simplified proof of the result (a) of [Zha] was presented. The relation (3.1.1) used here was derived in [Sch2], motivated also by the result of [SaS1] quoted above. Finally here we are building on the techniques of [SaS2], when we prove (3.1.4) and use (3.1.1) in order to prove Theorem 3.1.2. A guiding principle which helped us follow the path just described was the heuristic idea that complete convergence is for the contact process what uniqueness of the infinite cluster is for percolation.

Acknowledgments: As explained above, the techniques used in this paper were motivated by techniques and results on the contact process obtained in the papers [SaS1] and [SaS2]; it is a pleasure to thank Marcia Salzano for her collaboration in that project and for discussions of the current paper. I am very grateful to Itai Benjamini and Oded Schramm for having made a number of important suggestions which led to improvements to the presentation of this paper. It is also a pleasure to thank Maury Bramson and Rick Durrett for organizing the conference and volume in honor of Harry Kesten for which this paper was written. This work was partially supported by the N.S.F. through grant DMS-9703814.

REFERENCES

[AKN] M. Aizenman, H. Kesten, and C.M. Newman. Uniqueness of the infinite cluster and continuity of connectivity functions in short- and long-range percolation. *Communications in Mathematical Physics*, **111** (1987), 505–532.

[AN] M. Aizenman and C.M. Newman. Tree graph inequalities and critical behavior in percolation models. *Journal of Statistical Physics*, **36** (1984), 107–143.

[BB] E. Babson and I. Benjamini. Cut sets and normed cohomology with application to percolation. *Proceedings of the American Mathematical Society*, **127** (1999), 589–597.

[BA] D. Barsky and M. Aizenman. Percolation critical exponents under the triangle condition *Communications in Mathematical Physics*, **19** (1991), 1520–1536.

[BGN] D. J. Barsky, G.R. Grimmett, and C.M. Newman. Percolation in half spaces: equality of critical probabilities and continuity of the percolation probability. *Probability Theory and Related Fields*, **90** (1991), 111–148.

[BLPS1] I. Benjamini, R. Lyons, Y. Peres, and O. Schramm. Group-invariant percolation on graphs. Preprint, 1997.

[BLPS2] I. Benjamini, R. Lyons, Y. Peres, and O. Schramm. Critical percolation on any non-amenable group has no infinite clusters. *The Annals of Probability*, to appear.

[BS1] I. Benjamini and O. Schramm. Percolation beyond $\mathbb{Z}^d$, many questions and a few answers. *Electronic Communications in Probability*, **1** (1996), 71–82.

[BS2] I. Benjamini and O. Schramm. (Paper in preparation.)

[BG] C. Bezuidenhout and G. Grimmett. The critical contact process dies out. *The Annals of Probability*, **18** (1990), 1462–1482.

[BHW] G.R. Brightwell, O. Häggström, and P. Winkler. Nonmonotonic behavior in hard-core and Widom-Rowlinson models. *Journal of Statistical Physics*, to appear.

[BK] R.M. Burton and M. Keane. Density and uniqueness in percolation. *Communications in Mathematical Physics*. **121** (1989), 501–505.

[Dur1] R. Durrett. *Lecture Notes on Interacting Particle Systems and Percolation*. Wadsworth & Brooks/Cole Publ. Co., Belmont, CA, 1988.

[Dur2] R. Durrett. *Probability: Theory and Examples*, Second edition. Duxbury Press, Belmont, CA, 1996.

[DG] R. Durrett and D. Griffeath. *Supercritical contact processes on* $\mathbb{Z}$, *The Annals of Probability*, **11** (1983), 1–15.

[DS] R. Durrett and R.B. Schinazi. Intermediate phase for the contact process on a tree. *The Annals of Probability* **23** (1995), 668–673.

[Geo] H.-O. Georgii. *Gibbs Measures and Phase Transitions*. Walter de Gruyter, Berlin, 1988.

[Gri] G.R. Grimmett. *Percolation*. Springer-Verlag, New York–Berlin, 1989.

[GN] G.R. Grimmett and C.M. Newman. Percolation in $\infty + 1$ dimensions. In *Disorder in Physical Systems*. G. R. Grimmett and D. J. A. Welsh, editors. Claremond Press, Oxford, 1990, 219–240.

[HP] O. Häggström and Y. Peres. Monotonicity of uniqueness for percolation on Cayley graphs: all infinite clusters are born simultaneously. Preprint, 1997.

[HPS] O.Häggström, Y. Peres, and R.H. Schonmann. Percolation on transitive graphs as a coalescent process: relentless merging followed by simultaneous uniqueness. Preprint, 1998.

[HS] T. Hara and G. Slade. Mean-field critical behavior for percolation in high dimensions. *Communications in Mathematical Physics*, **128** (1990), 333–391.

[JS] J. Jonasson and J.E. Steif. Amenability and phase transition in the Ising model. *Journal of Theoretical Probability*, to appear.

[Kes1] H. Kesten. The critical probability of bond percolation on the square lattice equals 1/2. *Communications in Mathematical Physics*, **74** (1980), 41–59.

[Kes2] H. Kesten. *Percolation Theory for Mathematicians*. Birkhäuser, Boston–Basel–Stuttgart, 1982.

[Lal1] S.P. Lalley. Percolation on Fuchsian groups. *Annales de L'Institut Henri Poincaré (Probability and Statistics)*, **34** (1998), 151–177.

[Lal2] S. Lalley. Growth profile and invariant measures for the weakly supercritical contact process on a homogeneous tree. Preprint, 1997.

[LS] S. Lalley and T. Sellke. Limit set of a weakly supercritical contact process on a homogeneous tree. *The Annals of Probability*, **26** (1998), 644–657.

[Lig1] T.M. Liggett. *Interacting Particle Systems*. Springer-Verlag, New York, 1985.

[Lig2] T.M. Liggett. Multiple transition points for the contact process on the binary tree. *The Annals of Probability*, **24** (1996), 1675–1710.

[Lig3] T.M. Liggett. Branching random walks and contact processes on homogeneous trees. *Probability Theory and Related Fields*, **106** (1996), 495–519.

[Lyo1] R. Lyons. The Ising model and percolation on trees and tree-like graphs. *Communications in Mathematical Physics*, **125** (1989), 337–353.

[Lyo2] R. Lyons. Random walk and percolation on trees. *The Annals of Probability*, **18** (1990), 931–958.

[MS] N. Madras and R.B. Schinazi. Branching Random Walk on trees. *Stochastic Processes and Their Applications*, **42** (1992), 255–267.

[MSZ] G.J. Morrow, R.B. Schinazi, and Y. Zhang. The critical contact process on a homogeneous tree. *Journal of Applied Probability*, **31** (1994), 250–255.

[New] C.M. Newman. Some critical exponent inequalities for percolation. *Journal of Statistical Physics*, **45** (1986), 359–368.

[NS] C.M. Newman and L.S. Schulman. Infinite clusters in percolation models. *Journal of Statistical Physics*, **26** (1981), 613–628.

[NW] C.M. Newman and C.C. Wu. Markov fields on branching planes. *Probability Theory and Related Fields*, **85** (1990), 539–552.

[Pem] R. Pemantle. The contact process on trees. *The Annals of Probability*, **20** (1992), 2089–2116.

[Rus] L. Russo. On the critical percolation probabilities. *Zeitschrift für Wahrscheinlichkeitstheorie und Verwandte Gebiete*, **56** (1981), 229–237.

[SaS1] M. Salzano and R.H. Schonmann. The second lowest extremal invariant measure of the contact process. *The Annals of Probability*, **25** (1997), 1846–1871.

[SaS2] M. Salzano and R.H. Schonmann. A new proof that for the contact process on homogeneous trees local survival implies complete convergence. *The Annals of Probability*, **26** (1998), 1251–1258.

[SaS3] M. Salzano and R.H. Schonmann. The second lowest extremal invariant measure of the contact process II.

[Sch1] R.H. Schonmann. The triangle condition for contact processes on homogeneous trees. *Journal of Statistical Physics*, **90** (1998), 1429–1440.

[Sch2] R.H. Schonmann. Stability of infinite clusters in supercritical percolation. *Probability Theory and Related Fields*, **113** (1999), 287–300.

[ST] R.H. Schonmann and N.I. Tanaka. Lack of monotonicity in ferromagnetic Ising model phase diagrams. *The Annals of Applied Probability*, **8** (1998), 234–245.

[SeS] C.M. Series and Ya. G. Sinai. Ising models on the Lobachevsky Plane. *Communications in Mathematical Physics*, **128** (1990), 63–76.

[Sta] A.M. Stacey. The existence of an intermediate phase for the contact process on trees. *The Annals of Probability*, **24** (1996), 1711–1726.

[Wu1] C.C. Wu. Critical behavior of percolation and Markov fields on branching planes. *Journal of Applied Probability*, **30** (1993), 538–547.

[Wu2] C.C. Wu. The contact process on a tree – behavior near the first transition. *Stochastic Processes and Their Applications*, **57** (1995), 99–112.

[Wu3] C.C. Wu. Ising models on hyperbolic graphs. *Journal of Statistical Physics*, **85** (1996), 251–259.

[Zha] Y. Zhang. The complete convergence theorem of the contact process on trees. *The Annals of Probability*, **24** (1996), 1408–1443.

Mathematics Department
University of California at Los Angeles
Los Angeles, CA 90095
rhs@math.ucla.edu

4

Percolation on Transitive Graphs as a Coalescent Process: Relentless Merging Followed by Simultaneous Uniqueness

Olle Häggström, Yuval Peres, and Roberto H. Schonmann

ABSTRACT Consider i.i.d. percolation with retention parameter p on an infinite graph G. There is a well known critical parameter $p_c \in [0, 1]$ for the existence of infinite open clusters. Recently, it has been shown that when G is quasi-transitive, there is another critical value $p_u \in [p_c, 1]$ such that the number of infinite clusters is a.s. ∞ for $p \in (p_c, p_u)$, and a.s. one for $p > p_u$. We prove a simultaneous version of this result in the canonical coupling of the percolation processes for all $p \in [0, 1]$. Simultaneously for all $p \in (p_c, p_u)$, we also prove that each infinite cluster has uncountably many ends. For $p > p_c$ we prove that all infinite clusters are *indistinguishable by robust properties*. Under the additional assumption that G is unimodular, we prove that a.s. for all $p_1 < p_2$ in (p_c, p_u), every infinite cluster at level p_2 contains *infinitely many* infinite clusters at level p_1. We also show that any Cartesian product G of d infinite connected graphs of bounded degree satisfies $p_u(G) \leq p_c(\mathbf{Z}^d)$.

Keywords: Percolation, transitive graphs, uniqueness of infinite cluster, ends of infinite clusters, indistinguishability of infinite clusters, critical points.

AMS Subject Classification: 60K35.

4.1 Introduction

We consider i.i.d. bond percolation with retention parameter $p \in [0, 1]$ on an infinite locally finite connected graph $G = (V, E)$. This means that each edge is independently assigned the value 1 (open) with probability p, and the value 0 (closed) with probability $1 - p$. We write $\mathbf{P}_p^G$, or simply $\mathbf{P}_p$, for the resulting probability measure on $\{0, 1\}^E$. All our results and proofs may be adapted to site percolation as well.

Percolation theory deals with the structure of the connected components of open edges, especially infinite connected components (clusters). By Kolmogorov's zero-one law, the existence of *at least one* infinite cluster has probability 0 or 1, and one defines

$$p_c(G) = \inf\{p \in [0, 1] : \mathbf{P}_p^G(\exists \text{ an infinite cluster}) = 1\}.$$

Following Benjamini and Schramm [9], we also define

$$p_u(G) = \inf\{p \in [0,1] : \mathbf{P}_p^G(\exists \text{ a unique infinite cluster}) = 1\}.$$

For $G = \mathbf{Z}^d$, Aizenman, Kesten and Newman [1] showed that whenever an infinite cluster exists, it is a.s. unique, so that $p_c = p_u$; subsequently, shorter proofs were given in [14] and [11]. (With the usual abuse of notation, we write $\mathbf{Z}^d$ for the graph whose vertex set is $\mathbf{Z}^d$ and whose edge set consists of the pairs of Euclidean nearest neighbors.) For general graphs uniqueness no longer holds, but for a large class of graphs, including the *quasi-transitive* ones (see Definition 4.1.1 below), the number of infinite clusters is an a.s. constant (depending on p) which may be either 0, 1 or ∞. As noted independently by several authors, this follows from the arguments of Newman and Schulman [27].

The pioneering paper of Grimmett and Newman [16] revealed that surprising new phenomena appear when one goes beyond lattices in Euclidean space. The work of Benjamini and Schramm [9] indicated the right level of generality to study these phenomena, and was the impetus for much of the recent progress in percolation theory. Nevertheless, as we shall see in Section 4.8, certain deep results for percolation in $\mathbf{Z}^d$ (uniqueness in orthants, and estimates of p_c) have significant implications beyond the Euclidean setting.

Write $\mathrm{Aut}(G)$ for the group of graph automorphisms of the graph G.

Definition 4.1.1 A graph $G = (V, E)$ is called *transitive* if for any $x, y \in V$ there exists a $\gamma \in Aut(G)$ which maps x to y. The graph G is called *quasi-transitive* if V can be partitioned into finitely many sets (orbits) $V_1, \ldots, V_k$, so that for $x \in V_i$ and $y \in V_j$, there exists $\gamma \in Aut(G)$ mapping x to y iff $i = j$.

Clearly, a transitive graph is quasi-transitive.

Benjamini and Schramm [9] conjectured that when G is quasi-transitive, a.s. uniqueness of the infinite cluster holds for all $p > p_u$. This was proved for Cayley graphs (and, more generally, for quasi-transitive unimodular graphs; see Definition 4.6.1) by Häggström and Peres [18], and in full generality by Schonmann [29].

Theorem 4.1.1 ([18], [29]) *Consider bond percolation on a connected, infinite, locally finite, quasi-transitive graph G. Then $\mathbf{P}_p^G$-a.s., the number N of infinite clusters satisfies*

$$N = \begin{cases} 0 & \text{if} \quad p \in [0, p_c) \\ \infty & \text{if} \quad p \in (p_c, p_u) \\ 1 & \text{if} \quad p \in (p_u, 1] \,. \end{cases}$$

The parameter space $[0, 1]$ is thus split into three qualitatively different intervals, separated by the two critical values p_c and p_u. Some of the intervals may be degenerate or empty (e.g., for $\mathbf{Z}^d$ we have $p_c = p_u$, and for trees we have $p_u = 1$). Grimmett and Newman [16] presented the first example of a transitive graph where all three regimes are nondegenerate: the product of a regular tree and $\mathbf{Z}$. Other examples were given by Benjamini and Schramm [9] and Lalley [22].

There is a natural way to couple the percolation processes for all p simultaneously. Equip the edges of G with i.i.d. random variables $\{U(e)\}_{e\in E}$, uniform in $[0, 1]$, and write Ψ^G for the resulting product measure on $[0, 1]^E$. For each p, the edge set $\{e \in E : U(e) \leq p\}$ has the same distribution as the set of open edges under $\mathbf{P}_p^G$. This yields a *coalescent process* which has turned out to be a fruitful object of study in Erdős–Rényi random graph theory (see, e.g., [19, 2]) and which has recently attracted more attention also in percolation theory (e.g., [10]): When $p = 0$ every vertex is its own connected component. As the parameter p is increased, more and more edges become open, causing connected components to coalesce, until finally, when $p = 1$ all edges are open.

By Theorem 4.1.1 and Fubini's Theorem, we have Ψ^G-a.s. that the number of infinite clusters is ∞ for (Lebesgue-)a.e. $p \in (p_c, p_u)$, and 1 for a.e. $p \in (p_u, 1]$. However, it is not obvious that the quantifier "a.e." can be strengthened to "every" in these statements. Alexander [3] demonstrated that this strengthening holds for $G = \mathbf{Z}^d$ and other Euclidean lattices, and Häggström and Peres [18] handled the case where G is quasi-transitive and unimodular. Here we prove the simultaneous version of Theorem 4.1.1 for all quasi-transitive graphs.

Theorem 4.1.2 *Let G be an infinite, locally finite, connected, quasi-transitive graph, and let p_c and p_u be as in Theorem 4.1.1. Consider the coupling Ψ^G of the percolation processes on G for all $p \in [0, 1]$ simultaneously, and let $N(p)$ be the number of infinite clusters determined by the edge set $\{e \in E : U(e) \leq p\}$. With Ψ^G-probability 1, we then have*

$$N(p) = \begin{cases} 0 & \textit{for all} \quad p \in [0, p_c) \\ \infty & \textit{for all} \quad p \in (p_c, p_u) \\ 1 & \textit{for all} \quad p \in (p_u, 1]\,. \end{cases}$$

This is an immediate consequence of the following result in conjunction with Theorem 4.1.1.

Theorem 4.1.3 *Let G be an infinite, locally finite, connected, quasi-transitive graph. With Ψ^G-probability 1, for all $p_1 < p_2$ in $(p_c, 1]$, every infinite p_2-cluster contains an infinite p_1-cluster.*

This sharpens a result of Schonmann [29], which gives the same assertion except that the order of the quantifiers "with Ψ^G-probability 1" and "for all $p_1 < p_2$" is interchanged.

Theorems 4.1.2 and 4.1.3 imply that as the parameter p increases, infinite clusters are "born" only at, or immediately after, level p_c. For larger p, infinite clusters grow and merge, but no new ones are formed from finite clusters. Our next result shows that infinite clusters "merge relentlessly" in the intermediate regime (p_c, p_u). We can only prove this result for quasi-transitive graphs under the additional assumption of unimodularity (see Definition 4.6.1), but we believe it holds for all quasi-transitive graphs.

Theorem 4.1.4 *Let G be an infinite, locally finite, connected, quasi-transitive unimodular graph, and let p_c and p_u be as in Theorem 4.1.1. Then, with Ψ^G-*

probability 1, *for any* $p_1 < p_2$ *in* (p_c, p_u), *any infinite cluster at level* p_2 *contains infinitely many infinite clusters at level* p_1.

The next result also concerns the intermediate regime (p_c, p_u). Say that two infinite self-avoiding paths ξ_1 and ξ_2 in the same infinite cluster $\mathcal{C}$ are *equivalent* if for any finite set $\{e_1, \dots, e_n\}$ of edges in $\mathcal{C}$, both paths are eventually in the same connected component of $\mathcal{C} \setminus \{e_1, \dots, e_n\}$. Equivalence classes of self-avoiding paths in $\mathcal{C}$ are called *ends* of $\mathcal{C}$. The following theorem is proved in Section 4.4.

Theorem 4.1.5 *Let G be an infinite, locally finite, connected, quasi-transitive graph. Then,* Ψ^G*-a.s., for all* $p \in (p_c, p_u)$ *every infinite p-cluster has precisely* $2^{\aleph_0}$ *many ends.*

The proof extends to the case $p = p_u$ if there are multiple infinite clusters at that level; thus, this theorem confirms Conjecture 5 of Benjamini and Schramm [9] (except for the case $p = p_c$, if infinitely many infinite clusters can exist there). The fixed-p unimodular case was first proved in an early version of [18].

When there are infinitely many infinite clusters, can they be qualitatively different? To make this question precise, we need some definitions.

Definition 4.1.2 Let $G = (V, E)$ be a quasi-transitive graph. By a *subgraph* of G, we mean a collection of edges. A set Q of subgraphs of G is called a *property* if for every $p \in (0, 1)$ and every vertex x, the event that the open cluster of x at level p belongs to Q is $\mathbf{P}_p$-measurable.

- Q is an *invariant* property if for every $\gamma \in \mathrm{Aut}(G)$ and $E_0 \in Q$, necessarily $\gamma(E_0) \in Q$.
- Q is *monotone* if whenever $E_1 \in Q$ and $E_1 \subset E_2$, then also $E_2 \in Q$.
- Q is *robust* if for for every infinite connected subgraph $\mathcal{C}$ of G and every edge $e \in \mathcal{C}$, we have the equivalence: $\mathcal{C} \in Q$ iff there is an infinite connected component of $\mathcal{C} \setminus \{e\}$ that satisfies Q.

Suppose that Q is a robust property and $\mathcal{C}$ is an infinite cluster satisfying Q. If an edge adjacent to $\mathcal{C}$ is opened, then the resulting cluster will satisfy Q, and if an edge in $\mathcal{C}$ is closed, then at least one of the resulting infinite clusters will satisfy Q.

Transience (for simple random walk) is a robust, monotone, invariant property of subgraphs that has been studied extensively. An invariant property of interest, that is robust but not monotone, is

$$\{\mathcal{C} : \exists \text{ infinitely many encounter points in } \mathcal{C}\}$$

(see [11], [8], or the end of the present paper for the definition and significance of encounter points). A monotone invariant property for which robustness is not known is $\{\mathcal{C} : p_c(\mathcal{C} \times \mathbf{Z}) < p_0\}$, where $p_0 < 1$ is fixed.

Following a question of O. Schramm (personal communication), Häggström and Peres [18] showed that for G quasi-transitive and unimodular, if Q is any monotone invariant property, then $\mathbf{P}_p$-a.s, infinite clusters with and without Q cannot coexist, except possibly at one value of p. This result was substantially improved by Lyons and Schramm [24], who showed there is no exceptional p, and the monotonicity assumption on Q can be dropped. Thus on quasi-transitive unimodular graphs, [24] shows that infinite clusters are *indistinguishable by invariant properties.* As noted there, this strong result fails without unimodularity; see Section 4.6.

Nevertheless, on any quasi-transitive graph, infinite clusters cannot be distinguished by *robust* invariant properties. The following theorem is proved in Section 4.5.

Theorem 4.1.6 *Let G be an infinite, locally finite, connected, quasi-transitive graph, and let $p \in (p_c, p_u]$. If Q is a robust invariant property of subgraphs of G such that $\mathbf{P}_p(\exists$ an infinite cluster satisfying $Q) > 0$, then $\mathbf{P}_p$-a.s., all infinite clusters in G satisfy Q.*

Next, we present an upper bound on p_u for products of infinite graphs. For d graphs $\{G_i = (V_i, E_i)\}_{i=1}^d$, define the *product graph* $G = G_1 \times \cdots \times G_d$ as the graph with vertex set $V = V_1 \times \cdots \times V_d$, and edge set E consisting of pairs $(x_1, \ldots, x_d)$ and $(y_1, \ldots, y_d)$ such that x_i and y_i are neighbors in G_i for exactly one coordinate $i \in \{1, \ldots, d\}$, and $x_j = y_j$ for all other coordinates j. Clearly, a product of two or more quasi-transitive graphs is quasi-transitive.. Some of the most natural examples (such as the Grimmett–Newman example) arise this way.

Theorem 4.1.7 *Let $G_1, \ldots, G_d$ be infinite connected graphs with bounded degree, and let G be their product $G_1 \times \cdots \times G_d$. Then, for bond percolation on G with parameter $p > p_c(\mathbf{Z}^d)$, we have $\mathbf{P}_p^G$-a.s. that the number of infinite clusters is exactly 1. Moreover, in the coupling Ψ^G, uniqueness of the infinite cluster holds a.s. simultaneously for all $p > p_c(\mathbf{Z}^d)$.*

In particular, if $G_1, \ldots, G_d$ are infinite connected graphs with bounded degree, then

$$p_u(G_1 \times \cdots \times G_d) \leq p_c(\mathbf{Z}^d).$$

The rest of the paper is organized as follows. In the next section we state an extension of Theorem 4.1.3. We prove this extension in Section 4.3, by combining the approach of Schonmann [29] with invasion percolation ideas. Theorem 4.1.5 on ends is established by similar means in Section 4.4. We prove Theorem 4.1.6 (indistinguishability by robust properties) in Section 4.5. In Section 4.6, we define unimodularity and recall a technique known as the mass-transport method, which we then use in Section 4.7 to prove Theorem 4.1.4. In Section 4.8 we prove Theorem 4.1.7, building on classical results for percolation in $\mathbf{Z}^d$, and a result in [29]. Lower bounds on p_u are also discussed there.

Section 4.9 contains examples, remarks, and unsolved problems.

4.2 Uniform Percolation and Semi-Transitive Graphs

In this section we will extend Theorem 4.1.3. To state this extension, we will need the notion of *uniform percolation* from [29]. The ball $B(x, R)$ of radius R centered at $x \in V$, is defined as the set of edges in G which have both endpoints within (graph-theoretic) distance R from x.

Definition 4.2.1 A graph $G = (V, E)$ exhibits *uniform percolation at level p* if

$$\lim_{R\to\infty} \inf_{x\in V} \mathbf{P}_p(\text{some infinite } p\text{-cluster intersects } B(x, R)) = 1\,. \tag{4.2.1}$$

It is easy to see that any quasi-transitive graph exhibits uniform percolation at all levels $p > p_c$. In fact, this holds in the larger class of *semi-transitive* graphs.

Definition 4.2.2 A graph $G = (V, E)$ is called *semi-transitive* if there is a finite set $V_F \subset V$ such that for any vertex $x \in V$, there is a vertex $y \in V_F$ and an injective graph homomorphism of G that maps y to x.

The simplest examples of semi-transitive graph that are not quasi-transitive are the nearest-neighbor graph on the positive integers $\mathbf{Z}_+$, and d-ary trees where the root has degree d and all other vertices have degree $d + 1$. More generally, the "super-periodic" trees that discussed in Lyons and Peres [23] are semi-transitive; an example is the subtree of the binary tree consisting of all vertices such that the path from the root to v has at least as many left turns as right turns. These trees are closely related to the "super self-similar" sets studied by Falconer [13]. A class of graphs, mentioned in [29], which are semi-transitive but not quasi-transitive, are products $G \times \mathbf{Z}_+$, where G is quasi-transitive.

The next result extends Theorem 4.1.3.

Theorem 4.2.1 *Let G be an infinite connected graph with bounded degree, that exhibits uniform percolation at level p_*. With Ψ^G-probability 1, for all $p_1 < p_2$ in $(p_*, 1]$, every infinite p_2-cluster contains some infinite p_1-cluster. In particular, there is Ψ^G-a.s. a unique infinite cluster at level p for all $p > \max(p_u, p_*)$.*

This will be proved in the next section using invasion percolation. Here, we show how it implies a generalization of Theorem 4.1.3.

Proof of Theorem 4.1.3, generalized to semi-transitive G. Let $p > p_c(G)$. Since the existence of an infinite cluster has $\mathbf{P}_p$-probability 1, we have for each fixed $x \in V$ that

$$\lim_{R\to\infty} \mathbf{P}_p(\text{some infinite } p_1\text{-cluster intersects } B(x, R)) = 1\,.$$

The infimum in (4.2.1) is attained for some y in the finite set V_F specified in Definition 4.2.2, and it follows that (4.2.1) holds for any $p > p_c(G)$. Invoking Theorem 4.2.1 completes the proof. □

Theorem 4.1.2 may fail in the semi-transitive setting, because there exist semi-transitive graphs where with positive probability, the number of infinite clusters is finite but greater than one. (An example of this, due to O. Schramm, is described in the final section.) Nevertheless, Theorem 4.1.2 does extend to semi-transitive graphs G where $\mathrm{Aut}(G)$ has an infinite orbit (e.g., $G = G_1 \times G_2$ where G_1 is quasi-transitive and G_2 is semi-transitive), since a standard argument shows that in such graphs G, for each parameter p the number of infinite clusters is 0, 1 or ∞ a.s.

4.3 Invasion Hits Infinite Percolation Clusters

A key idea in proving Theorem 4.2.1 is to use *invasion percolation*, which is a sequential construction based on the same uniform random variables $\{U(e)\}_{e\in E}$ as the canonical coupling of the ordinary percolation processes. Here we give only a brief description of invasion percolation; we refer to Chayes, Chayes and Newman [12] for a general introduction to the model, and to [25, 28] for some interesting recent applications in statistical mechanics.

The invasion cluster of a vertex $x \in V$ is built up sequentially by constructing an increasing sequence of edge sets $I_1^x \subset I_2^x \subset \cdots$ as follows. Let I_1^x consist of the single edge e which minimizes $U(e)$ among all edges incident to x. When I_i^x is constructed, I_{i+1}^x is taken to be $I_i^x \cup \{e\}$, where e is the edge which minimizes $U(e)$ among all edges e that are not in I_i^x but are adjacent to some edge in I_i^x. The invasion cluster of x is the edge set

$$I_\infty^x = \bigcup_{i=1}^{\infty} I_i^x .$$

Proposition 4.3.1 *Let $G = (V, E)$ be an infinite, connected graph with bounded degrees. If G exhibits uniform percolation at level p_*, then Ψ^G-a.s. for any $p > p_*$ and any $x \in V$, the invasion cluster I_∞^x intersects some infinite p-cluster.*

This proposition was proved by Chayes, Chayes and Newman [12] for $\mathbf{Z}^d$, by Alexander [4] for other Euclidean graphs, and by O. Schramm (personal communication) for transitive unimodular graphs. Before proving Proposition 4.3.1, we explain how it implies Theorem 4.2.1.

PROOF OF THEOREM 4.2.1. For $p \in [0, 1]$ and a vertex x, let $\mathcal{C}(x, p)$ denote the cluster at level p containing x. Also let $\Omega_{x,p}$ denote the event that (i) all the edges in G are assigned distinct labels $U(e)$, and (ii) the invasion cluster I_∞^x hits some infinite p-cluster. Fix $p > p_*$ and an edge labeling $\{U(e)\}_{e\in E}$ in $\Omega_{x,p}$. For any parameter $p_2 > p$ such that the cluster $\mathcal{C}(x, p_2)$ is infinite, it must contain the invasion cluster I_∞^x, and hence $\mathcal{C}(x, p_2)$ must intersect some infinite p-cluster $\mathcal{C}(y, p)$. Obviously, $\mathcal{C}(x, p_2)$ then intersects some infinite p_1-cluster for any $p_1 \in [p, p_2)$. Proposition 4.3.1 ensures that $\Psi^G(\cap_{x,p}\Omega_{x,p}) = 1$, where the intersection ranges over all $x \in V$ and all rational $p > p_*$, and this proves the theorem. □

The proof of Proposition 4.3.1 is based on an adaptation to invasion percolation of the proof of the main result in [29]. The following lemma is needed.

Lemma 4.3.1 *Let $G = (V, E)$ be an infinite, connected graph with bounded degrees, and let $R > 0$ be an integer. With Ψ^G-probability 1, the invasion cluster I_∞^x contains a ball of radius R.*

PROOF. Denote by D the maximum degree of vertices in G. The standard inequality

$$p_c(G) \geq \frac{1}{D-1} > 0 \tag{4.3.1}$$

(see, e.g., [15]), will be used at the end of the proof. Let $(v_1, v_2, \ldots)$ be an arbitrary enumeration of the vertex set V. For $n = 1, 2, \ldots$, set $L_n = n(2R+1)$, and define

$$\tau_n := \min\{k : I_k^x \text{ comes within distance } R \text{ from some } y \in V \setminus B(x, L_n)\}.$$

Since the invasion cluster I_∞^x is infinite, τ_n is a.s. finite for every n. For each n, define y_n to be the vertex in $V \setminus B(x, L_n)$ at minimal distance from $I_{\tau_n}^x$ (in case of a tie, y_n is the one with minimal index in the above enumeration). Finally, consider the events

$$A_n = \{U(e) < p_c \text{ for all } e \in B(y_n, R)\}.$$

Since there are a.s. no infinite p-clusters for $p < p_c$, on A_n the invasion cluster I_∞^x must contain the ball $B(y_n, R)$. Thus it suffices to prove that $\Psi^G\left(\cap_{i=1}^\infty A_i^c\right) = 0$. Note that $I_{\tau_n}^x$ and $B(y_n, R)$ "touch" but do not intersect, and that the invasion process up to time τ_n gives no information about edges in $B(y_n, R)$. The conditional probability of A_n given $A_1^c, \ldots, A_{n-1}^c$ and the invasion process up to time τ_n, is therefore at least $p_c^{D^{R+1}}$. Therefore

$$\Psi^G\left(\cap_{i=1}^n A_i^c\right) \leq \left(1 - p_c^{D^{R+1}}\right)^n,$$

and the right-hand side tends to 0 as $n \to \infty$ by (4.3.1). □

PROOF OF PROPOSITION 4.3.1. Fix p_*, p and x as in the proposition. Define the random variable $\xi_{p_*}^x$ as the number of edges that have one endpoint in I_∞^x and the other in some infinite p_*-cluster. Our proof consists of first showing that

$$\Psi^G(\xi_{p_*}^x = \infty) = 1 \tag{4.3.2}$$

and then showing that for $p > p_*$,

$$\Psi^G(I_\infty^x \text{ intersects some infinite } p\text{-cluster} \mid \xi_{p_*}^x = \infty) = 1. \tag{4.3.3}$$

Letting $p \downarrow p_*$ through a countable sequence then proves the proposition.

By the uniform percolation assumption, we can, for any $\varepsilon > 0$, pick an R so large that

$$\inf_{y \in V} \Psi^G(\text{some infinite } p_*\text{-cluster intersects } B(y, R)) \geq 1 - \varepsilon. \tag{4.3.4}$$

Let τ denote the smallest k for which I_k^x contains a ball of radius R; by Lemma 4.3.1, $\tau < \infty$ a.s.

For an edge set $E_0 \subset E$, set

$$V(E_0) = \{y \in V : y \text{ is an endpoint of some } e \in E_0\}.$$

If E_0 is finite and contains some ball of radius R, then by (4.3.4) we have with probability at least $1-\varepsilon$ that some vertex in $V(E_0)^c$ at distance 1 from $V(E_0)$ has an open path to infinity at level p_* via vertices in $V(E_0)^c$ only. Since the invasion cluster up to time τ gives no information about the set of edges not adjacent to I_τ^x, we may apply the above reasoning with $E_0 = I_\tau^x$ to deduce that the conditional probability that there is some infinite p_*-cluster within distance 1 from I_τ^x is at least $1-\varepsilon$. This shows that $\Psi^G(\xi_{p_*}^x = 0) \leq \varepsilon$, and since ε was arbitrary we have

$$\Psi^G(\xi_{p_*}^x = 0) = 0\,. \tag{4.3.5}$$

The next step is to rule out the possibility of having $\xi_{p_*}^x = n$ for any finite n. Note that on the event $(\xi_{p_*}^x = n)$ we can move into the event $(\xi_{p_*}^x = 0)$ by changing the status of finitely many edges. It is easy to see that this implies that if $\Psi^G(\xi_{p_*}^x = n) > 0$, then $\Psi^G(\xi_{p_*}^x = 0) > 0$ holds as well. But this would contradict (4.3.5), so we have

$$\Psi^G(\xi_{p_*}^x = n) = 0 \tag{4.3.6}$$

for any $n < \infty$, and (4.3.2) is established.

To prove (4.3.3), consider the following "coloring followed by invasion percolation" procedure. First mark every edge blue which is in some infinite p_*-cluster. Then mark every edge red which is not blue but is adjacent to some blue edge. Given the coloring information, start to build the invasion cluster at x in the usual way. For the event $(\xi_{p_*}^x = \infty)$ to happen, the invasion cluster has to meet (become adjacent to) infinitely many colored edges. If the invasion cluster ever meets any of the blue edges, then we are done (i.e., the invasion cluster intersects some infinite p-cluster), because the invasion cluster must then eventually contain the encountered blue edge unless it penetrates some infinite p^*-cluster elsewhere. Otherwise (still on the event $(\xi_{p_*}^x = \infty)$) the invasion cluster has to meet infinitely many red edges. Suppose now that a given red edge e is met for the first time. Then the conditional distribution of $U(e)$ (given the coloring information and the invasion cluster so far) is uniform on $(p_*, 1]$. Thus the event that $U(e) < p$ (which obviously implies that I_∞^x intersects some infinite p-cluster) has conditional probability $\frac{p-p_*}{1-p_*} > 0$. Since this conditional probability is the same every time a red edge is encountered by the invasion cluster for the first time, we have (4.3.3), and the proof is complete. □

4.4 Uncountably Many Ends

PROOF OF THEOREM 4.1.5. We shall prove that for any $p_1 < p_2$ in (p_c, p_u) we have

$$\Psi^G(\forall p \in [p_1, p_2],\ \text{all infinite } p\text{-clusters have } 2^{\aleph_0} \text{ ends}) = 1\,. \tag{4.4.1}$$

Sending $p_1 \downarrow p_c$ and $p_2 \uparrow p_u$ through countable sequences then proves the theorem.

Fix p_1 and p_2 as above, and set $p_0 = \frac{p_c+p_1}{2}$. To prove (4.4.1), it is (due to Theorem 4.1.3) enough to show, for any $x_0 \in V$, that

$$\Psi^G(H^{x_0}_{p_0,p_1,p_2}) = 0 \tag{4.4.2}$$

where $H^{x_0}_{p_0,p_1,p_2}$ is the event that x_0 is in an infinite p_0-cluster and for some $p \in [p_1, p_2]$ it is in an infinite p-cluster with less than $2^{\aleph_0}$ ends. Also define $\tilde{H}^{x_0}_{p_0,p_1,p_2}$ as the event that x_0 is in an infinite p_0-cluster and for some $p \in [p_1, p_2]$ it is in an infinite p-cluster with just one end. If a given realization $\eta \in [0,1]^E$ of the variables $\{U(e)\}_{e\in E}$ is in $H^{x_0}_{p_0,p_1,p_2}$, then (arguing as in Benjamini and Schramm [9], p. 76) one can change finitely many of the variables to obtain a realization η' which is in $\tilde{H}^{x_0}_{p_0,p_1,p_2}$. Thus, (4.4.2) follows easily once we show

$$\Psi^G(\tilde{H}^{x_0}_{p_0,p_1,p_2}) = 0\,. \tag{4.4.3}$$

Let $L_{x,R,k}$ be the event that there are at least k infinite p_2-clusters which contain infinite p_1-clusters that intersect $B(x, R)$. Fix $k \geq 1$. Since $p_2 \in (p_c, p_u)$, and each infinite p_2-cluster contains some infinite p_1-cluster Ψ^G-a.s., for every $x \in V$ we have $\lim_{R\to\infty} \Psi^G(L_{x,R,k})$. Thus given $\varepsilon > 0$, for every $x \in V$ there is an R such that

$$\Psi^G(L_{x,R,k}) \geq 1 - \varepsilon. \tag{4.4.4}$$

Since G is quasi-transitive, there exists an R that satisfies (4.4.4) for all $x \in V$. (R may depend on p_1, p_2, k and ε, but not on x.) Fix such an R, and grow the invasion cluster of x_0 until the first time τ for which $I^{x_0}_\tau$ contains some ball of radius R; Lemma 4.3.1 ensures that this happens a.s. for some finite τ. Let $\partial I^{x_0}_\tau$ be the set of edges in $E \setminus I^{x_0}_\tau$ that are adjacent to $I^{x_0}_\tau$. Using (4.4.4) and arguing as in the proof of Proposition 4.3.1, we have that

$$\Psi^G(A^{x_0}_k \mid \text{the invasion process up to time } \tau) \geq 1 - \varepsilon\,,$$

where $A^{x_0}_k$ is the event that the percolation process restricted to the edge set $E \setminus (I^{x_0}_\tau \cup \partial I^{x_0}_\tau)$ has at least k infinite p_1-clusters that

(i) are contained in separate p_2-clusters, and
(ii) contain some vertex at distance 1 from $I^{x_0}_\tau$.

If x_0 is in an infinite p_0-cluster, then

$$U(e) \leq p_0 \text{ for every } e \in I^{x_0}_\tau\,. \tag{4.4.5}$$

Given (4.4.5) and the invasion process up to time τ, each $e \in \partial I^{x_0}_\tau$ is open at level p_1 independently with conditional probability at least $\frac{p_1-p_0}{1-p_0}$. If $A^{x_0}_k$ happens, we may pick $e_1, \ldots e_k \in \partial I^{x_0}_\tau$ adjacent to k different p_1-clusters with the properties (i), (ii) above. These properties guarantee that if x_0 is in an infinite p_0-cluster and at least two of the edges $e_1, \ldots, e_k$ are open at level p_1, then x_0 is in an infinite p-cluster with at least two ends for all $p \in [p_1, p_2]$. Hence

$$\Psi^G(\tilde{H}^{x_0}_{p_0,p_1,p_2}) \leq \varepsilon + \left(\frac{1-p_1}{1-p_0}\right)^k + k\left(\frac{p_1-p_0}{1-p_0}\right)\left(\frac{1-p_1}{1-p_0}\right)^{k-1}.$$

Sending $\varepsilon \to 0$ and $k \to \infty$ proves (4.4.3), and thus also (4.4.2) and (4.4.1), so the proof is complete. □

We end this section by noting the following very simple corollary to Theorem 4.1.3. It is the natural analogue for the uniqueness regime $(p_u, 1)$ of Theorem 4.1.5.

Corollary 4.4.1 *Let G be an infinite, locally finite, connected, semi-transitive graph. Then, Ψ^G-a.s., for all $p \in (p_u, 1)$ the (unique) infinite p-cluster has a single end.*

PROOF. Suppose for contradiction that with positive Ψ^G-probability, there exists some $p \in (p_u, 1)$ for which the infinite cluster has more than one end. Any realization $\eta \in [0, 1]^E$ of the $\{U(e)\}_{e \in E}$ variables for which this happens at level p can be modified into a configuration η' in which uniqueness of the infinite cluster fails at level p, by changing the status of just finitely many edges. It follows that with positive Ψ^G-probability, there is some $p \in (p_u, 1)$ for which uniqueness of the infinite cluster fails, contradicting Theorem 4.1.3. □

4.5 Indistinguishability by Robust Properties

PROOF OF THEOREM 4.1.6. Fix $p_0 \in (p_c, p)$. Since, by Theorem 4.1.3, Ψ^G-a.s. any infinite p-cluster contains an infinite p_0-cluster, it suffices to show that for all $x \in V$,

$$\Psi^G[\mathcal{C}(x, p_0) \text{ is infinite and } \mathcal{C}(x, p) \notin \mathcal{Q}] = 0 . \tag{4.5.1}$$

Define the random variable ξ^x as the number of edges that are adjacent to, or contained in, I_∞^x and are also adjacent to, or contained in, some infinite p-cluster which satisfies $\mathcal{Q}$. We will establish (4.5.1) by proving the following two statements:

$$\Psi^G[\xi^x < \infty] = 0 , \tag{4.5.2}$$

and

$$\Psi^G[\mathcal{C}(x, p_0) \text{ is infinite, } \xi^x = \infty \text{ and } \mathcal{C}(x, p) \notin \mathcal{Q}] = 0 . \tag{4.5.3}$$

We first prove (4.5.2). By the 0-1 law for automorphism-invariant events,

$$\Psi^G[\exists \text{ an infinite } p\text{-cluster satisfying } \mathcal{Q}] = 1 .$$

Therefore, for any $\varepsilon > 0$, we can pick an R so that

$$\inf_{y \in V} \Psi^G\Big(\text{no infinite } p\text{-cluster with property } \mathcal{Q} \text{ intersects } B(y, R)\Big) < \varepsilon . \tag{4.5.4}$$

Let τ be the smallest m for which I_m^x contains a ball of radius R; by Lemma 4.3.1, $\tau < \infty$ a.s.

For an edge set $E_0 \subset E$, let ∂E_0 be the set of edges outside E_0 that are adjacent to E_0, and denote by $\mathcal{S}(E_0)$ the set of edges in ∂E_0 that are adjacent to an infinite connected component of $\{e \notin \partial E_0 : U(e) \leq p\}$ which has property Q.

If a finite edge set E_0 intersects an infinite p-cluster that has property Q, then robustness of Q implies that $\mathcal{S}(E_0) \neq \emptyset$. Therefore, any finite edge set E_0 that contains a ball of radius R, satisfies $\Psi^G\Big(\mathcal{S}(E_0) = \emptyset\Big) < \varepsilon$ by (4.5.4).

Since the invasion cluster I_τ^x gives no information about the labels on edges not in $I_\tau^x \cup \partial I_\tau^x$, we may apply the above reasoning with $E_0 = I_\tau^x$ to deduce that $\Psi^G\Big(\mathcal{S}(I_\tau^x) = \emptyset\Big) < \varepsilon$. In particular, $\Psi^G(\xi^x = 0) < \varepsilon$, and since ε was arbitrary, $\Psi^G(\xi^x = 0) = 0$. On the event $(\xi^x = n)$, we can move into the event $(\xi^x = 0)$ by changing the labels $U(e)$ on finitely many edges to be greater than p; it follows that $\Psi^G(\xi^x = n) = 0$ for any $n < \infty$, and (4.5.2) is established.

To prove (4.5.3), observe that if $\mathcal{C}(x, p_0)$ is infinite, then $I_\infty^x \subseteq \mathcal{C}(x, p_0)$. Therefore, if also $\xi^x = \infty$ and $\mathcal{C}(x, p) \notin Q$, then the following event, which we call F, must happen: $\mathcal{C}(x, p) \notin Q$ but there are infinitely many edges in $\partial\mathcal{C}(x, p_0)$ which are adjacent to p-clusters $\mathcal{C} \in Q$ such that $\mathcal{C} \cap \mathcal{C}(x, p_0) = \emptyset$. To establish (4.5.3), it suffices to show that $\Psi^G(F) = 0$. To prove this, write $\mathcal{C}_\ell(x, p_0)$ for the collection of edges that can be reached from x by a path which is open at level p_0 and is contained in the ball $B(x, \ell)$. Consider the event $F_{\ell,k}$ that $\mathcal{C}(x, p) \notin Q$ and $\partial\mathcal{C}_\ell(x, p_0)$ contains at least k edges that are adjacent to infinite p-clusters $\mathcal{C} \in Q$ such that $\mathcal{C} \cap \mathcal{C}(x, p_0) = \emptyset$. Clearly, for every k,

$$F \subseteq \cup_{\ell=1}^{\infty} F_{\ell,k}\,, \quad \text{whence } \Psi^G(F) \leq \lim_{\ell\to\infty} \Psi^G(F_{\ell,k})\,.$$

The proof will therefore be complete once we establish that for any ℓ, k,

$$\Psi^G(F_{\ell,k}) \leq \left(\frac{1-p}{1-p_0}\right)^k. \tag{4.5.5}$$

Clearly,

$$F_{\ell,k} \subseteq \Big\{\mathcal{C}(x, p) \notin Q \text{ and } |\mathcal{S}(\mathcal{C}_\ell(x, p_0))| \geq k\Big\}.$$

(In fact these events coincide, but we do not need this.) Therefore, by robustness of Q,

$$F_{\ell,k} \subseteq \Big\{|\mathcal{S}(\mathcal{C}_\ell(x, p_0))| \geq k \text{ and } \forall e \in \mathcal{S}(\mathcal{C}_\ell(x, p_0)) \quad U(e) \geq p\Big\}. \tag{4.5.6}$$

Denote by $\mathcal{F}_\ell$ the σ-field generated by $\mathcal{C}_\ell(x, p_0)$ and the labels $\{U(e) : e \notin \partial\mathcal{C}_\ell(x, p_0)\}$. Then $\mathcal{S}(\mathcal{C}_\ell(x, p_0))$ is $\mathcal{F}_\ell$-measurable, and the remaining labels $\{U(e) : e \in \partial\mathcal{C}_\ell(x, p_0)\}$ are conditionally independent and uniform on $[p_0, 1]$ given $\mathcal{F}_\ell$. Thus (4.5.5) follows from (4.5.6). □

4.6 Unimodularity and Mass Transport

For $x \in V$, define the *stabilizer* $S(x) = \{\gamma \in \mathrm{Aut}(G) : \gamma(x) = x\}$, and for $y \in V$, define $S(x)y = \{\gamma(y) : \gamma \in S(x)\}$. Let $|A|$ denote the cardinality of a set A.

Definition 4.6.1 A quasi-transitive graph G is called *unimodular* if for any two vertices x, y in the same orbit, we have $|S(x)y| = |S(y)x|$.

For equivalent definitions of unimodularity, see Trofimov [31] and Benjamini, Lyons, Peres and Schramm [7]. Most quasi-transitive graphs that come up naturally are unimodular. In particular, the Cayley graph of any finitely generated group is transitive and unimodular. A transitive graph $\widehat{T_d}$ which is *not* unimodular can be constructed by considering a regular tree T_d with degree $d \geq 3$, fixing an end ξ of T_d, and for each vertex x adding an edge between x and its ξ-grandparent; see [31] or [7]. In this example, $p_u = 1$, and for $p \in (p_c, 1)$ every infinite cluster $\mathcal{C}$ has a unique vertex $v(\mathcal{C}, \xi)$ that is "closest" to ξ. Thus, as noted in [24], the invariant property $\{\mathcal{C} : v(\mathcal{C}, \xi)$ has degree 1 in $\mathcal{C}\}$ distinguishes some infinite clusters in $\widehat{T_d}$ from others. By utilizing more of the local structure of a cluster $\mathcal{C}$ as seen from $v(\mathcal{C}, \xi)$, *any* two infinite clusters in $\widehat{T_d}$ may be invariantly distinguished.

The significance of unimodularity for us is that it allows a certain mass-transport technique, which was introduced in percolation theory in [17] and systematically developed in [7]. Central to the mass-transport method is Theorem 4.6.1 below, which was proved (in a more general setting) in [7]. For any graph G, every automorphism in $\mathrm{Aut}(G)$ acts as a measure-preserving transformation on the probability space $(\{0,1\}^E, \mathbf{P}_p^G)$. Let $m(x, y, \omega)$ be a nonnegative function of three variables: two vertices x, y in the same orbit of $\mathrm{Aut}(G)$, and $\omega \in \{0,1\}^E$. Intuitively, $m(x, y, \omega)$ represents the mass transported from x to y given the configuration ω. We suppose that $m(\cdot,\cdot,\cdot)$ is invariant under the diagonal action of $\mathrm{Aut}(G)$, i.e., $m(x, y, \omega) = m(\gamma x, \gamma y, \gamma\omega)$ for all x, y, ω and $\gamma \in \mathrm{Aut}(G)$.

Theorem 4.6.1 (The Mass-Transport Principle) *Let $G = (V, E)$ be a unimodular and quasi-transitive graph. Given $m(\cdot,\cdot,\cdot)$ as above, let*

$$M(x, y) = \int_{\{0,1\}^E} m(x, y, \omega)\, d\mathbf{P}_p^G(\omega)\,.$$

Then the expected total mass transported out of any vertex x equals the expected total mass transported into x, i.e.,

$$\forall x \in V \quad \sum_{y \in V} M(x, y) = \sum_{y \in V} M(y, x)\,. \tag{4.6.1}$$

We remark that (4.6.1) fails in the nonunimodular case; see [7]. The key element in a successful application of the mass-transport method is to make a suitable choice of the transport function $m(\cdot,\cdot,\cdot)$; examples can be found, e.g., in [17, 7, 8, 18], and also in Section 4.7 below.

4.7 Relentless Merging

The main step in proving Theorem 4.1.4 is showing that for $p \in (p_c, p_u)$, any infinite p-cluster will a.s. come within distance 1 from other infinite p-clusters in infinitely many places, as stated in the following proposition. The result assumes quasi-transitivity and unimodularity; we conjecture the latter condition to be removable.

Proposition 4.7.1 *Consider bond percolation on an infinite, locally finite, connected, quasi-transitive unimodular graph G with retention parameter $p \in (p_c, p_u)$. Then, $\mathbf{P}_p^G$-a.s., for any infinite cluster C there exist infinitely many closed edges e with one endpoint in C and the other endpoint in some other infinite cluster (which may depend on e).*

PROOF. Assume for contradiction that with positive $\mathbf{P}_p^G$-probability there is some infinite cluster C which comes within distance exactly 1 from the set of other infinite clusters in at most finitely many locations. As in the argument for (4.3.6) in the proof of Proposition 4.3.1, it follows that with positive probability there is some infinite cluster C in which exactly one vertex x is at distance 1 from the set of other infinite clusters. Call such an infinite cluster C a *kingdom*, call x its *king*, and consider the following mass transport. If a vertex y is in a kingdom, and its king is in the same orbit as y (recall Definition 4.1.1), then y sends unit mass to its king. If y is in a kingdom but not in the same orbit as the king, then y sends unit mass which it distributes equally among the vertices in its orbit which are closest in G to the king. Otherwise, no mass is sent from y. The expected mass sent from each vertex is then at most 1, whereas the expected mass received has to be ∞ for some vertices. By the Mass-Transport Principle (Theorem 4.6.1) we have the desired contradiction. □

PROOF OF THEOREM 4.1.4. We first prove the assertion of the theorem with the quantifiers interchanged, i.e., that

> for all $p_1 < p_2$ in (p_c, p_u), we have Ψ^G-a.s. that any infinite p_2-cluster contains infinitely many infinite p_1-clusters. (4.7.1)

We know from Theorem 4.1.1 that any infinite p_2-cluster contains some infinite p_1-cluster. Hence, it suffices to show that any infinite p_1-cluster C gets connected to infinitely many infinite p_1-clusters disjoint from C as we raise the percolation level to p_2. Fix a vertex x, let $C(x, p_1)$ be the p_1-cluster containing x, and assume that $C(x, p_1)$ is infinite. Call an edge e *pivotal* if it is closed at level p_1, has one endpoint in C and the other endpoint in some other infinite p_1-cluster. By Proposition 4.7.1, there are Ψ^G-a.s. infinitely many pivotal edges. As we raise the level to p_2, each pivotal edge gets turned on independently with probability $\frac{p_2 - p_1}{1 - p_1}$, whence at least one of them gets turned on a.s., so that $C(x, p_1)$ gets connected to at least one other infinite cluster a.s.

Now pick $k \geq 2$ and $q_1, \ldots, q_k$ such that $p_1 = q_1 < q_2 < \cdots < q_k = p_2$. The above reasoning shows that the infinite cluster $C(x, p_1)$ gets connected to at least one additional infinite p_1-cluster for each interval (q_i, q_{i+1}), so that $\mathcal{C}(x, p_2)$ contains at least $k-1$ infinite p_1-clusters. Since k was arbitrary, we have established (4.7.1).

It remains to change the order of the quantifiers "for all $p_1 < p_2$" and "Ψ^G-a.s." in (4.7.1). To do this, note first that we can apply (4.7.1) to all rational $p_1 < p_2$ in (p_c, p_u) simultaneously. The assertion of the theorem now follows easily using Theorem 4.1.2 and the observation that for any $p_1 < p_2$, we can find two distinct rational numbers between p_1 and p_2. □

4.8 Product Graphs and Estimates on p_u

Let us collect the results from the literature that are needed to prove Theorem 4.1.7. The first one concerns percolation on the orthant $\mathbf{Z}_+^d$. For $d = 2$, it follows from the work of Kesten [20], while for general d it was first obtained by Barsky, Grimmett and Newman [6].

Theorem 4.8.1 ([20], [6]) *For bond percolation on $\mathbf{Z}_+^d$, $d \geq 2$, we have*

(a) $p_c(\mathbf{Z}_+^d) = p_c(\mathbf{Z}^d)$, and

(b) for $p > p_c(\mathbf{Z}_+^d)$, there is $P_p^{\mathbf{Z}_+^d}$-a.s. a unique infinite cluster.

The following result is due to Schonmann [29]. There it was formulated in the setting of quasi-transitive graphs, but that proof goes through unchanged in the generality stated here.

Theorem 4.8.2 ([29]) *Let G be any bounded degree graph, and pick $p \in [0, 1]$. If*

$$\lim_{R\to\infty} \inf_{x,y\in V} \mathbf{P}_p^G(B(x, R) \leftrightarrow B(y, R)) = 1\,, \tag{4.8.1}$$

then for all $p' > p$, there is $\mathbf{P}_{p'}^G$-a.s. exactly one infinite cluster.

Remark The conclusion of this theorem can now be strengthened to "*Ψ^G-a.s. there is exactly one infinite cluster at each level $p' > p$.*" Indeed, (4.8.1) implies uniform percolation at level p, so Theorem 4.2.1 applies.

In the following proof, we shall work with more than one graph, and therefore write $B_G(x, R)$ for $B(x, R)$ to indicate in which graph the ball sits.

Proof of Theorem 4.1.7. We first prove uniqueness of the infinite cluster on G for fixed $p > p_c(\mathbf{Z}^d)$. By Theorem 4.8.2, it is sufficient to show that (4.8.1) holds for all $p > p_c(\mathbf{Z}^d)$. Fix such a p. Theorem 4.8.1 implies that

$$\lim_{R\to\infty} \mathbf{P}_p^{\mathbf{Z}_+^d}(B_{\mathbf{Z}_+^d}(0, R) \leftrightarrow \infty) = 1\,, \tag{4.8.2}$$

where $(B_{\mathbf{Z}_+^d}(0, R) \leftrightarrow \infty)$ is the event that there is an infinite open cluster intersecting $B_{\mathbf{Z}_+^d}(0, R)$. Pick $\varepsilon > 0$, and R large enough so that

$$\mathbf{P}_p^{\mathbf{Z}_+^d}(B_{\mathbf{Z}_+^d}(0, R) \leftrightarrow \infty) \geq 1 - \varepsilon .$$

Now let $x = (x_1, \dots, x_d)$ and $y = (y_1, \dots, y_d)$ be arbitrary vertices of G. For $i = 1, \dots, d$, let T_x^i be some infinite self-avoiding path in G_i starting in x_i. Also let T_y^i be some infinite self-avoiding path in G_i from y_i which eventually coincides with T_x^i. Such a path is easily seen to exist: just take a path from y_i to x_i, concatenate it with T_x^i, and erase any circuits. Finally, let z_i be the first vertex on T_x^i with the property that T_x^i and T_y^i coincide from z_i to infinity, and define T_z^i to be the self-avoiding path starting at z_i that is contained in T_x^i. Define the product graph $G_x^* = T_x^1 \times \cdots \times T_x^d$, and define G_y^* and G_z^* analogously. Note that G_x^*, G_y^* and G_z^* are all isomorphic to $\mathbf{Z}_+^d$, and furthermore that they are all subgraphs of G and that G_z^* is a subgraph both of G_x^* and of G_y^*.

Let $D_{R,x}$ be the event that some vertex in $B_{G_x^*}(x, R)$ has an open path to infinity in G_x^*. Define $D_{R,y}$ analogously, and set $D_{R,x,y} = D_{R,x} \cap D_{R,y}$. Using (4.8.2), we get

$$\mathbf{P}_p^G(D_{R,x,y}) \geq 1 - 2\varepsilon .$$

By Theorem 4.8.1, we have $\mathbf{P}_p^G$-a.s. that G_x^* and G_y^* each have a unique infinite cluster, and that both these infinite clusters contain the unique infinite cluster of G_z^*. Hence, we have $\mathbf{P}_p^G$-a.s. on the event $D_{R,x,y}$ that there is an open path in G connecting $B_G(x, R)$ and $B_G(y, R)$, so that

$$\mathbf{P}_p^G(B_G(x, R) \leftrightarrow B_G(y, R)) \geq 1 - 2\varepsilon .$$

Note that this is a uniform bound for all vertices x and y in G. Since ε was arbitrary we have (4.8.1), so the proof for fixed p is complete.

The asserted simultaneous uniqueness is implied by the remark following Theorem 4.8.2. □

Next, we discuss lower bounds for p_u. Benjamini and Schramm [9, Theorem 4] proved that any quasi-transitive graph G satisfies $p_u(G) \geq (D\rho_G)^{-1}$, where D is the maximal degree in G, and ρ_G is the spectral radius for simple random walk on G. Their proof was based on coupling the percolation process with a branching random walk. In fact, a simple counting argument yields a slightly better bound.

Given a locally finite connected graph G, let $A_{x,y}^n$ denote the number of paths of length n which connect x to y. It is easy to see that

$$\lambda_G := \limsup_{n\to\infty} (A_{x,y}^n)^{1/n}$$

does not depend on x and y.

Proposition 4.8.1 *Suppose that G is an infinite, locally finite, connected graph. Then for $p < \lambda_G^{-1}$, for any $x, y \in V$,*

$$\mathbf{P}_p^G[x \leftrightarrow y] \leq \frac{(p\lambda_G)^{d(x,y)}}{1 - p\lambda_G}.$$

If G is also quasi-transitive, then $p_u(G) \geq \lambda_G^{-1}$.

This bound on p_u coincides with the bound in [9] for quasi-transitive graphs with constant degree, and improves upon it if the degree is nonconstant.

PROOF. Clearly,

$$A_{x,x}^{m+n} \geq A_{x,y}^{m} A_{y,x}^{n} \tag{4.8.3}$$

for any two sites x, y, and any non-negative integers m, n. Since $A_{x,x}^{2k} > 0$, it follows (see, e.g., [21, Section 8]) that

$$\lim_{k\to\infty} (A_{x,x}^{2k})^{\frac{1}{2k}} = \tilde{\lambda}_G = \sup_{k\geq 1} (A_{x,x}^{2k})^{\frac{1}{2k}} .$$

Using symmetry and (4.8.3), we obtain

$$\forall x, y \in V \quad \forall k \geq 1 \quad A_{x,y}^{k} \leq \left(A_{x,x}^{2k}\right)^{1/2} \leq \tilde{\lambda}_G^{k} . \tag{4.8.4}$$

Note that (4.8.4) implies that $\lambda_G = \tilde{\lambda}_G$. Denote by $[x \leftrightarrow y]$ the event that there is an open path connecting the sites x and y, and let $[x \leftrightarrow \infty]$ denote the event that x belongs to an infinite p-cluster. Let $\mathcal{N}_{x,y}^{(k)}$ be the number of self-avoiding paths of length k which connect x to y. Then

$$\begin{aligned} \mathbf{P}_p[x \leftrightarrow y] &\leq \sum_{k=d(x,y)}^{\infty} \mathcal{N}_{x,y}^{(k)} p^k \leq \sum_{k=d(x,y)}^{\infty} p^k A_{x,y}^{k} \\ &\leq \sum_{k=d(x,y)}^{\infty} (p\lambda_G)^k = \frac{(p\lambda_G)^{d(x,y)}}{1 - p\lambda_G} , \end{aligned} \tag{4.8.5}$$

provided $p < \lambda_G^{-1}$.

Suppose now that G is quasi-transitive. In this case

$$\theta(p) = \inf_{z\in V} \mathbf{P}_p[z \leftrightarrow \infty] > 0$$

for any $p > p_c$. If $p > p_u$, then for all $x, y \in V$,

$$\mathbf{P}_p[x \leftrightarrow y] \geq \mathbf{P}_p[x \leftrightarrow \infty, y \leftrightarrow \infty] \geq \mathbf{P}_p[x \leftrightarrow \infty]\mathbf{P}_p[y \leftrightarrow \infty] = \theta(p)^2 > 0 , \tag{4.8.6}$$

where the second inequality is an instance of the Harris inequality. Comparing (4.8.5) to (4.8.6) gives $p_u \geq \lambda_G^{-1}$. □

Remark O. Schramm (personal communication) has obtained a sharper lower bound for p_u. He showed that $p_u \geq \gamma_G^{-1}$, where

$$\gamma_G := \sup_{x \in V} \limsup_{k \to \infty} \left(\mathcal{N}_x^k\right)^{\frac{1}{k}}$$

and $\mathcal{N}_x^k$ is the number of self-avoiding cycles that start and end at x.

The final topic of this section is the relation between the number of ends of a quasi-transitive graph and the critical parameters p_c and p_u. It is well known that a quasi-transitive graph G can only have 1, 2 or uncountably many ends (see, e.g., [26, Section 6]). In case the number of ends is more than 1, one can use the converse of Theorem 4.8.2 to show that $p_u = 1$. This converse states that for quasi-transitive graphs,

$$\forall p > p_u \qquad \lim_{R \to \infty} \inf_{x,y \in V} \mathbf{P}_p^G\left(B(x,R) \leftrightarrow B(y,R)\right) = 1 . \tag{4.8.7}$$

(This is a consequence of the Harris inequality; see [29, Theorem 3.1]). The removal of an appropriate finite set of sites and edges from a graph G which has more than 1 end breaks the graph into more than one infinite connected component. Therefore the limit in (4.8.7) cannot be 1 when $p < 1$, and we must have $p_u = 1$. If the number of ends of G is uncountable, then $p_c(G) < 1$, since then G has a positive Cheeger constant (see [26, Proposition 6.2] and [9, Theorem 2]). If G has 2 ends, then G is just a "finite extension" of $\mathbf{Z}$, so $p_c(G) = 1$. More precisely, by the proof of [26, Proposition 6.1], there is a doubly infinite sequence $(\ldots, A_{-2}, A_{-1}, A_0, A_1, A_2, \ldots)$ of pairwise disjoint and isomorphic finite subgraphs of G, such that any infinite self-avoiding path in G must intersect either all the graphs A_j with large enough j, or all the graphs A_{-j} with large enough j.

The case of a single end is more delicate. Babson and Benjamini [5] proved that $p_u(G) < 1$ if G is the Cayley graph of a finitely presented group with one end. The question stated in [9], whether $p_u(G) < 1$ whenever G is a quasi-transitive graph with one end, is still unsolved.

4.9 Examples and Questions

A variant of the following example was shown to us by O. Schramm.

Example: A semi-transitive graph where exactly 2 infinite clusters can coexist Let T be a binary tree with root ρ, i.e., T is the tree in which ρ is incident to exactly two edges, and all other vertices are incident to exactly three edges. Let H be the product graph $T \times \mathbf{Z}$ with an additional distinguished vertex v^* joined by a single edge to the vertex $(\rho, 0)$ of $T \times \mathbf{Z}$. Theorem 4.1.7 implies that $p_u(H) = p_u(T \times \mathbf{Z}) < 1$. Finally, let G consist of two copies H_1, H_2 of H, glued together at their distinguished vertices (so these vertices v_1^* and v_2^* are identified, and the resulting vertex of G is denoted w^*). It is easy to see that G is semi-transitive, with V_F consisting of the distinguished vertex w^* only. For all

$p > p_u(H)$, it follows that bond percolation on G can have one or two infinite clusters with positive probability: If at least one of the two edges incident to w^* is closed, then a.s. there are exactly two infinite clusters, one contained in H_1 and the other in H_2. On the other hand, clearly the infinite clusters of H_1 and H_2 can connect to each other with positive probability. □

Next, we discuss briefly yet another phase transition. Let G be a nonunimodular quasi-transitive graph, and denote by μ the left Haar measure on $\mathrm{Aut}(G)$. Recall the definition of stabilizer $S(x)$ from Section 4.6. Say that a cluster $\mathcal{C}$ with vertex set $V(\mathcal{C})$ is *heavy* if $\sum_{x \in V(\mathcal{C})} \mu[S(x)] = \infty$; otherwise, we say that $\mathcal{C}$ is *light*. Theorem 4.1.6 implies that heavy and light infinite clusters cannot coexist at any fixed level p, so it is natural to define

$$p_h := \inf \Big\{ p : \mathbf{P}_p[\text{ there is a heavy infinite cluster}] > 0 \Big\} . \tag{4.9.1}$$

The mass transport method can be used effectively in heavy clusters. For instance, Theorem 4.1.4 can be easily extended to nonunimodular graphs, provided that the parameters p_1, p_2 considered there are greater than p_h. For the nonunimodular example $\widehat{T_d}$ mentioned in Section 4.6 (a tree with additional edges leading to ξ-grandparents) it is easy to see that $p_c < p_h = p_u = 1$. On the other hand, let T_k denote the k-regular tree. For any graph G_0 with bounded degrees, if k is large enough, then $G = G_0 \times T_k$ satisfies

$$p_h(G) < p_u(G) . \tag{4.9.2}$$

The proof is similar to an argument in [9, Section 4]. Let D_0 be the maximal degree in G_0, and let ρ_G be the spectral radius for simple random walk on G. It is easy to see that if $p > p_c(T_k)$, then any infinite p-cluster in $G = G_0 \times T_k$ is heavy. Therefore,

$$\lambda_G \cdot p_h(G) \le (D_0 + k) \cdot \rho_G \cdot p_h(G) \le (D_0 + k) \cdot \rho_G \cdot p_c(T_k) = \frac{D_0 + k}{k - 1} \rho_G . \tag{4.9.3}$$

Since $\rho_G = \rho_{G_0 \times T_k} \to 0$ as $k \to \infty$, it follows from [9, Theorem 4], or from Proposition 4.8.1 above, that for large enough k (when the right hand side of (4.9.3) is less than 1) we have (4.9.2). We expect that there exist transitive graphs where $p_c < p_h < p_u < 1$, but we do not have an explicit example.

We end the paper with some *questions:*

1. Is $p_c < p_h$ for every nonunimodular quasi-transitive graph? (p_h is defined in (4.9.1).)
   What geometric properties of G guarantee that $p_h < p_u$?
2. Can one drop the unimodularity assumption made in Theorem 4.1.4 and Proposition 4.7.1?
3. A graph G is said to exhibit *cluster repulsion* at level p if, for i.i.d. percolation with retention parameter p on G, any two infinite clusters can come within unit distance from each other in at most finitely many places. Does a quasi-transitive graph necessarily exhibit cluster repulsion for any $p \in [0, 1]$?

It is not hard to show that cluster repulsion at level p follows if the pair connectivity function $p \mapsto \mathbf{P}_p^G[u \leftrightarrow v]$ is continuous at p for all u, v. Hence cluster repulsion can fail at most for countably many values of p. If cluster repulsion holds, then it follows that for every R, any two infinite clusters come within distance R from each other at most finitely often a.s. We have an example of a (non-semi-transitive) graph for which cluster repulsion fails for certain p.

4. Let G be a quasi-transitive graph. A site x is an *encounter point* of the cluster C if removing the edges incident to x from C yields at least three infinite connected components. For $p \in (p_c, p_u)$, is $\mathbf{P}_p[\exists$ an infinite cluster with infinitely many encounter points$] > 0$? By Theorem 4.1.6, this would imply that *every* infinite cluster has infinitely many encounter points $\mathbf{P}_p$-a.s. This question has a positive answer in the unimodular case (see [8] or [24]), which readily extends to heavy clusters in the nonunimodular case. A general answer would be a significant step toward determining whether

$$\mathbf{P}_{p_c}[\exists \text{ infinite open clusters }] = 0 \qquad (4.9.4)$$

for quasi-transitive graphs with a nonamenable automorphism group. (Under the additional assumption of unimodularity, (4.9.4) is proved in [8].)

Acknowledgments: We thank Itai Benjamini, Russ Lyons, and Jeff Steif for stimulating discussions. We are especially indebted to Oded Schramm for many helpful and insightful comments. The research of the first author was supported by a grant from the Swedish Natural Science Research Council. The research of the second author was partially supported by NSF grant # DMS-9803597. The research of the third author was partially supported by NSF grant # DMS-9703814.

References

[1] Aizenman, M., Kesten H., and Newman, C.M. (1987) Uniqueness of the infinite cluster and continuity of connectivity functions for short- and long-range percolation, *Commun. Math. Phys.* **111**, 505–532.

[2] Aldous, D. (1997) Brownian excursions, critical random graphs and the multiplicative coalescent, *Ann. Probab.* **25**, 812–854.

[3] Alexander, K. (1995) Simultaneous uniqueness of infinite clusters in stationary random labeled graphs, *Commun. Math. Phys.* **168**, 39–55.

[4] Alexander, K. (1995) Percolation and minimal spanning forests in infinite graphs. *Ann. Probab.* **23**, 87–104.

[5] Babson, E. and Benjamini, I. (1998) Cut sets and normed cohomology with application to percolation, *Proc. Amer. Math. Soc.*, to appear.

[6] Barsky, D.J., Grimmett, G.R., and Newman, C.M. (1991) Dynamic renormalization and continuity of the percolation transition in orthants, *Spatial Stochastic Processes*, pp 37–55, Birkhäuser, Boston.

[7] Benjamini, I., Lyons, R., Peres, Y., and Schramm, O. (1997) Group-invariant percolation on graphs, *Geom. Func. Analysis*, to appear.

[8] Benjamini, I., Lyons, R., Peres, Y., and Schramm, O. (1998) Critical percolation on any nonamenable group has no infinite clusters, *Ann. Probab.*, to appear.

[9] Benjamini, I. and Schramm, O. (1996) Percolation beyond $\mathbf{Z}^d$: many questions and a few answers, *Electr. Commun. Probab.* **1**, 71–82.

[10] Borgs, C., Chayes, J., Kesten, H., and Spencer, J. (1998) The birth of the infinite cluster: finite-size scaling in percolation, *in preparation.*

[11] Burton, R. and Keane, M. (1989) Density and uniqueness in percolation, *Commun. Math. Phys.* **121**, 501–505.

[12] Chayes, J.T., Chayes, L., and Newman, C.M. (1985) The stochastic geometry of invasion percolation, *Commun. Math. Phys.* **101**, 383–407.

[13] Falconer, K. (1997) *Techniques in Fractal Geometry*, Wiley, New York.

[14] Gandolfi, A., Grimmett, G.R., and Russo, L. (1988) On the uniqueness of the infinite open cluster in the percolation model, *Commun. Math. Phys.* **114**, 549–552.

[15] Grimmett, G.R. (1989) *Percolation*, Springer-Verlag, New York.

[16] Grimmett, G.R. and Newman, C.M. (1990) Percolation in $\infty + 1$ dimensions, *Disorder in Physical Systems* (G. R. Grimmett and D. J. A. Welsh, editors), 219–240, Clarendon Press, Oxford, UK.

[17] Häggström, O. (1997) Infinite clusters in dependent automorphism invariant percolation on trees, *Ann. Probab.* **25**, 1423–1436.

[18] Häggström, O. and Peres, Y. (1998) Monotonicity of uniqueness for percolation on Cayley graphs: all infinite clusters are born simultaneously, *Probab. Th. Rel. Fields*, **113**, 273–285.

[19] Janson, S., Knuth, D.E., Luczak, T., and Pittel, B. (1993) The birth of the giant component, *Rand. Struct. Alg.* **4**, 233–358.

[20] Kesten, H. (1980) The critical probability for bond percolation on the square lattice equals $\frac{1}{2}$, *Commun. Math. Phys.* **74**, 41–59.

[21] Kingman, J.F.C. (1963) The ergodic decay of Markov transition probabilities, *Proc. London Math. Soc.* **13**, 337–358.

[22] Lalley, S. (1996) Percolation on Fuchsian groups, *Ann. Inst. H. Poincaré, Probab. Stat.* **34**, 151–178.

[23] Lyons, R. and Peres, Y. (1998) *Probability on Trees and Networks.* Book in preparation, draft available at URL `http://php.indiana.edu/~rdlyons`

[24] Lyons, R. and Schramm, O. (1998) Indistinguishability of Percolation Clusters, *in preparation.*

[25] Machta, J., Choi, Y.S., Lucke, A., Schweizer, T., and Chayes, L. (1995) Invaded cluster algorithms for equilibrium critical points, *Phys. Rev. Lett.* **75**, 2792–2795.

[26] Mohar, B. (1991) Some relations between analytic and geometric properties of infinite graphs, *Discr. Math.* **95**, 193–219.

[27] Newman, C.M. and Schulman, L.S. (1981) Infinite clusters in percolation models, *J. Statist. Phys.* **26**, 613–628.

[28] Newman, C.M. and Stein, D.L. (1996) Ground-state structure in a highly disordered spin-glass model, *J. Statist. Phys.* **82**, 1113–1132.

[29] Schonmann, R.H. (1998) Stability of infinite clusters in supercritical percolation, *Probab. Th. Rel. Fields*, **113**, 287–300.

[30] Schonmann, R.H. (1998) Percolation in $\infty + 1$ dimensions at the uniqueness threshold, *this volume.*

[31] Trofimov, V.I. (1985) Automorphism groups of graphs as topological groups, *Math. Notes* **38**, 717–720.

Olle Häggström
Mathematical Statistics
Chalmers University of Technology
Gothenburg
olleh@math.chalmers.se

Yuval Peres
Department of Statistics
University of California
Berkeley
and
Department of Mathematics
Hebrew University
Jerusalem
peres@stat.berkeley.edu

Roberto H. Schonmann
Department of Mathematics
University of California
Los Angeles
rhs@math.ucla.edu

5

Inequalities and Entanglements for Percolation and Random-Cluster Models

Geoffrey R. Grimmett

ABSTRACT We discuss inequalities and applications for percolation and random-cluster models. The relevant areas of methodology concern the following two types of inequality: inequalities involving the probability of a general increasing event, and certain differential inequalities involving the percolation probability. We summarise three areas of application of such inequalities, namely strict inequality between the bond and site critical percolation probabilities of a general graph, the general study of entanglements in percolation, and strict inequalities for critical points of disordered random-cluster models.

Keywords: Percolation, random-cluster model, Ising model, Potts model, quenched system, phase transition, entanglement.

AMS Subject Classifications: 60K35, 82B20.

5.1 Introduction

Harry Kesten's achievements across probability theory continue to be enormously influential and stimulating, and nowhere more so than in the study of spatial random processes. The results reported in this paper have been inspired in part by Harry's beautiful work on percolation.

Inequalities are central to the mathematics of disordered physical systems such as percolation and random-cluster models. They occur in several different ways, some of which are discussed here.

The methodological uses of inequalities include applications of the FKG and BK inequalities; these inequalities are now well understood and appreciated (see [6]). Less well known is an inequality used in [9, 11] in order to study exponential decay in random-cluster models. We present this inequality in Section 5.3.1, together with an application to percolation entanglements in Section 5.4.

Our second 'methodological' inequality is more a frame of mind than a theorem, and concerns the problem of proving that enhancements of certain processes cause *strict* changes in the values of the critical point. We present a very brief account of the relevant methods of [2] in Section 5.3.2. This method will be applied in Section 5.5 to obtain a theorem concerning strict inequalities between critical points of 'disordered' (or 'quenched') random-cluster models.

Some of the results of this paper have appeared or will appear elsewhere, and therefore no proofs are included here, although references are listed. The results of Section 5.5 concerning disordered random-cluster models are however new, and proofs are included in that section.

5.2 Percolation and Random-Cluster Models

Let $\mathbb{L}^d$ denote the d-dimensional cubic lattice having vertex set $\mathbb{Z}^d$ and edge set $\mathbb{E}^d$, where $d \geq 2$. We consider bond percolation on $\mathbb{L}^d$. The appropriate sample space is $\Omega = \{0, 1\}^{\mathbb{E}^d}$, and the probability measure is the product measure P_p with density p, where $0 \leq p \leq 1$. As usual, we call an edge e *open* in the configuration ω $(\in \Omega)$ if $\omega(e) = 1$, and we call e *closed* otherwise.

A path in $\mathbb{L}^d$ is called *open* if and only if all its edges are open. For $A, B \subseteq \mathbb{Z}^d$, we write $A \leftrightarrow B$ if there exists an open path having one endpoint in A and the other in B. We write $A \leftrightarrow \infty$ if there exists some vertex in A which is the endpoint of an infinite open path. For $x \in \mathbb{Z}^d$, we write $C_x = \{y \in \mathbb{Z}^d : x \leftrightarrow y\}$ for the *open cluster* at x. The origin of $\mathbb{L}^d$ is denoted as 0, and we abbreviate C_0 to C.

We shall be particularly interested in the existence (or not) of infinite open clusters. The principal objects of study in percolation theory are the *percolation probability*

$$\theta(p) = P_p(|C| = \infty) = P_p(0 \leftrightarrow \infty), \tag{5.2.1}$$

together with the associated *critical probability*

$$p_c = \sup\{p : \theta(p) = 0\}. \tag{5.2.2}$$

See [6, 15] for detailed accounts of the percolation model.

Site percolation is a variant of the above model in which the vertices rather than the edges of $\mathbb{L}^d$ are designated either open or closed. The 'site' percolation probability is defined as in (5.2.1), with the difference that a path is called *open* if and only if all its *vertices* are open.

The random-cluster model of this paper will be defined in a slightly more general way than was percolation. Let $\mathbf{p} = (p_e : e \in \mathbb{E}^d)$ be a vector of numbers from the interval $[0, 1]$, and let $q \geq 1$. For a finite box Λ in $\mathbb{L}^d$, we write $\mathbb{E}_\Lambda$ for the set of edges induced by Λ. For $\xi \in \Omega = \{0, 1\}^{\mathbb{E}^d}$, we write Ω_Λ^ξ for the set of all configurations $\omega \in \Omega$ satisfying $\omega(e) = \xi(e)$ for all $e \notin \mathbb{E}_\Lambda$. The random-cluster measure $\phi_{\Lambda,\mathbf{p},q}^\xi$ on Ω_Λ^ξ is given by

$$\phi_{\Lambda,\mathbf{p},q}^\xi(\omega) = \frac{1}{Z_{\Lambda,\mathbf{p},q}^\xi} \left\{ \prod_{e \in \mathbb{E}_\Lambda} p_e^{\omega(e)} (1 - p_e)^{1-\omega(e)} \right\} q^{k(\omega,\Lambda)}, \qquad \omega \in \Omega_\Lambda^\xi, \tag{5.2.3}$$

where $Z_{\Lambda,\mathbf{p},q}^\xi$ is the appropriate normalising constant, and $k(\omega, \Lambda)$ is the number of open clusters of ω which intersect Λ. When $\xi = 0$ (respectively $\xi = 1$), this is called the 'free' (respectively 'wired') measure. For the purposes of this paper, it suffices to consider the wired measure, and we abbreviate henceforth $\phi^1_{\Lambda,\mathbf{p},q}$ to $\phi_{\Lambda,\mathbf{p},q}$. For a general guide to such random-cluster measures, see [8] and the references therein.

For $A \subseteq \mathbb{Z}^d$, the *surface* ∂A is the subset of A containing all vertices which have a neighbour in $\mathbb{L}^d$ not lying in A. We write $\Lambda_k = [-k, k]^d$ for the box of $\mathbb{L}^d$ having side-length $2k$.

The following facts are standard (see [8]):

(a) the limit measure $\phi_{\mathbf{p},q} = \lim_{\Lambda\to\mathbb{Z}^d} \phi_{\Lambda,\mathbf{p},q}$ exists in the sense of weak convergence,

(b) for any finite subset A of $\mathbb{Z}^d$,

$$\phi_{\Lambda,\mathbf{p},q}(A \leftrightarrow \partial\Lambda) \to \phi_{\mathbf{p},q}(A \leftrightarrow \infty) \quad \text{as } \Lambda \uparrow \mathbb{Z}^d,$$

(c) $\phi_{\Lambda,\mathbf{p},q}$ and $\phi_{\mathbf{p},q}$ satisfy the FKG inequality.

The random-cluster percolation probability is given by

$$\theta(\mathbf{p}, q) = \phi_{\mathbf{p},q}(0 \leftrightarrow \infty). \tag{5.2.4}$$

For reasons which will become clear in Section 5.5, we shall work not with $\theta(\mathbf{p}, q)$ but with the function $\psi(\mathbf{p}, q)$ defined by

$$\psi(\mathbf{p}, q) = \phi_{\mathbf{p},q}(I) \tag{5.2.5}$$

where I is the event that there exists at least one infinite open cluster.

We note that the random-cluster model with parameters $\mathbf{p}$, q reduces to the above bond percolation model when $q = 1$ and $\mathbf{p} = p$, the vector all of whose entries equal p. It is a standard fact concerning percolation that

$$\theta(p, 1) = 0 \text{ if and only if } \psi(p, 1) = 0.$$

Therefore, the percolation critical probability satisfies

$$p_c = \sup\{p : \psi(p, 1) = 0\}.$$

5.3 Inequalities

5.3.1 *An Inequality for Increasing Events*

There is a partial order on Ω given by $\omega \le \omega'$ if and only if $\omega(e) \le \omega'(e)$ for all $e \in \mathbb{E}^d$. A random variable X is called *increasing* if $X(\omega) \le X(\omega')$ whenever $\omega \le \omega'$. An event A is called *increasing* if its indicator function 1_A is increasing.

For any $\omega \in \Omega$ and any increasing event A, we define the 'distance' $F_A(\omega)$ from ω to A by

$$F_A(\omega) = \inf\left\{\sum_e \big(\omega'(e) - \omega(e)\big) : \omega' \ge \omega,\ \omega' \in A\right\}. \tag{5.3.1}$$

That is to say, $F_A(\omega)$ is the minimal number of extra edges which must be designated 'open' in order for A to occur.

Let E be a finite set of edges of $\mathbb{L}^d$, and let A be an event defined in terms of the states of edges belonging to E. Let

$$N(\omega) = \sum_{e \in E} \omega(e),$$

the total number of open edges of E in the configuration ω. The following proposition follows by Russo's formula and the FKG inequality, on noting that $F_A 1_A = 0$ and that $N + F_A$ is an increasing random variable.

Proposition 5.3.1 *Let $0 < p < 1$. For any non-empty increasing cylinder event A,*

$$\frac{d}{dp}\{\log P_p(A)\} \geq \frac{P_p(F_A)}{p(1-p)}. \tag{5.3.2}$$

(We write $\mu(X)$ for the mean of the random variable X under the probability measure μ.)

Proposition 5.3.1 relates the gradient of $\log P_p(A)$ to the mean of F_A. A different type of inequality is needed in order to bound this mean value below. In a way similar to the 'sprinkling' argument of [1], one may obtain the following.

Proposition 5.3.2 *For all p_1, p_2 satisfying $0 < p_1 < p_2 < 1$, there exist strictly positive numbers $a = a(p_1, p_2)$ and $b = b(p_1, p_2)$ such that, for any increasing cylinder event A,*

$$P_{p_1}(F_A) \geq -b - a \log P_{p_2}(A). \tag{5.3.3}$$

These two propositions do not appear to be sufficient by themselves in applications, and it is useful in practice to have recourse to the following additional general proposition.

Proposition 5.3.3 *Let $A, B_1, B_2, \ldots, B_m$ be increasing cylinder events such that $A \subseteq B_1 \cap B_2 \cap \cdots \cap B_m$ and such that the B_i are defined on disjoint sets of edges. Then*

$$F_A \geq \sum_{i=1}^{m} F_{B_i}. \tag{5.3.4}$$

Similar propositions are valid in the more general context of random-cluster models, and their proofs may be found in [9]. They may be applied to study the decay rates of the connectivity functions of subcritical random-cluster models. They also have applications to percolation, and such an application to entanglements in percolation is described in Section 5.4.

5.3.2 *Multiparameter Processes*

The following general situation occurs frequently. One encounters some random process having two (or more) real-valued parameters p, s, say. This random process has a phase transition, in the sense that some 'macroscopic function' $\theta = \theta(p, s)$

satisfies

$$\theta(p,s)\begin{cases} = 0 & \text{if } \phi(p,s) < 0 \\ > 0 & \text{if } \phi(p,s) > 0, \end{cases}$$

for some smooth function ϕ. The set of pairs (p, s) satisfying $\phi(p, s) = 0$ is sometimes called the 'critical surface' of the process. In many situations, one may be able to prove the existence of such a function ϕ, but its detailed properties, such as continuity or strict monotonicity, can be difficult to ascertain.

Here are two examples of questions which may be formulated in this way. First, one may ask whether or not there exists a critical probability p_c^{ent} for the existence of an infinite entanglement in bond percolation, and in addition whether or not p_c^{ent} differs from p_c. (See Section 5.4.) Secondly, if some of the strengths of interactions of a disordered ferromagnetic Ising or Potts model are increased, does the critical temperature necessarily change? We discuss this latter question further in Section 5.5.

A useful method for approaching and sometimes answering such questions has been described in [2]. In a broad class of situations including the two examples above, one may reformulate the question in the manner of the first paragraph of this section. One may then find a sequence $\theta_n(p, s)$, $n \geq 1$, of non-decreasing real-analytic functions satisfying $\theta_n(p, s) \to \theta(p, s)$, and in addition such that

$$\alpha \frac{\partial \theta_n}{\partial p} \leq \frac{\partial \theta_n}{\partial s} \leq \alpha^{-1} \frac{\partial \theta_n}{\partial p} \tag{5.3.5}$$

for some continuous function $\alpha = \alpha(p, s)$ which is strictly positive and finite on the interior of the parameter space. Such differential inequalities may be used to gain information about the gradient vector of θ_n, and this in turn implies certain properties of continuity and strict monotonicity for a natural parametrization of the critical surface of the process.

This approach has recently yielded, amongst other results, a solution to the problem of proving strict inequality between the critical probabilities of bond and site percolation on a given graph, and we state the relevant theorem here.

Let G be an arbitrary connected graph, and write $p_c^{\text{bond}}(G)$ (respectively $p_c^{\text{site}}(G)$) for the critical point of bond percolation (respectively site percolation) on G. The automorphism group of G acts on the vertices of G in a natural way, and we call G *finitely transitive* if this group action has only finitely many orbits. An edge e of G is called a *bridge* if its removal disconnects G; G is said to be *bridgeless* if it contains no bridges. We write $\Delta = \Delta(G)$ for the maximum vertex degree of a graph G.

Theorem 5.3.4 *Let G be an infinite connected graph with $\Delta = \Delta(G) < \infty$.*

(a) *If G is finitely transitive and bridgeless, then either*

(i) $p_c^{\text{bond}}(G) = p_c^{\text{site}}(G) = 1$, *or*
(ii) $0 < p_c^{\text{bond}}(G) < p_c^{\text{site}}(G) < 1$.

FIGURE 5.1. Sketches of two graphs. The first is entangled, the second is not.

(b) *We have that*

$$p_c^{\text{site}}(G) \le 1 - \left(1 - p_c^{\text{bond}}(G)\right)^{\Delta - 1}.$$

The proof may be found in [12]. This result generalises related inequalities valid for certain two-dimensional lattices; see [15] and elsewhere. For more details and applications of the general argument around (5.3.5), see [2, 9].

5.4 Entanglements in Percolation

The question was posed in [14] whether or not a percolation model on $\mathbb{Z}^3$ can contain large entangled clusters but no large connected clusters. Numerical work reported in [14] suggested the existence of an 'entanglement critical point' p_c^{ent} satisfying $p_c^{\text{ent}} \approx p_c - 1.8 \times 10^{-7}$. No formal definition of this critical point was presented, and indeed the discussion of this initial paper concerned the contents of finite boxes only, rather than the configuration on the entire infinite lattice. We summarise in this section recent progress towards a rigorous formulation of the problem of entanglements in percolation, and we present an application of Propositions 5.3.1–5.3.3, together with some open problems.

Here is some terminology. With each edge e of $\mathbb{E}^3$, we associate the closed straight line segment of $\mathbb{R}^3$ joining its endpoints. For $E \subseteq \mathbb{E}^3$, we write $[E]$ for the union of the corresponding line segments. A 'sphere' shall be taken to mean any subset of $\mathbb{R}^3$ which is homeomorphic to the unit sphere. The complement of a sphere S has two connected components, an unbounded component called the *outside* of S and denoted $\text{out}(S)$, and a bounded component called the *inside* and denoted $\text{ins}(S)$.

Let E be a finite subset of $\mathbb{E}^3$. We call E *entangled* if, for any sphere S not intersecting $[E]$, either $[E] \subseteq \text{ins}(S)$ or $[E] \subseteq \text{out}(S)$. This definition is illustrated in Figure 5.1.

There is more than one way of extending the notion of entanglement to an infinite set E of edges. Here are two such ways.

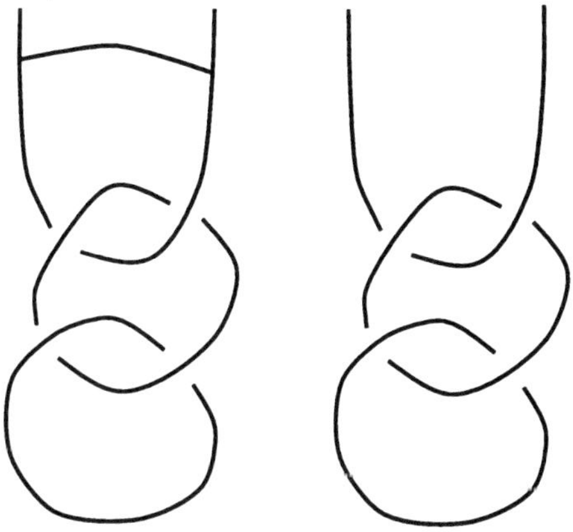

FIGURE 5.2. The four uppermost points lie in disjoint infinite paths not shown in this figure. The first graph is strongly entangled; the second graph is weakly entangled but not strongly entangled.

(a) We call E *strongly entangled* if, for every finite subset F of E, there exists a finite entangled subset F' of E satisfying $F \subseteq F'$.
(b) We call E *weakly entangled* if, for any sphere S not intersecting $[E]$, either $[E] \subseteq \text{ins}(S)$ or $[E] \subseteq \text{out}(S)$.

Such definitions are explored in [10, 13], where it is shown that E is weakly entangled whenever it is strongly entangled; the converse statement is false. See Figure 5.2.

Let J^{w} (respectively J^{s}) be the event that the origin is an endvertex of some edge lying in an infinite weakly (respectively strongly) entangled set E of open edges. It may be shown that J^{w} and J^{s} are indeed events, and it is clear that they are increasing. One may therefore define the *weak* and *strong entanglement probabilities*

$$\theta^{\text{w}}(p) = P_p(J^{\text{w}}), \quad \theta^{\text{s}}(p) = P_p(J^{\text{s}}), \tag{5.4.1}$$

and associated *entanglement critical points*

$$\begin{aligned} {}^{\text{w}}p_{\text{c}}^{\text{ent}} &= \sup\{p : \theta^{\text{w}}(p) = 0\}, \\ {}^{\text{s}}p_{\text{c}}^{\text{ent}} &= \sup\{p : \theta^{\text{s}}(p) = 0\}. \end{aligned} \tag{5.4.2}$$

It may be conjectured that

$$ {}^{\text{w}}p_{\text{c}}^{\text{ent}} = {}^{\text{s}}p_{\text{c}}^{\text{ent}}. \tag{5.4.3}$$

Article [10] contains a further discussion of types of entanglement, but no proof of this conjecture. In order to be more concrete in the remainder of this section, we concentrate henceforth on 'strong entanglement', and shall suppress further reference to the word 'strong'. Thus, for example, we write $p_{\text{c}}^{\text{ent}} = {}^{\text{s}}p_{\text{c}}^{\text{ent}}$.

Since all connected graphs are entangled, it is immediate that $p_{\text{c}}^{\text{ent}} \le p_{\text{c}}$. The technique of Section 5.3.2 was used in [2] in such a way as to imply the strict inequality $p_c^{\mathrm{ent}} < p_c$. Only recently was it proved in [13] that $p_c^{\mathrm{ent}} > 0$. We summarise these two facts in a theorem.

Theorem 5.4.1 *It is the case that* $0 < p_c^{\mathrm{ent}} < p_c$.

We consider next a further problem, namely to ascertain the manner of decay of the sizes of large finite entanglements. Let E_x be the maximal entanglement touching the vertex x, and write $E = E_0$; it is not hard to see that E_x is well defined for any x. For $n \geq 1$, let $B(n)$ be the box $[-n, n]^3$, and $\partial B(n) = B(n) \backslash B(n-1)$. It seems reasonable to believe that $P_p(E \cap \partial B(n) \neq \varnothing)$ should decay exponentially as $n \to \infty$, whenever $p < p_c^{\mathrm{ent}}$, and it is an open problem to prove this. The inequalities of Section 5.3.1 allow a little progress in the direction of estimating the decay rate of $P_p(E \cap \partial B(n) \neq \varnothing)$ as $n \to \infty$, as follows.

For a positive integer k, we write λ_k for the kth iterate of the natural logarithm function. More precisely, let

$$\lambda_1(x) = \log x,$$
$$\lambda_{k+1}(x) = \max\{1, \log \lambda_k(x)\} \quad \text{for } k \geq 1.$$

Theorem 5.4.2 *There exists* $p_0 > 0$ *such that, for* $p \in (0, p_0)$ *and* $k \geq 1$, *there exists* $\alpha_k(p) > 0$ *such that*

$$P_p\big(E \cap \partial B(n) \neq \varnothing\big) \leq \exp\left\{-\frac{\alpha_k(p)n}{\lambda_k(n)}\right\} \quad \textit{for all large } n. \tag{5.4.4}$$

We expect that the logarithmic term in (5.4.4) may be removed, and that the conclusion is valid for all p satisfying $p < p_c^{\mathrm{ent}}$. The proof of Theorem 5.4.2 exploits versions of Propositions 5.3.1–5.3.3, and may be found in [10].

5.5 Disordered Random-Cluster Models

The problem of proving strict inequality between two critical points occurs frequently in probability theory and statistical mechanics. The fundamental mechanism summarised in Section 5.3.2 for establishing such inequalities was applied in [2] to percolation and Ising models. This work was extended in [3, 7] to random-cluster models, thereby deriving an attractive methodology for Ising and Potts systems. There has been considerable interest recently in disordered (or 'quenched') systems, in which the interaction function is itself sampled at random from an appropriate ensemble, and it is the purpose of this section to explore strict inequalities for random-cluster models in this setting.

Perhaps the main motivation for the general study of disordered systems is the desire to understand phase transitions in spin glasses (see [18]). Indeed, it is currently unknown whether or not such phase transitions exist. Theorem 5.5.2 of this section has proved useful to recent work [4] intended to elucidate this question.

Here is a description of a disordered random-cluster model. Let $\mathbf{J} = \{J_e : e \in \mathbb{E}^d\}$ be a family of non-negative random variables governed by a probability measure $\mathbb{P}$; we allow the J_e to take values in the extended half-line $[0, \infty]$, and we define

$$p_e = 1 - e^{-\beta J_e}, \quad e \in \mathbb{E}^d,$$

where $0 < \beta < \infty$. Let q be a real number satisfying $q \geq 1$. The random-cluster measure $\phi_{\mathbf{p},q}$, defined as in Section 5.2, is a random probability measure.

Let I ($\subseteq \Omega$) be the event that there exists at least one infinite open cluster. Since I is an increasing event, we have by the FKG inequality that $\phi_{\mathbf{p},q}(I)$ is non-decreasing in β. We define the critical point $\beta_c(\mathbf{J})$ by

$$\beta_c(\mathbf{J}) = \sup\{\beta > 0 : \phi_{\mathbf{p},q}(I) = 0\},$$

with the convention that the supremum of the empty set is 0. It is more usual (see [8]) to define the critical point via the event $\{0 \leftrightarrow \infty\}$ rather than via I. Such a definition may be inappropriate whenever the J_e are permitted to take the value 0 with strictly positive probability, since there may exist (with strictly positive $\mathbb{P}$-probability) configurations $\mathbf{J}$ such that $\phi_{\mathbf{p},q}(I) > 0$ while $\phi_{\mathbf{p},q}(0 \leftrightarrow \infty) = 0$. It is not difficult to see however that the two such definitions are equivalent whenever $\mathbb{P}$ satisfies

$$\mathbb{P}\big(\phi_{\mathbf{p},q}(x \leftrightarrow y) > 0\big) = 1 \quad \text{for all } x, y \in \mathbb{Z}^d.$$

Let τ_i, $1 \leq i \leq d$, be the d fundamental lattice shifts of $\mathbb{L}^d$; that is, $\tau_i(x) = x + e_i$ where e_i is a unit vector in the direction of increasing ith coordinate. We recall that the *invariant σ-field $\mathcal{I}$* of the random field $\mathbf{J} = \{J_e : e \in \mathbb{E}^d\}$ is the σ-field of all events which are invariant under the natural shift operators on Ω induced by the τ_i. We call a σ-field *trivial* if all events therein have probability either 0 or 1.

Theorem 5.5.1 *If the family* $\mathbf{J}$ *has trivial invariant σ-field, then there exists a constant* $\beta_c = \beta_c(\mathbb{P})$ *satisfying* $0 \leq \beta_c \leq \infty$ *and*

$$\mathbb{P}\big(\beta_c(\mathbf{J}) = \beta_c\big) = 1.$$

PROOF. The quantity $\phi_{\mathbf{p},q}(I)$ is a function of $\mathbf{J}$ which is invariant under lattice shifts. Therefore it is measurable on the invariant σ-field, and is therefore a.s. constant. It follows that $\beta_c(\mathbf{J})$ is a.s. constant as claimed. □

We do not have useful necessary and sufficient conditions for the strict inequalities $0 < \beta_c < \infty$. Instead we note that, when $\mathbf{J}$ has trivial invariant σ-field, then

(a) $\beta_c = \infty$ if the edge set $\{e : J_e > 0\}$ possesses a.s. finite clusters only, and
(b) $\beta_c = 0$ if the edge set $\{e : J_e = \infty\}$ possesses a.s. one or more infinite clusters.

We shall prove that $\beta_c(\mathbb{P})$ is strictly monotone in $\mathbb{P}$, subject to certain conditions. Although the main applications of such a result are currently to situations where the J_e are *independent* random variables (see [4]), we shall consider here a more general setting, as follows.

Let $\mathbf{X} = \{X_e : e \in \mathbb{E}^d\}$ and $\mathbf{Y} = \{Y_e : e \in \mathbb{E}^d\}$ be families of non-negative random variables indexed by $\mathbb{E}^d$ and defined on the same probability space $(\Gamma, \mathcal{G}, \mathbb{P})$. We shall assume henceforth that

$$\mathbb{P}(\mathbf{X} \leq \mathbf{Y}) = 1, \tag{5.5.1}$$

which is to say that $\mathbb{P}(X_e \leq Y_e) = 1$ for all e. We propose to compare with one another the two random-cluster models having respective edge interactions $\mathbf{X}$ and $\mathbf{Y}$. It is evident (by the FKG inequality) that

$$\beta_c(\mathbf{X}) \geq \beta_c(\mathbf{Y}). \tag{5.5.2}$$

In order to prove a strict inequality, we shall require some sort of lower bound for the difference $\mathbf{Y} - \mathbf{X}$. Let $\eta \in \mathbb{E}^d$ and let $\mathbf{k} = (k_1, k_2, \ldots, k_d) \in \mathbb{Z}^d$ satisfy $k_j \neq 0$ for $1 \leq j \leq d$. The pair $(\eta, \mathbf{k})$ generates a periodic class

$$\Xi = \Xi(\eta, \mathbf{k}) = \{\eta + \mathbf{m}.\mathbf{k} : \mathbf{m} \in \mathbb{Z}^d\} \tag{5.5.3}$$

of edges, where $\mathbf{m}.\mathbf{k} = (m_1 k_1, m_2 k_2, \ldots, m_d k_d)$.

For $f \in \mathbb{E}^d$, let

$$\mathcal{F}_f = \sigma\big(\{X_e, Y_e : e \in \mathbb{E}^d,\ e \neq f\}\big)$$

denote the σ-field generated by the random variables X_e, Y_e for $e \neq f$. We shall require that there exists η, $\mathbf{k}$, and $\delta > 0$ such that

$$\mathbb{P}(Y_f - X_f \mid \mathcal{F}_f) \geq \delta \quad \text{a.s., for all } f \in \Xi(\eta, \mathbf{k}). \tag{5.5.4}$$

(The expression $\mathbb{P}(Z \mid \mathcal{F}_f)$ denotes the appropriate conditional expectation of the random variable Z.)

Let $0 \leq s \leq 1$, and set

$$J_e = J_e(s) = X_e + s(Y_e - X_e), \quad e \in \mathbb{E}^d. \tag{5.5.5}$$

The $J_e(s)$ interpolate between $J_e(0) = X_e$ and $J_e(1) = Y_e$. If the family $\{(X_e, Y_e) : e \in \mathbb{E}^d\}$ has trivial invariant σ-field, then so does $\mathbf{J}(s) = \{J_e(s) : e \in \mathbb{E}^d\}$, whence there exists by Theorem 5.5.1 a constant $\beta_c = \beta_c(\mathbb{P}, s)$ such that

$$\mathbb{P}\big(\beta_c(\mathbf{J}(s)) = \beta_c(\mathbb{P}, s)\big) = 1.$$

Our target is to identify conditions under which $\beta_c(\mathbb{P}, 0) > \beta_c(\mathbb{P}, 1)$.

Theorem 5.5.2 *Let $q > 1$. Assume that:*

(i) *$\mathbb{P}$, η, $\mathbf{k}$, δ are such that $\delta > 0$ and (5.5.4) holds,*
(ii) *there exist reals ρ, σ such that $0 < \rho \leq \sigma < \infty$ and*

$$\mathbb{P}(\rho \leq X_e \leq Y_e \leq \sigma) = 1 \quad \textit{for all } e \in \mathbb{E}^d, \tag{5.5.6}$$

(iii) *the invariant σ-field of the family $\{(X_e, Y_e) : e \in \mathbb{E}^d\}$ is trivial.*

We have that $\beta_c(\mathbb{P}, 0) > \beta_c(\mathbb{P}, 1)$.

It may be possible to relax condition (ii) while retaining the conclusion of this theorem. A similar result is valid when $q = 1$, but a different argument is needed;

see the relevant discussion in [7]. An inequality related to the above theorem, and derived independently of the present paper, will appear in [5].

We begin the proof of Theorem 5.5.2 with a preliminary lemma. Let $G = (V, E)$ be a finite graph; let $\mathbf{p} = (p_e : e \in E)$ be a vector of numbers in [0, 1], and let $q \geq 1$. We write ϕ_G for the random-cluster measure on $\{0, 1\}^E$ having edge parameters p_e and cluster-weighting factor q; that is,

$$\phi_G(\omega) = \frac{1}{Z_G}\left\{\prod_{e\in E} p_e^{\omega(e)}(1-p_e)^{1-\omega(e)}\right\}q^{k(\omega)}, \quad \omega \in \Omega_E = \{0, 1\}^E,$$

where $k(\omega)$ is the number of open clusters of ω. For $e \in E$, let J_e denote the event $\{\omega(e) = 1\}$. We denote by $G.e$ (respectively $G\backslash e$) the graph obtained from G by contracting (respectively deleting) the edge e.

Lemma 5.5.3 *Let $e \in E$, $q \geq 1$, and $0 < p_e < 1$.*

(a) *For any event A ($\subseteq \{0, 1\}^E$),*

$$\frac{d}{dp_e}\phi_G(A) = \frac{\phi_G(J_e)(1-\phi_G(J_e))}{p_e(1-p_e)}\Delta_G(A, e) \qquad (5.5.7)$$

where

$$\Delta_G(A, e) = \phi_{G.e}(A) - \phi_{G\backslash e}(A).$$

(b) *We have that*

$$\frac{1}{q} \leq \frac{\phi_G(J_e)(1-\phi_G(J_e))}{p_e(1-p_e)} \leq q. \qquad (5.5.8)$$

Proof. (a) By [3, Proposition 4],

$$\frac{d}{dp_e}\phi_G(A) = \frac{1}{p_e(1-p_e)}\Big\{\phi_G(A \cap J_e) - \phi_G(A)\phi_G(J_e)\Big\}.$$

Now,

$$\phi_G(A \cap J_e) - \phi_G(A)\phi_G(J_e) = \phi_G(J_e)\phi_G(\overline{J}_e)\Big\{\phi_G(A \mid J_e) - \phi_G(A \mid \overline{J}_e)\Big\},$$

and (5.5.7) follows by [8, Theorem 2.3].

(b) It is standard (see [8, equation (3.10)]) that

$$\frac{p_e}{p_e + (1-p_e)q} \leq \phi_G(J_e) \leq p_e$$

and the claim follows easily. □

Proof of Theorem 5.5.2. Assume the hypotheses of the theorem. Let $0 \leq s \leq 1$, and define $\mathbf{J}(s) = \{J_e(s) : e \in \mathbb{E}^d\}$ accordingly by (5.5.5). With $\mathbf{p} = \mathbf{p}(s)$ given by

$$p_e(s) = 1 - \exp\big(-\beta J_e(s)\big),$$

we write $I_{m,n} = \{\partial\Lambda_m \leftrightarrow \partial\Lambda_n\}$, and

$$\theta_{m,n} = \phi_{\Lambda_n,\mathbf{p},q}(I_{m,n}) \quad \text{for } m \leq n.$$

We have that

$$\begin{aligned}\frac{\partial\theta_{m,n}}{\partial s} &= \sum_e \frac{\partial\theta_{m,n}}{\partial p_e}\frac{\partial p_e}{\partial s}\\ &\geq \sum_{f\in\Xi} \frac{\partial\theta_{m,n}}{\partial p_f}\frac{\partial p_f}{\partial s}\\ &\geq \sum_{f\in\Xi} \frac{\partial\theta_{m,n}}{\partial p_f}\beta(Y_f - X_f)e^{-\beta\sigma}, \end{aligned} \tag{5.5.9}$$

where $\Xi = \Xi(\eta, \mathbf{k})$. Similarly,

$$\begin{aligned}\frac{\partial\theta_{m,n}}{\partial \beta} &= \sum_e \frac{\partial\theta_{m,n}}{\partial p_e}\frac{\partial p_e}{\partial \beta}\\ &\leq \sum_e \frac{\partial\theta_{m,n}}{\partial p_e}\sigma e^{-\beta\rho}. \end{aligned} \tag{5.5.10}$$

These two sums may be compared with one another via the forthcoming Lemma 5.5.4.

Let $e \in \mathbb{E}_{\Lambda_n}$ and let $f = f(e)$ be the edge of $\Xi \cap \mathbb{E}_{\Lambda_{n-1}}$ which is closest to e. (That is, the midpoint of f is closest to the midpoint of e, according to some given norm, say L^∞, on $\mathbb{R}^d$. If two or more such edges f exist, we pick one of them according to some predetermined rule.) We note that, for any given edge $f \in \Xi$, there exist at most $K = d2^d k_1 k_2 \dots k_d$ edges e with $f(e) = f$.

Lemma 5.5.4 *There exist a positive integer N and a function $\zeta = \zeta(\beta)$, continuous and finite when $0 < \beta < \infty$, such that*

$$\frac{\partial\theta_{m,n}}{\partial p_e} \leq \zeta\frac{\partial\theta_{m,n}}{\partial p_{f(e)}} \quad \textit{for all } e \in \mathbb{E}_{\Lambda_n},\ m \leq n, \textit{ and } n \geq N. \tag{5.5.11}$$

Note that (5.5.11) is an inequality between random variables. The proof of this lemma is given later.

We deduce from (5.5.10)–(5.5.11) that

$$\frac{\partial\theta_{m,n}}{\partial \beta} \leq K\zeta\sigma e^{-\beta\rho}\sum_{f\in\Xi}\frac{\partial\theta_{m,n}}{\partial p_f} \quad \text{for } n \geq N. \tag{5.5.12}$$

By (5.5.9) and Lemma 5.5.3,

$$\frac{\partial\theta_{m,n}}{\partial s} \geq \frac{\beta e^{-\beta\sigma}}{q}\sum_{f\in\Xi}\Delta_{m,n}(f)(Y_f - X_f), \tag{5.5.13}$$

where

$$\Delta_{m,n}(f) = \phi_{\Lambda_n\cdot f,\mathbf{p},q}(I_{m,n}) - \phi_{\Lambda_n\backslash f,\mathbf{p},q}(I_{m,n}).$$

Note that $\Delta_{m,n}(f)$ does not depend on the random variable $p_f(s)$, and is therefore $\mathcal{F}_f$-measurable. It follows that

$$\begin{aligned}
\mathbb{P}\left(\frac{\partial\theta_{m,n}}{\partial s}\right) &\geq \frac{\beta e^{-\beta\sigma}}{q}\sum_{f\in\Xi}\mathbb{P}\Big(\Delta_{m,n}(f)(Y_f - X_f)\Big) \\
&= \frac{\beta e^{-\beta\sigma}}{q}\sum_{f\in\Xi}\mathbb{P}\Big(\Delta_{m,n}(f)\mathbb{P}(Y_f - X_f \mid \mathcal{F}_f)\Big) \\
&\geq \frac{\beta e^{-\beta\sigma}}{q}\sum_{f\in\Xi}\delta\,\mathbb{P}\Big(\Delta_{m,n}(f)\Big) && \text{by (5.5.4)} \\
&\geq \frac{\beta\delta e^{-\beta\sigma}}{q^2K\zeta\sigma e^{-\beta\rho}}\mathbb{P}\left(\frac{\partial\theta_{m,n}}{\partial\beta}\right) && \text{for } n \geq N,
\end{aligned}$$

where we have used Lemma 5.5.3 and (5.5.12) at the last step.

In summary, there exists $\zeta'(\beta)$, continuous and finite when $0 < \beta < \infty$, such that

$$\mathbb{P}\left(\frac{\partial\theta_{m,n}}{\partial\beta}\right) \leq \zeta'\mathbb{P}\left(\frac{\partial\theta_{m,n}}{\partial s}\right) \quad \text{for } n \geq N.$$

It follows that $\Gamma_{m,n} = \mathbb{P}(\theta_{m,n})$ satisfies

$$\frac{\partial\Gamma_{m,n}}{\partial\beta} \leq \zeta'\frac{\partial\Gamma_{m,n}}{\partial s} \quad \text{for } n \geq N.$$

Now,

$$\begin{aligned}
\Gamma_{m,n} &\to \mathbb{P}\Big(\phi_{\mathbf{p}(s),q}(\partial\Lambda_m \leftrightarrow \infty)\Big) && \text{as } n \to \infty \\
&\to \mathbb{P}\Big(\phi_{\mathbf{p}(s),q}(I)\Big) && \text{as } m \to \infty,
\end{aligned}$$

by the dominated convergence theorem. Furthermore, by Theorem 5.5.1,

$$\mathbb{P}\Big(\phi_{\mathbf{p}(s),q}(I)\Big) = \begin{cases} 0 & \text{if } \beta < \beta_c(\mathbb{P}, s) \\ 1 & \text{if } \beta > \beta_c(\mathbb{P}, s), \end{cases}$$

where we have used assumption (iii) of the theorem. It follows as in [2, 9] (see also Section 5.3.2 of the current article) that $\beta_c(\mathbb{P}, s)$ is strictly decreasing in s, which implies the claim of the theorem. □

Proof of Lemma 5.5.4. This is very similar to the proof of [3, Theorem 1], and we therefore omit many of the details. The first step is to express the two derivatives in (5.5.11) in terms of two coupled Markov processes R_t, S_t on the common state space $\Omega^1_{\Lambda_n}$, satisfying $R_t \leq S_t$, and whose respective equilibrium distributions are $\phi_{\Lambda_n,\mathbf{p},q}(\cdot)$ and $\phi_{\Lambda_n,\mathbf{p},q}(\cdot \mid I_{m,n})$; such representations are easily derived as in (5.5.9)–(5.5.10) from [3, Proposition 5], which states that

$$\frac{\partial\theta_{m,n}}{\partial p_e} = \frac{\theta_{m,n}}{p_e(1-p_e)}\lim_{t\to\infty}\big\{P(R_t(e) = 0,\ S_t(e) = 1)\big\}. \tag{5.5.14}$$

(Here, P denotes the appropriate probability measure for the processes R, S.) Utilising the argument of [3], particularly the proof of inequality (4.4) there, one may obtain in the following way the required (5.5.11). The only difference of significance arises in the definition of the events V^1, V^2, V^3 of [3].

Let $e \in \mathbb{E}_{\Lambda_n}$ and let $f = f(e)$; let u, v be the endvertices of e. We may assume that e has at least one endvertex, u say, belonging to Λ_{n-1}; if, on the contrary, $u, v \in \Lambda_n \setminus \Lambda_{n-1}$ then it is a consequence of our assumption of wired boundary conditions that

$$\frac{\partial \theta_{m,n}}{\partial p_e} = 0,$$

and inequality (5.5.11) is trivial in this case.

Let C be a circuit of edges in $\mathbb{E}_{\Lambda_n}$ containing both e and f. The set $C \setminus \{e, f\}$ is the union of two paths π_u, π_v, where π_u (respectively π_v) is the path containing u (respectively v). Let $C_n(e)$ be a shortest such circuit with the property that π_u contains no vertices in $\Lambda_n \setminus \Lambda_{n-1}$. The following statement constitutes an easy piece of graph theory. There exists a constant M, depending only on d and $\mathbf{k}$, such that $C_n(e)$ exists, and furthermore every vertex therein belongs to $e + \Lambda_M$.

Let $\langle B_e \rangle$ denote the collection of all edges of $\mathbb{E}^d$ having both endvertices in $(e + \Lambda_{M+1}) \cap \Lambda_n$. Assume that the event $V_t = \{R_t(e) = 0,\ S_t(e) = 1\}$ occurs. We define the following further events V^1, V^2, V^3:

(i) V^1 is the event that: during the time-interval $(t, t+1]$, all edges in $\langle B_e \rangle$ which are present in R_t are removed, and no edges in $\langle B_e \rangle$ are added to R; e remains present in S,

(ii) V^2 is the event that: during $(t+1, t+2]$, all edges in $C_n(e) \setminus \{e\}$ are added to R, but no other edges in $\langle B_e \rangle$ are added to R; e remains present in S,

(iii) V^3 is the event that: during $(t+2, t+3]$, the edge f is removed from R but not from S.

We note that

$$V_t \cap V^1 \cap V^2 \cap V^3 \subseteq \big\{R_{t+3}(f) = 0,\ S_{t+3}(f) = 1\big\}.$$

It may be shown as in [3] that

$$P(V^1 \cap V^2 \cap V^3 \mid V_t) \geq \nu(\beta)$$

for some $\nu(\beta)$ which is continuous and strictly positive on $(0, \infty)$. With the above definitions of V^1, V^2, V^3, the proof of [3] goes through in the present situation, and yields (5.5.11) by way of (5.5.14). □

References

[1] Aizenman, M., Chayes, J.T., Chayes, L., Fröhlich, J., and Russo, L. On a sharp transition from area law to perimeter law in a system of random surfaces. *Communications in Mathematical Physics*, **92** (1983), 19–69.

[2] Aizenman, M. and Grimmett, G.R. Strict monotonicity for critical points in percolation and ferromagnetic models. *Journal of Statistical Physics*, **63** (1991), 817–835.

[3] Bezuidenhout, C.E., Grimmett, G.R., and Kesten, H. Strict inequality for critical values of Potts models and random-cluster processes. *Communications in Mathematical Physics*, **158** (1993), 1–16.

[4] Campanino, M. Strict inequality for critical percolation value in frustrated random cluster models. *Markov Processes and Related Fields*, **4** (1998), 395–410.

[5] Gandolfi, A. Inequalities for critical points in disordered ferromagnets. 1998, to appear.

[6] Grimmett, G.R. *Percolation*, 2nd edition. Springer-Verlag, New York, 1999.

[7] Grimmett, G.R. Potts models and random-cluster processes with many-body interactions. *Journal of Statistical Physics*, **75** (1994), 67–121.

[8] Grimmett, G.R. The stochastic random-cluster process and the uniqueness of random-cluster measures. *Annals of Probability*, **23** (1995), 1461–1510.

[9] Grimmett, G.R. Percolation and disordered systems. In *Ecole d'Eté de Probabilités de Saint Flour* XXVI–1996, P. Bernard, ed., Lecture Notes in Mathematics 1665. Springer-Verlag, Berlin, 1997, 153–300.

[10] Grimmett, G.R. and Holroyd, A.E., Entanglement in percolation. 1998, to appear.

[11] Grimmett, G.R. and Piza, M.S.T. Decay of correlations in subcritical Potts and random-cluster models. *Communications in Mathematical Physics*, **189** (1997), 465–480.

[12] Grimmett, G.R. and Stacey, A.M. Critical probabilities for site and bond percolation models. *Annals of Probability*, **26** (1998), 1788–1812.

[13] Holroyd, A.E. Existence of a phase transition for entanglement percolation. 1998, to appear.

[14] Kantor, T. and Hassold, G.N. Topological entanglements in the percolation problem, *Physical Review Letters*, **60** (1988), 1457–1460.

[15] Kesten, H. *Percolation Theory for Mathematicians*. Birkhäuser, Boston, 1982.

[16] Kesten, H. Aspects of first-passage percolation. In *Ecole d'Eté de Probabilités de Saint Flour* XIV-1984, P. L. Hennequin, ed., Lecture Notes in Mathematics 1180. Springer-Verlag, Berlin, 1986, 125–264.

[17] Kesten, H. Percolation theory and first-passage percolation. *Annals of Probability*, **15** (1987), 1231–1271.

[18] Newman, C.M. Disordered Ising systems and random cluster representations. In *Probability and Phase Transition*, G. R. Grimmett, ed. Kluwer, Dordrecht, 1994, 247–260.

Statistical Laboratory
DPMMS
University of Cambridge
16 Mill Lane, Cambridge CB2 1SB, United Kingdom
G.R.Grimmett@statslab.cam.ac.uk

6

From Greedy Lattice Animals to Euclidean First-Passage Percolation

C. Douglas Howard and Charles M. Newman

ABSTRACT We discuss some new results concerning infinite geodesics in models of Euclidean First-Passage Percolation (FPP) on $\mathbb{R}^d$ introduced previously by the authors. In particular, for any d, with probability one, every semi-infinite geodesic has an asymptotic direction and every direction has at least one semi-infinite geodesic (starting from each Poisson particle). These results, which have not been proved for lattice FPP, are nevertheless based on an extension to the Euclidean case of lattice FPP estimates due to Kesten and Alexander. A key technical ingredient that we prove here is a large deviation estimate for greedy lattice animals applied to Bernoulli percolation.

Keywords: Greedy lattice animals, first-passage percolation, large deviations.

AMS Subject Classifications: Primary 60F10, 60K35; Secondary 82B41, 82D30.

6.1 Introduction

In the general area of percolation theory and in the specific subject of first-passage percolation, Harry Kesten has played a leading role for many years. It should thus come as no surprise to the reader to see the strong influence of his previous work on the results reported in this paper. Indeed, he has been a major influence, both personally and through his papers, on the research of both of us. This applies not only to percolation, but also to such areas as random matrices and random walks with scenery. It is with great pleasure that we participate in this volume in honor of Harry Kesten.

The standard first-passage percolation (FPP) model [HW, K1] is constructed on the nearest neighbor graph $(\mathbb{Z}^d, \mathbb{E}^d)$ using non-negative i.i.d. random variables $(\tau(e) : e \in \mathbb{E}^d)$, with common distribution F. If F is continuous, then a.s., for any $u, v \in \mathbb{Z}^d$, the minimum sum $T(u, v)$ of $\tau(e)$'s along (self-avoiding) paths between u and v is achieved by a unique path $M(u, v)$. Then $T(\cdot, \cdot)$ is a.s. a metric on $\mathbb{Z}^d$ and the $M(u, v)$'s are its finite geodesics.

An infinite path is called a geodesic if every finite sub-path is a finite geodesic. As we shall immediately explain, it is easy to see that there must be at least one infinite geodesic starting from every u. The only stronger result of this sort about $\mathbb{Z}^d$ lattice FPP models of which we are aware is that for $d = 2$ and F an exponential distribution, there is, for any deterministic u, positive probability of at least two infinite geodesics starting from that u [HP]. The existence of one infinite geodesic from u follows from the simple fact that $R(u) \equiv \cup_{v\in\mathbb{Z}^d} M(u, v)$ (i.e., the graph with vertex set $\mathbb{Z}^d$ and all edges from $\cup_v M(u, v)$) is an *infinite* tree, since it spans all of $\mathbb{Z}^d$, and hence has at least one infinite path from every vertex. Infinite paths from u are necessarily geodesics.

It is natural to conjecture that (for fairly general F's), a.s. every infinite geodesic from u has an asymptotic direction $\hat{x}$ (a unit vector in $\mathbb{R}^d$) and that to every $\hat{x}$ there is at least one such infinite $\hat{x}$-geodesic from u. The main purpose of this paper, after recalling the existing partial proof of this conjecture for lattice FPP [N1], is to discuss recent developments which lead to a complete proof of the analogous conjecture (and to some related results) for certain non-lattice Euclidean FPP models originally introduced in [HN1].

These Euclidean FPP models replace the lattice $\mathbb{Z}^d$ as the vertex set by the set Q of particle locations of a homogeneous Poisson process on $\mathbb{R}^d$, nearest neighbor edges of $\mathbb{E}^d$ by *all* pairs $\{q, q'\}$ from Q, and the random $\tau(e)$ by the deterministic (given Q) $|q - q'|^\alpha$, where $\alpha > 1$ is a parameter of the model and $|\cdot|$ denotes Euclidean length. We remark that other FPP models based on Poisson processes were introduced earlier by [VW1, VW2] and studied further by [S]; it is likely that the results discussed here should be valid for these other models but, at least on a technical level, our proofs are not directly applicable. Other Poisson-process-based random metrics are studied in [Sz].

The partial proof in [N1] of the above-mentioned conjecture for lattice FPP is Theorem 2.1 there. The problem with that theorem, which states that the conjecture is valid if three hypotheses are satisfied, and the reason it provides only a partial proof, concerns the third hypothesis. The first two hypotheses, that F is continuous and has an exponential tail ($E(\exp(\beta\tau(e)) < \infty$ for some $\beta > 0$), are unobjectionable; the third hypothesis, that the "asymptotic shape" B_0 is uniformly curved, may very well be true but unfortunately has not been verified for any F, and seems to be well beyond our current state-of-the-art. The asymptotic shape B_0, a convex subset of $\mathbb{R}^d$ with nonempty interior, arises in the celebrated Shape Theorem [R, CD, K1], and uniformly curved means, in a sense, that its surface curvature is bounded away from zero. The Shape Theorem states, roughly speaking, that $\tilde{B}_t$, the ball of radius t based on the the metric $T(\cdot,\cdot)$ and centered at the origin, behaves a.s. as $tB_0 + o(t)$ as $t \to \infty$. A related description of B_0 is that it is the unit ball of a deterministic norm $g(\cdot)$ on $\mathbb{R}^d$ determined by $T(\cdot,\cdot)$ according to

$$T(0, v)/g(v) \to 1 \text{ as } |v| \to \infty \text{ (a.s. and in } L^1). \tag{6.1.1}$$

Since the homogeneous Poisson process is statistically isotropic, it is clear that a shape theorem for Euclidean FPP cannot possibly be valid unless B_0 is a Euclidean ball and correspondingly $g(v) = \mu|v|$ for some $\mu \in (0, \infty)$. Indeed, such a shape theorem was proved in [HN1] for all $d \geq 2$ and all $\alpha > 1$. At first glance, it appears that since this removed the basic impediment of [N1, Theorem 2.1], the conjecture about infinite geodesics for Euclidean FPP should have quickly followed. But alas, there was a technical obstruction whose resolution, which took some time, is reported in this paper.

The technical difficulty concerned the following large deviation estimate for lattice FPP due to Kesten and Alexander [K2, A] (see also [T]) which was a major ingredient in the proof of [N1, Theorem 2.1]: for any $\epsilon > 0$, there exist $C_1, C_2 \in (0, \infty)$ such that

$$P[|T(0,v) - g(v)| \geq \lambda] \leq C_1 \exp(-C_2\lambda/|v|^{1/2}) \text{ for } |v|^{\frac{1}{2}+\epsilon} \leq \lambda \leq |v|^{\frac{3}{2}-\epsilon}. \tag{6.1.2}$$

The difficulty was that their results were not applicable to Euclidean FPP. In the remainder of this paper we report on our extension of the Kesten-Alexander estimate and its implications. In Section 6.2, we state precisely some of the new results for Euclidean FPP, including the infinite geodesic conjecture as a theorem. In Section 6.3, we state precisely our extension of the Kesten-Alexander estimate to Euclidean FPP, discuss briefly how it leads to the results of Section 6.2, and sketch the proof of our extension including the role played in the proof by greedy lattice animals in a Bernoulli percolation context. Finally, in Section 6.4 we prove the greedy lattice animal large deviation estimates needed in Section 6.3; these estimates are an elaboration on the original lattice animal results of Cox-Gandolfi-Griffin-Kesten and may be of independent interest. They presumably also lead to an extension of the Kesten-Alexander results to lattice FPP models with a subexponential tail for F.

6.2 Euclidean FPP: Main Results

Let Q denote the set of particle locations of a homogeneous (density one) Poisson point process on $\mathbb{R}^d$. The dimension d is fixed and, to avoid the trivial case, is assumed to be at least two. We also take a fixed parameter $\alpha > 1$. For q and q' points in Q, we denote by $\mathcal{P}(q, q')$ the set of finite sequences $q_1, \ldots, q_k$ of distinct points in Q (for arbitrary $k \geq 2$) with $q_1 = q$ and $q_k = q'$. Thinking of these as paths from q to q', we then define

$$T(r) = \sum_{j=1}^{k-1} |q_{j+1} - q_j|^\alpha \text{ for } r = (q_1, \ldots, q_k) \in \mathcal{P}(q, q'), \tag{6.2.1}$$

and

$$T(q, q') = \begin{cases} \inf\{T(r) : r \in \mathcal{P}(q, q')\} & \text{if } q \neq q' \\ 0 & \text{if } q = q'. \end{cases} \tag{6.2.2}$$

We note that $T(\cdot, \cdot)$ is almost surely a metric on Q. A geodesic from q to q' is a path $r^* \in \mathcal{P}(q, q')$ such that $T(q, q') = T(r^*)$; it almost surely exists and is unique and we denote it by $M(q, q')$. For $q \in Q$, we denote by $\mathcal{P}(q)$ the set of semi-infinite sequences $(q_1, q_2, \ldots)$ of distinct points in Q with $q_1 = q$ and we define $\mathcal{M}(q)$ to be the set of $r = (q_1, q_2, \ldots) \in \mathcal{P}(q)$ such that, for every $k \geq 2$, $(q_1, \ldots, q_k)$ is a geodesic from $q_1 = q$ to q_k. Then $\mathcal{M}(q)$ is the set of semi-infinite geodesics starting from q.

For any unit vector $\hat{x} \in \mathbb{R}^d$ and $r = (q_1, q_2, \ldots) \in \mathcal{M}(q)$, we say that r has direction $\hat{x}$ if $q_n/|q_n| \to \hat{x}$ as $n \to \infty$. We say that r has a direction if it has direction $\hat{x}$ for some unit vector $\hat{x}$. Our main result about semi-infinite geodesics concerns the existence of directions for geodesics and the existence of geodesics for directions.

Theorem 6.2.1 *Let D denote the event that for every $q \in Q$, every $r \in \mathcal{M}(q)$ has a direction. Let D_E denote the event that for every unit vector $\hat{x} \in \mathbb{R}^d$ and for every $q \in Q$ there is at least one $r \in \mathcal{M}(q)$ with direction $\hat{x}$. Then for any $d \geq 2$ and $\alpha > 1$, D and D_E occur with probability one. Thus, a.s. $\mathcal{M}(q)$ is uncountable for every $q \in Q$.*

Remark For a discussion of the measurability of D and D_E, see [N2, Appendix A].

The proof of this theorem is based on the improbability of a long finite geodesic starting off in one direction and then "noticeably" changing course. This is closely related to the unlikelihood of a geodesic between q and q' (for $|q - q'|$ large) departing from the straight line between q and q' by a distance of order $|q - q'|^{\xi+\epsilon}$ for some (conjectured to exist) wandering exponent $\xi = \xi(d)$ and for any $\epsilon > 0$. Heuristically, Theorem 6.2.1 should follow from $\xi < 1$. In [NP], a weak version of $\xi \leq 3/4$ was proved for lattice FPP and in [N1] a stronger version (that really would yield a lattice version of Theorem 6.2.1) was proved, but only under the unverified assumption that the asymptotic shape B_0 is uniformly curved. We next state a strong Euclidean version of $\xi \leq 3/4$ that does yield Theorem 6.2.1.

Theorem 6.2.2 *For any $d \geq 2$, $\alpha > 1$, and $\epsilon > 0$, the following is true with probability one. For every $q \in Q$, there are only finitely many $q' \in Q$ such that there is some $q'' \in Q$ with $M(q, q'')$ passing through q' and such that the angle (in $[0, \pi]$) between $q'' - q$ and $q' - q$ exceeds $|q' - q|^{\frac{3}{4}+\epsilon}/|q' - q|$.*

There is a second fluctuation exponent χ defined by the conjecture that, roughly speaking, the fluctuations of $T(q, q')$ are of order $|q - q'|^{\chi}$ for $|q - q'|$ large. More precisely, one may define, for any x and y in $\mathbb{R}^d$, $T(x, y) = T(q(x), q(y))$ where $q(x)$ and $q(y)$ are the closest points in Q to x and y with some deterministic rule for breaking ties and then define χ as the minimum value such that for every $\epsilon > 0$,

$$\mathrm{Var}\,(T(x, y)) = O(|x - y|^{2\chi+\epsilon}) \text{ as } |x - y| \to \infty. \tag{6.2.3}$$

A small part of the Kesten-Alexander large deviation estimate (6.1.2) was Kesten's proof [K2] for lattice FPP that $\chi \leq 1/2$. This is also the case for our extension of (6.1.2) to Euclidean FPP.

6.3 Euclidean FPP: Large Deviations

The function $T(x, y) = T(q(x), q(y))$ does not quite define a metric on $\mathbb{R}^d$ (it is a pseudometric since $T(x, y) = 0$ whenever $q(x) = q(y)$), but still it is useful to consider the random ball of radius t centered at x: $\tilde{B}_t^x = \{y \in \mathbb{R}^d : T(x, y) \leq t\}$. The essential ingredient in deriving a result, such as Theorem 6.2.2, that $\xi \leq 3/4$, is some estimate of how far away from the straight line between x and y the geodesic $M(q(x), q(y))$ can wander when $|x - y|$ is large.

Suppose q^* is on the geodesic $M(q(x), q(y))$. Then there must be t_1 and t_2 such that $\tilde{B}_{t_1}^x$ and $\tilde{B}_{t_2}^y$ barely overlap with just the one Poisson particle $q^* \in \tilde{B}_{t_1}^x \cap \tilde{B}_{t_2}^y$. We know from the Shape Theorem of [HN1] that to leading order, as $t_1 \to \infty$, $\tilde{B}_{t_1}^x \approx x + t_1 B_0$ with B_0 a Euclidean ball of radius $1/\mu(d, \alpha)$ and similarly for $\tilde{B}_{t_2}^y$. If these were exact formulae for finite t_1 and t_2, then the overlap would be exactly on the straight line and ξ would be zero. To show that $\xi \leq 3/4$, one basically needs to show that, with probability rapidly approaching one as $t \to \infty$, $\tilde{B}_t \equiv \tilde{B}_t^0$ is within $O(t^{\frac{1}{2}+\epsilon})$ of tB_0; the number $3/4$ results from the Pythagorean Theorem.

In the case of lattice FPP such a result about $\tilde{B}_t$, due to Kesten and Alexander [K2, A], is a direct consequence of the large deviation estimate (6.1.2). Using that result and an argument like the one just outlined, a lattice version of $\xi \leq 3/4$ can be obtained [NP], but one for which the vector $y - x$ is restricted to directions where the surface of B_0 is curved. Thus, obtaining the lattice analogue of Theorem 6.2.2 (this is [N1, Proposition 3.2]), which has no directional restrictions, seems to require a uniform curvature hypothesis on B_0—one that has not been verified for lattice models. For Euclidean FPP, since B_0 is a Euclidean ball, an analogue of (6.1.2) is all that is needed to obtain Theorem 6.2.2 By Euclidean invariance, we may restrict our attention to a large deviation estimate for $T_\ell \equiv T(0, \ell\hat{e}_1)$ where $\hat{e}_1$ is the unit vector $(1, 0, 0, \ldots, 0)$. Our analogue of the Kesten-Alexander estimate is as follows.

Theorem 6.3.1 *Let $d \geq 2$ and $\alpha > 1$ be fixed. There exists $\mu > 0$ so that for any $\epsilon > 0$, there exists $C_1, C_2 \in (0, \infty)$ such that*

$$P[|T_\ell - \mu\ell| \geq \lambda] \leq C_1 \exp\Big(-C_2\big(\lambda/\sqrt{\ell}\big)^{\kappa_1}\Big) \text{ for } \ell^{\frac{1}{2}+\epsilon} \leq \lambda \leq \ell^{\kappa_2+\frac{1}{2}-\epsilon} \tag{6.3.1}$$

where $\kappa_1 = \min(1, d/\alpha)$ and $\kappa_2 = 1/(4\alpha + 3)$.

The estimate for deviations of T_ℓ from the asymptotic expression $\mu\ell$ is itself a consequence of an estimate for deviations from the mean: There exist $C_0, C_1, C_2 \in (0, \infty)$ such that for all $\ell \geq 0$,

$$P[|T_\ell - ET_\ell| > w\sqrt{\ell}] \leq C_1 \exp(-C_2 w^{\kappa_1}) \text{ for } w \leq C_0 \ell^{\kappa_2}. \tag{6.3.2}$$

We finish this section by sketching the derivation of (6.3.2) and in particular explaining its relation to a large deviation estimate for greedy lattice animals in Bernoulli site percolation. This heuristic sketch ignores various technical complications which involve truncations and other approximations.

First we observe that the exact analogue of Kesten's result in [K2], namely that (6.3.2) holds with $\kappa_1 = \kappa_2 = 1$, cannot possibly be true in general. To see what goes wrong, let A_ℓ denote the event that: (1) there is a Poisson particle in $\mathcal{B}(0, 1)$ (the Euclidean ball centered at 0 with radius 1); (2) there is also a particle in $\mathcal{B}(\ell\hat{e}_1, 1)$; and (3) there are no particles in the annulus $\mathcal{B}(0, \ell^\rho) \setminus \mathcal{B}(0, 1)$ (where $\rho \in (0, 1)$). Then, for large ℓ, on A_ℓ we have $T_\ell \geq (\ell^\rho - 1)^\alpha \geq \ell^{\rho\alpha}/2^\alpha$ since the optimal path encounters no particles as it crosses the annulus, a distance of $\ell^\rho - 1$. On the other hand, $P[A_\ell] \geq (1-e^{-\theta})^2 \exp(-\theta \ell^{\rho d})$ where θ is the d-dimensional volume of a unit ball. If $\alpha > 3d/2$, then, by taking ρ close to $3/(2\alpha)$, we see that this contradicts (6.3.2) with $\kappa_1 = \kappa_2 = 1$. (Here we use that ET_ℓ is of order ℓ.) Indeed, a more basic statement, true for lattice FPP when edge times have an exponentially bounded tail, fails in our setting. In particular, based on lattice FPP results, one might hope that

$$P[T_\ell > x] \leq C_1 \exp(-C_2 x^{\kappa_1}) \text{ for } x \geq C_3 \ell \tag{6.3.3}$$

holds with $\kappa_1 = 1$. Yet this same construction with a different choice of ρ shows that (6.3.3) cannot hold unless $\kappa_1 \leq d/\alpha$. In fact, (6.3.3) is true [HN2] with $\kappa_1 = \min(1, d/\alpha)$. The basic strategy in proving this is to produce an algorithm to construct a suboptimal path $r^{**}(\ell)$ such that $T(r^{**}(\ell))$ enjoys property (6.3.3). (The bound (6.3.3) is an improvement of [HN1, Lemma 1].)

Let $r^* = M(q(0), q(\ell\hat{e}_1))$, so $T_\ell = T(r^*)$, and, replacing α by 2α in the formula for $T(r^*)$, put

$$S_\ell = \sum_j |r^*_{j+1} - r^*_j|^{2\alpha}. \tag{6.3.4}$$

Then an argument in which $T_\ell - ET_\ell$ is represented as the (a.s. and L^2) limit of a martingale yields that $\operatorname{Var} T_\ell \leq CES_\ell$. (Our martingale representation is a variation of the approach introduced by Kesten in [K2] for the standard FPP setting.) To get from a bound on $\operatorname{Var} T_\ell$ to an estimate such as (6.3.2), a crucial step is obtaining a large deviation estimate, analogous to (6.3.3), for S_ℓ. Such a bound on the tail of S_ℓ is an ingredient in a deviation-from-the-mean result for martingale limits (such as $T_\ell - ET_\ell$), proved in [HN2], that closely resembles [K2, Theorem 3]. At first glance, it might appear that (6.3.3) should apply directly to S_ℓ. After all, α is an arbitrary parameter and (6.3.3) should hold with α replaced by 2α. The problem is that r^* optimizes (6.2.2) for α, not for 2α. Nevertheless, it turns out that (6.3.3) is useful here. For some $\delta > 0$ (thought of as large), we write

$$S_\ell = \sum_{|r^*_{j+1} - r^*_j| < \delta} |r^*_{j+1} - r^*_j|^{2\alpha} + \sum_{|r^*_{j+1} - r^*_j| \geq \delta} |r^*_{j+1} - r^*_j|^{2\alpha} \equiv S_\ell^{\text{small}} + S_\ell^{\text{big}}.$$

Then $S_\ell^{\text{small}} \leq \delta^\alpha T_\ell$, so S_ℓ^{small} has a subexponential tail in the sense of (6.3.3).

It is in controlling the tail of S_ℓ^{big} that large deviations for greedy lattice animals in Bernoulli percolation play a role. For $a, b \in \mathbb{R}^d$, define

$$W(a,b) = \{x \in \mathbb{R}^d : |a-x|^\alpha + |x-b|^\alpha < |a-b|^\alpha\}.$$

The region $W(a,b)$ is convex, invariant with respect to rotations about the line segment $\overline{ab}$, and homogeneous in the sense that $W(a,b) = a + |b-a|UW(0,\hat{e}_1)$ where U is any unitary transformation with $U\hat{e}_1 = (b-a)/|b-a|$. For example, when $\alpha = 2$, $W(a,b)$ is a Euclidean ball with a and b on its surface and antipodal to each other. Note also that the regions $W_j \equiv W(r_j^*, r_{j+1}^*)$ are devoid of Poisson particles since r^* is a geodesic. For some $\lambda > 0$ (to be presently determined), we partition $\mathbb{R}^d$ into "λ-boxes" whose vertices are collectively $\lambda \cdot ((\frac{1}{2}, \ldots, \frac{1}{2}) + \mathbb{Z}^d)$ and whose centers are collectively $\lambda \cdot \mathbb{Z}^d$. We identify any collection of the λ-boxes with a corresponding collection of sites in $\mathbb{Z}^d$ in the natural way and we note that two λ-boxes are adjacent (i.e., share a $(d-1)$-dimensional face) if their corresponding $\mathbb{Z}^d$ sites are nearest neighbors. We choose λ sufficiently large that the probability that a particular λ-box is unoccupied by a Poisson particle is less than the critical probability for Bernoulli site percolation on the $\mathbb{Z}^d$ lattice.

Letting $\#W_j$ denote the number of λ-boxes completely contained in W_j, we see that $\#W_j \approx C|r_{j+1} - r_j|^d$, for large $|r_{j+1} - r_j|$, where $C = \text{volume}\,(W(0,\hat{e}_1))/\lambda^d$. It follows, then, that

$$\begin{aligned} \tfrac{1}{\ell} S_\ell^{\text{big}} \text{ is large} &\Longrightarrow \tfrac{1}{\ell} \textstyle\sum_{|r_{j+1}^* - r_j^*| \geq \delta} (\#W_j)^{2\alpha/d} \text{ is large} \\ &\Longrightarrow \tfrac{1}{\ell} \textstyle\sum_{\nu \in A^*} |\mathcal{C}_\nu|^{2\alpha/d} \text{ is large,} \end{aligned}$$

where A^* is the lattice animal of λ-boxes that covers r^* and $|\mathcal{C}_\nu|$ is the size of the cluster $\mathcal{C}_\nu$ of unoccupied λ-boxes at ν. (Any subset $A \subset \mathbb{Z}^d$ has an associated nearest neighbor graph whose vertices are the sites in A and whose edges are $\{\{\nu_1, \nu_2\} : \nu_i \in A \text{ and } |\nu_1 - \nu_2| = 1\}$. The sites A form a lattice animal if its associated nearest neighbor graph is connected.) A Peierls-type argument shows that, with large probability, the number $|A^*|$ of λ-boxes in A^* is of order ℓ. Thus, in a sense that is useful here,

$$\tfrac{1}{\ell} S_\ell^{\text{big}} \text{ is large} \Longrightarrow \tfrac{1}{|A^*|} \textstyle\sum_{\nu \in A^*} |\mathcal{C}_\nu|^{2\alpha/d} \text{ is large.} \tag{6.3.5}$$

Theorem 6.4.2 of the next section is precisely what is needed to control the likelihood of the right hand side of (6.3.5); it essentially replaces A^* by the "greedy" lattice animal (of the same size) that maximizes the sum.

6.4 Tail Estimates for Greedy Lattice Animals

We let $\mathcal{A}_0(n)$ denote the set of lattice animals of n vertices one of which is the origin and put $\mathcal{A}_0 = \cup_{n \geq 1} \mathcal{A}_0(n)$. Throughout this section, $(X_\nu : \nu \in \mathbb{Z}^d)$ represents an i.i.d. family of non-negative random variables and we put

$$N_n = \max_{A \in \mathcal{A}_0(n)} \sum_{\nu \in A} X_\nu. \tag{6.4.1}$$

Lattice animals that achieve the maximum in (6.4.1) are called greedy. In [CGGK] it is shown among other things that:

Theorem 6.4.1 (Cox, Gandolfi, Griffin, and Kesten) *If, for some $a > 0$,*

$$E\big[X_\nu^d (\log^+ X_\nu)^{d+a}\big] < \infty \tag{6.4.2}$$

then, almost surely,

$$N \equiv \limsup_{n\to\infty} \frac{N_n}{n} < \infty. \tag{6.4.3}$$

Furthermore, if $E\big[X_\nu^d\big] = \infty$, then, almost surely, $N = \infty$.

The second part of the theorem shows that (6.4.2) is nearly optimal. In [GK] it is shown that the lim sup in (6.4.3) is actually a limit both almost surely and in L^1. A key ingredient in their arguments is:

Proposition 6.4.1 ([CGGK]) *Assume* (6.4.2) *holds and put $b = 1 + a/d$. Then there is a continuous, strictly increasing function $\gamma(x)$ and positive finite constants c_0 and $\bar{c}$ such that:*

$$\frac{\gamma(x)(\log x)^b}{x} \to 0 \quad \textit{and} \quad \frac{\gamma(x)(\log x)^b \log\log x}{x} \to \infty \quad \textit{as} \quad x \to \infty, \tag{6.4.4}$$

and such that for all $c \geq \bar{c}$ and $n \geq 2$,

$$P\Bigg[\max_{A\in\mathcal{A}_0(n)} \frac{1}{n}\sum_{\nu\in A}(X_\nu \wedge \gamma(n)) > 1 + cc_0\Bigg] \leq \exp\Big(-\frac{c}{8}(\log n)^b\Big). \tag{6.4.5}$$

The special properties in (6.4.4) are used in [GK], but, in fact, the following also holds:

Proposition 6.4.1′ *Assume* (6.4.2) *holds. Then there are positive constants $c_0 < \infty$ (depending on d and the distribution of the X_ν) and $\bar{c} < \infty$ (depending only on d) such that for all $c \geq \bar{c}$, $n \geq 2$, and $0 < \gamma \leq n$:*

$$P\Bigg[\max_{A\in\mathcal{A}_0(n)} \frac{1}{n}\sum_{\nu\in A}(X_\nu \wedge \gamma) > 1 + cc_0\Bigg] \leq \Big(1 + \frac{\log n}{\log 2}\Big)\exp\Big(-\frac{c}{4}\frac{n}{\gamma}\Big). \tag{6.4.6}$$

Indeed, the proof of Proposition 6.4.1′ is identical to that of Proposition 6.4.1 except that starting at (3.9) in [CGGK] and ending at (3.26), $\gamma(n)$ is replaced with γ and $(\log n)^b$ is replaced with n/γ.

Presently we use Proposition 6.4.1′ to obtain large deviation estimates for N_n/n when assumptions stronger than (6.4.2) are made about the tail of the distribution of the X_ν. The basic inequality exploited is:

$$P\left[\frac{1}{n}N_n > y\right] \le P\left[\max_{\nu\in[-n,n]^d} X_\nu > \gamma\right] + P\left[\max_{A\in\mathcal{A}_0(n)} \frac{1}{n}\sum_{\nu\in A}(X_\nu \wedge \gamma) > y\right] \tag{6.4.7}$$

$$\le 3^d n^d P[X_\nu > \gamma] + 3\log n \exp\left(-\frac{\bar{c}}{4y_0}\frac{ny}{\gamma}\right), \tag{6.4.8}$$

where (6.4.7) holds for any positive n and γ and (6.4.8) holds for $n \ge 2, 0 < \gamma \le n$, and $y \ge y_0 \equiv 1 + \bar{c}c_0$. Here we use that $(y-1)/c_0 \ge \bar{c}y/y_0$. Our strategy, of course, is to choose γ a function of y and n to make both summands in (6.4.8) the same order of magnitude. In the following propositions the hypotheses regarding the tail of the X_ν's is made progressively stronger:

Proposition 6.4.2 *Suppose $P[X_\nu > x] \le C_1 x^{-\lambda}$ for some $\lambda > d$. Then there are finite constants C_2, C_3, and y_0 such that*

$$P\left[\frac{1}{n}N_n > y\right] \le C_2\frac{(\log n)^\lambda}{n^{\lambda-d}y^\lambda} \quad \textit{whenever} \quad y_0 \le y \le C_3\log n.$$

In particular, for some other constant $\tilde{C}_2$,

$$P\left[\frac{1}{n}N_n > y_0\right] \le \tilde{C}_2\frac{(\log n)^\lambda}{n^{\lambda-d}} \quad \textit{for} \quad n \ge 2.$$

Proposition 6.4.3 *Suppose $P[X_\nu > x] \le C_4\exp(-C_5x^\lambda)$ for some $\lambda > 0$. Then, with $\kappa = \lambda/(1+\lambda)$, there are finite constants C_6, C_7, and y_0 such that*

$$P\left[\frac{1}{n}N_n > y\right] \le C_6\exp\left(-C_7(ny)^\kappa\right) \quad \textit{whenever} \quad y_0 \le y \le n^\lambda.$$

In particular, for some other constants $\tilde{C}_6$ and $\tilde{C}_7$,

$$P\left[\frac{1}{n}N_n > y_0\right] \le \tilde{C}_6\exp\left(-\tilde{C}_7n^\kappa\right) \quad \textit{for} \quad n \ge 2.$$

We note that, in general, the exponent $\kappa = \lambda/(1+\lambda)$ will not be optimal. Indeed, the arguments in [CGGK] following (2.12) illustrate that for $\lambda \ge 1$ we may take $\kappa = 1$. Proposition 6.4.3 is therefore of interest for $0 < \lambda < 1$.

PROOFS. For both propositions we take $y_0 = 1 + \bar{c}c_0$, where $\bar{c}$ and c_0 are appropriate to the setting (see Proposition 6.4.1′). Proposition 6.4.2 follows from (6.4.8) by taking

$$\gamma = \frac{\bar{c}ny}{4y_0[(\lambda-d)\log n + \lambda\log y - (\lambda-1)\log\log n]},$$

which is meaningful for $n \geq 2$. If C_3 is any constant strictly less than $4y_0(\lambda - d)/\bar{c}$, then for some n^* we will have $\gamma \leq n$ when $n \geq n^*$ with $y \leq C_3 \log n$. The case $n < n^*$ with $y \leq C_3 \log n$ is handled by adjusting C_2.

Proposition 6.4.3 follows from (6.4.8) by taking $\gamma = (ny)^{1/(1+\lambda)}$. Here we have $\gamma \leq n$ provided $y \leq n^\lambda$. The choice of γ is less delicate in this case because the n^d and $\log n$ factors in (6.4.8) can be absorbed by making C_7 small. Trying to optimize C_7 is not worthwhile since κ is probably not optimal. □

As promised, we demonstrate a result about cluster sizes in the standard Bernoulli site percolation model on $\mathbb{Z}^d$. In this model, $(\eta_\nu : \nu \in \mathbb{Z}^d)$ is a family of i.i.d. $\{0, 1\}$-valued random variables with $P[\eta_\nu = 1] = p$. A site ν is "occupied" if $\eta_\nu = 1$. We consider sub-critical p, i.e., values of p below the critical probability for occupied site percolation. Let $\mathcal{C}_\nu$ denote the cluster of occupied sites at ν and $|\mathcal{C}_\nu|$ denote the number of sites in $\mathcal{C}_\nu$. We have:

Theorem 6.4.2 *For any $\rho > 0$ there is a $y_0 > 0$ and constants C_8 and C_9 such that:*

$$P\left[\sup_{m \geq n} \max_{A \in \mathcal{A}_0(m)} \frac{1}{m} \sum_{\nu \in A} |\mathcal{C}_\nu|^\rho > y_0\right] \leq C_8 \exp(-C_9 n^{1/(\rho+2)}) \text{ for all } n.$$

With an application of the Borel-Cantelli Lemma, it follows easily that:

Corollary 6.4.1 *For any $\rho > 0$,*

$$\sup_{A \in \mathcal{A}_0} \frac{1}{|A|} \sum_{\nu \in A} |\mathcal{C}_\nu|^\rho < \infty \text{ almost surely.} \tag{6.4.9}$$

This corollary generalizes a result of Fontes and Newman (see [FN, Theorem 4]) that establishes (6.4.9) for $\rho = 1$. We remark that Corollary 6.4.1 has a more direct proof along the lines of [FN] that does not yield Theorem 6.4.2 along the way.

Proof of Theorem 6.4.2. Let $(C_\nu^* : \nu \in \mathbb{Z}^d)$ be i.i.d. random subsets of $\mathbb{Z}^d$ equidistributed with $\mathcal{C}_0$, the random cluster at the origin in the site percolation model. Put $\tilde{\mathcal{C}}_\nu = \nu + C_\nu^*$ and define

$$U_\nu = \sup\{|\tilde{\mathcal{C}}_u| : u \in \mathbb{Z}^d, \nu \in \tilde{\mathcal{C}}_u\}, \tag{6.4.10}$$

where the supremum is 0 if the set on the right in (6.4.10) is empty. As in [FN], we have the stochastic dominance relation:

$$(|\mathcal{C}_\nu| : \nu \in \mathbb{Z}^d) \ll (U_\nu : \nu \in \mathbb{Z}^d). \tag{6.4.11}$$

Now for some constant $\phi > 0$,

$$P[|\mathcal{C}_0| > x] \leq \exp(-\phi x) \text{ for all } x, \tag{6.4.12}$$

(see, e.g., [Gr]). It follows from an application of the Borel-Cantelli Lemma that $U_\nu < \infty$ almost surely and, hence, if the set on the right in (6.4.10) is not empty, that

$$U_\nu = |\tilde{\mathcal{C}}_{z(\nu)}| \text{ for some } z(\nu) \in \mathbb{Z}^d \text{ with } \nu \in \tilde{\mathcal{C}}_{z(\nu)}.$$

If the set on the right of (6.4.10) is empty, take $z(\nu) = \nu$ and we still have that $U_\nu = |\tilde{C}_{z(\nu)}|$. Under this scheme, it follows that for any lattice animal A, there is a possibly larger $A' \supset A$ such that

$$\frac{1}{|A|}\sum_{\nu\in A} U_\nu^\rho = \frac{1}{|A|}\sum_{\nu\in A} |\tilde{C}_{z(\nu)}|^\rho \le \frac{2}{|A'|}\sum_{\nu\in A'} |\tilde{C}_\nu|^{\rho+1}. \tag{6.4.13}$$

The inequality in (6.4.13) follows substantially as in [FN, Lemma 1], where the special case $\rho = 1$ is proved. Combining (6.4.11) and (6.4.13), we see that

$$P\left[\sup_{m\ge n}\max_{A\in\mathcal{A}_0(m)} \frac{1}{m}\sum_{\nu\in A}|C_\nu|^\rho > y_0\right] \le P\left[\sup_{m\ge n}\max_{A\in\mathcal{A}_0(m)} \frac{1}{m}\sum_{\nu\in A}|\tilde{C}_\nu|^{\rho+1} > \frac{y_0}{2}\right]. \tag{6.4.14}$$

By (6.4.12),

$$P[|\tilde{C}_\nu|^{\rho+1} > x] \le \exp(-\phi x^{1/(\rho+1)}) \text{ for all } x,$$

and, since the $|\tilde{C}_\nu|^{\rho+1}$ are i.i.d., Proposition 6.4.3 yields that

$$P\left[\max_{A\in\mathcal{A}_0(m)} \frac{1}{m}\sum_{\nu\in A}|\tilde{C}_\nu|^{\rho+1} > \frac{y_0}{2}\right] \le \tilde{C}_8\exp(-\tilde{C}_9 m^{1/(\rho+2)}) \text{ for all } m \tag{6.4.15}$$

for some $y_0 > 0$ for appropriate constants $\tilde{C}_8$ and $\tilde{C}_9$.

In general, if $0 < a$, $0 < b < 1$ and $c = \sup_{x\ge 0} x^{1-b}\exp(-\frac{a}{2}x^b)$, then $c < \infty$ and

$$\begin{aligned}\sum_{m\ge n}\exp(-am^b) &\le \exp(-an^b) + \int_n^\infty \exp(-ax^b)\,dx \\ &\le \exp(-an^b) + c\int_n^\infty x^{b-1}\exp(-\frac{a}{2}x^b)\,dx \\ &\le \left(1+\frac{2c}{ab}\right)\exp(-\frac{a}{2}n^b).\end{aligned}$$

Applying this to (6.4.14) and (6.4.15) yields that

$$\begin{aligned}P\left[\sup_{m\ge n}\max_{A\in\mathcal{A}_0(m)} \frac{1}{m}\sum_{\nu\in A}|C_\nu|^\rho > y\right] &\le \sum_{m\ge n}\tilde{C}_8\exp(-\tilde{C}_9 m^{1/(\rho+2)}) \\ &\le C_8\exp(-C_9 n^{1/(\rho+2)}),\end{aligned}$$

for $C_9 = \tilde{C}_9/2$ and appropriate C_8. □

Acknowledgments: The research of the first author was partially supported by NSF Grant DMS-98-15226. The research of the second author was partially supported by NSF Grants DMS-95-00868 and DMS-98-03267.

REFERENCES

[A] Alexander, K.S. Approximations of subadditive functions and convergence rates in limiting-shape results. *Ann. Prob.* **25** (1997), 30–55.

[CD] Cox, J.T. and Durrett, R. Some limit theorems for percolation processes with necessary and sufficient conditions. *Ann. Prob.* **9** (1981), 583–603.

[CGGK] Cox, J.T., Gandolfi, A., Griffin, P.S., and Kesten, H. Greedy Lattice Animals I: Upper Bounds. *Ann. App. Prob.* **3** (1993), 1151–1169.

[DL] Durrett, R. and Liggett, T.M. The shape of the limit set in Richardson's growth model. *Ann. Prob.* **9** (1981), 186–193.

[FN] Fontes, L. and Newman, C.M. First Passage Percolation for Random Colorings of $\mathbb{Z}^d$. *Ann. App. Prob.* **3** (1993), 746–762.

[GK] Gandolfi, A. and Kesten, H. Greedy Lattice Animals II: Linear Growth. *Ann. App. Prob.* **4** (1994), 76–107.

[Gr] Grimmett, G. *Percolation.* Springer-Verlag, New York, 1989.

[HN1] Howard, C.D. and Newman, C.M. Euclidean Models of First Passage Percolation. *Prob. Th. Rel. Fields* **108** (1997), 153–170.

[HN2] Howard, C.D. and Newman, C.M. In preparation.

[HP] Häggström, O. and Pemantle, R. First-passage percolation and a model for competing spatial growth. *J. of App. Probab.*, **35** (1998), 683–692.

[HW] Hammersley, J.M. and Welsh, D.J.A. First-passage percolation, subadditive processes, stochastic networks and generalized renewal theory. In *Bernoulli, Bayes, Laplace Anniversary Volume*, J. Neyman and L. LeCam, eds. Springer-Verlag, Berlin, 1965, 61–110.

[K1] Kesten, H. Aspects of first-passage percolation. In *Lecture Notes in Math.* 1180. Springer-Verlag, Berlin, 1986, 125–264.

[K2] Kesten, H. On the speed of convergence in first-passage percolation. *Ann. Appl. Prob.* **3** (1993), 296–338.

[N1] Newman, C.M. A Surface View of First-Passage Percolation. In *Proceedings of the International Congress of Mathematicians*, S.D. Chatterji, ed. Birkhäuser, Basel, 1995, 1017–1023.

[N2] Newman, C.M. *Topics in Disordered Systems.* Birkhäuser, Basel, 1997.

[NP] Newman, C.M. and Piza, M.S.T. Divergence of shape fluctuations in two dimensions. *Ann. Prob.* **23** (1995), 977–1005.

[R] Richardson, D. Random growth in a tesselation. *Proc. Cambridge Philos. Soc.* **74** (1973), 515–528.

[S] Serafini, H. First-Passage Percolation in the Delaunay Graph of a d-Dimensional Poisson Process. Ph.D. Dissertation, Courant Institute of Mathematical Sciences, 1997.

[Sz] Sznitman, A.-S. Distance fluctuations and Lyapunov exponents. *Ann. Prob.* **24** (1996), 1504–1530.

[T] Talagrand, M. Concentration of measure and isoperimetric inequalities in product spaces *Publ. Math. I.H.E.S.* **81** (1995), 73–205.

[VW1] Vahidi-Asl, M.Q. and Wierman, J.C. First-passage percolation on the Voronoi tessellation and Delaunay triangulation. In *Random Graphs* '87, M. Karoński, J. Jaworski, and A. Ruciński, eds. Wiley, New York, 1990, 341–359.

[VW2] Vahidi-Asl, M.Q. and Wierman, J.C. A shape result for first-passage percolation on the Voronoi tessellation and Delaunay triangulation. In *Random Graphs* '89, A. Frieze and T. Łuczak, eds. Wiley, New York, 1992, 247–262.

C. Douglas Howard
Baruch College
Box G0930
17 Lexington Avenue
New York, NY 10010
howard@math.baruch.cuny.edu

Charles M. Newman
Courant Institute of Mathematical Sciences
New York University
251 Mercer Street
New York, NY 10012
newman@cims.nyu.edu

7

Reverse Shapes in First-Passage Percolation and Related Growth Models

Janko Gravner and David Griffeath

ABSTRACT Over the past 40 years, it has been observed that many of the simplest random and deterministic local growth dynamics expand at a linear rate in each radial direction, and attain an asymptotic geometry. Shape theorems to this effect have been proved in several instances. In a similar manner, initially very large holes within supercritical local dynamics may be expected to attain a characteristic shape as they shrink, a while before disappearing. We describe a general theory of reverse shapes which formalizes this phenomenology, and then apply it to first-passage percolation and related deterministic and stochastic growth models. As an application, we analyze the last holes of such models started from sparse product measures.

Keywords: First-passage percolation, growth model, cellular automaton, shape theory.

AMS Subject Classifications: Primary 60K35, 52A10.

7.1 Introduction

Since the pioneering work of Broadbent and Hammersley [BH] and Eden [Ede] in the late 1950s, there has been a steady stream of research on *shape theory*, the principles whereby "crystals" formed from many of the simplest spatial growth algorithms attain an asymptotic geometry as they spread. Starting from a bounded set A of occupied sites, if the dynamics are local, homogeneous and supercritical, then one typically finds that the occupied crystal A_t at time t satisfies

$$t^{-1}A_t \xrightarrow{H} \mathcal{L}, \tag{7.1.1}$$

($\xrightarrow{H}$ denotes convergence in the Hausdorff metric), for some *asymptotic shape* $\mathcal{L}$ which is independent of the initial seed A. The first rigorous result of type (7.1.1) was obtained by Richardson [Ric], using subadditivity arguments, for various random growth models on the d-dimensional integers $\mathbb{Z}^d$. The subadditivity approach to shape theory was subsequently extended to a variety of stochastic interacting particle systems (IPS) in [BrG1, BrG2, DL], and [DG]. See [Lig] and [Dur] for authoritative accounts of subadditivity and its application to additive growth models, and [BoG] for an extension to some random systems which are not additive. In the context of first-passage percolation [HW, CD], if $\tau(x)$ denotes the passage time from the origin to site x, then under quite general conditions $A_t = \{x : \tau(x) \leq t\}$ obeys the limit (7.1.1). Indeed, some of Richardson's processes may be viewed as special cases of first-passage growth, having passage times which are geometric with parameter p, while Eden's original crystal model can be interpreted as the asynchronous limit of such systems as $p \to 0$. Recent results for first-passage growth appear in [NP].

Shape theory for cellular automata (CA) and related deterministic dynamics was initiated by Willson [Wil], and more recently has been studied in considerable detail by the authors [GG1, GG2, GG3, GG4]. Some particularly simple growth rules evolve recursively, and subadditivity applies as in the random case, but the most illuminating route to (7.1.1) for cellular automata is based on a polar transform representation of $\mathcal{L}$ in terms of half-space velocities, the so-called "Wulff construction" (cf. [TCH]). In the course of our earlier work on Excitable CA systems [FGG1, FGG2], and in the study of various first-passage and dynamic tiling problems [GG2, GG3], it became evident that rare nucleating crystals pass through three stages as they fill space: initially individual droplets spread with characteristic shape $\mathcal{L}$; later on large droplets collide to create wedge-shaped interstices often governed by new forces; and then eventually the last isolated empty pockets are filled. An analysis of convex corners and their complements, as in [GG2], is crucial to understanding the intermediate time regime. One goal of this paper is to formulate a general *reverse shapes* approach to the final process of filling in large but bounded holes.

Although the Wulff recipe for asymptotic shape also seems quite broadly applicable to IPS models, only in a few cases (cf. [Sep]) are there techniques currently available to handle the difficult issue of stochastic interface fluctuations, and thereby establish the polar representation of $\mathcal{L}$ rigorously. For Richardson models which admit a first-passage representation, as for a large class of well-behaved first-passage percolation models, Kesten's groundbreaking work [Kes2] supplies the needed large deviation estimates. The polar formula is well-suited for Monte Carlo simulation (cf. [GG4, Section 7]), but otherwise offers little advantage over subadditivity unless half-space velocities can be computed effectively, which is virtually never the case with random dynamics. However, subadditivity does not seem directly applicable to the second and third stages of crystallization described above, in which case half-space analysis can provide genuinely new insights. As motivating examples, we consider two problems—reverse shapes and last holes—for the following one parameter family of systems.

Synchronously, at each discrete update, if x is any empty site which has at least one of eight nearest neighbors $(y : \|y - x\|_\infty = 1)$ occupied, then x becomes occupied with probability p, independently of all other sites. This is *Richardson's model*, as studied in [Ric, DL, Dur], except that we choose the box instead of diamond $(\|\cdot\|_1)$ neighborhood, since we will also briefly discuss *Threshold Growth* rules in which "one of eight" is replaced by "two of eight," "three of eight," or "four of eight."

The problem of reverse shapes starts such dynamics from a large hole of prescribed shape, i.e., $A_0 = (mA)^c$ for some fixed A and large m. We will establish a reverse limit shape $\mathcal{R}$ for the shrinking hole, ϵm time units before it vanishes, as $m \to \infty$ and then $\epsilon \to 0$. An important feature distinguishes reverse limits from the "forward" result (7.1.1): since the initial hole mA becomes arbitrarily large, $\mathcal{R}$ typically depends on the geometry of A.

For the last holes problem, begin by populating $\mathbb{Z}^d$ according to Bernoulli product measure ν_δ, with density δ of occupied sites. For small δ, the vast majority of initially occupied sites are isolated singletons which spawn autonomous spreading crystals that approximate the shape $\mathcal{L}$ of (7.1.1) as they grow. Later, on a time scale of δ^{-1}, these droplets interact and combine, leaving isolated pockets of unoccupied sites which eventually shrink and disappear. What, then, is the *shape* of the last remaining holes a while before they vanish? To be precise, fix attention on the origin, say, denote the occupation time of $\mathbf{0}$ by T, and let C_n be the connected cluster, in the usual site percolation sense, of unoccupied sites containing $\mathbf{0}$. Thus $\{C_n = \emptyset\} = \{T \le n\}$. Denote the law of $\frac{1}{\epsilon n} C_{t-\epsilon n}$ under the conditional measure $P(\cdot | T > n)$ as $\mu_{\delta,\epsilon,n}$. Then a random set $\mathcal{H}_\delta$ with associated distribution $\mu_{\mathcal{H}_\delta}$ is called the last hole if

$$\lim_{\epsilon \to 0} \limsup_{n \to \infty} d(\mu_{\delta,\epsilon,n}, \mu_{\mathcal{H}_\delta}) = 0 \tag{7.1.2}$$

(Here d is the Prohorov distance between random compact sets corresponding to the Hausdorff metric d_H on compact subsets of $\mathbb{R}^2$.)

Using the reverse shapes formalism, we will see that for symmetric additive deterministic dynamics, and for any $\delta \in (0, 1)$, *$\mathcal{H}_\delta$ equals the forward shape $\mathcal{L}$* of (7.1.1) with probability one. For example, this result applies to the deterministic Richardson model above with $p = 1$, in which case the forward shape and last holes are squares. Our analysis will also show that for certain quasi-additive systems (to be defined later), $\mathcal{H}_\delta$ depends on δ but converges to $\mathcal{L}$ as $\delta \to 0$. For instance, the "two of eight" solidification CA has last holes which are approximately diamond shaped for small δ. More generally, e.g., for CA rules such as Bootstrap Percolation [AL], or "three of eight" solidification, $\mathcal{H}_\delta$ most likely consists of a random assortment of holes, possibly nonconvex or with volume 0. In additive random models, the last hole as defined in (7.1.2) turns out to be $\{\mathbf{0}\}$, but for first passage times n with $\delta^{-\frac{1}{d}} \ll n \ll \delta^{-\frac{1}{d-1}}$, holes of shape $\mathcal{L}$ are observed. The nature of $\mathcal{H}_\delta$ for general random dynamics remains an enigma, as the event $\{T > n\}$ seems very difficult to understand in detail.

The rest of the paper is organized as follows. Section 7.2 begins with an abstract framework in $\mathbb{R}^d$, convenient for development of the basic theory of deterministic growth. Then we formulate precisely the general problem of reverse shapes, state and prove two reverse limit theorems, give corresponding formulas for the limit, and derive various consequences of those formulas. In Section 7.3 we offer six examples of Threshold Growth automata, on box neighborhoods of various ranges ρ, and with various thresholds θ, to illustrate some of the subtleties of reverse shape theory. These CA rules are chosen for their relative ease of computation, but the reader should understand that many of the same phenomena can be expected to arise in stochastic growth models. Section 7.4 applies the results of Section 7.2 to the problem of last holes for deterministic growth started from random seedings ν_δ. Finally, Section 7.5 details the modifications to our arguments which are needed to prove reverse shape and last hole results for additive stochastic solidification dynamics such as Richardson's model and first-passage percolation.

7.2 Reverse Shape Theory for Deterministic Growth

Let $\mathcal{B}$ denote the σ-algebra of Borel subsets of $\mathbb{R}^d$. A map $\mathcal{T} : \mathcal{B} \to \mathcal{B}$ is a crystal growth transformation (CGT) if it satisfies the following five properties:

(a) *absorption*: $\mathcal{T}\emptyset = \emptyset$;
(b) *translation invariance*: for all $\mathcal{A} \in \mathcal{B}$ and $x \in \mathbb{R}^d$, $\mathcal{T}(x + A) = x + \mathcal{T}A$;
(c) *locality*: there exists a *neighborhood radius* $R > 0$ so that for every $\mathcal{A} \in \mathcal{B}$,

$$\mathcal{T}A = \{x \in \mathbb{R}^d : x \in \mathcal{T}(a \cap B_2(x, r))\};$$

(We use the standard notation for Euclidean balls: $B_p(x, r) = \{y \in \mathbb{R}^d : \|y - x\|_p \le r\}$. When $d = 2$, we will refer to the shapes $B_\infty(0, 1)$, $B_2(0, 1)$, and $B_1(0, 1)$ as *box*, *circle*, and *diamond*, respectively.)
(d) *monotonicity*: for every $A, B \in \mathcal{B}$, $A \subset B \Rightarrow \mathcal{T}A \subset \mathcal{T}B$;
(e) *solidification*: for every $A \in \mathcal{B}$, $A \subset \mathcal{T}A$.

The iterates of a crystal growth transformation, given by $\mathcal{T}^n A_0$ for a fixed initial set A_0 are called *crystal growth dynamics*. Write $\mathcal{T}^\infty A_0 = \bigcup_{n \ge 0} \mathcal{T}^n A_0$. Such dynamics are *supercritical* if $\mathcal{T}^\infty B_2(0, R) = \mathbb{R}^d$ for a sufficiently large R, *subcritical* if $\mathcal{T}^\infty B_2(0, R)^c \ne \mathbb{R}^d$ for a sufficiently large R, and *critical* otherwise. *Completely symmetric* sets are invariant with respect to permutation of coordinates and switching signs of coordinates. Also, $\mathcal{T}$ is *completely symmetric* if it commutes with all such operations on coordinates.

Threshold growth dynamics, with parameters μ, θ, where μ is a measure on $\mathbb{R}^d$ with compact support, and $\theta \ge 0$ is the threshold, are given by $\mathcal{T}A = \{x \in \mathbb{R}^d : \mu(A - x) \ge \theta\}$. *Additive dynamics* with parameter a bounded set , are defined simply by $\mathcal{T}A = A + \mathcal{N}$. These are the two basic classes of crystal growth transformations we have in mind, but many other examples may be obtained by taking unions and intersections Indeed, if $\mathcal{T}_1, \mathcal{T}_2, \ldots$ have the same neighborhood radius, then one can define new CGTs $\bigcup_n \mathcal{T}_n$ and $\bigcap_n \mathcal{T}_n$ by

$$\left(\bigcup_n \mathcal{T}_n\right) = \bigcup_n (\mathcal{T}_n A), \qquad \left(\bigcap_n \mathcal{T}_n\right) = \bigcap_n (\mathcal{T}_n A).$$

Note that any crystal growth transformation $\mathcal{T}$ must translate half-spaces: if $u \in S^{d-1}$ is a unit vector and $H_u^- = \{x \in \mathbb{R}^d : \langle x, u \rangle \le 0\}$, then there exists a *speed* $w(u) \ge 0$ such that either $\mathcal{T}H_u = H_u^- + [0, w(u))u$, or $\mathcal{T}H_u = H_u^- + [0, w(u)]u$.

Now, introduce the (possibly unbounded) set $K_{1/w} = \bigcup_u [0, w(u)]u \subset \mathbb{R}^d$. Locality of $\mathcal{T}$ implies that $w : S^{d-1} \to \mathbb{R}$ is Lipschitz continuous. Conversely, it is not hard to show that any such Lipschitz w is the speed function of a CGT, and that if $K_{1/w}$ is completely symmetric, then $\mathcal{T}$ can be chosen completely symmetric as well.

Continuity of w immediately implies that supercriticality is equivalent to $w > 0$ [GG1]. On the other hand, subcriticality is not so easy to characterize. To see this, note that threshold growth with μ uniform on *box* and $\theta = 2$ is critical, while it has $w \equiv 0$. On the other hand, we have proved in [GG2] that the two-dimensional discrete threshold model, in which μ is a counting measure on a finite subset of $\mathbb{Z}^2$ is subcritical if and only if $w \equiv 0$.

In the supercritical case, assuming that the iterates of a bounded initial A_0 eventually cover any compact set, the *(Forward) Shape Theorem* [GG1] states that

$$\frac{1}{n}\mathcal{T}^n A \to \mathcal{L} = K^*_{1/w} \quad \text{as } n \to \infty, \tag{7.2.1}$$

where $K^* = \{y : \langle x, y\rangle \leq 1 \text{ for every } x \in K\}$ is the polar transform of the set K.

We now proceed to the formulation and proof of two *Reverse Shape Theorems*, beginning with the simpler case. Since we will be concerned, in part, with infinite limit sets, let us specify precisely what is meant by convergence. In the present context, it is most convenient to say that $\Lambda_n \subset \mathbb{R}^d$ converges to a closed set $\Lambda \subset \mathbb{R}^d$ if $\Lambda_n \cap B_2(0, R)$ converges in the Hausdorff metric d_H to $\Lambda \cap B_2(0, R)$ for all sufficiently large $R > 0$. For notational convenience, we write $\lim_{\epsilon\to 0}\lim_{m\to 0}\Lambda_{\epsilon,m} = \Lambda$ if for every R,

$$\lim_{\epsilon\to 0}\limsup_{m\to 0} d_H(\Lambda_{\epsilon,m} \cap B_2(0, R), \Lambda \cap B_2(0, R)) = 0.$$

Throughout the paper, A will be denote a prescribed compact set which contains a neighborhood of $\mathbf{0}$. As in Section 7.1, set $A_0 = (mA)^c$ for some large m. Let $T = T(m)$ be the first time the origin is occupied by the dynamics:

$$T = \inf\{n : \mathbf{0} \in \mathcal{T}^n((mA)^c)\}.$$

Our first result establishes a reverse limit $\mathcal{R}(A)$ in the case of *convex* A.

Theorem 7.2.1 *Assume that* $w \geq 0$ *is not identically* 0 *on* S^{d-1}, *and that* A *is convex. Then there is a unique nonempty convex proper subset* $\mathcal{R}(A)$ *of* $\mathbb{R}^d$ *such that*

$$\lim_{\epsilon\to 0}\lim_{m\to 0}\frac{1}{\epsilon m}\mathcal{T}^{T-\epsilon m}(mA^c) = \mathcal{R}(A)^c. \tag{7.2.2}$$

If $\mathcal{T}$ *and* A_0 *are completely symmetric, then* $\mathcal{R}(A)$ *is bounded.*

Remark The weak double limit formulation in (7.2.2) for the reverse shape property holds quite generally for CGTs, and also for stochastic growth models, as we shall see in Section 7.5. In deterministic lattice dynamics the rapid convergence methods developed by [Wil] imply that ϵm can be replaced by a large M and the double limit by $\lim_{M\to 0}\lim_{m\to 0}$.

PROOF. If the boundary of has small curvature, then $\mathcal{T}^n((mA)^c)$ is well-approximated by the union of iterates over all exterior half-spaces [GG1, GG2]. Hence, for m and n large,

$$\mathcal{T}^n((mA)^c) \approx \bigcup \left\{ \left(\left(m\alpha_A(u) - n\frac{w(v)}{\langle u, v\rangle} \right) u + H_{-v}^- \right) : u, v \in S^{d-1}, v \in \nu_A(u) \right\}. \tag{7.2.3}$$

Here $A = \{\lambda u : u \in S^{d-1}, 0 \le \lambda \le \alpha_A(u)\}$, and $\nu_A(u)$ is the set of exterior normals to A at $\alpha_A(u)u$. The right side of (7.2.3) defines a transformation $\overline{\mathcal{T}}_{n,m}(A)$. For $u \in S^{d-1}$, $v \in \nu_A(u)$, define

$$\phi(u, v) = \frac{w(v)}{\alpha_A(u)\langle u, v\rangle} = w(v)\alpha_{A^*}(v),$$
$$M = \max_{u,v} \phi(u, v).$$

Then, if one chooses $n = (M^{-1} - \delta)m$ in (7.2.3), one gets

$$\lim_{m\to\infty} \frac{T(m)}{m} = \frac{1}{M}, \quad \text{and}$$
$$\mathcal{R}(A) = \bigcap \{M\alpha_A(u)u + H_v^- : v \in \nu_A(u), \phi(u, v) = M\}. \tag{7.2.4}$$

The proof of [GG2, (3.2)] shows that R(A) is invariant for $\overline{\mathcal{T}}_{n,m}$, i.e.,

$$\overline{\mathcal{T}}_{n,m}(\mathcal{R}(A)) = (m - Mn)(\mathcal{R}(A)^c),$$

and so a good candidate for the reverse shape.

Rigorous justification of (7.2.4) hinges on three observations. First, the $\supset$ inclusion in (7.2.3) follows from monotonicity. Second, one can approximate *all of the* iterates by sets of uniformly small curvature uniformly within ϵn. As in [GG1, Section 4], this implies that $\mathcal{T}$ is well-approximated by $\overline{\mathcal{T}}$. Third,

$$\lim_{\epsilon\to 0} \lim_{m\to\infty} \frac{\overline{\mathcal{T}}_{\epsilon m,m}(A)}{\epsilon m} = \mathcal{R}(A)^c$$

by direct computation, since only u, v which maximize ϕ determine either side. □

Another characterization of $\mathcal{R}(A)$ follows from (7.2.4).

Corollary 7.2.1 *Let*

$$t_0 = \max\{t : tA^* \subset K_{1/w}\}.$$

Then

$$\mathcal{R}(A) = (t_0 A^* \cap \partial K_{1/w})^*. \tag{7.2.5}$$

In particular, starting from circle, *with* $\overline{w} = \max\{w(u) : u \in S^{d-1}\}$,

$$\mathcal{R}(\text{circle}) = \bigcap \{\overline{w}u + H_u^- : w(u) = \overline{w}\}. \tag{7.2.6}$$

PROOF. Since $\alpha_{A^*}(v) = (\alpha_A(u)\langle u, v\rangle)^{-1}$, (7.2.4) can be rewritten as

$$\mathcal{R}(A) = \bigcap \left\{ \frac{1}{\alpha_{A^*}(v)} v + H_v^- : v \in \nu_A(u), \alpha_{A^*}(v) = \frac{M}{w(v)} \right\}.$$

The condition $v \in \nu_A(u)$ is superfluous, and (7.2.5) follows. In the case A = *circle*, note that $M = \overline{w}$. □

One unsatisfactory feature of Theorem 7.2.1 is the rather severe restriction to convex initial A. Unfortunately, nonconvex holes typically arise in disordered systems, and the general mechanism by which they shrink is quite complicated. As an illustration of the added complexity with nonconvex initial sets, consider additive dynamics with $\mathcal{N}$ = *box*, and the completely symmetric stars A_x with boundary consisting of 8 line segments, one of which connects (1, 1) with any point $(x, 0)$, where $1 \le x \le \infty$. All of these sets are invariant, and all except the square are nonconvex. We call the dynamics $\mathcal{T}$ *quasi-additive* if $K_{1/w}$ is convex, a property enjoyed by additive $\mathcal{T}$, since $K_{1/w} = \mathcal{N}^*$ in that case, but otherwise not often satisfied (cf. [GG1, Gra1]). Only the two-dimensional quasi-additive case is simple enough to permit a thorough understanding of reverse shapes starting from fairly general sets. The construction leading to the following result is rather involved, so we will only provide a sketch in the additive case with convex $\mathcal{N}$.

Theorem 7.2.2 *Assume $d = 2$, that $\mathcal{T}$ is quasi-additive, that $w > 0$ on S^1, and that the boundary of A consists of a finite number of piecewise differentiable curves. Then the double limit in* (7.2.2) *exists, and $\mathcal{R}(A)$ is nonempty, although it need not be convex.*

SKETCH OF THE PROOF. We begin with a recipe for $\mathcal{R}(A)$. For additive dynamics, the first-passage time of the origin can be represented as

$$T = \sup\{t : T\mathcal{N} \cap \partial A = \emptyset\}.$$

For each $x \in t\mathcal{N} \cap \partial A$, we define a wedge or half-space W_x as follows. By convention, none of the wedges or half-spaces in the following construction include $\mathbf{0}$. If ∂A is differentiable at x, then W_x is the tangent half-plane to ∂A at x. If ∂A is nondifferentiable at x, then there are two rays tangent to ∂A at x. Denote the wedge determined by those rays as W_x'. The ray from x through $\mathbf{0}$ intersects $W_x' + \mathcal{N}$ at some point y. W_x is the half-plane or wedge at x which is parallel to the half-plane or wedge tangent to $\partial(W_x' + \mathcal{N})$ at y. Finally, we claim that

$$\mathcal{R}(A) = \left(\bigcup_{x \in T\mathcal{N} \cap \partial A} W_x \right)^c. \tag{7.2.7}$$

To justify this formula, note first that only an infinitesimal neighborhood of $T\mathcal{N} \cap \partial A$ plays a role in the reverse shape since other points of A lag linearly behind. Also, the prescription is correct if $T\mathcal{N} \cap \partial A$ consists of a single point. Finally, the general case follows by additivity.

Formula (7.2.7) also holds for any $\mathcal{T}$ and initial A satisfying the hypotheses of the theorem provided $\mathcal{N}$ is replaced by $\mathcal{L}$ in the construction, but the proof is somewhat more involved so we omit it. The formula shows that $\mathcal{R}(A)$ is convex for every A if and only if $\mathcal{L}$ is C^1. □

Key to our analysis of last holes for symmetric, additive and quasi-additive CA crystals in Section 7.4 is the following special case of Corollary 7.2.1, which asserts that $\mathcal{L}$ is *invariant*, by which we mean that the forward asymptotic shape is its own reverse limit. In other words, the hole obtained by growing a large crystal A_t, and then reversing the occupied and empty sites of $\mathbb{R}^d$, retains its shape as it shrinks.

Corollary 7.2.2 *If $\mathcal{T}$ is quasi-additive, then $\mathcal{R}(\mathcal{L}) = \mathcal{L}$.*

PROOF. Since $K_{1/w}$ is convex, $A^* = \mathcal{L}^* = K_{1/w}$, so that in (7.2.5) evidently $t_0 = 1$ and $t_0 A^* \cap \partial K_{1/w}$. Hence $\mathcal{R}(\mathcal{L}) = (\partial K_{1/w})^* = \mathcal{L}$, as claimed. □

For the remainder of this section we assume that $d = 2$. Let us first try to answer the following question: is there a way to center convex initial A so that $\mathcal{R}(A)$ is bounded? More precisely, call $x_0 \in \mathbb{R}^2$ a *$\mathcal{T}$-center* for A if $\mathcal{R}(x_0 + A)$ is bounded. If A is not lattice symmetric, then the center need not exist. To see this, write $e_1 = (1, 0)$, and consider the case with $K_{1/w} = B_\infty(e_1, 1) \cup B_\infty(e_{-1}, 1)$, corresponding to additive dynamics with $\mathcal{N} = \{(\pm 1, 0), (0, \pm\frac{1}{2})\}$. Choose $A =$ *box*, in which case different speeds in the horizontal and vertical directions produce increasingly oblong shrinking holes. However such anomalies cannot occur in completely symmetric cases.

Proposition 7.2.1 *Under the assumptions of Theorem* 7.2.1, *and complete symmetry of $\mathcal{T}$ and A, the origin is the unique $\mathcal{T}$-center for A.*

PROOF. Since w is finite, and positive in some direction u, t_0 in Corollary 7.2.1 must be positive and finite. By symmetry, $t_0 A^* \cap \partial K_{1/w}$ must contain at least 4 symmetrically situated points. Hence the dual of this set, $\mathcal{R}(A)$, is bounded with nonempty interior containing the origin. Thus $\mathbf{0}$ is a $\mathcal{T}$-center. For $x_0 \neq \mathbf{0}$, the dynamics from $x_0 + A$ are the translate by x_0 of the dynamics from A. So when $\mathcal{T}^n((m(x_0 + A))^c)$ hits the origin, its diameter is bounded below by a constant multiple of m. □

For simplicity, then, we will assume complete symmetry of $\mathcal{T}$, and complete symmetry and convexity of all sets A, in the following analysis of shape dependence on initial seeds. For a typical $K_{1/w}$, different A yield many different reverse shapes. As we will see, most of them are unstable, since small perturbations produce large differences in $\mathcal{R}(A)$. It is important to note that the following notion of stability is formulated *only* within the restricted class of convex and completely symmetric sets A.

Definitions A has a *weakly stable* reverse shape $\mathcal{R}(A)$ if for every $\epsilon > 0$ there exists a $\delta > 0$ such that $d_H(A, A') < \delta$ implies $d_H(\mathcal{R}(A), \mathcal{R}(A')) < \epsilon$. $\mathcal{R}(A)$ is stable if there exists a $\delta > 0$ such that $d_H(A, A') < \delta$ implies $d_H(\mathcal{R}(A), \mathcal{R}(A'))$ $= 0$.

In the stability criterion which follows,

$$\Omega = \{(x_1, x_2) : 0 \le x_2 \le x_1\}, \qquad R = \{(x_1, x_2) : x_1 > 0,\ x_1 + x_2 > 0\} \cup \{\mathbf{0}\}.$$

Proposition 7.2.2 *The reverse shape $\mathcal{R}(A)$ is weakly stable if and only if $|t_0 A^* \cap \partial K_{1/w}|$ is either 4 or 8 (in which case $\mathcal{R}(A)$ has 4 or 8 sides, respectively). $\mathcal{R}(A)$ is stable if and only if it is weakly stable and, for any $x_0 \in t_0 A^* \cap \partial K_{1/w}$ which lies in the closed first octant, there is a $\delta > 0$ such that $B_2(x_0, \delta) \cap \partial K_{1/w} \Omega \subset x_0 + R$.*

In other words, stability means that in the first octant, $t_0 A^*$ only meets parts of the boundary of $K_{1/w}$ which lie inside the translate of cone R. Note that in quasi-additive cases the forward shape $\mathcal{L}$, which is its own reverse shape by Corollary 7.2.2, can *never* be even weakly stable.

PROOF. We apply Corollary 7.2.1, using the fact that small perturbations of A produce small changes in t_0. Assume first that $t_0 A^*$ intersects $K_{1/w}$ in 8 points, one in the interior of each octant by symmetry. Fix an $\epsilon > 0$, and let B_ϵ denote the ϵ-neighborhood of these 8 points. There exists an $\eta > 0$ such that $d_H(B_\epsilon^c \cap K_{1/w}, t_0'(A')^*) > \eta/2$. Therefore, for A' sufficiently close to A, the corresponding t_0' is close to t_0, and $d_H(B_\epsilon^c \cap K_{1/w}, t_0(A)^*) > \eta$. By symmetry, $t_0'(A')^* \cap \partial K_{1/w}$ contains at least one point in each octant. It follows that the d_H-distance between $t_0'(A')^* \cap \partial K_{1/w}$ and $t_0 A^* \cap \partial K_{1/w}$ is at most ϵ. Continuity of the polar transform establishes weak stability. The remaining cases of 4 or 8 point intersections are similar. Conversely, if $\Omega \backslash \{x_1 = x_2\}$ contains two points of intersection, then the point in the interior of Ω can be removed by an arbitrarily small perturbation.

As before, we consider only the case of one intersection point x interior to Ω. If $t_0 A^*$ only meets $\partial K_{1/w}$ in $x + R$, then instability of $\mathcal{R}(A)$ would imply the existence of 8 symmetrically located points in $t_0 A^*$, having x interior to their convex hull. This contradicts the definition of t_0. Conversely, if the stability condition is violated, then the octagon with one vertex at x is not stable. □

A particularly strong form of stability is *uniqueness*. The rare cases in which it holds are as follows.

Proposition 7.2.3 *All reverse shapes $\mathcal{R}(A)$ agree if $\mathcal{R}$(diamond) $=$ $\mathcal{R}$(box). Moreover, if the reverse shape is unique, then it must be a diamond, a box, or an octagon.*

PROOF. By Corollary 7.2.1, if the reverse shapes for diamond and box agree, then there exist t_0 and t_0' such that $\partial K_{1/w}$ meets t_0 *diamond* $\cup\, t_0'$ *box* only in $S = t_0 \partial$ *diamond* $\cap\, t_0' \partial$ *box*. Any completely symmetric convex set A such that $S \subset \partial A$ is included in t_0 *diamond* $\cup\, t_0'$ *box*. This proves uniqueness. The reverse shape is then determined by S, which consists of 4 points in a box, 4 points in a diamond, or 8 points in an octagon. □

Necessary and sufficient conditions for this unique $\mathcal{R}$ to be a diamond or box are $\mathcal{R}$(*diamond*) $=$ *box* and $\mathcal{R}$(*box*) $=$ *diamond*, respectively. For example, let $\mathcal{T} = \mathcal{T}_1 \cap \mathcal{T}_2$, where $\mathcal{T}_1, \mathcal{T}_2$ correspond to additive dynamics with $\mathcal{N}_1 = \{(\pm 1, 0), (0, \pm\frac{1}{3})\}$ and $\mathcal{N}_2 = \{(0, \pm 1), (\pm\frac{1}{3}, 0)\}$, respectively. Then

$K_{1/w} = \{0, (\pm 2, 0), (0, \pm 2)\} + box$, and *diamond* is the unique shape. Rotate by 45° to get a (rescaled) box as the unique shape. Example 7.3.1 of the next section illustrates octagonal uniqueness.

7.3 Examples

Let us now present five illustrative examples of Theorem 7.2.2 for deterministic Threshold Growth on the two dimensional integers. Such dynamics have already been studied elsewhere in considerable detail (see [Boh, GG1, GG2, GG3, GG4]). Within our general framework, these cases have μ a counting measure on the range ρ lattice box $\mathcal{N} = B_\infty(0, \rho) \cap \mathbb{Z}^2$, and θ an integer-valued threshold. In words, an empty site x joins the occupied set if it sees at least θ occupied cells within its range ρ box neighborhood. Such dynamics are supercritical for $\theta \leq \rho(2\rho + 1)$, and critical for $\rho(2\rho + 1) < \theta \leq 2\rho(2\rho + 1)$. The discrete deterministic setting makes computation of the half-space velocities $w(u)$ straightforward, and hence, starting from completely symmetric convex initial data A, a complete analysis of the possible reverse shapes $\mathcal{R}(A)$ follows from Corollary 7.2.1. The solidification process called *Hickerson's Diamoeba* [GG4] provides a beautiful example of distinct subsequential reverse shapes when CGT axiom (d) is lacking.

For each of the following examples we specify the range ρ, threshold θ, vertices of $K_{1/w}$ in the first octant $\{0 \leq y \leq x\}$ (the others are determined by symmetry), and the maximal number of sides in any (polygonal) reverse shape $\mathcal{R}$. In each case we show the graph of $K_{1/w}$, including the largest inscribed diamond, circle, and square. Their points of contact with $K_{1/w}$ in (7.2.5) characterize the dual of the reverse shape for a square, diamond and circle, respectively (recall that the diamond and square are dual and the circle self-dual). A second graphic for all but one of the examples shows the evolution of a shrinking hole, in an alternating black and white palette, for a representative initial A. Finally, a brief description summarizes the distinctive features of each case. Octagons play a central role in the stability analysis of our examples, so we denote by O_m the completely symmetric shape with sides of slope $\pm m, \pm m^{-1}$. For simplicity we identify all dilations of $\mathcal{R}$ in the discussion, ignoring the *size* (as opposed to *shape*) of the limit. Thus the notation $\mathcal{R} \propto O_m$ means that $\mathcal{R} = cO_m$ for a suitable constant $c > 0$ which we choose not to compute. However c is easily determined from $K_{1/w}$.

Some "diophantine" aspects of these deterministic lattice examples presumably do not arise in more general growth models, especially those with random dynamics. But other features should be shared by many nonlinear systems, deterministic or stochastic, with space and time either discrete or continuous, since it would seem common for anisotropic neighborhoods to generate nonconvex $K_{1/w}$. We shall return to this point in Sections 7.4 and 7.5.

Example 7.3.1 (see Figure 7.1) $\rho = 1, \theta = 4$: $K_{1/w}$ has vertices $(\frac{1}{0}, 0), (2, 1), (\frac{1}{0}, \frac{1}{0})$. $\mathcal{R}$ is always a rescaling of O_2.

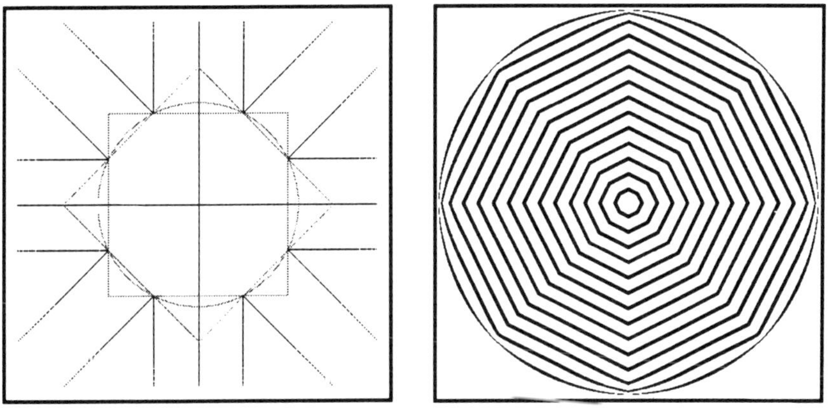

FIGURE 7.1.

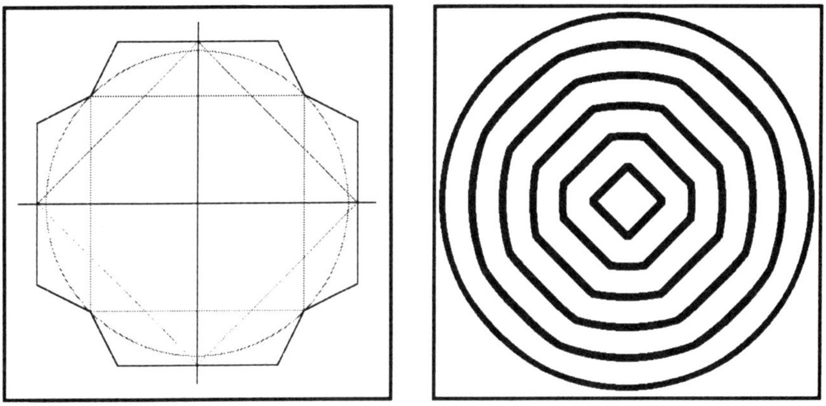

FIGURE 7.2.

This critical rule seems to be the only lattice example of reverse shape uniqueness with box neighborhood, for any range ρ. Note that $K_{1/w}$ is infinite in the horizontal, vertical, and diagonal directions, a manifestation of criticality. The figure on the right shows convergence to $\propto O_2$ from a lattice circle, after 530 steps from "radius" 250).

Example 7.3.2 (see Figure 7.2) $\rho = 2$, $\theta = 3$: $K_{1/w}$ has vertices $(\frac{1}{2}, 0)$, $(\frac{1}{2}, \frac{1}{4})$, $(\frac{1}{3}, \frac{1}{3})$. $\mathcal{R}$ has at most 12 sides.

O_m octagons are invariant (up to rescaling) and weakly stable for $m \geq 3$, but converge to a multiple of *diamond* for $1 \leq m < 3$. The right figure shows $\mathcal{R}(circle) \propto diamond$. (A hole of "radius" 250 fills at time 118.)

Example 7.3.3 (see Figure 7.3) $\rho = 2$, $\theta = 6$: $K_{1/w}$ has vertices $(1, 0)$, $(\frac{2}{3}, \frac{1}{6})$ $(\frac{3}{4}, \frac{1}{4})$, $(\frac{2}{3}, \frac{1}{3})$, $(\frac{3}{4}, \frac{1}{2})$, $(\frac{1}{2}, \frac{1}{2})$. $\mathcal{R}$ has at most 12 sides.

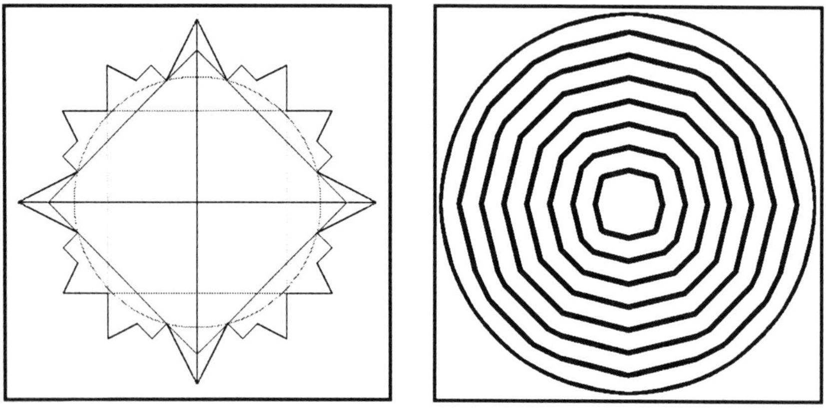

FIGURE 7.3.

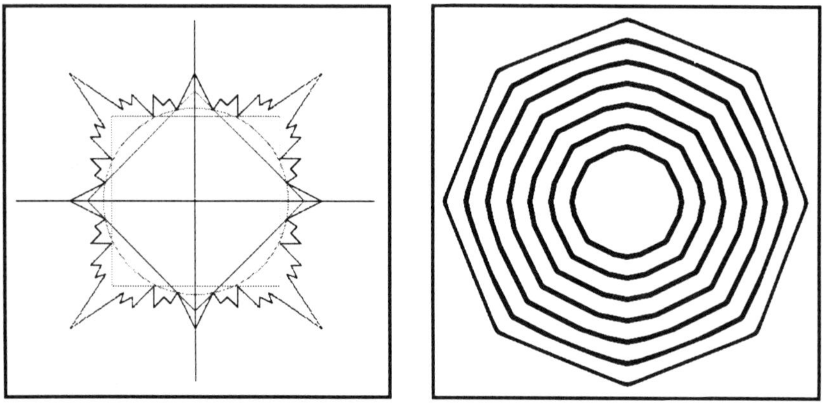

FIGURE 7.4.

O_2 is unstable with three distinct limit shapes in its neighborhood: $\propto$ *diamond* (stable), $\propto O_4$ (stable), and a rescaling of the unstable 12-gon with vertices $(51, 0)$, $(44, 28)$. The right side of Figure 7.3 illustrates convergence to $\propto O_4$ from a lattice circle (at time 150 from "radius" 250).

Example 7.3.4 (see Figure 7.4) $\rho = 3$, $\theta = 16$: $K_{1/w}$ has vertices $(\frac{5}{7}, \frac{1}{7})$, $(\frac{2}{3}, \frac{1}{3})$, and many others irrelevant to reverse shapes. $\mathcal{R}$ has at most 16 sides.

O_2 and O_5 are invariant up to rescaling, and stable. O_m tends to $\propto O_2$ for $1 \leq m \leq \frac{5}{2}$, to $\propto O_5$ for $m > \frac{5}{2}$. $O_{\frac{5}{2}}$ tends to a rescaling of the unstable 16-gon with vertices $(234, 0)$, $(215, 95)$, $(175, 175)$. The left side of Figure 7.4 shows convergence of $O_{\frac{5}{2}}$ to the unstable 16-gon (at time 125 from "principal radius" 245).

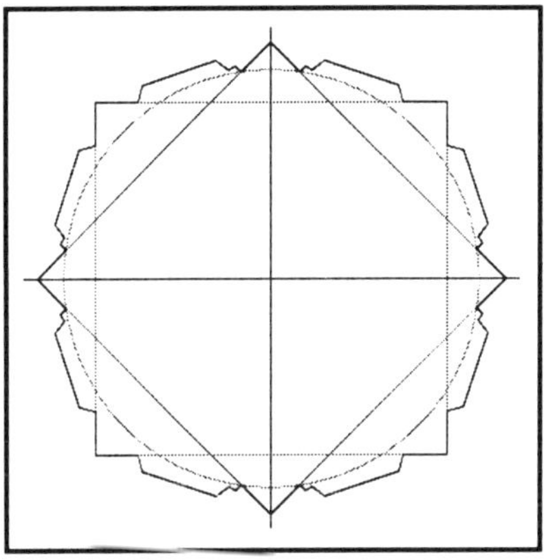

FIGURE 7.5.

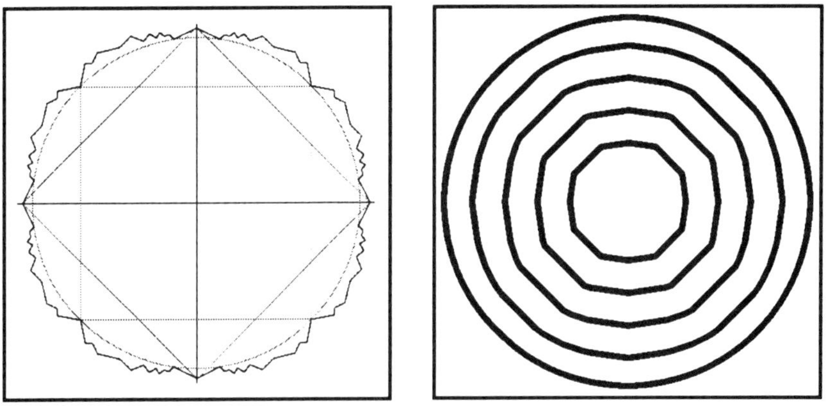

FIGURE 7.6.

Example 7.3.5 (see Figure 7.5) $\rho = 4$, $\theta = 11$: $K_{1/w}$ has vertices $(\frac{1}{3}, 0)$, $(\frac{7}{24}, \frac{1}{24})$, $(\frac{3}{10}, \frac{1}{20})$, $(\frac{5}{17}, \frac{1}{17})$, $(\frac{4}{13}, \frac{1}{13})$, $(\frac{3}{11}, \frac{2}{11})$, $(\frac{3}{16}, \frac{1}{4})$, $(\frac{1}{4}, \frac{1}{4})$. $\mathcal{R}$ has at most 32 sides.

Flat edges of $K_{1/w}$ give rise to a continuum of unstable reverse shapes; here an example of an unstable 32-gon, invariant up to rescaling, has polar determined by $(7/24, 1/24)$, $(9/31, 4/31)$, $(2/7, 1/7)$, $(3/16, 1/4)$.

Example 7.3.6 (see Figure 7.6) $\rho = 5$, $\theta = 15$: $K_{1/w}$ has vertices $(\frac{7}{30}, \frac{1}{30})$, $(\frac{2}{9}, \frac{1}{9})$, $(\frac{3}{13}, \frac{1}{13})$, $(\frac{1}{6}, \frac{1}{6})$, and *many* more irrelevant to reverse shapes. $\mathcal{R}$ has at most 24 sides.

From a circle, the reverse limit shape $\mathcal{R}(circle)$ is an unstable rescaling of the 12-gon with vertices (150, 0), (140, 70). To obtain a reverse shape with more than 8 sides starting from a circle, by (7.2.6) there must be two points x_1 and x_2 in the first octant of the boundary of $K_{1/w}$ such that $\|x_1\|_2 = \|x_2\|_2$, as in the present case where $\|(\frac{7}{30}, \frac{1}{30})\|_2 = \|(\frac{1}{6}, \frac{1}{6})\|_2$. Since edge speeds are a nonlinear function of system parameters, such examples are very rare; this is the only case with small range, and perhaps the only one for any choice of ρ and θ. The right side of Figure 7.6 shows convergence to the unstable 12-gon from a lattice circle (at time 42 from "radius" 250).

7.4 Last Holes for Deterministic Growth on $\mathbb{Z}^d$

We now turn to the study of last holes (7.1.2), as described in Section 7.1. Let $\Pi = \Pi(\delta)$ be a random configuration of occupied sites in $\mathbb{Z}^d$ distributed according to the Bernoulli product measure ν_δ with density δ. Recall that T is the occupation time of $\mathbf{0}$ starting from Π. Throughout this section, let $\mathcal{T}$ be a CGT satisfying (a)–(e) of Section 7.2, and also

(f) *symmetry*: for every A, $\mathcal{T}(-A) = -\mathcal{T}A$. (In the additive case, $\mathcal{N} = -\mathcal{N}$.)

For additive supercritical transformations, we are able to identify the shape of last holes starting from any nontrivial $\Pi(\delta)$. Roughly, if the origin is not occupied at time n, then the set $n\mathcal{L}$ initially has no seeds, but if $n^d \gg \delta^{-1}$, then there are lots of seeds everywhere around the boundary. Since $\mathcal{L}$ is an invariant reverse shape by Corollary 7.2.2, the hole around $\mathbf{0}$ at time $T - \epsilon n$ is approximately $\epsilon n\mathcal{L}$.

Theorem 7.4.1 *For any* $\delta \in (0, 1)$, *additive supercritical growth satisfies* (7.1.2), *with the last hole* $\mathcal{H}_\delta$ *equal to the (forward) asymptotic shape* $\mathcal{L}$ *of* (7.1.1). *That is, for every* $0 < \delta < 1$, *and every* $\eta > 0$,

$$\lim_{\epsilon \to 0} \limsup_{n \to \infty} \left(d_H \left(\frac{C_{T-\epsilon n}}{\epsilon n}, \mathcal{L} \right) > \eta \mid T > n \right) = 0. \tag{7.4.1}$$

PROOF. Writing $A_n = \mathcal{T}^n\{\mathbf{0}\}$, symmetry implies that

$$\{T > n\} = \{A_n \cap \Pi = \emptyset\}. \tag{7.4.2}$$

Conditioning on $\{T > n\}$ therefore has no effect on sites in A_n^c. By (7.2.1), for large n and any $\eta_1 > 0$,

$$(1 - \eta_1)n\mathcal{L} \subset A_n \subset (1 + \eta_1)n\mathcal{L}.$$

In particular, the event $\{T > n\}$ and events involving sites outside $(1 + \eta_1)n\mathcal{L}$ are independent. With overwhelming probability, for every x belonging to $(1 + 2\eta_1)n\mathcal{L}\backslash(1 + \eta_1)n\mathcal{L}$,

$$\Pi \cap B_\infty(x, \eta_1 n) \neq \emptyset.$$

Moreover, on $\{T > n\}$, $\Pi \cap (1 - \eta_1)n\mathcal{L} = \emptyset$. The last two conditions, together with Corollary 7.2.2, imply that, on $\{T > n\}$, $C_{T-\epsilon n}$ differs from $\epsilon n\mathcal{L}$ by $(c\eta_1 + o(\epsilon))n$ for some constant $C = C(\mathcal{N})$. Choosing $\eta_1 = o(\epsilon)$, (7.4.1) follows. □

An elaboration of the last argument shows that the last holes of *quasi-additive* dynamics from sparse product measures have shape approximately $\mathcal{L}$. For simplicity we assume $d = 2$. For instance, our next result applies to "two of eight" solidification, or to Threshold Growth on F^2 with $\theta = 2$ and any range ρ box neighborhood [GG2]. New complications arise due to the presence of failed nucleation centers around the origin (e.g., isolated singletons in our $\theta = 2$ examples) which slightly alter the process whereby holes are filled in. For this reason, $\mathcal{L}$ is only achieved in the limit as $\delta \to 0$. To ensure that failed centers are well-behaved, in addition to (a)–(f) we assume

(g) *strong omnivoracity*: Let R be such that $\mathcal{T}^\infty B_\infty(0, R) = \mathcal{R}^d$. There exists an N so that for every A_0 such that $\mathcal{T}^{N+1} A_0 \neq \mathcal{T}^N A_0$, $B_\infty(x, R) \subset \mathcal{T}^N A_0$ for some x.

This uniform version of the *omnivorous* assumption in [GG2] holds for all the threshold 2 models mentioned above, and can be verified by computer for other simple crystals.

Theorem 7.4.2 *If $\mathcal{T}$ is quasi-additive supercritical and strongly omnivorous, then the last hole converges to $\mathcal{L}$ as $\delta \to 0$, in the sense that, for every $\eta > 0$,*

$$\lim_{\delta\to 0} \limsup_{\epsilon\to 0} \limsup_{n\to\infty} P_\delta \left(d_H \left(\frac{\mathcal{C}_{T-\epsilon n}}{\epsilon n}, \mathcal{L} \right) > \eta \mid T > n \right) = 0.$$

PROOF. Call site x is a *nucleus* if at the time N of (g), $B_2(x, R) \subset \mathcal{T}^N \Pi$. Given $R > 0$, introduce comparison dynamics $\mathcal{T}_\delta$ as on [GG2, p. 1771]. This process supposes a nucleus at a single site, but nowhere else, and models growth in a sparse random environment with effects of range at most R. [GG2, Lemma 6.4] shows that for every $\eta_1 > 0$, there is a δ sufficiently small that

$$P(\mathcal{T}_\delta^n \{x\} \not\subset (1 + \eta_1) n\mathcal{L}) \leq e^{-n}.$$

Let G be the following event, which depends on $\Pi|_{(1+2\eta_1)n\mathcal{L}}$. On G, impose a boundary condition of all 1's outside $(1 + 2\eta_1)n\mathcal{L}$ and require that every nucleus x is such that $\mathcal{T}_\delta^k\{x\} \cap (1 + 2\eta_1)n\mathcal{L} \subset x + (1 + \eta_1)k\mathcal{L}$ for every $k > \eta_1 n$. By the FKG inequality,

$$P(G^c \mid T > n) \leq P(G^c) \leq 8C^2 n^3 e^{-\eta_1 n}.$$

Since the dynamics are deterministic, on $\{T > n\}$ there is no nucleus within $(1 - \eta_1)n\mathcal{L}$. Moreover, on G no points outside $(1 + 2\eta_1)n\mathcal{L}$ can effect $\{T > n\}$.

From this point on, the proof is a minor modification of the additive argument. Write

$$G_1 = \{\forall x \in (1 + 4\eta_1)n\mathcal{L} \backslash (1 + 3\eta_1)n\mathcal{L},\ B_\infty(x, \eta_1 n) \text{ contains a nucleus}\},$$
$$G_2 = \{\text{there is no nucleus in } (1 - \eta_1)n\mathcal{L}\}.$$

If n is very large compared to δ^{-1} (say, of order δ^{-8R^2}) then $P(G_1^c)$ is very small. Hence, since G_1^c and $\{T > n\}$ are conditionally independent given G, $P(G_1^c \mid T > n)$ is also very small for large n. Moreover, G_2 occurs deterministically on $\{T > n\}$. On $G \cap G_1 \cap G_2$, as in the additive case, $d_H(\mathcal{C}_{T-\epsilon n}, \epsilon n\mathcal{L}) \leq (C\eta_1 + o(\epsilon))n$. □

Without quasi-additivity a host of new issues arise, so we conclude this section with speculation about two of the simplest cases.

Example 7.4.1 For "three of eight" solidification, forward shape $\mathcal{L}$ is the completely symmetric octagonal region with vertices $(\frac{1}{2}, 0)$, $(\frac{1}{3}, \frac{1}{3})$, and $\mathcal{R}(\mathcal{L})$ is the square $B_\infty(0, \frac{1}{2})$. But there are five holes A smaller than that square, hence more prevalent in ν_δ, which are invariant and such that mA reaches the origin at the same time T. These candidates for last holes are the nonconvex completely symmetric star with vertices $(\frac{1}{2}, 0)$, $(\frac{1}{6}, \frac{1}{6})$, and the asymmetric nonconvex quadrilateral with vertices (counterclockwise) $(\frac{1}{2}, 0)$, $(-\frac{1}{2}, -\frac{1}{2})$, $(0, -\frac{1}{2})$, $(\frac{1}{4}, \frac{1}{4})$ and its rotations by 90°, 180°, and 270°, all with area $\frac{1}{3}$. We suspect that no smaller invariant hole has the same T, and hence conjecture that

$$\lim_{\delta\to 0} \limsup_{\epsilon\to 0} \limsup_{n\to\infty} d(\mu_{\delta,\epsilon,n}, \mu_{\mathcal{H}_\delta}) = 0$$

(in the manner of (7.1.2)), where $\mu_{\mathcal{H}_\delta}$ is some measure concentrated on the above five shapes. Several hurdles would need to be overcome in order to establish this result, and even so, the relative weight of last holes assigned to stars vs. quadrilaterals would pose an extremely difficult problem.

Example 7.4.2 "Four of eight" solidification is a *critical* CGT, convex-confined like its *bootstrap* [AL] relatives on the range 1 diamond neighbor set. The last hole is most likely, for every δ, a random set of dimension one. This is probably quite difficult to prove, but [AMS] provides good evidence: for one variant of bootstrap percolation, in which a site becomes occupied if it has one occupied nearest neighbor in the horizontal direction and one in the vertical direction, that paper shows that as $n \to \infty$,

$$\frac{\log P(T > N)}{2n \log(1-\delta)} \to 1. \tag{7.4.3}$$

The denominator is the logarithm of the probability that an interval of $2n$ sites is empty. This should mean that conditioning on $\{T > n\}$ is equivalent to conditioning on such a vacant interval around $\mathbf{0}$. For every large n (much larger than $e^{c/\delta}$), the rest of space is filled way before the interval shortens significantly from its ends. The last stage then consists of filling the interval and the shape is obvious, but lattice symmetry dictates a random distribution of orientations. Of course, whether or not such heuristics can be substantiated, the last hole of the [AMS] model cannot possibly be a set with positive area, since that would violate (7.4.3).

7.5 Reverse Shapes for Additive Random Crystals

Our setting for stochastic growth on $\mathbb{Z}^d$ begins with a finite set $\mathcal{N} \subset \mathbb{Z}^d$ (the *neighborhood* of the origin) with $\mathbf{0} \in \mathcal{N}$, and a table of local transition probabilities $\pi : 2^{\mathcal{N}} \to [0, 1]$. We assume that π is a *monotone solidification* map, i.e., $A \subset B \subset \mathcal{N}$ implies $\pi(A) \le \pi(B)$, $\pi(\emptyset) = 0$, and $\pi(\{\mathbf{0}\}) = 1$. Then a random crystal growth model $\xi_n \in \{0, 1\}^{\mathbb{Z}^d}$ is constructed as follows. Identifying ξ_n with $\{\xi_n = 1\}$, at every update, each site x looks at the configuration of its neighborhood $x + \mathcal{N}$, and decides independently, with probability $\pi((\xi_n - x) \cap \mathcal{N})$, whether to become occupied. Given any $\Lambda \subset \mathbb{R}^d$, let ξ_n^Λ denote the dynamics started from $\Lambda \cap \mathbb{Z}^d$.

Although propagation of stochastic interfaces is a notoriously difficult subject, at least the formulation of half-space speeds is straightforward. For $u \in S^{d-1}$, let $\xi_0 = H_u^- \cap \mathbb{Z}^d$, and introduce

$$V_n(u) = \left\{\lambda \in \mathbb{R} : \lambda u \in \xi_n + B_\infty\left(\mathbf{0}, \frac{1}{2}\right)\right\}.$$

Call $w(u)$ the *half-space velocity in direction u* if

$$w(u) = \lim_{n\to\infty} n^{-1} \max V_n(u) = \lim_{n\to\infty} n^{-1} \min(\mathbb{R}\backslash V_n(u))$$

exists with probability one. In order to apply the deterministic methods of Section 7.2 to random crystals, one needs a large deviations inequality of the type established in [Kes2] for some first-passage percolation models to be mentioned below. We formalize the required estimate as follows.

Definition Say that ξ_n is *Kesten* if w exists and there is a strictly positive function $\Gamma : S^{d-1} \times (0, \infty) \to (0, \infty)$ such that for every $u \in S^{d-1}$ and every $\epsilon > 0$,

$$P(|\max V_n(u) - nw(u)| + |\min(\mathbb{R}\backslash V_n(u)) - nw(u)| > \epsilon n) \le e^{-\Gamma(u,\epsilon)n}. \tag{7.5.1}$$

In [BGG] we show that Kesten random crystals grow with an asymptotic shape $\mathcal{L}$ given by the polar formula in (7.2.1). The main result of this section establishes corresponding reverse shapes given by (7.2.3). In the statement of the next theorem, a.s. convergence refers to the basic coupling which constructs processes $\xi_n^{mA^c}$ for every m on a common probability space. Note that coupling and monotonicity imply continuity of $w : S^{d-1} \to [0, \infty)$ whenever it exists.

Theorem 7.5.1 *Assume that ξ_n is Kesten, and that $w > 0$. Given convex $A \subset \mathbb{R}^d$, there is a unique nonempty convex proper subset $\mathcal{R}(A)$ of $\mathbb{R}^d$ such that almost surely,*

$$\lim_{\epsilon\to 0} \lim_{m\to\infty} \frac{1}{\epsilon m} \xi_{T-\epsilon m}^{mA^c} = \mathcal{R}(A)^c. \tag{7.5.2}$$

PROOF. Introduce

$$V_n(u, x) = \left\{\lambda \in \mathbb{R} : \lambda u + x \in \xi_n + B_\infty\left(\mathbf{0}, \frac{1}{2}\right)\right\}.$$

Then by (7.5.1),

$$\begin{aligned} &P(|\max V_n(u, x) - nw(u)| + |\min(\mathbb{R}\backslash V_n(u, x)) - nw(u)| \le \epsilon n \\ &\quad \text{for every } x \text{ such that } \langle x, u\rangle = 0 \text{ and } \|x\|_2 \le n^2) \\ &\quad \le Cn^{2(d-1)} e^{-\Gamma(u,\epsilon)n}. \end{aligned}$$

Assume that A is a convex set with C^2 boundary having curvature bounded above and below. In other words, there is a positive constant C_1 such that

$$C_1^{-1}\|u_2 - u_1\| \le \|\nu_A(u_2) - \nu_A(u_1)\| \le C_1\|u_2 - u_1\|. \tag{7.5.3}$$

For a fixed $\eta > 0$, we will select $v_1, \ldots, v_k \in S^{d-1}$ depending on η and C_1, and set

$$\gamma = \frac{1}{2} \min_i \Gamma(v_i, \eta).$$

We claim that the v_i can be chosen so that, for large m, $\xi_{\eta m}^{mA^c}$ differs from

$$\bigcup \{((m\alpha_A(u) - \eta m w(\nu_A(u)))\nu_A(u) + H_{\nu_A(u)}) : u \in S^{d-1}\}$$

(the right side of (7.2.3) with $n = \eta m$) by at most $C_2\eta^2 m$, with probability at least $1 - e^{-\gamma\eta m}$. Here C_2 depends only on C_1 of (7.5.3).

In order to verify the claim, first choose the (unique) direction u_i such that $\nu_A(u_i) = v_i$. Then there is a constant C_3 (depending only on C_1) such that within distance ηm of $m\alpha_A(u_i)u_i$ the boundary of mA lies within a cylinder of height $C_3\eta^2 m$. Moreover, there is a C_4 (depending only on $\mathcal{N}$) such that only sites within distance ηm influence how far sites within distance $C_4^{-1}\eta m$ advance. Hence we simply choose the v_i so that $\alpha_A(u_{i+1})u_{i+1}$ is within the cylinder with height $C_3\eta^2$ and base consisting of points on the boundary of $\alpha_A(u_i)u_i + H_{v_i}^-$ within distance $C_4^{-1}\eta$.

To finish the proof, we now apply the claim repeatedly, $\frac{M-\epsilon}{\eta}$ times in all, using smooth approximations as in the proof of Theorem 7.2.1. □

Example 7.5.1 One class of Kesten random dynamics to which Theorem 7.5.1 applies comprises any additive growth with neighborhood $\mathcal{N}$ in which $\pi(A) = 0$ if $A = \emptyset$, $= 1$ if $0 \in A$, and $= p$ otherwise. Such systems permit a first-passage interpretation, and then the martingale difference method of [Kes2] yields (7.5.1). More generally, Kesten's analysis and the techniques of the present paper apply to the growth of any first-passage percolation model in which the distribution function F of the time it takes to cross a site (bond) satisfies

$$F(0) < p_c, \text{ the critical value for the site (bond) percolation, and}$$
$$\int e^{hx} dF(x) < \infty \text{ for some } h > 0.$$

Example 7.5.2 (see Figure 7.7) Another instance of Theorem 7.5.1 is "barely supercritical random threshold growth" [Gra2, KeSc]. In this case, $\mathcal{N}$ consists of $\mathbf{0}$ and its four closest neighbors, $\pi(A) = 1$ if $0 \in A$ or $|A| \geq 2$, $= p$ if A is a singleton other than $\mathbf{0}$, $= 0$ otherwise. Estimate (7.5.1) is obtained from a last passage representation (cf. [GK, Gra2]). As p decreases from 1 to 0, $K_{1/w}$ changes from $B_\infty(\mathbf{0}, 1)$ to the *cross* $\{\|x_1\| \leq 1 \text{ or } \|x_2\| \leq 1\}$. Hence the reverse shape of any convex, completely symmetric A other than $B_1(\mathbf{0}, 1)$ (which is always invariant) makes a transition from a square or octagon at $p = 1$ to a diamond at $p = 0$. In particular, the reverse shape of a circle changes from square to diamond, perhaps through some intermediate regime of octagons. Level sets of one simulated sample path of the reverse dynamics, starting from a lattice circle of "radius" 250, at times 0, 50, 100, ..., 250 are shown below for parameter values $p = .7$ (left side of Figure 7.7) and .3 (right side of Figure 7.7).

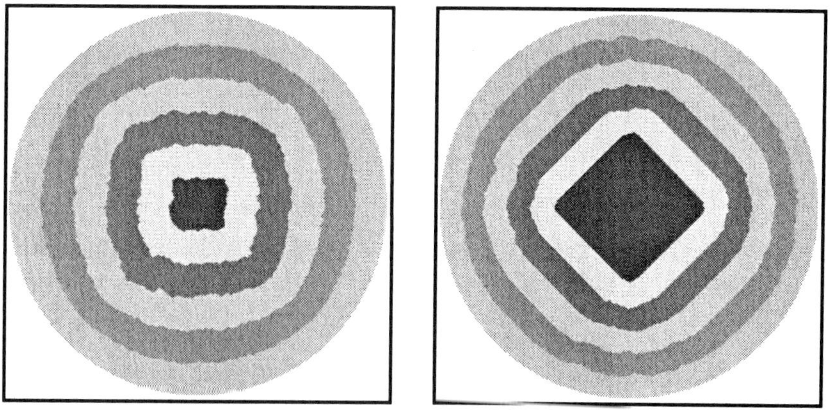

FIGURE 7.7.

Remark For p close to 1 or 0, this last example illustrates weak stability of CA crystals with small random "error." Results about the forward shape $\mathcal{L}$ under perturbations appear in [DL] (the additive case) and [BGG] (nonconvex $K_{1/w}$). A representative general result for reverse shapes is the following. Given a CGT $\mathcal{T}$, let $\mathcal{T}_p$ denote the random dynamics which adjoin sites with probability p whenever they are added deterministically under $\mathcal{T}$, and let $\mathcal{R}_p(\cdot)$ be the reverse shapes for $\mathcal{T}_p$. If A is any completely symmetric, convex set which is weakly stable for $\mathcal{T}$, then as p tends to 1, $\mathcal{R}_p(A)$ can be made arbitrarily close to $\mathcal{R}(A)$.

We conclude by analyzing the last holes problem for additive random crystals with symmetric neighborhood $\mathcal{N}$ and update probability p, started from ν_δ. (Essentially the same analysis applies to the first-passage models of Example 7.5.1.) For large n, the most likely scenario on $\{\mathcal{T} > n\}$ is that $\mathbf{0}$ refuses to be affected by its surroundings, leading to a trivial last hole.

Theorem 7.5.2 *For additive, symmetric random growth, $\mathcal{H}_\delta = \{\mathbf{0}\}$.*

PROOF. Let A_n be the random crystal started from $\{\mathbf{0}\}$, in the standard additive graphical representation, so that (7.4.2) holds. By the method used to prove Theorem 7.4.1, the last hole must be trivial if the conditional distribution of A_n concentrates on balls of radius $o(n)$ around $\{\mathbf{0}\}$, i.e.,

$$P_\delta(\mathrm{diam}(A_n) \geq \eta n \mid T > n) \to 0 \quad \text{as } n \to \infty \tag{7.5.4}$$

for any $\eta > 0$. Indeed, this implies that T is less than $(1+\eta)n$ with high probability, in which case $\mathcal{C}_{t-\epsilon n}$ has diameter at most ηn.

The first step in checking (7.5.4) is to show, for a suitable $M < \infty$,

$$P_\delta(|A_n| \geq Mn \mid T > n) \to 0 \quad \text{as } n \to \infty. \tag{7.5.5}$$

By definition, this last probability equals

$$\frac{E_\delta[(1-\delta)^{|A_n|} 1_{\{|A_n| \geq Mn\}}]}{P_\delta(T > n)}.$$

The denominator here is at least $(1-\delta)(1-p)^{|\mathcal{N}\setminus\{\mathbf{0}\}|n}$, corresponding to the event that the origin is initially empty and $A_n = \{\mathbf{0}\}$. Of course the numerator is at most $(1-\delta)^{Mn}$, so (7.5.5) follows by choosing M large enough that $(1-\delta)^M \le (1-p)^{|\mathcal{N}\setminus\{\mathbf{0}\}|}$.

The second step asserts that given any $\eta > 0$, there is a $c > 0$ such that

$$P_\delta(\text{diam}(A_n) \ge \eta n,\ |A_n| \le Mn) \le e^{-cn^2} \tag{7.5.6}$$

for n large. As noted in the previous paragraph, $\{T > n\}$ has at least exponentially small probability in n, so conditioning on that event does not appreciably alter (7.5.6). In combination with (7.5.5), this yields (7.5.4). Finally, to check (7.5.6), note that the total number of deterministic sequences $A_0 = \{\mathbf{0}\}, A_1, \ldots, A_n$ which lead to a possible configuration of size $\le Mn$ is at most $(M'n)^n$. (A_n is $\mathcal{N}$-connected, and every site in A_n has at most n choices of when to first become occupied.) If $\text{diam}(A_n) \ge \eta n$, then linearly many (in n) of $A_1, \ldots, A_n$ must have linear diameter, and hence linear boundary. Thus the probability of a prescribed trajectory entering into (7.5.6) is at most $(\max\{p, 1-p\})^{c'n^2}$. Summing over trajectories finishes the proof. We note that the same estimate precludes $\text{diam}(A_n) \ge n^{\frac{1}{2}+\eta}$ on $\{T > n\}$. □

In spite of the previous theorem, a "last nontrivial hole" with shape $\mathcal{L}$ arises by restricting the window of observation times n to a shorter horizon as the seeding density δ decreases.

Theorem 7.5.3 *Under the same hypotheses, if* $n = n_\delta \in [\delta^{-\frac{1}{d}-\alpha}, \delta^{-\frac{1}{d-1}+\alpha}]$ *for some* $\alpha > 0$, *then*

$$\lim_{\epsilon\to 0} \limsup_{\delta\to 0} P_\delta\left(d_H\left(\frac{C_{T-\epsilon n}}{\epsilon n}, \mathcal{L}\right) > \eta \mid T > n\right) = 0. \tag{7.5.7}$$

PROOF. For a suitable constant $C = C_p$, with δ sufficiently small and n in the prescribed range,

$$P_\delta(T > n) \ge \exp\{-Cn^d\delta\} \gg e^{-C'n}$$

(any C'). Hence, the large deviation bounds established in [Kes2] for the forward limit shape $\mathcal{L}$ give

$$P_\delta((1-\eta)n\mathcal{L} \subset A_n \subset (1+\eta)n\mathcal{L} \mid T > n) \to 0 \quad \text{as } \delta \to 0.$$

Also for n in the prescribed range, with overwhelming probability there are initially lots of occupied sites everywhere around the boundary of $n\mathcal{L}$, as in the argument for Theorem 7.4.1, and the rest of the proof is essentially the same as for that case of additive deterministic dynamics. We remark that (7.5.7) also holds with the conditioning event changed to $\{T = n\}$, perhaps a more natural formulation. □

Acknowledgment: We extend our profound gratitude to Harry Kesten for all he has taught us about percolation, random growth, and stochastic processes over the years, and for providing us with a preeminent role model as master of mathematical probability.

References

[AL] M. Aizenman and J. Lebowitz. Metastability effects in bootstrap percolation. *J. Phys. A: Math. Gen.*, **21** (1988), 3801–3813.

[AMS] E. Andjel, T. Mountford, and R. Schonmann. Equivalence of exponential decay rates for bootstrap percolation like cellular automata. *Ann. Inst. H. Poincaré*, **31** (1995), 13–25.

[Big] J. Biggins. The asymptotic shape of the branching random walk. *Adv. Appl. Prob.*, **10** (1978), no. 1, 62–84.

[Boh] T. Bohman. Discrete Threshold Growth dynamics are omnivorous for box neighborhoods. *Trans. Amer. Math. Soc.*, **351** (1999), 947–983.

[BoG] T. Bohman and J. Gravner. Random threshold growth models. *Random Structures and Algorithms*, 1999, to appear.

[BGG] T. Bohman, J. Gravner, and D. Griffeath. Asymptotic shapes for random threshold growth models, 1998, in preparation.

[BrG1] M. Bramson and D. Griffeath. On the Williams-Bjerknes Tumour Growth Model I. *Ann. Probability*, **9** (1981), 173–185.

[BrG2] M. Bramson and D. Griffeath. On the Wiliams-Bjerknes Tumour Growth Model II. *Math. Proc. Cambridge Philos. Soc.*, **88** (1980), 339–357.

[BH] S. Broadbent and J. Hammersley. Percolation processes. I. Crystals and mazes. *Proc. Cambridge Philos. Soc.*, **53** (1957), 629–641.

[CD] J.T. Cox and R. Durrett. Some limit theorems for percolation processes with necessary and sufficient conditions. *Ann. Probability*, **9** (1981), 583–603.

[Dur] R. Durrett. *Lecture Notes on Particle Systems and Percolation*. Wadsworth & Brooks/Cole, Belmont, CA, 1988.

[DG] R. Durrett and D. Griffeath. Contact processes in several dimensions. *Z. Wahrscheinlichkeitstheorie verw. Gebiete*, **59** (1982), 535–552.

[DKS] R. Dobrushin, R. Kotecky, and S. Shlosman. *Wulff Construction, A Global Shape from Local Interaction*. American Mathematical Society, Providence, RI, 1992.

[DL] R. Durrett and T. Liggett. The shape of the limit set in Richardson's growth model. *Ann. Probability*, **9** (1981), 186–193.

[Ede] M. Eden. Page 359 of *Symp. on Information Theory in Biology*, H. Yockey, ed. Pergamon Press, New York, 1958.

[FGG1] R. Fisch, J. Gravner, and D. Griffeath. Threshold-range scaling of excitable cellular automata. *Statistics and Computing*, **1** (1991), 23–39.

[FGG2] R. Fisch, J. Gravner, and D. Griffeath. Metastability in the Greenberg-Hastings model. *Ann. Appl. Prob.*, **3** (1993), 935–967.

[GK] A. Gandolfi and H. Kesten. Greedy lattice animals. II. Linear growth. *Ann. Appl. Prob.*, **4** (1994), 76–107.

[Gra1] J. Gravner. The boundary of iterates in Euclidean growth models. *T.A.M.S.*, **348** (1996), 4549–4559.

[Gra2] J. Gravner. Recurrent ring dynamics in two-dimensional cellular automata. *J. Appl. Prob.*, 1999, to appear.

[GG1] J. Gravner and D. Griffeath. Threshold Growth dynamics. *T.A.M.S.*, **340** (1993), 837–870.

[GG2] J. Gravner and D. Griffeath. First-passage times for discrete Threshold Growth dynamics. *Ann. Probability*, **24** (1996), 1752–1778.

[GG3] J. Gravner and D. Griffeath. Multitype Threshold Growth: convergence to Poisson-Voronoi tessellations. *Ann. Appl. Prob.*, **7** (1997), 615–647.

[GG4] J. Gravner and D. Griffeath. Cellular Automaton Grown on $\mathbb{Z}^2$: theorems, examples and problems. *Adv. Appl. Math.*, **21** (1998), 241–304.

[Gri] D. Griffeath. Primordial Soup Kitchen. `http://psoup.math.wisc.edu/kitchen.html`.

[HW] J. Hammersley and D. Welsh. First-passage percolation, subadditive processes, stochastic networks, and generalized renewal theory. In *Proc. Internat. Res. Semin.* (Statist. Lab., Univ. California, Berkeley). Springer-Verlag, New York, 1965, 61–110.

[Kes1] H. Kesten. First-passage percolation and a higher-dimensional generalization. In *Particle Systems, Random Media and Large Deviations*, R. Durrett, ed. American Mathematical Society, Providence, RI, 1984.

[Kes2] H. Kesten. On the speed of convergence in first-passage percolation. *Ann. Appl. Prob.*, **4** (1994), 76–107.

[KeSc] H. Kesten and R. Schonmann. On some growth models with a small parameter. *Probab. Th. Rel. Fields*, **101** (1995), 435–468.

[KrSp] J. Krug and H. Spohn. Kinetic roughening of growing surfaces. In *Solids Far from Equilibrium*, C. Godreche, ed. Cambridge University Press, Cambridge, UK, 1992, 479–582.

[Lig] T. Liggett. *Interacting Particle Systems*. Springer-Verlag, New York, 1985.

[NP] C. Newman and M. Piza. Divergence of shape fluctuations in two dimensions. *Ann. Probability*, **23** (1995), 977–1005.

[Ric] D. Richardson. Random growth in a tesselation. *Proc. Camb. Phil. Soc.*, **74** (1973), 515–528.

[Sep] T. Seppäläinen. Hydrodynamic scaling, convex duality and asymptotic shapes of growth models. *Markov Process. Rel. Fields*, **4** (1998), 1–26.

[TCH] J. Taylor, J. Cahn, and C. Handwerker. Geometric models of crystal growth (Overview no. 98-1). *Acta Met.*, **40** (1992), 1443–1474.

[Wil] S. Willson. On convergence of configurations. *Discrete Math.*, **23** (1978), no. 3, 279–300.

Janko Gravner
Mathematics Department
University of California
Davis, CA 95616
`gravner@feller.ucdavis.edu`

David Griffeath
Mathematics Department
University of Wisconsin
Madison, WI 53706
`griffeat@math.wisc.edu`

8

Double Behavior of Critical First-Passage Percolation

Yu Zhang

ABSTRACT We consider standard first passage percolation on the Z^d lattice. Let $\{x(e) : e \text{ an edge of } Z^d\}$ be an i.i.d. family of random variables with distribution F. Denote by $c_{0,n}$ the first passage time from the origin to the boundary of $[-n, n]^d$. For $d = 2$ we show that there exist two curves F_a and G_b both with $F_a(0) = G_b(0) = p_c$ such that $\lim_{n\to\infty} Ec_{0,n}$ exists whenever $F(0) = p_c$ and $F \geq F_a$ or blows up whenever $F(0) = p_c$ and $F \leq G_b$, respectively. We also can obtain the corresponding results for the passage times $a_{0,n}$ and $b_{0,n}$. Furthermore, we will investigate the behavior of $\lim_{n\to\infty} E(c_{0,n})$ when $F(0)$ is near p_c. For a large d, we show the lower bound for $Ea_{0,n}$ is larger than $C \log\log n$ for some constant $C > 0$, and discuss the existence of routes for $a_{0,n}$ and $b_{0,n}$ when $F(0) = p_c$.

Keywords: First passage percolation, criticality.

AMS Subject Classifications: 60K, 35.

8.1 Introduction and Statement of Results

We consider Z^d as a graph with an edge connecting each pair of vertices $x, y \in Z^d$ with $\|x - y\| = 1$, where $\|u\| = \max_{1\leq i\leq d}\{|u_i|\}$ for $u = (u_1, \ldots, u_d)$. Denote by E^d the edge set of Z^d. Let $\Omega = [0, 1]^{E^d}$ and P be a probability measure on Ω. Expectation with respect to P is denoted by E. Let $\{U(e)\}_{e\in E^d}$ be a sequence of independent random variables with a common uniform distribution on $[0, 1]$. For a distribution F define its inverse in the usual way:

$$F^{-1}(t) = \inf\{x : F(x) \geq t\} \qquad 0 \leq t \leq 1.$$

By a well-known transformation, $\{F^{-1}(U(e))\}$ is an i.i.d. sequence of random variables with a common distribution F. For simplicity, for a given distribution F we use $\{X_F(e)\}$ to represent $\{F^{-1}(U(e))\}$. We also write $X_F(e, w)$ for the value of $X_F(e)$ at w for $w \in \Omega$. For any two distributions F_1 and F_2, if $F_1 \leq F_2$, then

$$X_{F_1}(e) \geq X_{F_2}(e) \text{ for any } e \in E^d.$$

Furthermore, we assume that for any $m \geq 1$

$$F(0^-) = 0 \text{ and } EX_F(e)^m < \infty. \tag{8.1.1}$$

For any vertices u and v, a path γ from u to v is an alternating sequence $(v_0, e_1, v_1, \ldots, e_n, v_n)$ for some integer $n \geq 1$, where v_i $(0 \leq i \leq n)$ is a vertex in Z^d, and e_i is an edge in E^d which connects v_{i-1} to v_i, and $v_0 = u$ and $v_n = v$. If $v = u$, the path is called the circuit. Define the passage time of γ to be

$$t(\gamma) = \sum_{i=1}^{n} X_F(e_i).$$

For any two sets A and B of vertices of Z^d we define the first passage time from A to B as

$$T(A, B) = \inf\{t(\gamma) : \gamma \text{ a path from some vertex in } A \text{ to some vertex in } B\}.$$

Let

$$\begin{aligned} a_{m,n} &= T((m, 0, \ldots, 0), (n, 0, \ldots, 0)), \\ b_{m,n} &= \inf\{T((m, 0, \ldots, 0), (n, k_2, \ldots, k_d)) : k_2, \ldots, k_d \in Z\} \\ c_{x,n} &= T(x, x + \partial B(n)), \end{aligned}$$

where $B(n) = \{x \in Z^d : \|x\| \leq n\}$, and $\partial B(n) = \{x \in Z^d : \|x\| = n\}$, and $x + \partial B(n) = \{x + y : y \in \partial B(n)\}$. It is well-known (see [7, Theorem 2.18]) that

$$\lim_{n\to\infty} \frac{1}{n} a_{0,n} = \lim_{n\to\infty} \frac{1}{n} b_{0,n} = \mu \text{ a.s. and in } L^1. \tag{8.1.2}$$

On the other hand, the same proof of [7, Theorem 2.18] can be adapted to show

$$\lim_{n\to\infty} \frac{1}{n} c_{0,n} = \mu \text{ a.s. and in } L^1, \tag{8.1.3}$$

where $c_{0,n}$ is defined to be $c_{x,n}$ for $x = (0, \ldots, 0)$. Here (see [7, Theorem 6.1]) $\mu = 0$ if and only if $F(0) \geq p_c(d)$, where $p_c(d)$ is the critical probability for Bernoulli (bond) percolation on Z^d. On the other hand, by the monotonicity of $\{c_{0,n}\}$

$$\lim_{n\to\infty} c_{0n} = \rho(F) \text{ a.s.} \tag{8.1.4}$$

for some random variable $\rho(F)$. When $F(0) > p_c$, [9] proved that for any integer $m > 0$

$$E\rho^m(F) \text{ is finite}$$

and

$$\lim_{n\to\infty} Ea_{0,n} = 2E\rho(F) \text{ and } \lim_{n\to\infty} Eb_{0n} = E\rho(F). \tag{8.1.5}$$

In fact, for any distribution F it is easy to show that

$$\rho(F) < \infty \text{ with probability one or zero.} \tag{8.1.6}$$

Here we say that the system has subcritical behavior if $\rho(F) = \infty$ and that it has supercritical behavior if $\rho(F) < \infty$. It follows from (8.1.3) and (8.1.5) that

$$E\rho(F) \begin{cases} = \infty & \text{if } F(0) < p_c \\ < \infty & \text{if } F(0) > p_c. \end{cases} \tag{8.1.7}$$

One immediate question raised by (8.1.7) is the behavior of $\rho(F)$ when $F(0) = p_c$. It was shown by Chayes, Chayes and Durrett (see [3]) that there exist two positive constants C_1 and C_2 such that

$$C_2 \log n \le Ec_{0,n} \le C_1 \log n \tag{8.1.8}$$

for the Z^2 lattice when F is a Bernoulli distribution with $F(0) = p_c$. In this paper there are many such constants like C_1 and C_2 above. Throughout this paper, C or C_i stands for a strictly positive constant which may depend on p but not n, whose value is of no significance to us. In fact, the value of C or C_i may change from appearance to appearance. It follows from (8.1.8) and the monotone convergence theorem that

$$E(\rho(F)) = \infty \tag{8.1.9}$$

if F is a Bernoulli distribution with $F(0) = p_c$. It might be believed that $E\rho(F) = \infty$ for all F with $F(0) = p_c$ by the evidence of (8.1.8). However, we find that $E\rho(F) = \infty$ is not always true for some $F(x)$ with $F(0) = p_c$. Let us first focus on the two dimensional lattice. For each $0 < a < \infty$ we consider

$$F_a(x) = \begin{cases} 1 & \text{if } (1 - p_c)^{1/a} \le x \\ x^a + p_c & \text{if } 0 \le x \le (1 - p_c)^{1/a} \\ 0 & \text{if } x < 0 \end{cases}$$

as a special distribution. With the definition of F_a we can prove the following theorem.

Theorem 8.1.1 *Assume that $d = 2$. For any integer $m > 0$ there exists $0 < a_m < \infty$ such that $E\rho^m(F_a) < \infty$ if $a < a_m$.*

Corollary 8.1.1 *It follows from Theorem 8.1.1 that $\rho(F_a) < \infty$ a.s. if $a < a_m$.*

Remark Clearly, $E\rho(F) < \infty$ will imply that $\rho(F) < \infty$ a.s., but does $\rho(F) < \infty$ imply that $E\rho(F) < \infty$? We are unable to answer this question.

It follows from (8.1.9) that $E\rho(F) = \infty$ if $F(x)$ is a Bernoulli distribution with $F(0) = p_c$. Then it is natural to guess that there is a curve such that $E\rho(F) = \infty$ if F is below the curve. Here we give a poor, but non-trivial, bound for such a curve as follows. For $0 < b < \infty$ let

$$G_b(x) = \begin{cases} 1 & \text{if } \exp(-1/x^b) > 1 - p_c \\ \exp(-1/x^b) + p_c & \text{if } 0 \le \exp(-1/x^b) \le (1 - p_c) \\ 0 & \text{if } x < 0. \end{cases} \tag{8.1.10}$$

Clearly, $G_b(0) = p_c$. In fact, $G_b(p)$ converges to p_c too fast to have a positive derivative of any order at 0. With G_b we have the following theorem.

Theorem 8.1.2 *Assume that $d = 2$ and $X(e)$ has a distribution G_b. Then $Ec_{0,n} \geq C \log^{1-1/b} n$ when $b > 1$.*

It follows from Theorem 8.1.2 that one might guess that $\rho(G_b) = \infty$ a.s. Here we would like to answer the question.

Theorem 8.1.3 *Assume that $d = 2$ and $X(e)$ has a distribution G_b with $b > 1$. Then $\rho(G_b) = \infty$ a.s.*

It follows from the proofs of Theorems 8.1.1 and 8.1.2 that the finiteness of $\rho(F)$ depends only on the behavior of the value of $F(x)$ when x is near zero. More precisely, for any two distributions F and G we say that

$$F \preceq G \text{ if there exists } \epsilon > 0 \text{ such that } F(x) \leq G(x) \text{ for } x \leq \epsilon.$$

With this definition we have the following theorem.

Theorem 8.1.4 *Assume that $d = 2$. If $\rho(F) = \infty$ a.s. and $G \preceq F$, then $\rho(G) = \infty$ a.s. On the other hand, if $\rho(F) < \infty$ a.s. and $F \preceq G$, then $\rho(G) < \infty$ a.s.*

By the "monotonicity" in Theorem 8.1.4 one may consider the family $\{a > 0 : \rho(F_a) < \infty\}$. Clearly, from Theorem 8.1.1, the above set is not empty. Let

$$\beta = \sup\{a : \rho(F_a) < \infty\}.$$

One immediate question is whether $\beta < \infty$ or $\beta = \infty$. In fact, we believe the following conjecture.

Conjecture $\beta < \infty$.

If we focus on the processes $a_{0,n}$ and $b_{0,n}$, we can show the following theorem.

Theorem 8.1.5 *Assume that $d = 2$. If $E\rho(F) = \infty$, then*

$$\lim_{n\to\infty} Eb_{0,n} = \infty \text{ and } \liminf_{n\to\infty} Ea_{0,n} = \infty.$$

On the other hand, if $E\rho(F) < \infty$, then

$$\lim_{n\to\infty} Eb_{0,n} = E\rho(F) \text{ and } \lim_{n\to\infty} Ea_{0,n} = 2E\rho(F).$$

We also investigate the behavior of $E\rho(F)$ when $F(0)$ is near p_c. For $d = 2$ the behavior of μ has been studied in [3] when $F(0) \uparrow p_c$. Here we also consider the behavior of our $\rho(F)$ when $F(0) \downarrow p_c$. Denote the connectivity probability by

$$\tau_{0,n}(F(0)) = P((n, 0, \ldots, 0) \in C(\mathbf{0}), |C(\mathbf{0})| < \infty),$$

where the cluster of the origin, $C(\mathbf{0})$, consists of all vertices which are connected by nearest neighbor paths such that for each e in the paths, $X(e) = 0$ and $|A|$ is the cardinality of A. Then the correlation length is defined to be

$$\frac{1}{\xi^f(F(0))} = -\lim_{n\to\infty} n^{-1} \log \tau_{0,n}(F(0)).$$

It is known (see [5]) that $\xi^f(F(0)) < \infty$ if $F(0) \neq p_c$. It is also proved (see [3]) that there exist two positive constants C_1 and C_2 which do not depend on F such that

$$C_1 \xi^f(F(0)) \leq \mu(F) \leq C_2 \xi^f(F(0)) \tag{8.1.11}$$

if F is a Bernoulli distribution with $F(0) < p_c$ and $d = 2$. On the other hand, we can show the following theorem.

Theorem 8.1.6 *Suppose that $d = 2$. For a Bernoulli distribution F with $F(0) > p_c$, then there exist two positive constants C_1 and C_2 which do not depend on F such that*

$$C_1 \log \xi^f(F(0)) \leq E\rho(F) \leq C_2 \log \xi^f(F(0)).$$

Remark For $d = 2$, it has been proved (see [5]) that there exist $\delta > 0$, C_1 and C_2 such that

$$C_1 (p - p_c)^{-1} \leq \xi^f(p) \leq C_2 (p - p_c)^{-\delta}$$

for $p > p_c$. By (8.1.11) and Theorem 8.1.6 there exist C_1 and C_2 which do not depend on F such that

$$-C_1 \log(F(0) - p_c) \leq E\rho(F) \leq -C_2 \log(F(0) - p_c), \tag{8.1.12}$$

if F is a Bernoulli distribution with $F(0) > p_c$.

Unfortunately, we have to restrict ourselves to $d = 2$ in order to obtain the results above, even though we believe that these results also hold for $d > 2$. In fact, for $d > 2$, there is only very limited information when $F(0) = p_c$. Clearly, one of the most important problems is to find the asymptotic behavior of $Ec_{0,n}$ when $F(0) = p_c$. [2] shows that $Ec_{0,n} = o(n^\epsilon)$ for every fixed $\epsilon > 0$ when F is a Bernoulli distribution with $F(0) = p_c$, and $d > 2$. For the Bethe tree, which is usually regarded as $d = \infty$, [1] shows that $Ec_{0,n} = O(\log \log n)$. By the last result, it might be believed that $Ec_{0,n} = O(\log \log n)$ if $d > d_c = 6$, where $d_c = 6$ is the critical dimension of the bond percolation. Here we will give the lower bound for $Ec_{0,n}$. Before giving the lower bound, we introduce some basic percolation results. Indeed, it is believed widely that for a Bernoulli distribution F with $F(0) = p_c$

$$\begin{aligned} P(\exists \text{ a path } r \text{ from the origin to } \partial B(n) \text{ such that} \\ X(e) = 0 \ \forall e \in r) = O(n^{-c}), \end{aligned} \tag{8.1.13}$$

where c is a constant. (8.1.13) is called the power law hypothesis. Unfortunately, as far as we know, (8.1.13) is not proved for a general d. However, [6] proved that (8.1.13) holds with $\rho = 2$ for some d larger than 92. Intuitively, with assumption (8.1.13) there are many "bubbles" with $X(e) = 1$ surrounding the origin. Any path from the origin to $\partial B(n)$ for a large n has to pass these bubbles. To pass through each such bubble, it will cost passage time 1. Clearly, $c_{0,n}$ has to be very large for a large n if there indeed exist many such bubbles. More precisely, we will establish a more precise lower bound for $Ec_{0,n}$ in the following theorem.

Theorem 8.1.7 *Assume that $d > 2$, F is a Bernoulli distribution with $F(0) = p_c$, and* (8.1.13) *holds. Then there exists a constant $C > 0$ such that*

$$Ec_{0,n} \geq C \log \log n \tag{8.1.14}$$

and

$$\rho(F) = \infty \ a.s. \tag{8.1.15}$$

Clearly, this implies that

$$Ea_{0,n} \geq Eb_{0,n} \geq Ec_{0,n} \geq C \log \log n.$$

Remark If we only assume that

$$\lim n \to \infty P(\exists \text{ a path } r \text{ from the origin to } \partial B(n) \text{ such that}$$
$$X(e) = 0 \, \forall e \in r) = 0, \tag{8.1.16}$$

for a Bernoulli distribution F with $F(0) = p_c$, then it is easy to show that $\rho(F) = \infty$ a.s. by using the same method as for (8.2.12) below.

As an application of Theorem 8.1.7 let us consider the existence of "route problem." Let r be a path from A to B. If $t(r) = T(A, B)$, we say r is a route of $T(A, B)$. An essential question in first passage percolation is to discuss the existence of routes for both $a_{0,n}$ and $b_{0,n}$. When $F(0) < p_c$, [7] proved the existence of such routes for both $a_{0,n}$ and $b_{0,n}$. When $F(0) > p_c$, [9] also proved the existence of such routes. Now we discuss the existence of route at criticality.

Theorem 8.1.8 *Assume that $\rho(F) = \infty$ a.s. There exist routes for both $a_{0,n}$ and $b_{0,n}$ almost surely.*

Remark Clearly, by Theorems 8.1.7 and 8.1.8, there exist routes for both $a_{0,n}$ and $b_{0,n}$ if F is a Bernoulli distribution with $F(0) = p_c$ and (8.1.13) holds. We are unable to prove the existence of routes for $a_{0,n}$ and $b_{0,n}$ if $F(0) = p_c$ and $\rho(F) < \infty$. Note that the existence of routes for $a_{0,n}$ and $b_{0,n}$ has been proved for any F when $d = 2$ (see [8]).

8.2 Proofs

In this section we will prove all theorems mentioned in section 8.1. We first need some notation. For any two sets of vertices A and B with $A \cap B = \emptyset$ let $A \to B$ be the event that there exists a path with $X(e) = 0$ from a vertex of A to a vertex of B.

Now we begin to show Theorem 8.1.1. Let f be a function mapping E^2 into R^1 (real numbers). For any $e \in E^2$, let e be f-open or f-closed if $X_F(e) \leq f(e)$ or $X_F(e) > f(e)$. An f-open path here is a nearest neighbor path on Z^2, whose bonds are f-open. The f-open cluster of the vertex x consists of all vertices which are connected to x by an f-open path on Z^d. We write C_f for one of the infinite f-open clusters if it exists. For each $e \in E^2$, let $\langle e \rangle = \max\{\|x\|, \|y\|\}$, where

x and y are the two vertices of e. With these definitions, the following lemma is proved by [4] (see [4, Theorem 2 and (19)]).

Lemma 8.2.1 *Let $d = 2$ and F_a be the distribution defined in section 8.1. If $f(e) = \frac{1}{<e>}$, then there is $0 < \alpha < \infty$ such that with probability one there exists the unique $\mathcal{C}_f$ if $a \leq \alpha$. Furthermore, under the same assumption there exist C_1, C_2 and $\delta = \delta(\alpha) > 0$ such that*

$$P(\mathcal{C}_f \cap B(i) \neq \emptyset) \geq 1 - C_1 \exp(-C_2 i^\delta).$$

Clearly, for F_α and $f(e) = \frac{1}{\langle e \rangle}$, e is f-open with probability

$$P(X_{F_\alpha}(e) \leq f(e)) = p_c + (\frac{1}{\langle e \rangle})^\alpha.$$

PROOF OF THEOREM 8.1.1. If

$$F_a(y) = p_c + (\frac{1}{n})^\alpha$$

for some large n, then

$$y = (\frac{1}{n})^{\alpha/a}.$$

We take $a < 1$ small such that $\frac{\alpha}{a} > 2$. Now we consider F_a and $g(e) = (\frac{1}{\langle e \rangle})^{\alpha/a}$. Note that e is g-open with the following probability:

$$P(X_{F_a}(e) \leq g(e)) = P(X_{F_a}(e) \leq (\frac{1}{\langle e \rangle})^{\alpha/a}) = p_c + (\frac{1}{\langle e \rangle})^\alpha. \tag{8.2.1}$$

By Lemma 8.2.1 and (8.2.1), $\mathcal{C}_g$ exists and is unique almost surely. Let

$$E_i = \{\mathcal{C}_g \cap B(i) \neq \emptyset\} \text{ and } U = \min\{i : E_i \text{ occurs}\}.$$

We also write I_l for the indicator of the event that there does not exist a g-open path from $B(l)$ to ∞. Clearly, I_l is decreasing (see the definitions of decreasing and increasing functions in [4]) and

$$\{U = l + 1\} \subset \{I_l = 1\}.$$

By Lemma 8.2.1 and (8.2.1),

$$P(I_l = 1) \leq C_1 \exp(-C_2 l^\delta). \tag{8.2.2}$$

Clearly, for each integer $m \geq 1$ and j, the m-th moment of the sum of $X_{F_a}(e)$ for all $e \in B(j)$ is less than $(2(2j+1)^2)^m E X^m_{F_a}(e)$ and $(\sum_{e \in B(m)} X_{F_a}(e))^m$ is increasing. On the other hand, note that the number of edges which have a vertex with $\|x\| = k$ is less than $16k$ so that in the event $\{U = k\}$

$$t(r) \leq 4(\frac{1}{k-1})^{\frac{\alpha}{a}-1} + 4(\frac{1}{k})^{\frac{\alpha}{a}-1} + 4(\frac{1}{k+1})^{\frac{\alpha}{a}-1} + \cdots \tag{8.2.3}$$

for any path $r \subset \mathcal{C}_g$ from a vertex of $\partial B(k)$ to ∞. Since $\frac{\alpha}{a} > 2$, $t(r) < \infty$. Let

$$M = 4\sum_{k=1}^{\infty}\left(\frac{1}{k}\right)^{\frac{\alpha}{a}-1}.$$

Finally, from (8.2.3) for each $m \geq 1$,

$$\begin{aligned} E\rho^m(F_a) &= \lim_{n\to\infty} Ec_{0,n}^m = \lim_{n\to\infty}\sum_{i=1}^{\infty} E(c_{0,n}^m; U = i) \\ &\leq \sum_{i=1}^{\infty} E((M + \sum_{e\in B(i)} X_{F_a}(e))^m; U = i). \end{aligned} \tag{8.2.4}$$

For each $j \leq m$,

$$\begin{aligned} &\sum_{i=1}^{\infty} E((\sum_{e\in B(i)} X_{F_a}(e))^j; U = i) \\ &\leq \sum_{i=1}^{\infty} E(\sum_{e\in B(i)} X_{F_a}(e))^j I_{i-1}) \\ &\leq \sum_{i=1}^{\infty} E(\sum_{e\in B(i)} X_{F_a}(e))^j (E(I_{i-1})) \\ &\quad \text{(by the FKG inequality (see [5]))} \\ &\leq \sum_{i=1}^{\infty} (2(2i+1)^2)^j EX_{F_a}^j(e) C_1 \exp(-C_2(i-1)^\delta). \end{aligned}$$

Then Theorem 8.1.1 follows from (8.2.4) and the binomial theorem. □

To show Theorem 8.1.2 we need to introduce the following basic knowledge of percolation. For each $e \in E^2$, set e to be red or black with probability p or $1-p$. We define a left-right (respectively top-bottom) red crossing of a rectangle B to be a red path in B which joins some vertex on the left (respectively upper) side of B to some vertex on the right (respectively lower) side of B but which uses no bonds joining two vertices in the boundary of B. Let

$$\sigma(n, p) = P_p(\exists \text{ a left-right red crossing on } [-n, n]^2)$$

and let

$$L(p) = L(p, \delta) = \begin{cases} \min\{n : \sigma(n, p) \geq 1 - \delta\} \text{ if } p > p_c, \\ \min\{n : \sigma(p, n) \leq \delta\} \text{ if } p < p_c \end{cases}$$

for some strictly positive constant δ whose precise value is not important. $L(p)$ is also called the correlation length. It follows from [5] that

$$L(p_c + \eta, \delta) \approx L(p_c - \eta, \delta), \tag{8.2.5}$$

where $f(x) \approx g(x)$ means that there exist C_1 and C_2 such that for all x

$$C_1 g(x) \le f(x) \le C_2 g(x).$$

With this definition, it is also proved in [3] that there exists C_1 such that for $p > p_c$

$$L(p) \ge C_1(p - p_c)^{-1/2}. \tag{8.2.6}$$

Then by the FKG inequality and the RSW lemma we can construct a black circuit (see [4]) for more details) in an annulus with a positive probability. More precisely, for $p > p_c$ and $2^i \ge L(p) \ge 2^{i-1}$ there exists $C > 0$ such that

$$P(\exists \text{ a black circuit in } B(2^n) \setminus B(2^{n-1})) \ge C \tag{8.2.7}$$

for $n = 1, \ldots, i-1$. Clearly,

$$i \ge C_1 \log L(p) \ge -C_2 \log(p - p_c). \tag{8.2.8}$$

Finally, we introduce the duality of planar graphs. Define Z^* as the dual graph of Z^2 with vertices $\{v + (\frac{1}{2}, \frac{1}{2})\}$ and bonds joining all pairs of vertices which are unit distance apart. For any bond set $A \subset Z^2$, we write $A^* \subset Z^*$ for the corresponding bonds of the dual graph of A. For each bond $e^* \in Z^*$, we declare it is red or black if e is red or black. In other words, if e^* crosses a red (black) bond in Z^2, then e^* is red (black). With this definition, we can obtain (see [6] for detail) that if there exists a black dual circuit D^* surrounding some set A^* on $B(n)^* \subset Z^*$, then any path on Z^2 from A to $B(n)$ has to use at least a black edge of D. We now begin to show Theorem 8.1.2.

PROOF OF THEOREM 8.1.2. Consider distribution G_b with $b > 1$. For each m, if

$$P(X_{G_b}(e) \le y) = p_c + \frac{1}{m},$$

then

$$y = C(1/\log^{1/b} m)$$

for some constant C. Define e to be red or black if $\{X_{G_b}(e) \le y\}$ or $\{X_{G_b}(e) > y)\}$. Clearly,

$$P(e \text{ is red}) = P(X_{G_b}(e) \le y) = p_c + \frac{1}{m}$$

$$P(e \text{ is black}) = P(X_{G_b}(e) > y) = 1 - (p_c + \frac{1}{m}).$$

By (8.2.6) we have

$$L(p_c + \frac{1}{m}) \ge C_1 m^{\frac{1}{2}}.$$

Then it follows from (8.2.7) that

$$P(\exists \text{ a black circuit in } B(2^n) \setminus B(2^{n-1})) \ge C_2$$

for $n = 1, \ldots, i$, where i is the integer such that $2^i \le L(p_c + \frac{1}{m}) \le 2^{i+1}$. It follows from (8.2.8) that

$$i \ge C \log m.$$

In other words,

$$\begin{aligned} &P(\exists \text{ a circuit with } X_{G_b}(e) > y \text{ in } B(2^n)^* \setminus B(2^{n-1})^*) \\ &= P(\exists \text{ a circuit with } X_{G_b}(e) > \tfrac{1}{\log^{1/b} m} \text{ in } B(2^n)^* \setminus B(2^{n-1})^*) \\ &\ge C_2. \end{aligned}$$

Clearly, the expected number of such black dual circuits is at least $C \log m$ and any path from the origin to $\partial B(m)$ on Z^2 crossing such a black dual circuit will cost time at least $\frac{1}{\log^{1/b} m}$. Therefore,

$$Ec_{0,m} \ge C(\log m)(\frac{1}{\log^{1/b} m}) \ge C \log^{1-1/b} m.$$

Theorem 8.1.2 is proved. □

PROOF OF THEOREM 8.1.3. Consider G_b with $b > 1$. Let I_n be the indicator of the event that there exists a dual circuit with $X_{G_b}(e) \ge \frac{1}{\log^{1/b} n}$ in $B(C_1 n^{1/2}) \setminus B(C_1 \frac{n^{1/2}}{2})$, where C_1 is the constant defined in (8.2.6). Here we assume that both $C_1 n^{1/2}$ and $C_1 \frac{n^{1/2}}{2}$ are integers, otherwise we can always use $\lfloor C_1 n^{1/2} \rfloor$ and $\lfloor C_1 \frac{n^{1/2}}{2} \rfloor$ instead of $C_1 n^{1/2}$ and $C_1 \frac{n^{1/2}}{2}$. By (8.2.7) and the same argument in the proof of Theorem 8.1.2,

$$P(I_n = 1) \ge C.$$

Clearly, I_i and I_j are independent if $i > j$. By the law of large number theorem

$$\lim_{n \to \infty} \frac{I_2 + I_{2^2} + \cdots + I_{2^n}}{n} \ge C \text{ a.s.}$$

In other words, given $\epsilon > 0$, there exists N such that

$$P(I_2 + \cdots + I_{2^k} \ge Ck) \ge 1 - \epsilon \tag{8.2.9}$$

for any $k > N$. If $I_{2^k} = 1$, i.e., there exists a dual circuit in $B(C_1 2^k) \setminus B(C_1 2^{k-1})$, then for $2^k \le m^{1/2}$, any path on Z^2 from the origin to $\partial B(m)$ crossing the dual circuit will cost at least time $\frac{1}{\log^{1/b} 2^k}$. For $w \in \Omega$ and a large m,

$$c_{0,m} \ge \sum_{k=1}^{i} \frac{1}{\log^{1/b} 2^k} I_{2^k},$$

where i is an integer such that $2^{i+1} \ge m^{1/2} \ge 2^i$. Clearly,

$$i \ge C \log m.$$

Note that $\frac{1}{\log^{1/b} m} \leq \frac{1}{\log^{1/b} 2^k}$ if $2^k \leq m^{1/2}$ so that by (8.2.9) for any $C \log m > N$ with a probability larger than $1 - \epsilon$

$$\begin{aligned} c_{0,m} &\geq \sum_{k=1}^{i} \frac{1}{\log^{1/b} 2^k} I_{2^k} \\ &\geq \frac{1}{\log^{1/b} m} (\sum_{k=1}^{i} I_{2^k}) \\ &\geq C \log^{1-1/b} m. \end{aligned}$$

Let r_1 be a route of $T(\partial B(m), \partial B(m_1))$ for some $m_1 > m$. Since $c_{0,m_1} \geq C \log^{1-1/b} m_1$ with a probability larger than $1 - \epsilon$ for any m_1 and

$$c_{0,m_1} \leq t(r_1) + \sum_{e \in B(m)} X_{G_b}(e),$$

we can choose m_1 large such that

$$t(r_1) \geq 1 \tag{8.2.10}$$

with a probability larger than $1 - \epsilon$. By the same argument of (8.2.10), we can construct a sequence

$$T(\partial B(m_1), \partial B(m_2)), \ldots, T(\partial B(m_k), \partial B(m_{k+1})), \ldots$$

such that

$$P(T(\partial B(m_k), \partial B(m_{k+1})) \geq 1) \geq 1 - \epsilon. \tag{8.2.11}$$

Note that

$$\sum_{k=1}^{\infty} P(T(\partial B(m_k), \partial B(m_{k+1})) \geq 1) = \infty,$$

and $T(\partial B(m_k), \partial B(m_{k+1}))$ and $T(\partial B(m_{k+1}), \partial B(m_{k+2}))$ are independent so that by the Borel-Cantelli lemma

$$\sum_{k=1}^{\infty} T(\partial B(m_k), \partial B(m_{k+1})) = \infty \text{ a.s.} \tag{8.2.12}$$

Then Theorem 8.1.3 follows from

$$\rho(F) \geq \sum_{k=1}^{\infty} T(\partial B(m_k), \partial B(m_{k+1})) = \infty \text{ a.s.}$$ □

PROOF OF THEOREM 8.1.4. Clearly, by (8.1.6), to show Theorem 4 we only need to show that if $F \preceq G$, then

$$\rho(G) < \infty \text{ a.s. if } \rho(F) < \infty \text{ a.s.}$$

Let F and G be two distributions satisfying $F(0) = G(0) = p_c$ and $F \preceq G$, i.e, there exists ϵ such that

$$F(x) \leq G(x) \text{ for } 0 \leq x \leq \epsilon. \tag{8.2.13}$$

For each $w \in \Omega$ with $\rho(F, w) < \infty$, note that $L(F(0)) = \infty$ so that by (8.2.7), there exist infinitely many disjoint circuits about the origin with $X_F(e, w) = 0$. We denote by $D_1, \ldots, D_n, \ldots$ such circuits, where D_n is surrounded by D_{n+1}. Let $r_0, r_1, \ldots, r_n, \ldots$ be paths from D_0 to D_1, D_1 to $D_2, \ldots, D_n$ to $D_{n+1}, \ldots$ such that $t(r_n) = T_F(D_n, D_{n+1})$, where $D_0 = (0, 0)$ and

$$T_F(D_n, D_{n+1}) = \inf\{\sum_{e \in r} X_F(e, w) : r \text{ a path from } D_n \text{ to } D_{n+1}\}.$$

Note that travel along D_n costs no time and $\lim c_{0,n}$ exists so that

$$\sum_{n=0}^{\infty} t(r_n) = \rho(F, w). \tag{8.2.14}$$

Since $\rho(F, w) < \infty$, there exists N such that

$$X_F(e, w) \leq t(r_n) \leq \epsilon$$

for all $e \in r_n$ when $n \geq N$. By (8.2.13) and the definitions of X_F and X_G

$$X_G(e, w) \leq X_F(e, w) \tag{8.2.15}$$

for $e \in r_n$ when $n \geq N$. On the other hand, by (8.2.13), $X_G(e, w) = 0$ for $e \in D_n$ for $n \geq N$. Let r'_n be the paths from D_n to D_{n+1} such that $t(r'_n) = T_G(D_n, D_{n+1})$. Similarly,

$$\sum_{n=0}^{\infty} r'_n = \rho(G, w). \tag{8.2.16}$$

By (8.2.15),

$$t(r'_n) \leq t(r_n) \text{ for } n \geq N. \tag{8.2.17}$$

It follows from (8.2.17) that

$$\sum_{n \geq N} t(r'_n) \leq \sum_{n \geq N} t(r_n). \tag{8.2.18}$$

Therefore $\rho(G, w) < \infty$ follows from (8.2.18) and (8.2.16). Theorem 8.1.4 is proved. □

PROOF OF THEOREM 8.1.5. Clearly,

$$a_{0,n} \geq b_{0,n} \geq c_{0,n}.$$

Therefore, if $E\rho(F) = \infty$, then

$$\lim Eb_{0,n} = \liminf Ea_{0,n} = \infty.$$

Assume that $E\rho(F) < \infty$. Note that any path from the origin to $(n, 0)$ has to intersect $\partial B(n/2)$ and $(n, 0) + \partial B(n/2)$ so that

$$T((0,0), \partial B(n/2)) + T((n,0), (n,0) + \partial B(n/2)) \leq a_{0,n}.$$

This implies that

$$2E\rho(F) \leq \liminf Ea_{0,n}. \tag{8.2.19}$$

On the other hand, by (8.2.7) we can find m such that there exists a circuit D_m with $X(e) = 0$ inside $B(m) \setminus B(n)$ surrounding both $(0, 0)$ and $(n, 0)$ with a probability larger than $1 - \exp(-n)$. Let r_1 and r_n be two routes of $c_{0,m}$ and $T((n, 0), \partial B(m))$, respectively. Then r_1 and r_n must intersect D_m if D_m exists. We now can construct a path from (0,0) to $(n, 0)$ by using r_1 and r_n and D_m. Note that $t(D_m) = 0$ by the definition of D_m so that in the event of the existence of D_m

$$a_{0,n} \leq t(r_1) + t(r_n).$$

On the other side, for each w,

$$a_{0,n}(w) \leq \sum_{e \in r} X_F(e, w),$$

where r is the path following the X-axis from $(0, 0)$ to $(n, 0)$. Denote by E_m the event that there exists D_m defined above. Then by the translation invariance

$$\begin{aligned} Ea_{0,n} &= E(a_{0,n}; E_m) + E(a_{0,n}; E_m^C) \\ &\leq Ec_{0,m} + Ec_{0,m+n} + (E\sum_{e \in r} X_F(e))(E(E_m^C)) \\ &\quad \text{(note that } E_m \text{ only depends on the } X_F(e) \text{ for } e \in B(m) \setminus B(n)) \\ &\leq 2E\rho(F) + n\exp(-n)EX_F(e). \end{aligned} \tag{8.2.20}$$

Therefore, $\lim_{n\to\infty} Ea_{0,n} = 2E\rho(F)$ follows from (8.2.19) and (8.2.20). To show that $\lim Eb_{0,n} = E\rho(F)$, we let

$$\lim_{n\to\infty} b_{0,n} = \rho(F) \text{ and } \lim_{n\to\infty} Eb_{0,n} = E\rho_b(F).$$

Note that existence of the above limits follows from the monotonicity of $\{b_{0,n}\}$. By the same argument above we can show that

$$2E\rho_b(F) = \lim_{n\to\infty} Ea_{0,n}. \tag{8.2.21}$$

Therefore,

$$\lim Eb_{0,n} = E\rho(F),$$

since $b_{0,n} \geq c_{0,n}$ and (8.2.21). Theorem 8.1.5 is proved. □

PROOF OF THEOREM 8.1.6. We first show the lower bound of Theorem 8.1.6. Since we only deal with a Bernoulli distribution F with $F(0) = p$, we say e is open (closed) if $X(e) = 0$ ($X(e) = 1$). Furthermore, we use P_p and E_p to replace

P and E in the proof of Theorem 8.1.6. Let N be the number of disjoint closed circuits on the dual surrounding the origin in $B^*(L(p))$. Then by (8.2.7)

$$E_p N \geq C \log L(p).$$

Clearly, any path on Z^2 crossing such a dual circuit in $B^*(L(p))$ will cost at least time 1. Therefore, for $m > L(p)$,

$$E_p c_{0,m} \geq E_p N \geq C(\log L(p)).$$

The lower bound of Theorem 8.1.6 is proved.

To show the upper bound we first introduce two results as the following lemmas.

Lemma 8.2.2 *Assume that $F(0) = p_c$ and $d = 2$. For any n, there exists C such that*

$$E_{p_c} c_{0,n} \leq C \log n.$$

PROOF. See the proof of [3, Theorem 3.3]. □

Let $Z(n)$ be the event that there exists an open circuit surrounding $B(2n) \setminus B(n)$ and there also exists an open path from $\partial B(n)$ to ∞. Then by adapting the same proofs of [10, Lemmas 2 and 3], we have the following result.

Lemma 8.2.3 *Assume that $d = 2$ and $F(0) = p > p_c$. Then for any $n \geq (L(p))^2$, there exist C and $\sigma > 0$ such that*

$$P_p(Z(n)) \geq 1 - C \exp(-n^\sigma). \tag{8.2.22}$$

With Lemmas 8.2.2 and 8.2.3, we will show the upper bound of Theorem 6. Let J_n be the indicator of the event that $Z(n)$ occurs and $Z(n-1)$ does not occur. By (8.2.22)

$$\sum_{i=1} J_i = 1.$$

Clearly, in the event $J_i = 1$

$$c_{0,m} \leq 16 i^2.$$

On the other hand, by Lemma 8.2.2

$$E_p c_{0,L^2(p)} \leq E_{p_c} c_{0,L^2(p)} \leq C \log L(p). \tag{8.2.23}$$

Therefore, for any m

$$\begin{aligned} E_p(c_{0,m}) &= \sum_{i=1}^{\infty} E_p(c_{0,m}) J_i \\ &\leq \sum_{i=1}^{L^2(p)} E_p(c_{0,m}) J_i + \sum_{i=L^2(p)}^{\infty} E_p(c_{0,m}) J_i \end{aligned}$$

$$\leq \sum_{i=1}^{L^2(p)} E_p(c_{0,L^2(p)})J_i + \sum_{i=L^2(p)}^{\infty} E_p(c_{0,i})J_i$$

$$\leq E_p(c_{0,L^2(p)}) \sum_{i=1}^{L^2(p)} J_i + \sum_{i=L^2(p)}^{\infty} 16i^2 P_p(J_i)$$

$$\leq E_p(c_{0,L^2(p)}) + \sum_{i=L^2(p)}^{\infty} 16i^2 C_1 \exp(-C_2 i^{\sigma}) \text{ (by Lemma 3)}$$

$$\leq E_{p_c}(c_{0,L^2(p)}) + C_3 \exp(-C_4(L(p))^{C_5})$$

$$\leq C_6 \log L(p) \text{ (by (39))}.$$

The upper bound of Theorem 8.1.6 is also proved. □

Before proving Theorem 8.1.7 we first show a lemma.

Lemma 8.2.4 *Let F be a Bernoulli distribution with $F(0) = p_c$ and* (8.1.13) *holds. We can show that for any d, there exist constants $\eta > 1$ and $\delta > 0$ which do not depend on n such that*

$$P(\exists \text{ a path with } X(e) = 0 \text{ from } \partial B(n) \text{ to } \partial B(n^{\eta})) < 1 - \delta.$$

Proof. Clearly,

$$1 - P(\exists \text{ a path with } X(e) = 0 \text{ from } \partial B(n) \text{ to } \partial B(n^{\eta}))$$

$$= P(\nexists v \in \partial B(n) \text{ such that } v \to B(n^{\eta}))$$

$$\geq (P((0, \ldots, 0) \not\to \partial B(n^{\eta}))^{C_1 n^d}$$

(by the FKG inequality)

$$\geq (1 - n^{-C\eta})^{C_1 n^d} \text{ (by (16))}$$

$$\geq \delta \text{ (by taking } \eta \text{ large)}.$$

Lemma 8.2.4 is proved. □

Proof of Theorem 8.1.7. We divide $B(n)$ into some smaller boxes as follows.

$$B(2), B(2^{\eta}), \ldots, B(2^{\eta^k}),$$

where k is the largest integer such that $2^{\eta^k} \leq n$. Clearly, $k \geq C_1 \log\log n$. Let

$$E_i = \{\nexists \text{ a path with } X(e) = 0 \text{ from } \partial B(2^{\eta^i}) \text{ to } \partial B(2^{\eta^{i+1}})\}$$

and let I_i be the indicator of E_i. Then by Lemma 8.2.4

$$P(I_i = 1) \geq \delta.$$

Clearly, if E_i occurs, then there is no path with $X(e) = 0$ from $\partial B(2^{\eta^i})$ to $\partial B(2^{\eta^{i+1}})$. In other words, every path from $\partial B(2^{\eta^i})$ to $\partial B(2^{\eta^{i+1}})$ has at least passage time larger than 1. Since routes of $c_{0,m}$ have to pass $B(2^{\eta^i}) \setminus B(2^{\eta^{i-1}})$ for $i \leq k$, then

$$Ec_{0,n} \geq E \sum_{i=1}^{k} I_i \geq C_1 \delta \log \log n. \tag{8.2.24}$$

Finally, we can also show (8.1.15) by adapting the same proof of Theorem 8.1.3. Theorem 8.1.7 is proved. □

PROOF OF THEOREM 8.1.8. For each $w \in \Omega$ with $\rho(w) = \infty$, let $t(r) = M$, where r is the path from the origin to $(n, 0, \ldots, 0)$ along the line $x_i = 0$ for $i = 2, 3, \ldots, d$. Since $\rho(F, w) = \infty$, there exists N such that $c_{0,m} > 2M$ for $m \geq N$. Suppose that there does not exist a route of $a_{0,n}$ inside of $B(N)$. Then there should exist a path r' from the origin to $(n, 0, \ldots, 0)$ with passage time smaller than M such that some of vertices of r' are outside of $B(N)$, otherwise we can find a route inside $B(N)$. On the other hand,

$$t(r') \geq c_{0,N} \geq 2M$$

which is a contradiction. Similarly, we can show the existence of a route for $b_{0,n}$. Theorem 8.1.8 is proved. □

Acknowledgments: This research was supported in part by NSF grant DMS 9400467.

REFERENCES

[1] M. Bramson, Minimal displacement of branching random walk, *Z. Wahrsch. verw. Geb.*, **45** (1978) 89–108.

[2] L. Chayes, On the critical behavior of the first passage time in $d \geq 3$, *Helv. Phys. Acta*, **64** (1991), 1055–1069.

[3] J. Chayes, L. Chayes, and R. Durrett, Critical behavior of two-dimensional first passage time, *J. Statist. Phys.*, **45** (1986), 933–951.

[4] J. Chayes, L. Chayes, and R. Durrett, Inhomogeneous percolation problems and incipient infinite clusters, *J. Phys. A: Math. Gen.*, **20** (1987) 1521–1530.

[5] G. Grimmett, *Percolation*, Springer-Verlag, New York, 1989.

[6] T. Hara and G. Slade, Mean-field critical behaviour for percolation in high dimensions, *Comm. Math. Phys.*, **128** (1990), 333–391.

[7] H. Kesten, Aspect of first-passage percolation, in *Lecture Notes in Mathematics* 1180, Springer-Verlag, Berlin, 1986.

[8] J. Wierman and W. Reh, On conjecture in first passage percolation theory, *Ann. Probab.*, **6** (1978), 388–397.

[9] Y. Zhang, Supercritical behaviors in first-passage percolation, *Stoch. Proc. Appl.*, **59** (1995), 251–266.

[10] Y. Zhang, The fractal volume of the two dimensional invasion percolation cluster, *Comm. Math. Phys.*, **167** (1995), 237–254.

Department of Mathematics
University of Colorado
Colorado Springs, CO 80933
yzhang@vision.uccs.edu

9

The van den Berg-Kesten-Reimer Inequality: A Review

C. Borgs, J.T. Chayes, and D. Randall

ABSTRACT We present a variant of Reimer's proof of the van den Berg-Kesten conjecture.

Keywords: Percolation, correlation inequalities, van den Berg-Kesten inequality.

AMS Subject Classification: 82B43.

9.1 Introduction

In this note, in honor of Harry Kesten's $66\frac{2}{3}$rds birthday, we give an expository treatment of a result that is near and dear to his heart, namely the van den Berg-Kesten (BK) inequality. Specifically, we give a variant of Reimer's proof of the BK inequality for percolation.

The BK inequality provides an upper bound on the probability of the disjoint occurrence of two events in terms of the product of their probabilities. The inequality was first proved by van den Berg and Kesten in 1985 [BK] for the case in which the two events in question are both increasing or both decreasing in the sense of Fortuin, Kasteleyn and Ginibre [FKG]; the BK proof holds for a class of measures called strongly new better than used (SNBU) which includes the product measure. Van den Berg and Kesten conjectured that the inequality holds for all events in percolation.

During the next decade, there were many attempts to prove the BK conjecture. Van den Berg and Fiebig [BF] refined the conjecture and showed that it holds whenever each of the events in question is the intersection of an FKG increasing and an FKG decreasing event. Finally, in 1994, a general proof by Reimer [Re] confirmed the belief that the inequality holds for all events in percolation.

A version of the Reimer proof was presented by one of us in some lectures given at the Institute for Advanced Study in 1996 [CPS]. The proof given there was based on a copy of a preliminary manuscript and some notes by D. Reimer [Re], as well as on a lecture by J. Kahn, and on some comments by C. Borgs, H. Kesten and P. Deligne. The proof in [CPS] modified some of Reimer's notation and added a few details to the proofs previously seen, but the main proof presented there was very similar to that given by Reimer. In particular, the proof of the main lemma in [CPS] used the notion of butterflies, introduced by Reimer, as the principal construct.

Here we give another treatment of Reimer's proof. Although our treatment follows quite closely that of [CPS], it differs in a number of important respects: First, we review the proofs of both van den Berg and Fiebig [BF] and Fishburn and Shepp [FS], on which Reimer's proof depends. The proof presented here is therefore entirely self-contained. Second, in contrast to [CPS], we use the more familiar notions of cylinders and subcubes, rather than butterflies. We hope that the reader will find it easier to understand the proof in this more familiar language. Finally, we have streamlined and generalized aspects of the proof of Reimer's main lemma, see Lemmas 9.5.1 and 9.5.2 and the remarks in Section 9.5.

9.2 Formal Statement of the BKR Inequality

We consider a finite probability space $(\Omega, \mathcal{F}, \mu)$, where Ω is a product of finite sets $\Omega_1, \Omega_2, \ldots, \Omega_n$,

$$\Omega = \Omega_1 \times \Omega_2 \times \cdots \times \Omega_n,$$

$\mathcal{F} = 2^\Omega$ is the set of all subsets of Ω, and μ is a product of n probability measures $\mu_1, \mu_2, \ldots, \mu_n$,

$$\mu = \mu_1 \times \mu_2 \times \cdots \times \mu_n.$$

As usual, elements $\omega = (\omega_1, \omega_2, \ldots, \omega_n) \in \Omega$ are called configurations, and two configurations $\omega = (\omega_1, \omega_2, \ldots, \omega_n) \in \Omega$ and $\tilde{\omega} = (\tilde{\omega}_1, \tilde{\omega}_2, \ldots, \tilde{\omega}_n) \in \Omega$ are said to be *equal on* $S \subset [n] := \{1, 2, \ldots, n\}$ if $\omega_i = \tilde{\omega}_i$ for all $i \in S$.

Definition Let $S \subset [n]$, and let $S^c = [n] \setminus S$. An event $A \in \mathcal{F}$ is said to *occur on the set* S in the configuration ω if A occurs using only the random variables over S, i.e., if A occurs independent of the values of $\{\omega_i\}_{i \in S^c}$. We denote the collection of all such ω by $A|_S$:

$$A|_S = \{\omega : \forall \tilde{\omega},\ \tilde{\omega} = \omega \text{ on } S \Rightarrow \tilde{\omega} \in A\}. \tag{9.2.1}$$

Two events $A_1, A_2 \in \Omega$ are said to *occur disjointly*, denoted by $A_1 \circ A_2$, if there are two disjoint sets on which they occur:

$$A_1 \circ A_2 = \{\omega : \exists S_1, S_2 \subset [n],\ S_1 \cap S_2 = \emptyset,\ \omega \in A_1|_{S_1} \cap A_2|_{S_2}\}. \tag{9.2.2}$$

BK Conjecture *Let $n \in \mathbb{N}$, let Ω_i be finite sets and μ_i be probability measures on Ω_i, $i \in [n]$. Let $\Omega = \Omega_1 \times \Omega_2 \times \cdots \times \Omega_n$, $\mu = \mu_1 \times \mu_2 \times \cdots \times \mu_n$ and let $\mathcal{F} = 2^\Omega$. Then*

$$\mu(A \circ B) \leq \mu(A)\mu(B) \tag{9.2.3}$$

for all $A, B \in \mathcal{F}$.

Theorem 9.2.1 (Reimer [Re]) *The BK conjecture holds.*

9.3 Equivalent Forms of the Inequality

First we reformulate the disjoint occurrence event $A \circ B$ in terms of cylinders. Given a configuration $\omega \in \Omega$ and a set of points $S \subset [n]$, we define the *cylinder* $[\omega]_S$ by

$$[\omega]_S = \{\tilde{\omega} : \tilde{\omega}_i = \omega_i \quad \forall i \in S\}. \tag{9.3.1}$$

We say that a set $A \subset \Omega$ is a cylinder set if $\exists\, \omega \in \Omega$ and $S \subset [n]$ such that $A = [\omega]_S$. With this definition, we may rewrite $A \circ B$ as

$$A \circ B = \{\omega : \exists\, S = S(\omega) \subset [n],\ [\omega]_S \subset A,\ [\omega]_{S^c} \subset B\}. \tag{9.3.2}$$

The first simplification of the BK conjecture (9.2.3) was due to van den Berg and Fiebig [BF] who showed that it is sufficient to prove the inequality for $\Omega = \{0,1\}^n$ and the uniform measure on Ω, i.e., for the pure percolation problem at density $1/2$. This was a significant simplification because it turned the inequality into a purely combinatorial one.

Proposition 9.3.1 (van den Berg-Fiebig [BF]) *The BK conjecture holds if for all $n \in \mathbb{N}$ it holds for the uniform measure on $\{0,1\}^n$, i.e., if for all $n \in \mathbb{N}$ and for all $A, B \subset \{0,1\}^n$*

$$|A \circ B|\, 2^n \le |A|\,|B|. \tag{9.3.3}$$

Proof. Assume that (9.3.3) holds for all $n \in \mathbb{N}$ and for all $A, B \subset \{0,1\}^n$. Let $\tilde{n} \in \mathbb{N}$, let $\tilde{\Omega}_i = \{\omega_{i1}, \omega_{i2}, \dots, \omega_{im_i}\}$ and let μ_i be probability measures on $\tilde{\Omega}_i$, $i \in [\tilde{n}]$. Let $\tilde{\Omega} = \tilde{\Omega}_1 \times \tilde{\Omega}_2 \times \cdots \times \tilde{\Omega}_{\tilde{n}}$, $\mu = \mu_1 \times \mu_2 \times \cdots \times \mu_{\tilde{n}}$ and let $\tilde{A}, \tilde{B} \subset \tilde{\mathcal{F}} = 2^{\tilde{\Omega}}$. We then have to show that

$$\mu(\tilde{A} \circ \tilde{B}) \le \mu(\tilde{A})\mu(\tilde{B}). \tag{9.3.4}$$

In order to prove (9.3.4), we will approximate the measure $\mu = \mu_1 \times \mu_2 \times \cdots \times \mu_{\tilde{n}}$ by a measure $\tilde{\mu} = \tilde{\mu}_1 \times \tilde{\mu}_2 \times \cdots \times \tilde{\mu}_{\tilde{n}}$ for which (9.3.4) can be reduced to (9.3.3). To this end, we approximate the probability measures μ_i on $\tilde{\Omega}_i$ by probability measures $\tilde{\mu}_i = \tilde{\mu}_i^{(K)}$ on $\tilde{\Omega}_i$ with the property that

$$k_{ij} := 2^K \tilde{\mu}_i(\omega_{ij}) \in \mathbb{N} \cup \{0\} \tag{9.3.5}$$

for all $i \in [\tilde{n}]$ and all $j \in [m_i]$. Obviously, the sequence $\tilde{\mu}^{(K)}$ can be chosen in such a way that $\tilde{\mu}^{(K)}$ converges weakly to μ. As a consequence, it is enough to show (9.3.4) for measures $\tilde{\mu} = \tilde{\mu}_1 \times \tilde{\mu}_2 \times \cdots \times \tilde{\mu}_{\tilde{n}}$ which obey the condition (9.3.5).

To prove (9.3.4) for measures $\tilde{\mu}$ which obey the condition (9.3.5), let $n = K\tilde{n}$,

$$\Omega_i = \{0,1\}^K, \qquad \text{and} \qquad \Omega = \Omega_1 \times \cdots \times \Omega_{\tilde{n}} = \{0,1\}^n. \tag{9.3.6}$$

Let

$$\Omega_i = \bigcup_{j=1}^{m_i} \Omega_{ij} \tag{9.3.7}$$

be an arbitrary decomposition of Ω_i into disjoint sets Ω_{ij} with

$$|\Omega_{ij}| = k_{ij}, \tag{9.3.8}$$

and let $f : \Omega \to \tilde{\Omega}$ be defined by

$$f = f_1 \times \cdots \times f_{\tilde{n}} \quad \text{with} \quad f_i : \Omega_i \to \tilde{\Omega}_i \quad \text{given by} \quad f_i(\omega) = \omega_{ij} \quad \text{if} \quad \omega \in \Omega_{ij}. \tag{9.3.9}$$

Then

$$\tilde{\mu}(\cdot) = \mu_0(f^{-1}(\cdot)), \tag{9.3.10}$$

where μ_0 is the uniform measure on $\Omega = \{0, 1\}^n$. Defining

$$A = f^{-1}(\tilde{A}) \qquad \text{and} \qquad B = f^{-1}(\tilde{B}), \tag{9.3.11}$$

we then may use (9.3.3) to conclude that

$$\tilde{\mu}(\tilde{A})\tilde{\mu}(\tilde{B}) = \mu_0(A)\mu_0(B) \geq \mu_0(A \circ B). \tag{9.3.12}$$

Next, we claim that

$$f^{-1}(\tilde{A} \circ \tilde{B}) \subset f^{-1}(\tilde{A}) \circ f^{-1}(\tilde{B}). \tag{9.3.13}$$

Indeed, let $x \in f^{-1}(\tilde{A} \circ \tilde{B})$, i.e., let $f(x) \in \tilde{A} \circ \tilde{B}$. By the definition (9.3.2) of disjoint occurrence, this implies that there exists a set $\tilde{S} \subset [\tilde{n}]$ such that $[f(x)]_{\tilde{S}} \subset \tilde{A}$ and $[f(x)]_{\tilde{S}^c} \subset \tilde{B}$. Therefore

$$f^{-1}([f(x)]_{\tilde{S}}) \subset f^{-1}(\tilde{A}) \qquad \text{and} \qquad f^{-1}([f(x)]_{\tilde{S}^c}) \subset f^{-1}(\tilde{B}).$$

Defining the set $S \subset [n]$ as $S = \{i \in [n] : \lceil i/K \rceil \in \tilde{S}\}$, we then have

$$[x]_S \subset \bigcup_{y \in f^{-1}(f(x))} [y]_S = f^{-1}([f(x)]_{\tilde{S}}) \subset f^{-1}(\tilde{A}) \tag{9.3.14}$$

and

$$[x]_{S^c} \subset \bigcup_{y \in f^{-1}(f(x))} [y]_{S^c} = f^{-1}([f(x)]_{\tilde{S}^c}) \subset f^{-1}(\tilde{B}). \tag{9.3.15}$$

The only subtle step in (9.3.14) and (9.3.15) are the equalities, which follow from the fact that $f : \Omega \to \tilde{\Omega}$ and the set S are defined to respect the product structure (9.3.6) of Ω. By the definition of disjoint occurrence, (9.3.14) and (9.3.15) imply that $x \in f^{-1}(\tilde{A}) \circ f^{-1}(\tilde{B})$ which in turn implies (9.3.13).

Combining (9.3.10) with (9.3.13), (9.3.11) and (9.3.12), we get that

$$\begin{aligned} \tilde{\mu}(\tilde{A} \circ \tilde{B}) &= \mu_0(f^{-1}(\tilde{A} \circ \tilde{B})) \\ &\leq \mu_0(f^{-1}(\tilde{A}) \circ f^{-1}(\tilde{B})) \\ &= \mu_0(A \circ B) \\ &\leq \tilde{\mu}(\tilde{A})\tilde{\mu}(\tilde{B}). \end{aligned} \tag{9.3.16}$$

This gives (9.3.4) for all product measures obeying the condition (9.3.5). Choosing a sequence of measures $\tilde{\mu} = \tilde{\mu}^{(K)}$ that converges weakly to μ, we obtain the BK inequality (9.3.4) for general product measures μ. □

Fishburn and Shepp [FS] derived yet another way of expressing the BK inequality, and it was their form that was ultimately proved by Reimer [Re]. While Fishburn and Shepp stated their inequality in the special case of the uniform measure on $\{0, 1\}^n$, we will state it here in the general context of the full BK conjecture.

We need some notation: Let $X \subset \Omega$, and let $S : X \to 2^{[n]} : x \mapsto S(x) \subset [n]$ be an arbitrary map from X into $2^{[n]}$. We then define

$$[X]_S = \bigcup_{x \in X} [x]_{S(x)} \tag{9.3.17}$$

and

$$[X]_{S^c} = \bigcup_{x \in X} [x]_{S(x)^c}, \tag{9.3.18}$$

where, as before, $S(x)^c = [n] \setminus S(x)$.

Proposition 9.3.2 (Fishburn-Shepp [FS]) *Let $n \in \mathbb{N}$, let Ω_i be finite sets and μ_i be probability measures on Ω_i, $i \in [n]$. Let $\Omega = \Omega_1 \times \Omega_2 \cdots \Omega_n$, $\mu = \mu_1 \times \mu_2 \times \cdots \times \mu_n$ and let $\mathcal{F} = 2^{\Omega}$. Then the following two statements are equivalent:*

(i) *For all $A, B \in \mathcal{F}$,*

$$\mu(A \circ B) \leq \mu(A)\mu(B). \tag{9.3.19}$$

(ii) *For all $X \subset \Omega$ and all $S : X \to 2^{[n]}$,*

$$\mu(X) \leq \mu([X]_S)\, \mu([X]_{S^c}). \tag{9.3.20}$$

PROOF.

(i) $\Longrightarrow$ (ii): Let $X \subset \Omega$, and let $S : X \to 2^{[n]}$. Consider

$$A = [X]_S = \bigcup_{x \in X} [x]_{S(x)} \quad \text{and} \quad B = [X]_{S^c} = \bigcup_{x \in X} [x]_{S(x)^c}.$$

Then

$$X \subset A \circ B.$$

Hence, by (i),

$$\mu(X) \leq \mu(A \circ B) \leq \mu(A)\, \mu(B) = \mu([X]_S)\, \mu([X]_{S^c}),$$

which shows that (i) implies (ii).

(ii) $\Longrightarrow$ (i): Let $A, B \subset \Omega$. Let $X = A \circ B$. By the definition (9.3.2) of $A \circ B$, for each $x \in X$ there is an $S(x)$ such that $[x]_{S(x)} \subset A$ and $[x]_{S(x)^c} \subset B$. We therefore have

$$[X]_S = \bigcup_{x \in X} [x]_{S(x)} \subset A$$

and

$$[X]_{S^c} = \bigcup_{x \in X} [x]_{S(x)^c} \subset B.$$

So, by (ii),

$$\mu(A \circ B) = \mu(X) \leq \mu([X]_S)\, \mu([X]_{S^c}) \leq \mu(A)\, \mu(B),$$

which shows that (ii) implies (i). □

9.4 Reduction of the BKR Inequality to Reimer's Main Lemma

We now come to the main lemma in the proof of the BKR inequality. As noted before, it is enough to show the BKR inequality (9.2.3), and hence the Fishburn-Shepp inequality (9.3.20), in the special case in which μ is the uniform measure on $\Omega = \{0, 1\}^n$. From now on, we will restrict ourselves to this case.

We will use the notation $\bar{x}$ to denote the bitwise complement of a configuration $x \in \Omega$, i.e., $\bar{x}_i = 1 - x_i$. For a cylinder $A = [y]_\Lambda$ and $x \in \Omega$, we define $\bar{x}^{(A)}$ to be the complement of x in A, i.e.,

$$\bar{x}_i^{(A)} = \begin{cases} x_i & \text{if} \quad i \in \Lambda \\ \bar{x}_i & \text{if} \quad i \notin \Lambda. \end{cases} \tag{9.4.1}$$

For a set T, we write $\bar{T} = \bigcup_{x \in T} \bar{x}$, and $\bar{T}^{(A)} = \bigcup_{x \in T} \bar{x}^{(A)}$.

Lemma 9.4.1 (Reimer's Main Lemma) *Let $n \in \mathbb{N}$, $X \subset \Omega = \{0, 1\}^n$ and $S : X \to 2^{[n]} : x \mapsto S(x)$. Let*

$$U = [X]_S \qquad \text{and} \qquad V = [X]_{S^c}. \tag{9.4.2}$$

Then

$$\left|U \cap \bar{V}\right| = \left|\bar{U} \cap V\right| \geq |X|. \tag{9.4.3}$$

In this section, we will show that the above lemma implies the BKR inequality. It turns out that the sufficiency of the main lemma was already known independently to van den Berg [Be] and Talagrand [Ta]; however, they did not have a proof of the lemma. As noted earlier, it suffices to show the Fishburn-Shepp form (9.3.20) of the BK inequality for the uniform measure on $\Omega = \{0, 1\}^n$. For $x, y \in \Omega$, let $\langle x, y\rangle$ be the cylinder $\langle x, y\rangle = \{z \in \Omega : z_i = x_i \text{ whenever } x_i = y_i\}$. Then

$$\begin{aligned} |U|\,|V| &= |\{(u, v) \in U \times V\}| \\ &= \sum_A |\{(u, v) \in U \times V : \langle u, v\rangle = A\}| \\ &= \sum_A |\{(u, v) \in (U \cap A) \times (V \cap A) : \langle u, v\rangle = A\}| \end{aligned} \tag{9.4.4}$$

where the sum runs over all cylinder sets $A \subset \Omega$. Defining

$$U_A = U \cap A \qquad \text{and} \qquad V_A = V \cap A \tag{9.4.5}$$

and observing that that $\langle u, v\rangle = A$ if and only if $u \in A$ and $v = \bar{u}^{(A)}$, we get that

$$|U|\,|V| = \sum_A |\{(u, v) \in U_A \times V_A : v = \bar{u}^{(A)}\}| = \sum_A |U_A \cap \bar{V}_A^{(A)}|. \tag{9.4.6}$$

We claim that Reimer's Main Lemma can be used to show that for each cylinder set $A \subset \Omega$

$$|U_A \cap \bar{V}_A^{(A)}| \geq |X_A|, \qquad \text{where} \qquad X_A = X \cap A. \tag{9.4.7}$$

Indeed, let $A = [z]_\Lambda$ for some $\Lambda \subset [n]$ and some $z \in \Omega$. Let $\Omega_A = \{0, 1\}^{\Lambda^c}$, let $f : \Omega_A \to A : \omega \mapsto f(\omega)$ be the bijection

$$f(\omega)_i := \begin{cases} \omega_i & \text{if} \quad i \in \Lambda^c \\ z_i & \text{if} \quad i \in \Lambda. \end{cases} \tag{9.4.8}$$

and let

$$\tilde{X}_A = f^{-1}(X_A). \tag{9.4.9}$$

Introducing

$$\tilde{S} : \tilde{X}_A \to 2^{\Lambda^c} : \tilde{x} \mapsto S(f(\tilde{x})) \cap \Lambda^c \tag{9.4.10}$$

and it's complement in Λ^c,

$$\tilde{S}^c : \tilde{X}_A \to 2^{\Lambda^c} : \tilde{x} \mapsto \Lambda^c \setminus (S(f(\tilde{x})) \cap \Lambda^c), \tag{9.4.11}$$

Reimer's Main Lemma now implies that

$$|X_A| = |\tilde{X}_A| \leq |[\tilde{X}_A]_{\tilde{S}} \cap [\overline{\tilde{X}}_A]_{\tilde{S}^c}| = |[X_A]_{S\cup\Lambda} \cap [\bar{X}_A^{(A)}]_{S^c\cup\Lambda}|, \tag{9.4.12}$$

where

$$[X_A]_{S\cup\Lambda} = \bigcup_{x\in X_A} [x]_{S(x)\cup\Lambda} \qquad \text{and} \qquad [\bar{X}_A^{(A)}]_{S^c\cup\Lambda} = \bigcup_{x\in \bar{X}_A^{(A)}} [x]_{S(x)^c\cup\Lambda}, \tag{9.4.13}$$

with, as before, $S^c(x) = [n] \setminus S(x)$.

Next we claim that

$$[X_A]_{S\cup\Lambda} \cap [\bar{X}_A^{(A)}]_{S^c\cup\Lambda} \subset U_A \cap \bar{V}_A^{(A)}. \tag{9.4.14}$$

Indeed, since $x \in A$ implies that $A = [x]_\Lambda$ which in turn implies that $[x]_{S(x)\cup\Lambda} = A \cap [x]_{S(x)}$, we have

$$[X_A]_{S\cup\Lambda} = \bigcup_{x\in X\cap A} [x]_{S(x)\cup\Lambda} = \bigcup_{x\in X\cap A} A \cap [x]_{S(x)} \subset \bigcup_{x\in X} A \cap [x]_{S(x)} = U_A. \tag{9.4.15}$$

In a similar way, one gets

$$[\bar{X}_A^{(A)}]_{S^c \cup \Lambda} \subset \bar{V}_A^{(A)}. \tag{9.4.16}$$

The relations (9.4.15) and (9.4.16) imply (9.4.14). Together with (9.4.14), the bound (9.4.12) now gives the bound (9.4.7).

Combining (9.4.6) and (9.4.7), we get

$$|U|\,|V| \geq \sum_A |X \cap A|. \tag{9.4.17}$$

An easy counting argument gives that the right hand side of (9.4.17) is equal to $|X||\Omega|$. Indeed,

$$\sum_A |X \cap A| = \sum_A \sum_{x \in X \cap A} 1 = \sum_{x \in X} \sum_{A \ni x} 1 = |X||\Omega|,$$

which, together with (9.4.16), implies that

$$|U|\,|V| \geq |X||\Omega|, \tag{9.4.18}$$

the Fishburn-Shepp inequality (9.3.20) for the uniform measure on $\Omega = \{0, 1\}^n$.

9.5 Proof of Reimer's Main Lemma

The first half of the statement of the main lemma just follows from the simple observation that $x \in U \cap \bar{V} \Longleftrightarrow \bar{x} \in \bar{U} \cap V$. We therefore have to show that $|U \cap \bar{V}| \geq |X|$. Using de Morgan's laws, this is equivalent to showing that

$$|U^c \cup \bar{V}^c| \leq |\Omega| - |X|, \tag{9.5.1}$$

or

$$|U^c| + |U \cap \bar{V}^c| + |X| \leq |\Omega| = 2^n. \tag{9.5.2}$$

Since $|\bar{U}^c| = |U^c|$, this is equivalent to

$$|\bar{U}^c| + |U \cap \bar{V}^c| + |X| \leq |\Omega| = 2^n. \tag{9.5.3}$$

To obtain (5.3), we will construct injective maps α, β and γ from $\bar{U}^c$, $U \cap \bar{V}^c$ and X into $\mathbb{R}^{2^n}$. We will show that the images of these maps are disjoint and that the union of the images is a set of linearly independent vectors in $\mathbb{R}^{2^n}$. This immediately implies that the number of elements in the union, and hence on the left hand side of (9.5.3), is bounded above by 2^n.

We begin by defining the maps α, β and γ. While these three maps were defined separately in Reimer's original proof, our treatment allows for a unified definition in terms of one function $\Phi : \Omega \times 2^{[n]} \to \mathbb{R}^{2^n} : (x, S) \mapsto \Phi(x, S)$. In particular, the oberservation that (9.5.2) is equivalent to (9.5.3) allows the definition of a single function. In terms of the (still to be defined) function Φ, the maps α, β and γ are defined as

$$\begin{aligned} \alpha : \bar{U}^c \to \mathbb{R}^{2^n} &: \quad x \mapsto \Phi(x, \emptyset) \\ \beta : U \cap \bar{V}^c \to \mathbb{R}^{2^n} &: \quad x \mapsto \Phi(x, [n]) \\ \gamma : X \to \mathbb{R}^{2^n} &: \quad x \mapsto \Phi(x, S(x)). \end{aligned} \tag{9.5.4}$$

To define $\Phi(\cdot, S)$, we first define functions $\varphi_i(\cdot, S)$ on a single bit x_i:

$$\varphi_i(x_i, S) = \begin{cases} (x_i, -1) & \text{if} \quad i \notin S \\ (1, x_i) & \text{if} \quad i \in S. \end{cases} \tag{9.5.5}$$

To define Φ on $\Omega = \{0, 1\}^{[n]}$, we must set some notation. Let $\oplus$ denote concatenation given by, $(a, b) \oplus (c, d) = (a, b, c, d)$. Let $\otimes$ be the tensor product given by $(a, b) \otimes v = av \oplus bv$ for $a, b \in \mathbb{R}$ and $v \in \mathbb{R}^m$. Equipping $\mathbb{R}^{2^n}$ with the standard inner product: $\langle v \mid w \rangle = \sum_{i=1}^{2^n} v_i w_i$, notice that an easy inductive proof yields

$$\left\langle \bigotimes_{i=1}^{n} v_i \mid \bigotimes_{i=1}^{n} w_i = \right\rangle = \prod_{i=1}^{n} \langle v_i \mid w_i \rangle \tag{9.5.6}$$

for $v_i, w_i \in \mathbb{R}^2$, $1 \leq i \leq n$. With this notation in hand, let

$$\Phi(x, S) = \bigotimes_{i=1}^{n} \varphi_i(x_i, S) \tag{9.5.7}$$

for each $x \in \Omega$.

It suffices to verify the following six statements to show linear independence:

(1) $\Phi(y, \emptyset) \perp \Phi(z, [n])$ for all $y \in \bar{U}^c$ and all $z \in U \cap \bar{V}^c$.
(2) $\Phi(y, \emptyset) \perp \Phi(x, S(x))$ for all $y \in \bar{U}^c$ and all $x \in X$.
(3) $\Phi(z, [n]) \perp \Phi(x, S(x))$, for all $z \in U \cap \bar{V}^c$ and all $x \in X$.
(4) $\{\Phi(x, S(x)) : x \in X\}$ is linearly independent.
(5) $\Phi(\bar{U}^c, \emptyset)$ is linearly independent.
(6) $\Phi(U \cap \bar{V}^c, [n])$ is linearly independent.

The function Φ has been defined so that most of this will be routine.

(1) $\Phi(y, \emptyset) \perp \Phi(z, [n])$ for all $y \in \bar{U}^c$ and all $z \in U \cap \bar{V}^c$.

If $y \in \bar{U}^c$ and $z \in U \cap \bar{V}^c$, then $\bar{y} \notin U$ and $z \in U$, so in particular $\bar{y} \neq z$. Then $y_i = z_i$ for some i and

$$\langle \varphi_i(y_i, \emptyset) \mid \varphi_i(z_i, [n]) \rangle = \langle (y_i, -1) \mid (1, z_i) \rangle = 0.$$

Recalling (9.5.6), we have that

$$\langle \Phi(y, \emptyset) \mid \Phi(z, [n]) \rangle = 0.$$

Since it is easy to see that neither $\Phi(y, \emptyset)$ nor $\Phi(z, [n])$ can be the zero vector, it follows that $\Phi(y, \emptyset) \perp \Phi(z, [n])$.

(2) $\Phi(y, \emptyset) \perp \Phi(x, S(x))$ for all $y \in \bar{U}^c$ and all $x \in X$.

If $y \in \bar{U}^c$ and $x \in X$, then $\bar{y} \notin U$ which implies there exists $i \in S(x)$ such that $y_i = x_i$. Thus, it follows that

$$\langle \varphi_i(y_i, \emptyset) \mid \varphi_i(x_i, S(x)) \rangle = \langle (y_i, -1) \mid (1, x_i) \rangle = 0.$$

Hence, $\Phi(y, \emptyset) \perp \Phi(x, S(x))$.

(3) $\Phi(z, [n]) \perp \Phi(x, S(x))$, for all $z \in U \cap \bar{V}^c$ and all $x \in X$.

If $z \in U \cap \bar{V}^c$ and $x \in X$, then $\bar{z} \notin V$ which implies there exists $i \in S(x)^c$ such that $z_i = x_i$. It follows that

$$\langle \varphi_i(z_i, [n]) \mid \varphi_i(x_i, S(x)) \rangle = \langle (1, z_i) \mid (x_i, -1) \rangle = 0.$$

Hence $\Phi(z, [n]) \perp \Phi(x, S(x))$.

(4) $\{\Phi(x, S(x)) : x \in X\}$ is a set of linearly independent vectors.

This statement is the core of Reimer's proof. For this argument, it is sufficient to prove the independence on $\mathbb{Z}_2^{2^n}$ rather than $\mathbb{R}^{2^n}$, and, as will become clear, it turns out to be much simpler for $\mathbb{Z}_2^{2^n}$. For the moment, simply note that, in $\mathbb{Z}_2^2$, if $x_i = 1$ then $\varphi_i(x_i, S) = (1, 1)$ whether or not $i \in S$. Notice that since $X \subseteq \Omega$, we can think of $S : x \mapsto S(x)$ as a function from $X \to 2^{[n]}$. We can extend this by defining $S(x) \in \Omega$ for all $x \in \Omega \setminus X$ arbitrarily. This in turn induces a function $x \mapsto \Phi(x, S(x)) : \Omega \to \mathbb{R}^{2^n}$ (or $\mathbb{Z}_2^{2^n}$) which coincides with γ when $x \in X$. In order to prove (4), it is therefore enough to prove that for all $S : \Omega \to 2^{[n]}$, the set $\{\Phi(x, S(x)) : x \in \Omega\}$ is a set of linearly independent vectors in $\mathbb{Z}_2^{2^n}$. This is the content of Lemma 9.5.1 below.

(5) $\Phi(\bar{U}^c, \emptyset)$ is linearly independent, and
(6) $\Phi(U \cap \bar{V}^c, [n])$ is linearly independent.

Although these can both be argued by completely elementary methods, both of these statements follow as a special case of the statement that for all $S : \Omega \to 2^{[n]} : x \mapsto S(x)$, the set $\{\Phi(x, S(x)) : x \in \Omega\}$ is a set of linearly independent vectors in $\mathbb{R}^{2^n}$ (choose the constant functions $S(x) \equiv \emptyset$ and $S(x) \equiv [n]$, respectively).

The proof of Reimer's Main Lemma is therefore reduced to the proof of the following:

Lemma 9.5.1 *Let $n \in \mathbb{N}$, and let $\Phi : \{0, 1\}^n \times 2^{[n]} \to \mathbb{R}^{2^n}$ be defined by* (9.5.5) *and* (9.5.7). *Let $S : x \mapsto S(x) \subset [n]$ be an arbitrary function from $\{0, 1\}^n$ into $2^{[n]}$. Then the vectors $\Phi(x, S(x))$, $x \in \{0, 1\}^n$, are linearly independent in $\mathbb{Z}_2^{2^n}$, and hence in $\mathbb{R}^{2^n}$.*

Proof. For $0 < k \leq 2^n$, let y^k be the configuration in Ω given by the binary representation of $k - 1$ so that $\Omega = \{y^k : 0 < k \leq 2^n\}$, with $k = 1$ corresponding to $y_i^k \equiv 0$, $k = 2$ corresponding to $y_n^k = 1$ and $y_i^k = 0$ for all $i \leq n - 1$, etc. For the configuration y^k in $\{0, 1\}^n$, we let $0y^k$ be the configuration corresponding to the binary representation of $k - 1$ in $\{0, 1\}^{n+1}$, and $1y^k$ be the configuration corresponding to the binary representation of $2^n + k - 1$ in $\{0, 1\}^{n+1}$.

If we let $A_S^{(n)}$ be the $2^n \times 2^n$ matrix formed by letting row k be the vector $\Phi(y^k, S(y^k))$,

$$A_S^{(n)}(k, \cdot) = \Phi(y^k, S(y^k)) \tag{9.5.8}$$

then it suffices to show that for all functions $S : \Omega \to 2^{[n]}$, the matrix $A_S^{(n)}$ satisfies

$$\det A_S^{(n)} = 1. \tag{9.5.9}$$

We will prove this using induction on n. The base case $n = 1$ is trivial to check. So suppose that for all $S : \{0, 1\} \to 2^{[n]}$ we have $\det A_S^{(n)} = 1$ by induction. Analyzing the case $n + 1$, let now $\Omega = \{0, 1\}^{n+1}$, and let S be a function from $\{0, 1\}^{n+1}$ into $2^{[n+1]}$. Note that the binary representation of each of the first 2^n configurations begins with 0. So $\varphi_1(y_1^k, S(y^k)) = (1, 0)$ or $(0, -1)$ (which equals $(0, 1)$ in $\mathbb{Z}_2$), depending on whether $1 \in S(y^k)$ or not. Therefore, defining S^0 : $\{0, 1\}^n \to 2^{[n]}$ by $S^0(y^k) = \{i \in [n] : i + 1 \in S(0y^k)\}$, we get that for each $0 \le k < 2^n$, either $1 \in S(y^k)$ and

$$\begin{aligned}
A_S^{(n+1)}(k, \cdot) &= (1, 0) \otimes \bigotimes_{i=2}^{n+1} \varphi_i(y_i^k, S(y^k)) \\
&= \bigotimes_{i=2}^{n+1} \varphi_i(y_i^k, S(y^k)) \oplus \bigoplus_{j=1}^{2^n} 0 \\
&= A_{S^0}^{(n)}(k, \cdot) \oplus \bigoplus_{j=1}^{2^n} 0
\end{aligned}$$

or $1 \notin S(y^k)$ and

$$\begin{aligned}
A_S^{(n+1)}(k, \cdot) &= (0, 1) \otimes \bigotimes_{i=2}^{n+1} \varphi_i(y_i^k, S(y^k)) \\
&= \bigoplus_{j=1}^{2^n} 0 \oplus \bigotimes_{i=2}^{n+1} \varphi_i(y_i^k, S(y^k)). \\
&= \bigoplus_{j=1}^{2^n} 0 \oplus A_{S^0}^{(n)}(k, \cdot).
\end{aligned}$$

Defining $\varepsilon_k = 1$ if $1 \in S(y^k)$ and $\varepsilon_k = 0$ if $1 \notin S(y^k)$, we therefore have that

$$A_S^{(n+1)}(k, \cdot) = \varepsilon_k A_{S^0}^{(n)}(k, \cdot) \oplus (1 - \varepsilon_k) A_{S^0}^{(n)}(k, \cdot).$$

Meanwhile, note that $(1, -1) = (1, 1)$ in $\mathbb{Z}_2^2$, so that $\varphi_1(y_1^k, S(y^k)) = (1, 1)$ if the binary representation of k starts with 1. Therefore, defining $S^1 : \{0, 1\}^n \to 2^{[n]}$

by $S^1(y^k) = \{i \in [n] : i+1 \in S(1y^k)\}$, we get that for each $2^n < k \le 2^{n+1}$

$$\begin{aligned} A_S^{(n+1)}(k,\cdot) &= (1,1) \otimes \bigotimes_{i=2}^{n+1} \varphi_i(y_i^k, S(y^k)) \\ &= \bigotimes_{i=2}^{n+1} \varphi_i(y_i^k, S(y^k)) \oplus \bigotimes_{i=2}^{n+1} \varphi_i(y_i^k, S(y^k)) \\ &= A_{S^1}^{(n)}(k,\cdot) \oplus A_{S^1}^{(n)}(k,\cdot). \end{aligned}$$

Hence

$$A_S^{(n+1)} = \begin{pmatrix} \varepsilon_k A_{S^0}^{(n)}(k,\cdot) & (1-\varepsilon_k) A_{S^0}^{(n)}(k,\cdot) \\ A_{S^1}^{(n)} & A_{S^1}^{(n)} \end{pmatrix}.$$

Although this matrix looks messy, a few column operations—actually 2^n of them—will improve things, without changing the determinant, of course. By adding column $k+1$ to column $k+1+2^n$ (for each $0 \le k < 2^n$) which, in $\mathbb{Z}_2$, is the same as subtracting column $k+1$ from column $k+1+2^n$, we can conclude that

$$\begin{aligned} \det A_S^{(n+1)} &= \det \begin{pmatrix} \varepsilon_k A_{S^0}^{(n)}(k,\cdot) & A_{S^0}^{(n)}(k,\cdot) \\ A_{S^1}^{(n)} & 0 \end{pmatrix} \\ &= \det A_{S^0}^{(n)} \det A_{S^1}^{(n)} \\ &= 1, \end{aligned}$$

where the final step follows by induction. □

This completes the proof of Reimer's Main Lemma, and hence the proof of the BK conjecture, Theorem 9.2.1.

Remarks

(i) The proof given above for independence of $\{\Phi(x, S(x)) : x \in \Omega\}$ follows quite closely that presented in [CPS], which was in turn based on the matrix proof given by Reimer. Alternatively, there is a more algebraic proof, similar to one presented by J. Kahn and also to one suggested to us by P. Deligne. At the end of this section we give a version of such a proof, originally presented by one of us (J.T.C.) in the Kac Seminars of 1995, and reviewed in [CPS].

(ii) A close analysis of the above proof of Lemma 9.5.1 shows that it holds in the more general case in which (5.5) is replaced by the definition

$$\varphi_i(x_i, S) = \begin{cases} \psi_0(x_i) & \text{if } i \notin S \\ \psi_1(x_i) & \text{if } i \in S, \end{cases}$$

provided the four vectors $\psi_0(0)$, $\psi_0(1)$, $\psi_1(0)$ and $\psi_1(1) \in \mathbb{Z}_2^2$ are chosen in such a way that for each pair of vectors $\psi(0) \in \{\psi_0(0), \psi_1(0)\}$ and $\psi(1) \in \{\psi_0(1), \psi_1(1)\}$, the set $\{\psi(0), \psi(1)\}$ is a basis for $\mathbb{Z}_2^2$. This is easy to see once it is realized that all possible cases can be reduced to the following three cases:

(a) $\psi_0(0) = (0, 1)$, $\psi_1(0) = (1, 0)$, and $\psi_0(1) = \psi_1(1) = (1, 1)$,

the case studied in the proof of Lemma 9.5.1;

(b) $\psi_0(0) = (0, 1)$, $\psi_1(0) = (1, 1)$, and $\psi_0(1) = \psi_1(1) = (1, 0)$,

and

(c) $\psi_0(0) = (1, 0)$, $\psi_1(0) = (1, 1)$, and $\psi_0(1) = \psi_1(1) = (0, 1)$.

Revisiting the proof of Lemma 9.5.1, it can be easily seen that the inductive proof works for cases (b) and (c) in a similar way to case (a). We therefore obtain the following generalization of Lemma 9.5.1:

Lemma 9.5.2 *Let $\psi_0(0)$, $\psi_0(1)$, $\psi_1(0)$ and $\psi_1(1) \in \mathbb{Z}_2^2$ be chosen in such a way that for each pair of vectors $\psi(0) \in \{\psi_0(0), \psi_1(0)\}$ and $\psi(1) \in \{\psi_0(1), \psi_1(1)\}$, the set $\{\psi(0), \psi(1)\}$ is a basis for $\mathbb{Z}_2^2$. Let $n \in \mathbb{N}$, let S be an arbitrary function from $\{0, 1\}^n$ into $2^{[n]}$, and let*

$$\varphi_i(x_i, S(x)) = \begin{cases} \psi_0(x_i) & if\ i \notin S(x) \\ \psi_1(x_i) & if\ i \in S(x). \end{cases} \tag{9.5.10}$$

and

$$\Phi_S(x) := \bigotimes_{i=1}^{n} \varphi_i(x_i, S(x)) . \tag{9.5.11}$$

Then the set $\{\Phi_S(x) : x \in \{0, 1\}^n\}$ is a set of linearly independent vectors in $Z_2^{2^n}$.

Remarks

(iii) If the set $S(x) \subset [n]$ is independent of $x \in \{0, 1\}^n$, the statement of the lemma just follows from the well known fact that whenever $\{v_1, \dots, v_n\}$ and $\{w_1, \dots, w_m\}$ are bases for two vector spaces V and W, then $\{v_i \otimes w_j\}_{(i,j)\in[n]\times[m]}$ is a basis for $V \otimes W$.

(iv) The generalization of Lemma 9.5.2 to $\mathbb{Z}^2$ or $\mathbb{R}^2$ is false. Indeed, taking $n = 2$ and choosing $\psi_0(0) = (0, 1)$, $\psi_1(0) = (1, 1)$, $\psi_0(1) = (1, 0)$, $\psi_1(1) = (-1, 1)$, $S(00) = S(11) = \{2\}$ and $S(01) = S(10) = \{1\}$, we find that the four vectors

$$\begin{aligned}
\Phi_S(00) &= \psi_0(0) \otimes \psi_1(0) = (0, 1) \otimes (1, 1) = (0, 0, 1, 1), \\
\Phi_S(01) &= \psi_1(0) \otimes \psi_0(1) = (1, 1) \otimes (1, 0) = (1, 0, 1, 0), \\
\Phi_S(10) &= \psi_1(1) \otimes \psi_0(0) = (1, -1) \otimes (0, 1) = (0, 1, 0, -1), \quad \text{and} \\
\Phi_S(11) &= \psi_0(1) \otimes \psi_1(1) = (1, 0) \otimes (1, -1) = (1, -1, 0, 0)
\end{aligned}$$

are linearly dependent, since $(0, 0, 1, 1)-(1, 0, 1, 0)+(0, 1, 0, -1)+(1, -1, 0, 0) = 0$.

We close this paper with the alternative proof of Lemma 9.5.1 mentioned in Remark (i) above.

Alternative Proof of Lemma 9.5.1. As in the above proof, we will establish independence on $\mathbb{Z}_2^{2^n}$ rather than $\mathbb{R}^{2^n}$. For convenience, we define $\mathbf{w} = (1, 0)$ and

$\mathbf{v} = (1, 1)$. Then $\varphi_i(x_i, S) = \mathbf{w}$ if $x_i = 0$ and $i \in S$, $\varphi_i(x_i, S) = \mathbf{v}$ if $x_i = 1$, and $\varphi_i(x_i, S) = \mathbf{w} + \mathbf{v}$ if $x_i = 0$ and $i \notin S$.

To show that the vectors in $\{\Phi(x, S(x)) : x \in \Omega\}$ are linearly independent, it suffices to expand them in a basis in $\otimes_{i=1}^n \mathbb{Z}_2^2$ and show that the coefficient matrix has nonzero determinant. To this end, let I be the set $I \subset [n]$. Our basis in $\otimes_{i=1}^n \mathbb{Z}_2^2$ will be $\{u_I \mid I \subset [n]\}$ where

$$u_I = \bigotimes_{j \notin I} \mathbf{v} \otimes \bigotimes_{j \in I} \mathbf{w}.$$

In order to expand $\Phi(x, S(x))$ in the $\{u_I\}$, we let $I_y = \{i \mid y_i = 0\}$. Then

$$\begin{aligned}
\Phi(x, S(x)) &= \bigotimes_{i=1}^{n} \varphi_i(x_i, S(x)) \\
&= \bigotimes_{i \notin I_y} \mathbf{v} \otimes \bigotimes_{i \in I} \begin{cases} \mathbf{w} & \text{if } x_i = 0 \\ \mathbf{w} + \mathbf{v} & \text{if } x_i = 1 \end{cases} \\
&= \bigotimes_{i \notin I_y} \mathbf{v} \otimes \bigotimes_{i \in I_y} (\mathbf{w} + x_i \mathbf{v}) \\
&= \sum_{J \subseteq I_y} \bigotimes_{i \notin I_y} \mathbf{v} \otimes \bigotimes_{i \in J} \mathbf{w} \otimes \bigotimes_{i \in I_y \setminus J} x_i \mathbf{v}.
\end{aligned}$$

Noting that $I_y^c \cup I_y \setminus J = J^c$, and defining $\epsilon(J) =: \prod_{I_y \setminus J} x_i \in \{0, 1\}$, we have

$$\begin{aligned}
\Phi(x, S(x)) &= \sum_{J \subseteq I_y} \epsilon(J) \bigotimes_{i \notin J} \mathbf{v} \otimes \bigotimes_{i \in J} \mathbf{w} \\
&= \sum_{J \subseteq I_y} \epsilon(J)\, u_J \\
&= u_{I_y} + \sum_{J \subset I_y} \epsilon(J)\, u_J,
\end{aligned}$$

where J is a proper subset of I_y in the final sum.

Now the above matrix is an upper triangular matrix with 1's along the diagonal. If the index set I_y were a totally ordered set, this would immediately imply the determinant of this matrix is one, and hence that the vectors $\Phi(x, S(x))$, $x \in \Omega$ are linearly independent. Since the index set is only partially ordered, this requires a little additional argument, which we leave to the reader. It is easy to verify using, e.g., the expansion of the determinant in minors. □

Acknowledgment: The research of the third author was partially supported by NSF Career Award No. CCR-9703206.

REFERENCES

[Be] J. van den Berg. Private communication.

[BF] J. van den Berg and U. Fiebig. On a combinatorial conjecture concerning disjoint occurrences of events. *Ann. Probab.*, **15** (1987), 354–374.

[BK] J. van den Berg and H. Kesten. Inequalities with applications to percolation and reliability. *J. Appl. Probab.*, **22** (1985), 556–569.

[CPS] J.T. Chayes, A. Puha, and T. Sweet. Independent and dependent percolation. In *IAS/Park City Mathematics Series*, Vol. 6. AMS, Providence, RI, 1998.

[FKG] C.M. Fortuin, P.W. Kasteleyn, and J. Ginibre. Correlation inequalities on some partially ordered sets. *Commun. Math. Phys.*, **22** (1971), 89–103.

[FS] P.C. Fishburn and L.A. Shepp. On the FKB conjecture for disjoint intersections. *Discrete Math.*, **98** (1991), 105–122.

[Re] D. Reimer. Preliminary manuscript. 1994.

[Ta] M. Talagrand. Some remarks on the Berg-Kesten inequality, In *Probability in Banach Spaces,* 9 (*Sandjberg,* 1993), *Progress in Probability* 35, Birkhäuser, Boston, 1994, 293–297.

Christian Borgs
Microsoft Research
One Microsoft Way
Redmond, WA 98052
borgs@microsoft.com
(on leave from Universität Leipzig, Germany)

Jennifer T. Chayes
Microsoft Research
One Microsoft Way
Redmond, WA 98052
jchayes@microsoft.com
(on leave from University of California, Los Angeles)

Dana Randall
School of Mathematics and College of Computing
Georgia Institute of Technology
Atlanta GA 30332-0160
randall@math.gatech.edu

10

Large Scale Degrees and the Number of Spanning Clusters for the Uniform Spanning Tree

Itai Benjamini

ABSTRACT We study large scale properties of the uniform spanning tree in $\mathbb{Z}^d$. It is shown that inside the n-cube of $\mathbb{Z}^d$, there are $o(n^d)$ vertices with large scale degree 2. In 2D the number of vertices with large scale degree 3 is uniformly bounded and there are no vertices with large scale degree bigger than 3. Also, it is shown that the number of spanning clusters for the uniform spanning tree in a square is tight as the square grows, and that this is not the case in dimensions 4 and higher. A similar transition is a known conjecture for critical Bernoulli percolation.

Keywords: Spanning trees, loop erased walks, percolation.

AMS Subject Classifications: Primary 60D05; Secondary 05C80, 05C05.

10.1 Introduction

The uniform spanning tree model has been recently getting some attention. In this note we would like to remark on large scale properties of the uniform spanning tree in $\mathbb{Z}^d$: large scale degrees and the number of spanning clusters in a large cube. Consider the uniform spanning tree of the square grid $\mathbb{Z}^d$, or the n-cube, denoted C_n^d, in $\mathbb{Z}^d$. Burton and Pemantle (1993) showed how to calculate the probability for a fixed edge to be in a tree, as well as any other cylindrical event. In particular the distribution of the degree of a fixed vertex can be calculated. Recently Schramm (1998), constructed an object "living" in $\mathbb{R}^2$, which is conjecturally the scaling limit of the uniform spanning tree in $\mathbb{Z}^2$. Having the proposed scaling limit in mind, here we are interested in the large scale degree of vertices. This is not a cylindrical event, and presents properties of the tree which are supposed to be inherited by the limiting object. Thus, different reasoning is used in the study of large scale degrees.

Roughly, the large scale degree of a vertex v, in a spanning tree, is the number of subtrees rooted at $v \in C_n^d$, which reach distance of order n from v in the $\mathbb{Z}^d$ graph metric.

We will show that the number of vertices with large scale degree 2 inside the n-cube is $o(n^d)$, in 2D we further show that the number of vertices with large scale degree 3, is uniformly bounded and that with probability going to 1, as n grows, there are no vertices with large scale degree bigger than 3. See below for the precise result.

We now turn to spanning clusters. Consider again C_n^d, the n-cube in $\mathbb{Z}^d$, i.e. the vertex set is $\{(x_1, \ldots, x_d) \in \mathbb{Z}^d \mid 0 \leq x_i \leq n\}$, and edges are between any two vertices with Euclidean distance 1. Assign each edge in the cube the value 0 or 1. A spanning cluster is a connected component in the 0, 1 configuration consisting only of 1's, which connect the face $x_1 = 0$ to its parallel face, $x_1 = n$. In particular, in our discussion, an edge is assigned 1 iff it is in the spanning tree. We would like to remark on the number of spanning clusters for the uniform spanning tree in $\mathbb{Z}^d$. Our motivation to look at this problem comes from percolation. We will see that the analogue of the conjectural picture for spanning clusters of critical Bernoulli percolation is in a larger part provable for the uniform spanning tree. It is conjectured that for critical Bernoulli percolation in a large cube, for $d \leq 6$, the number of disjoint spanning clusters is tight, as the size of the cube grows. Yet for $d > 6$, the expected number of spanning clusters grows to infinity, with the size of the cube. By the BK inequality the 2D case follows; the rest is still open. (See for instance Borgs, Chayes, Kesten and Spencer (1998), Aizenman (1997) and Cardy (1997)). The point of this note is that the qualitative transition between low and high dimensions is established for the uniform spanning tree (the exact picture is still open, in particular tightness in 3D). This can serve as additional support for the Bernoulli percolation conjecture.

10.2 Some Background and Formulation of the Results

The uniform spanning tree measure for a finite connected graph is just the uniform measure on the space of spanning trees of the graph. The uniform spanning tree measures for infinite graphs are weak limits of uniform spanning trees from finite subgraphs. See Pemantle (1991) or Benjamini, Lyons, Peres and Schramm (1998) for an introduction to and a study of uniform spanning tree measures.

Häggström (1995) showed that the uniform spanning tree measures arise as limits of (Fortuin-Kasteleyn) random cluster measures. Such measures generalize ordinary Bernoulli percolation, Ising and Potts models of statistical physics. Many questions that are difficult and often still unsolved for some of these models, become tractable for the uniform spanning tree model because of its close connection to potential theory and random walks. Conformal invariance as well as other properties, such as the distribution of spanning clusters discussed here, suggests that the uniform spanning tree behaves, like a critical model, although there is no parameter in the model.

Wilson's algorithm

Let G be a finite connected graph. Every connected graph has a spanning tree. It is not obvious how to choose one uniformly at random, and several sophisticated algorithms have been devised to do this. The fastest algorithm known, which we describe below, and then use, is due to Wilson (1996). To describe Wilson's algorithm, we define the loop erasure of a path. If γ is any finite path $\{v_0, v_1, \ldots, v_l\}$ in G, we define the loop erasure of γ, denoted $LE(\gamma)$, by erasing cycles in γ in the order they appear. A slightly different, and more precise, inductive description of the loop erasure $LE(\gamma) = \{u_0, u_1, \ldots, u_m\}$ of a path $\gamma = \{v_0, v_1, \ldots, v_l\}$ is as follows. The first vertex u_0 of $LE(\gamma)$ is the first vertex v_0 of γ. Suppose that u_j has been set. Let k be the last index such that $v_k = u_j$. Set $u_{j+1} := v_{k+1}$ if $k < l$; otherwise, let $LE(\gamma) := (u_1, \ldots, u_j)$. Note that $LE(\gamma)$ is still well defined when γ is an infinite path that visits no vertex infinitely often.

In order to generate a random spanning tree, first pick any vertex r to be the "root" of the tree. Then create a growing sequence of trees t_i $(i \geq 0)$ as follows. Choose any ordering $\{v_1, \ldots, v_n\}$ of the vertices. Let $t_0 := \{r\}$. Suppose that the tree t_i has been generated. Start an independent simple random walk at v_{i+1} and stop at the first time it hits t_i. (If $v_{i+1} \in t_i$, then the random walk will consist only of $\{v_{i+1}\}$.) Now create t_{i+1} by adding to t_i the loop erasure of this random walk. Then t_{i+1} is a tree. The output of Wilson's algorithm is the set of edges of the tree $T = t_n$. (The root is forgotten.)

Wilson (1996) proved that if G is a finite graph, then Wilson's method yields a uniform spanning tree.

Boundary conditions

In the study of spanning clusters for the cube, one can consider several uniform spanning tree measures. The most interesting is as follows. Pick a tree according to the uniform spanning tree measure in all of $\mathbb{Z}^d$, and look at the restriction to the n-cube. This measure will be used in the analysis of the number of spanning clusters in dimensions $d \geq 4$. Another natural measure is to start with a finite n-cube, ignoring the rest of the lattice, and pick a spanning tree uniformly, this is the free uniform spanning tree. For the study of spanning clusters in 2D, we will use a variant of this measure, that is, we will consider the uniform measure on spanning trees of the n-square, when all the vertices on the right side are identified (wired) to a single vertex. Call this measure, right wired. In this setup, two spanning clusters are disjoint if they intersect only in the rightmost wired vertex. The study of spanning clusters for the $\mathbb{Z}^2$ uniform spanning tree measure restricted to the square (the first measure above), is much harder. Regarding one possible approach in this setup, see the remark after the proof of Theorem 10.2.1.

With these measures in mind, we have,

Theorem 10.2.1 (number of spanning clusters) *For $d = 2$, with the right wired measure, the number of spanning clusters is tight as the size of the square grows. For $d \geq 4$, the expected number of spanning clusters grows to infinity with the size of the cube.*

Remarks

1) We can't decide the case $d = 3$, but **conjecture** tightness to hold.

2) Pemantle (1991), based on work by Lawler on loop erased random walk, proved that the uniform spanning forest of $\mathbb{Z}^d$ is a tree a.s., iff $d \leq 4$. Thus, the case $d = 4$, is interesting, since there is a unique infinite cluster, yet many spanning clusters.

3) For critical percolation in 2D, Aizenman (1997) proved that the probability of having k vertex disjoint crossings is bounded between $Ae^{-\alpha k^2}$ and $e^{-\alpha' k^2}$, for some constants A, α and α'. Cardy (1997) predicts that the exact exponent is $-(2\pi/3)k(k-1/2)$. Duplantier (private communication) predicts the exponent for the uniform spanning tree to be $-(\pi/2)k(k-1)$.

4) Consider the uniform spanning tree in $\mathbb{Z}^2$. Fix $c > 0$. What is the limit as n grows of the probability that there exists a spanning cluster for the $n \times cn$-rectangle? Cardy (1992) proposed an answer to this for critical Bernoulli percolation.

5) Recently Lawler (1998) gave a lower bound on the growth exponent for loop-erased random walk in two dimension, which is strictly bigger than 1. Aizenman and Burchard (1998), studied regularity properties of random curves, obtained as scaling limits. The theorem above and simple crossings bounds obtained by Wilson's algorithm provide the analogue of the "Russo-Seymour-Welsh theory" needed in order for their work to apply.

For a study of the large scale degrees, we will consider the $3n$-square in Z^2, with free boundary conditions. And we will define large scale degree, as follows:

Definition Consider the n-square centered in the $3n$-square. The large scale degree of a vertex v, in the n-square, denoted $lsd(v)$, is the maximal number of disjoint paths in the tree that connect v to the boundary of the $3n$-square.

In this setup, we have

Theorem 10.2.2 (large scale degree)

1. $\#\{v \in C_n^d \mid lsd(v) > 1\} = o(n^d)$ *in probability, and*
$$E\Big(\#\{v \in C_n^2 \mid lsd(v) > 1\}\Big) < Cn^{4/3}\log n.$$
2. $E(\#\{v \in C_n^2 \mid lsd(v) > 2\})$ *is uniformly bounded in n.*
3. $P(\exists v \in C_n^2, lsd(v) > 3)$ *goes to* 0, *with n.*

Remarks

1) See Figure 10.1 for an illustration of the theorem.

2) The 4/3 in item 1 should be improved to 5/4, which is believed to be the growth exponent for the loop-erased random walk (see Duplantier (1992)). Item 2 might be true also in dimension 3, and maybe item 3 holds in any dimension.

3) Since the number of vertices v with $lsd(v) > 2$ is uniformly bounded in n, it's limit distribution, if it exists, might serve as a natural conformal invariant for the annulus.

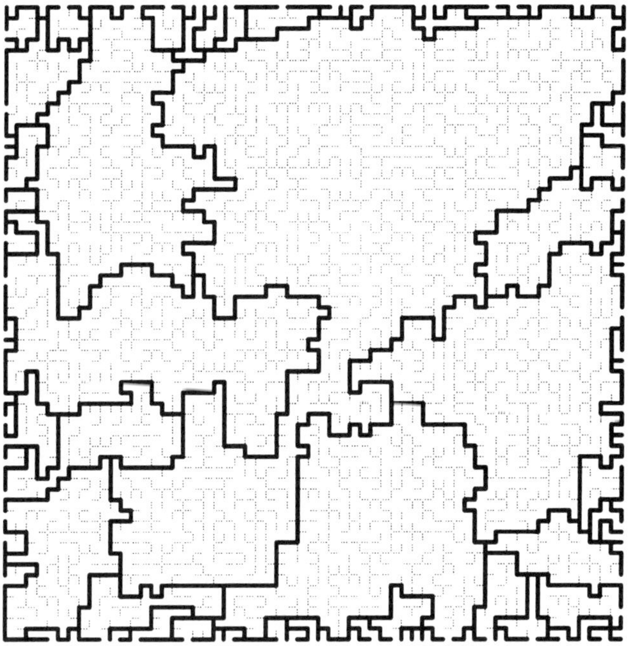

FIGURE 10.1. One picture (by Oded) is better than a thousand words (by ...).

10.3 Proofs and further comments

We start with spanning clusters.

PROOF OF THEOREM 10.2.1. We start with $d = 2$. Suppose that there is at least one crossing, and condition on the upper left-right crossing. The configuration of the uniform spanning tree in the domain below this crossing is independent from the configuration above, and is just the uniform spanning tree for the domain wired at the upper crossing and the right boundary. With probability at least $1/2$, simple random walk starting at the bottom left corner of the square will hit the upper crossing before hitting the right side of the square. Therefore, by Wilson's algorithm, with probability at least $1/2$ there is a path in the spanning tree, connecting the bottom left vertex to the upper crossing, which is disjoint from the right wired side. On this event, there is just one left-right crossing. By each time conditioning on the next upper crossing, for the remaining domain, and by the argument above, we get that the probability of having k disjoint crossings is bounded by 2^{-k}.

Remark Note that if you consider k independent simple random walks starting equally spaced on the left side of the square, then there is $c > 0$, so that with probability bounded below by c^{-k^2}, the k simple random walks will reach the right side without any intersection. By Wilson's algorithm, this gives a lower bound for the probability of having k disjoint spanning clusters.

We turn now to $d \geq 4$. As was pointed out in Benjamini, Lyons, Peres and Schramm (1998) Theorem 5.1, the uniform spanning tree measure for Z^d, $d \geq 3$, can be constructed via Wilson's method rooted at infinity, that is, start simple random walk at some vertex, and loop erase the infinite path. Then pick a vertex not on the path. Start simple random walk from the new vertex till it hits the first path or the root (infinity) and loop erase, and so on. So the case $d \geq 4$, for that measure, follows via Wilson's algorithm from standard properties of simple random walk in Z^d, $d \geq 4$. See Lawler (1991) for the fact that the probability that 2 and hence any k paths of k independent simple random walks, starting distance n apart, will intersect before reaching distance Cn (from one of the starting vertices), is going to 0, for any fixed $C > 0$, as n grows. In dimension 4, the probability that two random walks starting at distance n apart intersect before reaching distance Cn is of order $c/\log n$. In Benjamini, Kesten, Peres and Schramm (1998), it is shown that in $d > 4$, the expected size of the tree containing 0, in the uniform spanning forest, inside a ball of radius n, is of order Cn^4. Thus inside the n-cube one expects about n^{d-4}, large connected components, and hence, up to a constant, as many spanning clusters. □

Note that if we work with a measure as in the 2D, wired on the "right" face and ignoring the rest of the lattice, then the same proof works.

How many spanning clusters are there in 4D?

Remark Benjamini, Lyons, Peres and Schramm (1998) conjectured the following BK-Type Inequality: Denote by E the set of edges in a finite graph. We say that $A, B \subset 2^E$ occur disjointly for $F \subset E$ if there are disjoint sets $F_1, F_2 \subset E$ such that $F' \in A$ for every F' with $F' \cap F_1 = F \cap F_1$ and $F' \in B$ for every F' with $F' \cap F_2 = F \cap F_2$. Let $A, B \subset 2^E$ be increasing. Then the probability that A and B occur disjointly for the random spanning tree T is at most $P[T \in A]P[T \in B]$. (The BK inequality of van den Berg and Kesten says that the same is true when T is a random subset of E chosen according to any product measure on 2^E, see Grimmett (1989).)

If true, this will allow us to remove the wired boundary, and to consider the uniform spanning tree of $\mathbb{Z}^2$, restricted to an n-square. Also it will help in getting better tail estimates.

PROOF OF THEOREM 10.2.2. The first part of 1 follows from the fact that any tree in the uniform spanning forest of $\mathbb{Z}^d$, has a.s. one end. (See Pemantle for dimensions $2 \leq d \leq 4$, and Benjamini, Lyons, Peres and Schramm (1998), for a simple proof in 2D and for all other dimensions.) In particular for a fixed v the probability that v is in the middle of a path of length m, in the tree, goes to 0, with m. The uniform spanning tree measure on $\mathbb{Z}^d$ is a weak limit of the free uniform spanning tree measure on the n-cube. So first pick m so that in the uniform spanning tree of $\mathbb{Z}^d$, the origin has chance less than ϵ of connecting the origin in 2 disjoint paths to distance $m/2$, and then freeze m and pick n so that the chance of the cylinder event above in the free spanning tree in the n-cube is smaller than 2ϵ.

The sharper 2D case follows from duality. The dual of the free spanning tree on the $3n$-square is the wired spanning tree. See Benjamini, Lyons, Peres and Schramm (1998) for a discussion of the duality and applications. If $lsd(v) > 1$ then in the dual tree, on the dual lattice, there should be two vertices at distance 1 (neighbours of v in the dual) that are connected in the wired spanning tree only through the boundary. But by Lawler's (1991) upper bound on intersection probabilities, via Beurling's projection theorem which can be used here, this has probability bounded by $n^{-2/3} \log n$; multiplying by n^2 we get item 1.

Note that, since the set of vertices with $lsd(v) = 2$ contains a loop-erased path, which get to the boundary of the $3n$-square twice, we get that the number of vertices on such a path, inside the n-square, is bounded by $n^{4/3} \log n$, recovering another result in Lawler (1991).

We move to item 2. In Theorem 10.2.1, the exponential bound on the number of crossings was proved for a square with the right side wired, here we need a variant regarding crossings of an annulus, between the n-square and the $3n$-square, wired at the boundary of the $3n$-square. Following the argument in the proof of the first part of Theorem 10.2.1 we get that the probability for k crossings from, say, the top side of the n-square to the $3n$-square decays exponentially. Yet $P(\text{there are } k \text{ crossings of the annulus}) \leq 4P(\text{there are } k/4 \text{ crossings from the top side})$.

Now if there are k vertices in the n-square with $lsd(v) > 2$, then by the tree structure there are at least $k + 1$ disjoint crossings from the boundary of the n-square, to the boundary of the $3n$-square. Again by the duality between the free and the wired, there are $k+1$ or more disjoint crossings in the dual wired spanning tree, connecting the boundary of the n-square to the wired boundary of the $3n$-square. By the annulus version of the first part of Theorem 10.2.1, the probability of having such k crossings decays exponentially with k.

Remark Note that as n grows the mean number of vertices with $lsd(v) > 2$ is bounded away from 0. Using Wilson's algorithm it is not hard to see that with probability bounded away from 0, with n, the meeting vertex for the tree spanned by the mid-vertex on the top side of the $3n$-square, and the two mid-vertices on the left and right sides, will be inside the middle n-square.

To prove 3, start growing the tree in the $3n$-square via Wilson's algorithm. Assume that at some stage a vertex v, with $lsd(v) = 3$, was created. Part of the boundary of the $3n$-square can no longer connect to v, to increase it's large scale degree. For any vertex on the $3n$-boundary for which simple random walk can in principle hit v, before hitting the rest of the tree, the probability this will happen is going to zero, with n, by the discrete version of Beurling's projection theorem (see Kesten (1987)). Pick u_1 on the boundary to be the middle vertex of the part of the boundary that can hit the tree at v. Assume simple random walk, starting at u_1, failed to hit the tree at v. Pick u_2 as the middle of the boundary part, which can still hit the larger tree at v, and define u_i inductively in the same way, by each time picking the middle vertex of the part of the boundary that still can hit the previously constructed tree at v. Let p_i be the probability of a simple random walk reaching distance n from u_i, without hitting the part of the boundary, that can no longer "connect" to v. The p_i's decay exponentially with i, with a rate that is independent of n. By Beurling's projection the p_i's bound from above the probability that simple random walk starting at u_i will hit the tree, constructed so far, at v. We get that the probability that the large scale degree of v will increase to 4 is going to 0, with n. Since the probability that the first trails will hit v, is going to zero with n, and the probability the later trails will succeed is small by the uniform exponential decay. By 2, the number of vertices with $lsd(v) = 3$ is tight, and 3 follows. □

Acknowledgments: I would like to thank Bertrand Duplantier, Balint Toth and Wendelin Werner for useful discussions, Yuval Peres and Oded Schramm for simplifying some of the proofs and David Wilson for remarks on a preliminary version.

REFERENCES

[1] M. Aizenman, On the number of incipient spanning clusters, *Nucl. Phys.* B[FS] **485**, 551 (1997).
[2] M. Aizenman and A. Burchard, Hölder Regularity and dimension bounds for random curves, preprint (1998).
[3] I. Benjamini, H. Kesten, Y. Peres, and O. Schramm, Paper in preparation (1998).
[4] I. Benjamini, R. Lyons, Y. Peres, and O. Schramm, Uniform spanning forests, preprint (1998).
[5] C. Borgs, J. Chayes, H. Kesten, and J. Spencer, Uniform boundedness of critical crossing probabilities implies hyperscaling, preprint (1998).
[6] R. Burton and R. Pemantle, Local characteristics and limit theorems for spanning trees and domino tilings via transfer-impedances, *Ann. Prob.* **21**, 1329–1371 (1993).
[7] J. Cardy, Critical percolation in finite geometries, *J. Phys. A* **25**, 201–206 (1992).
[8] J. Cardy, The number of incipient spanning clusters in two-dimensional percolation, cond-mat/9705137 (1997).
[9] B. Duplantier, Loop-erased self-avoiding walks in 2D, *Physica A* **191**, 516–522 (1992).
[10] G. Grimmett, *Percolation*, Springer-Verlag, New York (1989).
[11] O. Häggström, Random-cluster measures and uniform spanning trees, *Stoch. Proc. Appl.* **59**, 267–275 (1995).
[12] H. Kesten, Hitting probabilities of random walks on $\mathbb{Z}^d$, *Stoc. Proc. and Appl.* **25**, 165–184 (1987).
[13] G. Lawler, *Intersections of Random Walks.* Birkhäuser, Boston (1991).
[14] G. Lawler, A lower bound on the growth exponent for loop-erase random walk in two dimension, preprint (1998).
[15] R. Pemantle, Choosing a spanning tree for the integer lattice uniformly, *Ann. Prob.* **19**, 1559–1574 (1991).

[16] D. Wilson, Generating random spanning trees more quickly than the cover time, *1996 ACM Sympos. Theory of Computing*, 296–303 (1996).
[17] O. Schramm, Scaling limits of loop-erased random walks and random spanning trees, preprint (1998).

Faculty of Computer Science
Weizmann Institute
Rehovot, Israel 76100
itai@wisdom.weizmann.ac.il

11

On the Absence of Phase Transition in the Monomer-Dimer Model

J. van den Berg

ABSTRACT Suppose we cover the set of vertices of a graph G by non-overlapping monomers (singleton sets) and dimers (pairs of vertices corresponding to an edge). Each way to do this is called a monomer-dimer configuration. If G is finite and $\lambda > 0$, we define the monomer-dimer distribution for G (with parameter λ) as the probability distribution which assigns to each monomer-dimer configuration a probability proportional to $\lambda^{|\text{dimers}|}$, where $|\text{dimers}|$ is the number of dimers in that configuration. If the graph is infinite, monomer-dimer distributions can be constructed in the standard way, by taking weak limits.

We are particularly interested in the monomer-dimer model on (subgraphs of) the d-dimensional cubic lattice. Heilmann and Lieb (1972) prove absence of phase transition, in terms of smoothness properties of certain thermodynamic functions. They do this by studying the location in the complex plane of the zeros of the partition function.

We present a different approach and show, by probabilistic arguments, that boundary effects become negligible as the distance to the boundary goes to ∞. This gives absence of phase transition in a related, but generally not equivalent sense as above. However, the decay of boundary effects appears to occur in such a strong way that, by results on general Gibbs systems of Dobrushin and Shlosman (1987) and Dobrushin and Warstat (1990), smoothness properties of thermodynamic functions follow. More precisely we show that, in their terminology, the model is *completely analytic*.

Keywords: Phase transition, monomer-dimer model, matchings, percolation.

AMS Subject Classifications: Primary 60K35; Secondary 62M40, 82B26, 05C70.

11.1 Introduction

One of the purposes of this paper is to give a clear example of the use of 'disagreement percolation' arguments in the study of spatial dependencies in Markov fields. These arguments were introduced in [2], and further developed in [3]–[5] and [10]. There are no totally new ideas in this paper, so in some sense it is a review-like paper. However, although the *proof* of the main result, Theorem 1, is (at least in spirit) 'almost' in [2] and [5] and in [13], the result itself has not been explicitly observed before. This, plus the fact that this result and its proof method may shed some new light on the classical Heilmann-Lieb results, are the main motivation for this paper.

I have tried to write the paper in such a way that it has something interesting for experts as well as beginners in the field of Gibbs measures and phase transitions. As to the latter group of readers, although the paper is not self-contained and part of it assumes some general knowledge about Gibbs measures, I hope that it gives at least an idea of the kind of problems and methods in this field, and that it is entertaining enough to stimulate further study. To avoid confusion: this paper is *not* about the *pure dimer* model (in which monomers are not allowed), which has in some sense a more interesting behaviour (see [14]).

Since percolation-like arguments play an important role in this paper, it is not out of place in a volume in honor of Harry Kesten. What made me hesitate is the level of difficulty of this paper, which may be somewhat low for his taste. However, 'not difficult' is not always the same as 'straightforward' and I hope that in this case the simplicity of the arguments has something unexpected and charming.

The monomer-dimer model originates from statistical physics: dimers represent molecules which occupy two adjacent vertices of a graph (crystal), and monomers molecules which occupy a single vertex (see the introduction of [11] for further background and references). Molecules are not allowed to overlap. The energy contribution of a dimer may be different from that of a monomer. By standard statistical-physics considerations it is then natural to assign to each monomer-dimer configuration a probability proportional to $\lambda_1^{|\text{monomers}|} \times \lambda_2^{|\text{dimers}|}$, where $|\text{monomers}|$ and $|\text{dimers}|$ denote the number of monomers and the number of dimers in the configuration respectively. Since the number of monomers plus twice the number of dimers equals the total number of vertices (hence does not depend on the configuration), the above is (assuming $\lambda_1 > 0$) exactly the same as assigning to each configuration a probability proportional to $\lambda^{|\text{dimers}|}$, where $\lambda = \lambda_2/\lambda_1^2$. The case $\lambda_1 = 0$ gives the pure dimer model which, as we mentioned before, will not be studied in this paper.

The monomer-dimer model also arises in operations research, combinatorial optimization and theoretical computer science. For instance, some vertices may represent tasks and the others persons, and the aim may be to match persons with tasks (not more than one task per person and vice versa). The edges indicate which person-task pairs are allowed. Since, for large λ, 'typical' configurations have many dimers, it makes sense to sample randomly from the monomer-dimer distribution with large λ in the hope to get a close to optimal matching. For rigorous results on how to do this 'efficiently' and related problems, see [12]. See also Remark 1 in Section 11.4.

More formally, the monomer-dimer model is described as follows: Let G be a finite graph. Denote the set of vertices of G by $V(G)$ and its set of edges by $E(G)$. We will often write just V and E respectively. If two vertices i and j are adjacent (i.e., share an edge), we write $i \sim j$. If e and e' are edges, $e \sim e'$ means that e and e' have a common endpoint. The *state space* Ω is given by $\Omega = \{0, 1\}^E$. Elements of Ω are called configurations and will typically be denoted by $\omega(= (\omega_i, i \in E))$, α etc. To denote a randomly chosen element from Ω we will mostly use the notation

σ. A configuration ω is called *feasible* if there are no edges e, e' with $e \sim e'$ and $\omega_e = \omega_{e'} = 1$.

Now consider the following probability distribution on Ω:

$$\mu(\omega) = \frac{\lambda^{|\omega|} I(\omega \text{ is feasible})}{Z}, \tag{11.1.1}$$

where $\lambda > 0$ is the parameter of the model, $|\omega|$ denotes $\sum_{e \in E} \omega_e$, and Z (which depends of course on λ) the normalizing constant (partition function). We call this the monomer-dimer distribution for G with parameter λ.

It should be clear that this corresponds with the more informal description given before. For instance, if $\omega_e = 1$, this means that the pair of endpoints of e is covered by a dimer. If, for a vertex i, each edge e having i as one of its endpoints has $\omega_e = 0$, then i is covered by a monomer. Feasibility means that no molecules overlap, $|\omega|$ is the number of dimers, etc.

Before we go on we need some more notation. If $\Lambda \subset E$, Ω_Λ denotes $\{0, 1\}^\Lambda$. If $\omega \in \Omega$, ω_Λ is the 'restriction' of ω to Λ, i.e., $\omega_\Lambda = (\omega_i, i \in \Lambda)$. Further, $\partial \Lambda$ denotes the boundary of Λ, that is, the set of all $e \in E \setminus \Lambda$ with $e \sim e'$ for some $e' \in \Lambda$. (If Λ has only one element, say i, we usually write ∂i instead of $\partial \{i\}$). If U and Λ are disjoint subsets of E, $\alpha \in \Omega_U$ and $\beta \in \Omega_\Lambda$, then $\alpha\beta$ denotes the 'concatenation' of α and β, i.e., the element of $\Omega_{U \cup \Lambda}$ for which $(\alpha\beta)_e$ equals α_e for all $e \in U$ and β_e for all $e \in \Lambda$. Finally, if A and B are sets, $A \subset\subset B$ means that A is a finite subset of B.

Now let G be an infinite, connected locally finite graph. Monomer-dimer distributions on G are defined in the standard way, as follows. Let $\lambda > 0$. Define, for each $\Lambda \subset E$ and each $\alpha \in \Omega_{\partial \Lambda}$ the monomer-dimer distribution for Λ with *boundary condition* α by

$$\mu_\Lambda^\alpha(\omega) = \frac{\lambda^{|\omega|} I(\omega \text{ is feasible w.r.t. } \alpha)}{Z_\Lambda}, \tag{11.1.2}$$

where the expression in the indicator function means that the configuration ω is feasible in the sense given before, and that, moreover, there are no $e \in \Lambda$ and $e' \in \partial \Lambda$ with $e \sim e'$ and $\omega_e = \alpha_e = 1$. A distribution μ on Ω is now called an (infinite-volume) monomer-dimer distribution for G (or infinite-volume Gibbs measure for the monomer-dimer model on G) if, for all $\Lambda \subset\subset E$, and all $\alpha \in \Omega_\Lambda$,

$$\mu(\sigma_\Lambda = \alpha \mid \sigma_e, e \in \Lambda^c) = \mu_\Lambda^{\sigma_{\partial \Lambda}}(\alpha), \ \mu - \text{a.s.}$$

Such distributions can be constructed in the standard way, as follows: take a nested sequence $\Lambda_1, \Lambda_2, \ldots$ of finite subsets of E with union E, and a sequence of boundary conditions $\alpha(i) \in \Omega_{\partial \Lambda_i}$ $i = 1, 2, \ldots$. Then take weak limits of subsequences of the sequence $\mu_1, \mu_2, \ldots$ on Ω, defined by

$$\mu_i(\sigma_{\Lambda_i} = \beta) = \mu_{\Lambda_i}^{\alpha(i)}(\beta), \ \beta \in \Omega_{\Lambda_i}, \ i = 1, 2, \ldots, \text{ and}$$

$$\mu_i(\sigma_{\Lambda_i^c} \equiv 0) = 1, \ i = 1, 2, \ldots.$$

A central question is whether there is a unique infinite-volume Gibbs measure. The above suggests that this is the case if, roughly speaking, the influence of the boundary condition becomes negligible as the boundary moves to infinity. This can indeed be stated in a precise way and is not difficult to prove. If the infinite-volume Gibbs measure is not unique, we say that there is a phase transition. An alternative notion of (absence of) phase transition is given in terms of smoothness properties of so-called thermodynamic functions, in particular the free energy function $\lim_{n\to\infty}(\log Z_{\Lambda_n})/|\Lambda_n|$ (this limit exists for sufficiently nice graphs and appropriate sequences (Λ_n)). For further study of these matters (for general Gibbs measures) see [9] and [8].

We will now concentrate on the d-dimensional cubic lattice, although many of the results hold in much more generality. This is the graph whose vertices are the elements of $\mathbf{Z}^d$ and where two vertices have an edge if their euclidian distance equals 1. With some abuse of notation we denote this graph simply by $\mathbf{Z}^d$.

Heilmann and Lieb (1972) proved in [11] (which has become a classical paper on the subject) absence of phase transition in terms of smoothness properties mentioned above. They did this by studying the zeros of the partition function (as a function on the complex plane). We follow a different, more probabilistic approach, and study the influence of boundary conditions. This is done by applying quite general percolation-like arguments introduced in [2]. In fact, the monomer-dimer model offers one of the nicest examples to illustrate these arguments: we will show that for every λ the above mentioned decay of boundary effects is so strong that one of the so-called *complete analyticity conditions* (namely condition III_c in [7]) of Dobrushin and Warstat (1990) holds. To state this condition we need a little more terminology and notation. If $\Lambda \subset\subset E$, $\Delta \subset \Lambda$ and $\alpha \in \Omega_{\partial\Lambda}$, $\mu^{\alpha}_{\Lambda,\Delta}$ denotes the marginal distribution on Ω_Δ of μ^{α}_{Λ}. Recall that if μ and ν are two probability distributions on a finite set A, then the *variational distance* between μ and ν is defined by $\mathrm{Var}(\mu, \nu) = 1/2 \sum_{i \in A} |\mu(i) - \nu(i)|$. This is also equal to $\max_{B \subset A} |\mu(B) - \nu(B)|$. The complete analyticity condition mentioned above can then be formulated as follows:

Dobrushin-Warstat condition III_c *There exist K and $\kappa > 0$ such that for all $\Lambda \subset\subset E$, $\Delta \subset \Lambda$, $e \in \partial\Lambda$, and all $\alpha, \beta \in \Omega_{\partial\Lambda}$ with $\alpha_s = \beta_s$, $s \neq e$, the following holds:*

$$Var(\mu^{\alpha}_{\Lambda,\Delta}, \mu^{\beta}_{\Lambda,\Delta}) \leq K \exp(-\kappa\, \rho(e, \Delta)), \tag{11.1.3}$$

where $\rho(e, \Delta)$ is the minimal length of a path which starts in e and ends in Δ.

Remarks

(i) The above is an obvious translation of condition III_c of Dobrushin and Warstat ([7]). In the original formulation the randomness is attached with the vertices instead of the edges, but this is immaterial.

(ii) It is easy to see that this condition implies uniqueness of the infinite-volume Gibbs measure. In fact, it is much stronger.

(iii) Dobrushin and Warstat state nine other conditions and prove that all ten conditions are 'essentially' equivalent (see the end of this remark). Some of those conditions are concerned with the zeros of the partition function, (and of the partition functions corresponding with small perturbations of the interactions) and imply that the free energy and related functions have analytic continuations (which partly explains the name complete analyticity).

The word 'essentially' in the beginning of this remark needs some explanation: In Section 2 of their paper Dobrushin and Warstat define a natural metric on the space of interactions (of bounded range). For each subset of interactions, the *main component* of that set is defined as its maximal connected open subset containing the 0 interaction. More precisely now, their main result is that for each of the ten conditions, the set of interactions satisfying that condition has the same main component. This common component is then called the class of completely analytic interactions.

(iv) The work of Dobrushin and Warstat is an extension of the results by Dobrushin and Shlosman in [6]. The Dobrushin-Warstat results are applicable to a larger class of interactions, including the monomer-dimer model.

Our main result is Theorem 11.1.1 below:

Theorem 11.1.1 *The monomer-dimer model on* $\mathbf{Z}^d$ *is completely analytic.*

Remark As the remarks after (11.1.3) indicate, complete analyticity is a very strong property, and, as far as I understand, does not follow from the results of Heilmann and Lieb (1972). In particular, the perturbations mentioned in Remark iii above are much more general than those only involving the 'ordinary' model parameters. To give one example, it also includes the possibility of adding an extra factor $x^{|\text{pairs}|}$ to (11.1.1), with the 'extra' parameter x sufficiently close to 1, where $|\text{pairs}|$ denotes the number of pairs of distinct edges $\{e, e'\}$ s.t. $\omega_e = \omega'_e = 1$ and for which there is an edge e^* with $e \sim e^* \sim e'$. On the other hand, although the analyticity notion in the Heilmann-Lieb paper is restricted to the 'ordinary' model parameters they have several detailed results which do not follow from ours.

In Section 11.2 we review the earlier mentioned percolation-like method for getting upper bounds for boundary effects. The main result in that section, Lemma 11.2.1, holds for all so-called Markov fields. Although it is, in spirit, already in [2], we state it here because the argument in [2] deals immediately with infinite-volume Gibbs measures, while here we need a more explicit expression for the boundary effects for finite graphs. In Section 11.3 we apply the Lemma to the special case of the monomer-dimer model, to prove Theorem 11.1.1. Section 11.4 gives a brief discussion and also mentions some other results in the literature.

11.2 Disagreement Percolation and Markov Fields

For those not familiar with the topic we explain the concept of Markov fields (on finite graphs): Let G be a finite graph, S a finite set and the state space $\Omega = S^V$, where $V = V(G)$ is the set of vertices of G. Note that the state space here involves, as is standard in the literature, the vertices instead of the edges, while in the formulation of the monomer-dimer model it was the other way round. We come back to this soon. We use the obvious analogs and generalizations of the notation in Section 11.1. For instance, if $\Lambda \subset V$, then $\partial\Lambda$ denotes $\{i \in \Lambda^c : \exists j \in \Lambda \text{ with } j \sim i\}$. And, if μ is a probability distribution on Ω, $\Lambda \subset V$ and $\alpha \in S^{\partial\Lambda}$ with $\mu(\sigma_{\partial\Lambda} = \alpha) > 0$, then μ_Λ^α denotes the conditional distribution of σ_Λ given $\sigma_{\partial\Lambda} = \alpha$. A probability distribution μ on Ω is called a Markov field if for all $\Lambda \subset V$ the conditional distribution of σ_Λ given σ_{Λ^c} depends only on $\sigma_{\partial\Lambda}$. Well-known examples in the literature are the Ising model and the hard-core lattice gas model. In the latter model we take $S = \{0, 1\}$ and

$$\mu(\omega) = \frac{\lambda^{|\omega|} I(\omega \text{ is feasible})}{Z},$$

where $|\omega| = \sum_{i \in V} \omega_i$ and 'ω is feasible' means that there are no vertices i, j with $i \sim j$ and $\omega_i = \omega_j = 1$. Comparison with (11.1.1) and a few moments reflection show that the monomer-dimer model on a graph G corresponds to the hard-core model on the so-called line graph of G. (The line graph of G is the graph whose vertices correspond to the edges of G, and in which two vertices have an edge if the corresponding two edges in G have a common endpoint). More about this correspondence below. It is easy to check that the hard-core model is indeed a Markov field. In particular we have the following observation: For each $i \in V$, the conditional probability that $\sigma_i = 1$ given the $\sigma-$values of all the other vertices, is $\lambda/(1+\lambda)$ if $\sigma_{\partial i} \equiv 0$, and 0 otherwise. This will be used in Section 11.3.

As we said before, we want to give an upper bound for boundary effects. So let $\Delta \subset \Lambda \subset V$, and $\alpha, \beta \in S^{\partial\Lambda}$. We want to estimate the variational distance between the marginal distribution on Ω_Δ for boundary condition α and that for boundary condition β. We now introduce the notion 'path of disagreement': If $\omega, \omega' \in S^\Lambda$ and $j \in \Lambda$, then we say that the pair (ω, ω') has a path of disagreement from j to Δ if there is a sequence $j = i_1 \sim i_2 \sim \cdots \sim i_n$, of vertices in Λ with $i_n \in \Delta$ and $\omega_{i_k} \neq \omega'_{i_k}$, $k = 1, \ldots, n-1$. Further, we say that (ω, ω') has a path of disagreement from $\partial\Lambda$ to Δ if there is a $j \in \partial\Lambda$ with $\alpha_j \neq \beta_j$, an $i \in \Lambda$ with $i \sim j$ and a path of disagreement from i to Δ. The following lemma is essentially in [2], but not explicitly stated there.

Lemma 11.2.1

$$Var(\mu_{\Lambda,\Delta}^\alpha, \mu_{\Lambda,\Delta}^\beta) \leq (\mu_\Lambda^\alpha \times \mu_\Lambda^\beta)(\exists \textit{ a path of disagreement from } \partial\Lambda \textit{ to } \Delta). \tag{11.2.1}$$

Remark In words, Lemma 11.2.1 says that the variational distance between the 'restrictions to Δ' of the two distributions μ_Λ^α and μ_Λ^β is smaller than or equal to the probability that, if we randomly and independently choose two realizations in Ω_Λ, one with the first distribution above, the other with the second distribution, this pair of realizations has a path of disagreement from $\partial\Lambda$ to Δ. The r.h.s. of (11.2.1) involves the product of μ_Λ^α and μ_Λ^β which can be considered as a trivial coupling of these distributions. A more clever coupling is given in [4]. However, it appears that for hard-core models these couplings are essentially the same.

PROOF OF LEMMA 11.2.1. Let $A \subset S^\Lambda \times S^\Lambda$ be the complement of the event in the r.h.s. of the inequality in the lemma. Define the "cluster of disagreement of Δ (w.r.t. the pair (ω, ω')", notation $C(\Delta)$, by

$$C(\Delta) = \Delta \cup \{i \in \Lambda : (\omega, \omega') \text{ has a path of disagreement from } i \text{ to } \Delta\}.$$

Let $T : \Omega_\Lambda \times \Omega_\Lambda \to \Omega_\Lambda \times \Omega_\Lambda$ be the map which exchanges ω and ω' on $C(\Delta)$. More precisely,

$$T(\omega, \omega') = (\omega'_{C(\Delta)}\omega_{\Lambda\setminus C(\Delta)}, \omega_{C(\Delta)}\omega'_{\Lambda\setminus C(\Delta)}),$$

where we have used the notation for concatenation given in Section 11.1. It is clear that T is a 1-1 map and that $(\omega, \omega') \in A \Leftrightarrow T((\omega, \omega')) \in A$. Moreover, it follows immediately from the definition of A that if $(\omega, \omega') \in A$ then $(\omega\alpha)_{\partial(C(\Delta))} = (\omega'\alpha)_{\partial(C(\Delta))}$. Combining these observations with the Markov property, we see that T preserves the measure $\mu_\Lambda^\alpha \times \mu_\Lambda^\beta$ on A. Now let $B \subset S^\Delta$. We have the following trivial identity:

$$\begin{aligned}
\mu_{\Lambda,\Delta}^\alpha(B) &= \mu_\Lambda^\alpha(\sigma_\Delta \in B)\\
&= (\mu_\Lambda^\alpha \times \mu_\Lambda^\beta)(\{\omega, \omega') \in \Omega_\Lambda \times \Omega_\Lambda : \omega_\Delta \in B\})\\
&= (\mu_\Lambda^\alpha \times \mu_\Lambda^\beta)(\{(\omega, \omega') \in \Omega_\Lambda \times \Omega_\Lambda : \omega_\Delta \in B,\ (\omega, \omega') \in A\})\\
&+ (\mu_\Lambda^\alpha \times \mu_\Lambda^\beta)(\{(\omega, \omega') \in \Omega_\Lambda \times \Omega_\Lambda : \omega_\Delta \in B,\ (\omega, \omega') \in A^c\}).
\end{aligned} \tag{11.2.2}$$

Similarly, we have

$$\begin{aligned}
\mu_{\Lambda,\Delta}^\beta(B) &= (\mu_\Lambda^\alpha \times \mu_\Lambda^\beta)(\{(\omega, \omega') \in \Omega_\Lambda \times \Omega_\Lambda : \omega'_\Delta \in B,\ (\omega, \omega') \in A\})\\
&+ (\mu_\Lambda^\alpha \times \mu_\Lambda^\beta)(\{(\omega, \omega') \in \Omega_\Lambda \times \Omega_\Lambda : \omega'_\Delta \in B,\ (\omega, \omega') \in A^c\}).
\end{aligned} \tag{11.2.3}$$

However, the image under T of the event in the first term in the r.h.s. of (11.2.2) is exactly the event in the first term of the r.h.s. of (11.2.3). Hence, since T is measure-preserving on A, the absolute value of the difference between the l.h.s. of (11.2.2) and (11.2.3) is smaller than or equal to the maximum of the second term in the r.h.s. of (11.2.2) and that in the r.h.s. of (11.2.3), which completes the proof of the lemma. □

The next step in [2] and [5] is a stochastic domination argument: Let, for $i \in \Lambda$, p_i be the maximum over all pairs $\alpha, \beta \in S^{\partial i}$ of $(\mu_i^\alpha \times \mu_i^\beta)(\{(s, s') \in S \times S : s \neq s'\})$. (In many models these values can be easily calculated explicitly). The r.h.s. of (11.2.1) is then bounded above by the probability that, for the independent site percolation model on G with parameters p_i, $i \in V$, there is an open path from $\partial\Lambda$ to Δ. In many cases, for instance when G is an $n \times n$ box of the square lattice, known results from percolation can then be used to prove asymptotic properties of the boundary influences.

The hard-core model, described in the beginning of this section, has the special feature that the vertices on a path of disagreement are alternatingly occupied in ω

(and hence empty in ω') and vice versa. Using this we can, for that model, refine the argument and replace the above given expression for p_i by the maximum over all $\alpha \in S^{\partial i}$ of $\mu_i^\alpha(1)$, which is simply equal to $\lambda/(1+\lambda)$. This leads then to the result of [5] that if this last expression is smaller than the critical probability for site percolation on an infinite graph G, then the hard-core model on G with parameter λ has a unique Gibbs measure. In fact, by using exponential-decay results for sub-critical percolation, which were first obtained by Kesten in his famous paper [15] for bond percolation on the square lattice, and later in more generality by Aizenman and Barsky [1] and Menshikov [17], we even get complete analyticity.

In the light of the earlier made observation that the monomer-dimer model corresponds to the hard-core model on the line graph, it is tempting to simply translate the above steps to the monomer-dimer model. This would, for instance, lead to the result that if $\lambda/(1+\lambda)$ is smaller than the critical probability for *bond* percolation on $\mathbf{Z}^d$, the monomer-dimer model on that graph is completely analytic. However, this is not sufficient for our purpose (to prove Theorem 11.1.1), because we want this result for *all* λ. This stronger result is obtained by being a little more patient and by observing that, in addition to the above mentioned special feature of the more general hard-core model, the monomer-dimer model has another simple but essential special feature. This will be explained in the next section.

11.3 Proof of Theorem 11.1.1

We now apply the quite general Lemma 11.2.1 to the special case of the monomer-dimer model, to prove Theorem 11.1.1. So let $\Delta \subset \Lambda \subset\subset \mathbf{E}^d$, $e \in \partial\Lambda$, and $\alpha, \beta \in \Omega_{\partial\Lambda}$ with $\alpha_s = \beta_s$ for all $s \neq e$. We will study for this model the l.h.s. of (11.1.3) by using Lemma 11.2.1. As mentioned before, the monomer-dimer model on a graph corresponds to the hard-core model on its line graph. So, strictly speaking, to apply the lemma we have to translate the monomer-dimer model in terms of the hard-core model, then apply the lemma and then translate the result back to the original monomer-dimer model. One can easily see that this gives an obvious 'edge-analog' of Lemma 11.2.1 (recall that Lemma 11.2.1 was formulated for models where the randomness involves the vertices rather than the edges). This edge analog of the lemma tells us that the l.h.s. of (11.1.3) is bounded above by

$$(\mu_\Lambda^\alpha \times \mu_\Lambda^\beta)(\exists \text{ a path of disagreement from } e \text{ to } \Delta), \qquad (11.3.1)$$

where the event above is formally given by

$$\{(\omega,\omega') \in \Omega_\Lambda \times \Omega_\Lambda : \exists \text{ a sequence } e_1 \sim e_2 \cdots \sim e_n \text{ of edges in } \Lambda, \text{ with } e_1 \sim e, e_n \in \Delta \text{ and } \omega_{e_i} \neq \omega'_{e_i}, i = 1, \ldots, n-1\}.$$

We now study the probability of this event. Let x and y be the two endpoints of e. A property which makes the monomer-dimer model so special within the more general hard-core model is that (see also Remark 1 in Section 11.4), for each integer $k > 0$, there is at most one path of disagreement $e_1, \ldots, e_k$ such that x is an endpoint of e_1. This is so, because of the following. For convenience, we define $e_0 = e_0^* = e$ and $\omega_e = \alpha_e$, $\omega'_e = \beta_e$. Suppose both $e_1, \ldots, e_k$ and $e_1^*, \ldots, e_k^*$ would be such paths. If they are not the same, take the smallest i, $1 \leq i \leq k$ for which $e_i \neq e_i^*$. Since both paths 'start' at the same point (namely x) it is easy to see that the three edges $e_{i-1}(= e_{i-1}^*)$, e_i and e_i^* are distinct and have a common endpoint. We also know that ω differs from ω' on each of these three edges. But then either two of these edges have ω value 1 or two of them have ω' value 1. In both cases we have a conflict with feasibility.

Now suppose $e_1, \ldots, e_k$ is a path of disagreement of length k with x an endpoint of e_1. Without loss of generality we assume that $\alpha_e = 1$. By the uniqueness property above, a path of disagreement of length $k+1$ starting in x (if one exists) must be of the form $e_1, \ldots, e_k, \tilde{e}$, where $\tilde{e}$ is an edge which has a common endpoint with e_k but not with e_{k-1}. There are clearly at most $2d - 1$ possible candidates for $\tilde{e}$. Suppose k is odd. It is easy to see (because of $\alpha_e = 1$ and $\beta_e = 0$) that then $\omega_{\tilde{e}}$ must be 1 and $\omega'_{\tilde{e}}$ must be 0. Hence, to have a path of disagreement of length $k+1$ starting at x, at least one of the above mentioned candidates must have ω-value 1. However, by the observation for the hard-core model in Section 2, the conditional probability that all of these have ω-value 0 is at least $(1+\lambda)^{-(2d-1)}$. From these arguments it is easy to conclude that the conditional probability that there is a path of disagreement of length $k+1$ starting at x, given that there is a path of disagreement of length k starting at x, is at most $1 - (1+\lambda)^{-(2d-1)}$. Since each path from e to Δ has length at least $\rho(e, \Delta)$, we finally see that the probability of having a path of disagreement from e to Δ is at most $2((1-(1+\lambda)^{-(2d-1)})^{\rho(e,\Delta)}$. (The factor 2 comes from the fact that the path could start at y instead of x). This shows that the monomer-dimer interaction satisfies the Dobrushin-Warstat condition III_c. To complete the proof of Theorem 11.1.1 we must (according to Remark (iii) after (11.1.3)) show that it is in the corresponding main component. This follows now by arguments which are quite standard and which we only sketch briefly: First of all, Dobrushin and Warstat have shown that condition III_c also has a so-called *constructive analog* involving boxes Λ of bounded size only. But for a fixed box simple continuity arguments can be applied. In this way it follows that the class of interactions satisfying III_c is open. Hence, by our result that III_c holds for the monomer-dimer model with arbitrary λ, changing λ keeps us in the same open connected component. However, for λ sufficiently small (so that we are in the so-called Dobrushin uniqueness region) it is well-known that the monomer-dimer model is in the main component. This completes the proof of Theorem 11.1.1.

11.4 Further Remarks and Discussion

1. An essential role in the proof in Section 11.3 was played by the uniqueness of certain disagreement paths, which made the monomer-dimer model so special. This property can also be formulated as follows: let ω and ω' be two feasible configurations. Then the set $\{e \in \Lambda : \omega_e \neq \omega'_e\}$ consists of pairwise disjoint paths (some of which may be cycles). ('Disjoint' means 'vertex-disjoint' here; so two paths are called disjoint if no edge in the first path has a common endpoint with any edge in the other path). Remarkably, this same property plays an essential role in work (on Markov chain Monte Carlo methods) by Jerrum and Sinclair (see, e.g., the interesting review paper [12] and the references given there), which apart from that is very different from ours. They use the property to construct so-called *canonical paths* from a configuration ω to a configuration ω'. (The word 'path' here refers to a sequence in the state space Ω, not in the graph). They then apply a quite general theorem to get an upper bound for the mixing time of a certain dynamics, which has the monomer-dimer distribution as its limit distribution. This dynamics is, roughly speaking, as follows: at each step an edge is chosen uniformly at random, and its value (0 or 1 in this case) is updated according to its conditional distribution given the current values of all the other edges. The mixing time is the number of steps after which the system is 'sufficiently close' to equilibrium.

One of their main results is, again roughly speaking, that the mixing time is polynomially bounded in the input parameters. One may wonder if these results for the '*dynamics*', applied to subgraphs of $\mathbf{Z}^d$, also give, for instance, uniqueness of the Gibbs measure. However, it seems that the bounds on the mixing time they obtain are not strong enough for that purpose. Conversely, one may wonder what our result for *spatial* dependencies yields for the mixing times. Using ideas from the literature (see, e.g., [16]) we can use our result to get upper bounds for the mixing time of a certain modified dynamics, a so-called block dynamics. Under this dynamics at each step in the procedure a whole box of edges, instead of a single edge, is updated simultaneously. Unfortunately, by lack of a certain monotonicity, it is not clear what these results tell us about the mixing time for the original (single-edge) dynamics. (Hard-core models on *bipartite* graphs do have the desired monotonicity properties, but the monomer-dimer model corresponds only in trivial cases with such a hard-core model). Moreover, the block size in the block dynamics depends on λ (the larger λ the larger the required block size), and even for moderate block sizes the simultaneous update of the edges within a block involves a huge amount of computations.

2. In the introduction we also mentioned the paper [13] by Kahn and Kayll. Although they too study spatial dependencies (and, e.g., refer to the percolation-like bound for boundary effects in [5]) their goals deviate considerably from ours. Their main motivation comes from combinatorics and graph theory, and the types of assumptions and conclusions in their main theorem (which is quite difficult) are different from those in statistical physics.

Note Added in Proof

As to the remarks on dynamics in part 1 of Section 11.4: recent work in cooperation with R. Brouwer shows how these results can indeed be applied to a single-edge update dynamics.

Acknowledgments: I thank A. van Enter and S. Shlosman for referring me to the Dobrushin-Warstat generalization of the Dobrushin-Shlosman theorem.

REFERENCES

[1] Aizenman, M. and Barsky, D.J. (1987). Sharpness of the phase transition in percolation models, *Comm. Math. Phys.* **108**, 489–526.
[2] van den Berg, J. (1993). A uniqueness condition for Gibbs measures, with application to the two-dimensional Ising antiferromagnet, *Comm. Math. Phys.* **152**, 161-166.
[3] van den Berg, J. (1997). A constructive mixing condition for 2-D Gibbs measures with random interactions. *Ann. Probab.*, **25**, 1316–1333.
[4] van den Berg, J. and Maes, C., (1994). Disagreement percolation in the study of Markov fields. *Ann. Probab.*, **22**, 749-763.
[5] van den Berg, J. and Steif, J.E., (1994). Percolation and the hard-core lattice gas model. *Stochastic Process. Appl.*, **49**, 179–197.
[6] Dobrushin, R.L. and Shlosman, S.B., (1986). Completely analytical Gibbs fields. In: *Statistical Physics and Dynamical Systems*, Birkhäuser, Boston.
[7] Dobrushin, R.L. and Warstat, V., (1990). Completely analytic interactions with infinite values. *Probab. Th. Rel. Fields*, **84**, 335–359.
[8] Ellis, R.S., (1985). *Entropy, large deviations and statistical mechanics,* Springer-Verlag, New York.
[9] Georgii, H., (1988). *Gibbs measures and phase transitions,* de Gruyter, New York.
[10] Gielis, G. and Maes, C. (1995). The uniqueness regime of Gibbs fields with unbounded disorder. *J. Statist. Phys.* **81**, 829–835.
[11] Heilmann, O.J. and Lieb, E.H., (1972). Theory of monomer-dimer systems. *Comm. Math. Phys.*, **25**, 190–232.
[12] Jerrum, M. and Sinclair, A., (1996). The Markov chain Monte Carlo method: an approach to approximate counting and integration. In: *Approximation Algorithms for NP-Hard Problems* (Dorit Hochbaum, ed.), PWS Publishing, Boston, 482–520.
[13] Kahn, J. and Kayll, P.M., (1997). On the stochastic independence properties of hard-core distributions, *Combinatorica*, **17**, 369–391.
[14] Kenyon, R., (1998). *Conformal invariance of domino tiling*. Preprint.
[15] Kesten, H., (1980). The critical probability of bond percolation on the square lattice equals $\frac{1}{2}$. *Comm. Math. Phys.* **74**, 41–59.
[16] Martinelli, F. and Olivieri, E., (1994). Approach to equilibrium of Glauber dynamics in the one phase region. I. The attractive case. *Comm. Math. Phys.*, **161**, 447–486.
[17] Menshikov, M.V. (1986). Coincidence of critical points in percolation problems. *Soviet Mathematics Doklady* **33**, 856–859.

CWI, Department PNA
Kruislaan 413
1098 SJ Amsterdam
The Netherlands
J.van.den.Berg@cwi.nl

12

Loop-Erased Random Walk

Gregory F. Lawler

ABSTRACT Loop-erased random walk (LERW) is a process obtained from erasing loops from simple random walk. This paper reviews some of the results and conjectures about LERW. In particular, we discuss the critical exponents for LERW, Wilson's algorithm for generating uniform spanning trees with LERW, and the role of conformal invariance in studying LERW in two dimensions.

Keywords: Loop-erased random walk, uniform spanning trees, conformal invariance.

AMS Subject Classification: 60J15.

12.1 Introduction

Loop-erased random walk on the integer lattice is a process obtained from simple random walk by erasing the loops chronologically. The process has self-avoiding paths. While the process was first studied [10] with the hope of getting some perspective on the problem of the usual self-avoiding walk (the uniform measure on random walk paths of a given length conditioned on no self-intersection), it has been found that the two processes are in different universality classes. Nevertheless, the loop-erased process is an interesting process in itself having many of the properties of interesting models from statistical physics:

- There is a critical dimension ($d = 4$) such that above the critical dimension the behavior is trivial (in this case, trivial means the same qualitative behavior as simple random walk, e.g., convergence to Brownian motion).
- The trivial behavior holds at the critical dimension with a logarithmic correction in the scaling.
- Below the critical dimension there is "non-mean field behavior"; in particular, the value of the growth exponent takes on values different from those of simple random walk.
- In two dimensions, the process is expected to have a continuum limit that is conformal, and nonrigorous conformal field theory gives an exact prediction of the exponent as a simple rational number. In three dimensions, there is no reason to believe that the exponent takes on a rational value.

While the loop-erased walk does not seem to be in the same universality class as the usual self-avoiding walk, it is closely related to other models: uniform spanning trees, domino tilings, and Q-state Potts model as $Q \to 0$. There has been a lot of recent interest in the behavior of uniform spanning trees, and this has brought about a renewed interest in the loop-erased walk. In particular, Wilson [21, 23] recently gave a beautiful and simple algorithm to generate uniform spanning trees using loop-erased walks from the simple random walk on a graph.

The goal of this paper is to summarize some of the results about loop-erased walk and to give some of the important open questions. We will not try to cover everything about loop-erased random walk (LERW), but rather focus on some of the key properties. We start by reviewing some of the basic facts about LERW. Most of this is can be found in [11]; however, we include a proof of a fact about two dimensional loop-erased walk that says essentially that (up to a uniform multiplicative constant) if we are interested in the LERW only up to the time it leaves the ball of radius n, we can consider the simple random walk up to the time it reaches the ball of radius $2n$ and erase loops on these finite paths. After the definitions, we focus on three areas. First, we will discuss what is known about the growth exponent for LERW on the integer lattice. We sketch arguments for the two relatively easy exponent equalities, the third moment estimate and the triple point estimate. The third moment estimate can be used to estimate the growth exponent as we describe. [As this paper was being prepared, Kenyon [8] announced that he has rigorously proved the major conjecture about the growth exponent in two dimensions: α_2, as described in Section 12.3, equals $3/4$.] Second, we give a short derivation of the validity of the Wilson algorithm. The basic idea of the proof is the same as the basic idea in [21, 23], but we think the presentation here is somewhat simpler. The main idea is that the distribution on paths from loop erasure is independent, in some sense, of the order in which the loops are erased. Finally, the last section, will be mainly heuristic and discuss the idea of conformal invariance and LERW in two dimensions. I would like to thank Richard Kenyon, Yuval Peres, Oded Schramm, Wendelin Werner, and David Wilson for assistance during the preparation of this paper.

12.2 Definition

If $\omega = [\omega_0, \omega_1, \ldots, \omega_n]$ is a finite sequence of points, its loop-erasure $L(\omega)$ is defined as follows. Let s_0 be the largest index such that $\omega_{s_0} = \omega_0$. Define s_m inductively by saying that if $s_{m-1} < n$, then s_m is the largest index j such that $\omega_j = \omega_{s_{m-1}+1}$. If l is the smallest index so that $s_l = n$, then

$$L(\omega) = [\omega_{s_0}, \ldots, \omega_{s_l}].$$

Note that $L(\omega)$ is a self-avoiding subpath of ω with the same starting and ending points. If $\omega = [\omega_0, \omega_1, \ldots]$ is an infinite sequence of points such that for each x,

$$\#\{j : \omega_j = x\} < \infty,$$

then $L(\omega)$ can be defined in a similar way, in this case producing an infinite self-avoiding path contained in ω with the same starting point.

Let $S(t)$ be a discrete time Markov chain on a countable state space X with transition probabilities $p(x, y)$. Let 0 denote a particular element of X. If $A \subset X$, let

$$\tau_A = \inf\{t \geq 0 : S(t) \in A\},$$

$$G(x, A) = \sum_{j=0}^{\infty} \mathbf{P}^x\{X(j) = x;\ X(t) \notin A,\ t = 0, 1, \ldots, j\}.$$

If S is transient, then we get a measure, μ_n, on n-step self-avoiding paths $\omega = [\omega_0 = 0, \omega_1, \ldots, \omega_n]$ by starting the Markov chain at 0, erasing loops from the infinite path, and considering the first n steps of the path remaining. Straightforward analysis of the loop-erasing procedure shows that

$$\mu_n(\omega) = [\prod_{j=1}^{n} p(\omega_{j-1}, \omega_j)][\prod_{j=0}^{n} G(\omega_j, A_{j-1})]\text{Es}(\omega_n, A_n),$$

where

$$A_j = \{\omega_0, \ldots, \omega_j\}, \quad A_{-1} = \emptyset,$$

$$\text{Es}(x, A) = \mathbf{P}^x\{X(t) \notin A,\ t = 1, 2, \ldots\}.$$

Similarly, if S is recurrent or transient, and $B \subset X$ with

$$\mathbf{P}^0\{\tau_B < \infty\} = 1,$$

we get a measure $\mu_{n,B}$ on n-step paths by starting the Markov chain at 0, letting it run until it reaches B, and then erasing loops. There is a chance that $\mu_{n,B}$ will not be a probability measure since the entire length of

$$L[S(0), \ldots, S(\tau_B)]$$

could be less than n. However, if B satisfies

$$\mathbf{P}^0\{n \leq \tau_B\} = 1,$$

then it will be a probability measure. In either case, if $\omega = [\omega_0 = 0, \ldots, \omega_n]$ is any self-avoiding path with $\omega_0, \ldots, \omega_{n-1} \notin B$,

$$\mu_n(\omega) = [\prod_{j=1}^{n} p(\omega_{j-1}, \omega_j)][\prod_{j=0}^{n} G(\omega_j, B \cup A_{j-1})]\text{Es}(\omega_n, A_n; B), \tag{12.2.1}$$

where

$$\text{Es}(x, A; B) = \mathbf{P}^x\{S(t) \notin A,\ 1 \leq t < \tau_B\}.$$

In particular, if $\omega_n \in B$, $\text{Es}(\omega_n, A_n; B) = 1$, and hence

$$\mathbf{P}^0\{L[S(0), \ldots, S(\tau_B)] = \omega\} = [\prod_{j=1}^{n} p(\omega_{j-1}, \omega_j)][\prod_{j=0}^{n} G(\omega_j, B \cup A_{j-1})]. \tag{12.2.2}$$

Note that the first product is just

$$\mathbf{P}^0\{[S(0), \ldots, S(n)] = \omega\}.$$

The second product is also interesting. A simple exercise in Markov chains is the following. Suppose S is irreducible, and either S is transient or $B \neq \emptyset$. Then for every $x, y \notin B$,

$$G(x, B)G(y, B \cup \{x\}) = G(x, B \cup \{y\})G(y, B).$$

By iterating this as many times as necessary, we see that

$$f(\omega_0, \ldots, \omega_n; B) = \prod_{j=0}^{n} G(\omega_j, B \cup A_{j-1}) \tag{12.2.3}$$

is a symmetric function of $\omega_0, \ldots, \omega_n$. This observation is crucial to the proof of the validity of Wilson's algorithm that we outline in Section 12.7.

If $d \geq 3$ and S is a simple random walk in $\mathbb{Z}^d$, the transience of S allows us to define the loop-erased random walk by erasing loops from the infinite path. For $d = 2$, we define the infinite LERW in a slightly different manner. Let

$$B_m = \{z \in \mathbb{Z}^2 : m \leq |z| < m + 1; \},$$

$$\tau_m = \tau_{B_m}.$$

Let Ω_n be the set of self-avoiding walks $\omega = [\omega_0 = 0, \omega_1, \ldots, \omega_k]$ in $\mathbb{Z}^2$ with

$$|\omega_j| < n, \quad j = 0, 1, 2, \ldots, k - 1;$$

$$|\omega_k| \geq n.$$

Then for each $n \leq m$, we can obtain a measure $\mu_{n,m}$ on Ω_n by considering the appropriate initial segment of

$$L[S(0), \ldots, S(\tau_m)].$$

The measures $\mu_{n,m}$ converge as $m \to \infty$ to a measure μ_n. If $\omega = [\omega_0, \ldots, \omega_k] \in \Omega_n$, then, by (12.2.1), if $m \geq n$,

$$\mu_{n,m}(\omega) = 4^{-k} G(0, B_m)[\prod_{j=1}^{k} G(\omega_j, A_{j-1} \cup B_m)]\mathrm{Es}(\omega_k, A_k; B_m).$$

It is known [11, Theorem 1.6.6] that

$$G(0, B_m) = \frac{2}{\pi} \log m + O(1),$$

and hence,

$$\mu_n(\omega) = \lim_{m \to \infty} \mu_{n,m}(\omega) = 4^{-k}[\prod_{j=1}^{k} G(\omega_j, A_{j-1})]a(\omega_k, A_j),$$

where

$$a(x, A) = \lim_{m\to\infty} (\frac{2}{\pi} \log m)\mathbf{P}^x\{S(t) \notin A,\ t = 1, 2, \ldots, \tau_m\}.$$

If ω is as above, then the discrete Harnack inequality can be used to show that for some c,

$$c^{-1}\mathrm{Es}(\omega_k, A_k; B_{2n}) \leq a(\omega_k, A_k) \leq c\mathrm{Es}(\omega_k, A_k; B_{2n}).$$

We now prove a lemma which can be considered an extenison of the Harnack inequality.

Lemma 12.2.1 *There is a $c > 0$ such that for every n and every $\omega = [\omega_0, \ldots, \omega_k] \in \Omega_n$,*

$$c^{-1}\mu_n(\omega) \leq \mu_{n,2n}(\omega) \leq c\mu_n(\omega).$$

PROOF. We have already seen that

$$\mu_{n,2n}(\omega) \asymp \mu_n(\omega)(\log n)^{-1}\frac{f(\omega_1, \ldots, \omega_k; \{0\})}{f(\omega_1, \ldots, \omega_k; \{0\} \cup B_{2n})}.$$

Here we are using $\asymp$ to mean that both sides are bounded above by a constant times the other side, where the constant is uniform over all n and all $\omega \in \Omega_n$. Hence it suffices to show for all $\omega \in \Omega_n$,

$$\frac{f(\omega_1, \ldots, \omega_k; \{0\})}{f(\omega_1, \ldots, \omega_k; \{0\} \cup B_{2n})} \asymp \log n,$$

or, equivalently,

$$\log \frac{f(\omega_1, \ldots, \omega_k; \{0\})}{f(\omega_1, \ldots, \omega_k; \{0\} \cup B_{2n})} = \log\log n + O(1). \qquad (12.2.4)$$

By symmetry of $f(\omega_1, \ldots, \omega_k; \cdot)$, we can order the points so that

$$|\omega_1| \leq |\omega_2| \leq \cdots \leq |\omega_k|.$$

Note for each j, although $[\omega_1, \ldots, \omega_j]$ may not be a path itself, $A_j = \{\omega_1, \ldots, \omega_j\}$ contains a path connnecting a neighbor of the origin with $B_{|\omega_j|-1}$. In particular, there is a positive number c such that if we start a random walker at radius less than $2|\omega_j|$, the probability that it reaches $B_{4|\omega_j|}$ without hitting A_j is at most $1 - c$.

Note that

$$G(\omega_j, A_{j-1} \cup B_{2n}) = G(\omega_j, A_{j-1})[1 - \mathrm{Es}(\omega_j, A_{j-1}; B_{2n})\tilde{H}(\omega_j, A_j; B_{2n})],$$

where

$$\tilde{H}(\omega_j, A_j; B_{2n}) = \mathbf{P}^{\omega_j}\{X(\tau_{A_j}) = \omega_j \mid \tau_{2n} < \tau_{A_j}\}.$$

By expanding the logarithm, we see that to establish (12.2.4), we need to show that

$$\sum_{j=1}^{k} \mathrm{Es}(\omega_j, A_{j-1}; B_{2n})\tilde{H}(\omega_j, A_j; B_{2n}) = \log\log n + O(1), \qquad (12.2.5)$$

where the $O(1)$ term is uniform over n and ω.

Let A be a finite subset of $\mathbb{Z}^2$ and

$$H_A(x,y) = \mathbf{P}^x\{S(\tau_A) = y\},$$

$$H_A(y) = \lim_{|x|\to\infty} H_A(x,y).$$

The limit is known to exist; moreover [11, Theorem 2.1.3], if A is contained in the ball of radius n,

$$H_A(x,y) \asymp H_A(y), \quad |x| \geq 2n, \quad y \in A,$$

and

$$H_A(x,y) = H_A(y)[1 + o(\sqrt{\frac{n}{|x|}})].$$

(There is a better estimate on the error, but this will do for our purposes.) Choose m so that $2^{m-1} < n \leq 2^m$. Let k_i be the largest index with $|\omega_{k_i}| < 2^i$, and let

$$\eta^i = \tau_{A_{k_i}}.$$

Let $z = (2n, 0)$, and let

$$p_i = p_i(\omega) = \mathbf{P}^z\{\eta^i < \tau_{4n};\ \tau_{\{0\}} > \tau_{4n}\}$$

Standard estimates (see [11, Exercise 1.6.8]) give for $1 \leq i \leq m-1$,

$$\begin{aligned} p_i &= \mathbf{P}^z\{\tau_{2^i} < \tau_{4n}\}\mathbf{P}^z\{\eta^i < \tau_{4n} \mid \tau_{2^i} < \tau_{4n}\} - \mathbf{P}^z\{\tau_{\{0\}} < \tau_{4n}\} \\ &= \frac{1}{m+1-i} + O(\frac{1}{(m+i-1)^2}) + O(\frac{1}{m}). \end{aligned}$$

Hence,

$$\sum_{i=1}^{m} p_i = \log m + O(1) = \log\log n + O(1).$$

Let

$$\sigma = \sigma_{n,\omega} = \inf\{j : \omega_j \in S[0, \tau_{4n}]\},$$

where $\sigma = \infty$ if no such j exists. Then

$$\{\sigma = j\} = \{\omega_j \in S[0,\tau_{4n}];\ \omega_l \notin S[0,\tau_{4n}], l = 0, 1, \ldots, j-1\}.$$

In order for σ to equal j, the random walk must first hit the set A_j at ω_j some time before reaching B_{4n}, and after this hit, the random walk must reach B_{4n} without hitting A_{j-1}. In other words,

$$\mathbf{P}^z\{\sigma = j\} = \mathbf{P}^z\{\tau_{A_j} < \tau_{4n}\}\mathbf{P}^z\{S(\tau_{A_j}) = \omega_j \mid \tau_{A_j} < \tau_{4n}\}\mathrm{Es}(\omega_j, A_{j-1}; B_{4n}).$$

Assume $2^{i-1} < |\omega_j| \leq 2^i$. Then,

$$\mathbf{P}^z\{\tau_{A_j} < \tau_{4n}\} = \frac{1}{m+1-i} + O(\frac{1}{(m+1-i)^2}),$$

and

$$\text{Es}(\omega_j, A_{j-1}; B_{4n}) = \text{Es}(\omega_j, A_{j-1}; B_{2n})[1 + O(\frac{1}{m+i-1})].$$

The estimates on the harmonic measure show that for any $|y| \geq 2n$,

$$H_{A_j}(y, \omega_j) = H_{A_j}(\omega_j)[1 + O(2^{(i-m)/2})].$$

In particular,

$$\begin{aligned}\mathbf{P}^z\{S(\tau_{A_j}) = \omega_j \mid \tau_{A_j} < \tau_{4n}\} &= H_{A_j}(\omega_j)[1 + O(2^{(i-m)/2})] \\ &= \tilde{H}(\omega_j, A_j, B_{2n})[1 + O(2^{(i-m)/2})].\end{aligned}$$

Hence,

$$\begin{aligned}p_i &= \sum_{j=1}^{k_i} \mathbf{P}\{\sigma = j\} \\ &= \frac{1}{m+1-i}\text{Es}(\omega_j, A_{j-1}; B_{4n})\tilde{H}(\omega_j, A_j, B_{2n})[1 + O(\frac{1}{m+1-i})],\end{aligned}$$

and

$$\sum_{i=1}^{m} p_i = \sum_{j=1}^{k} \text{Es}(\omega_j, A_{j-1}; B_{2n})\tilde{H}(\omega_j, A_j, B_{2n}) + O(1).$$

12.3 Growth Exponent

The growth exponent is a measure of how fast the loop-erased walk $\hat{S}(n)$ in $\mathbb{Z}^d$ grows. There are a few different ways to look at this exponent which should all be equivalent (although neither the existence of the exponent nor the equivalence of the formulations has been proven for all d). Let $p(n) = p_d(n)$ denote the probability that a random walk and the loop-erasure of another random walk do not intersect in n steps. More precisely, if S, S^1, S^2 denote independent simple random walks starting at the origin, let

$$p(n) = \mathbf{P}\{S^1[1,n] \cap L[S^2(0), \ldots, S^2(n)] = \emptyset\}.$$

The distribution from loop-erasing is the same whether one erases in the forward direction or the reverse direction [11, Lemma 7.2.1]. Hence, $p(n)$ also denotes the probability that the nth step of a random walk is not erased by time $2n$,

$$p(n) = \mathbf{P}\{L[S(0), \ldots, S(n)] \cap S[n+1, 2n] = \emptyset\}.$$

One expects the number of points in $L[S(0), \ldots, S(n)]$ to be on the order of $np(n)$. Since $\mathbf{E}[|S(n)|^2] = n$, we would expect

$$\mathbf{E}[|\hat{S}(np(n))|^2] \asymp n.$$

Or if we set

$$\sigma_n = \inf\{j : |\hat{S}(j)| \geq n\},$$

then

$$\mathbf{E}[\sigma_n] \asymp n^2 p(n^2).$$

If $d \geq 5$, by comparision to the probability that two simple random walks do not intersect, we see that there is a constant $c = c_d > 0$ such that for all n, $p(n) \geq c$. If $d = 4$ [13],

$$\mathbf{P}\{S[0, n] \cap S[n + 1, 2n] = \emptyset\} \asymp (\log n)^{-1/2}.$$

In the case of LERW, $p(n)$ also decays as a power of the logarithm, but a different power,

$$p(n) \asymp (\log n)^{-1/3}.$$

In the next section we will discuss where the $1/3$ comes from. If K_n denotes the number of points of the first n points of the random walk that remain after erasing loops, then

$$\mathbf{E}[K_n] \sim \begin{cases} c_d n, & d \geq 5 \\ c_d n (\log n)^{-1/3}, & d = 4. \end{cases}$$

In fact, if $d \geq 4$, $K_n/\mathbf{E}(K_n)$ converges in probability (almost everywhere for $d \geq 5$) to the constant 1 (see [11, Chapter 7]). In particular, the limiting distribution of $(b_n n)^{-1/2}\hat{S}(n)$ is Gaussian and there is convergence to Brownian motion, where $b_n = 1$ if $d \geq 5$ and $b_n = (\log n)^{1/3}$ for $d = 4$. We do not expect $K_n/\mathbf{E}(K_n)$ to converge to a constant random variable in $d < 4$.

For $d < 4$ we define the exponent $\alpha = \alpha_d$ by

$$p(n^2) \approx n^{-\alpha}.$$

We choose n^2 so that the probability a simple random walk and a loop-erased walk get to distance n without intersecting is of order $n^{-\alpha}$. We have no proof now that the exponent exists, so we should discuss

$$\underline{\alpha}_d = \liminf_{n\to\infty} \frac{p(n^2)}{-\log n}, \quad \overline{\alpha}_d = \limsup_{n\to\infty} \frac{p(n^2)}{-\log n}.$$

For shorthand we will write

$$a_1 \leq \alpha_d \leq a_2$$

for the statement

$$a_1 \leq \underline{\alpha}_d \leq \overline{\alpha}_d \leq a_2.$$

Assuming α exists,

$$\mathbf{E}[\sigma_n] \approx n^{2-\alpha},$$

and since n^2 steps of a simple random walk correspond to $n^{2-\alpha}$ steps of the loop-erased walk, m steps of the loop-erased walk correspond to $m^{2/(2-\alpha)}$ of a simple walk. This leads us to believe

$$\mathbf{E}[|\hat{S}(m)|^2] \approx m^{2/(2-\alpha)}.$$

We define the growth exponent ν

$$\nu = \nu_d = \frac{1}{2-\alpha},$$

so that

$$\mathbf{E}[|\hat{S}(m)|^2] \approx m^{2\nu}.$$

Since $\sigma_n \geq n$, we get the easy estimate $\alpha_d \preceq 1$. In the next section we will sketch the proof of a lower bound,

$$\alpha_d \geq \frac{4-d}{3}, \quad d = 2, 3,$$

which corresponds to a lower bound for ν,

$$\nu_d \geq \frac{6}{2+d}, \quad d = 2, 3.$$

The right hand side is exactly the Flory predictions for ν for the usual self-avoiding random walk [18]. However, as will be discussed in the next section, we do not expect this to be a sharp inequality so that the loop-erased walk grows faster than the usual self-avoiding walk. The exponent α is bounded above by the intersection exponent for simple random walk [11, Chapter 5], $\xi = \xi_d$, which is defined by

$$\mathbf{P}\{S[0, n^2] \cap S[n^2+1, 2n^2] = \emptyset\} \asymp n^{-\xi}.$$

It is known that $\xi_3 < 1$, and hence $\alpha_3 < 1$. If $d = 2, \xi_2 > 1$, and hence gives no information about α_2. However, it can be shown that $\alpha_2 < 1$; we discuss this somewhat in the next section.

Conformal field theory has been used to give the interesting conjecture

$$\alpha_2 = \frac{3}{4} \quad \nu_2 = \frac{4}{5}.$$

[As mentioned before, Kenyon [8] has recently proved this conjecture.] For $d = 3$, there is no reason to believe that α_3 or ν_3 takes any nice rational value; simulations [7] certainly suggest that $\nu_3 > 6/5$, and hence $\alpha_3 > 1/3$.

12.4 Long Range Intersection Probability

As a preliminary step to the estimates in the next section, we discuss the probability that a simple random walk and loop-erased walk starting a reasonable distance apart intersect. More precisely, let S and S^1 be independent simple random walks starting at the origin and $(n, 0, \ldots, 0)$, respectively, and let

$$q(n) = q_d(n) = \mathbf{P}\{S^1[0, n^2] \cap L[S(0), \ldots, S(n^2)] \neq \emptyset\}.$$

For two simple random walks it is known [11, Chapter 3] that ,

$$\mathbf{P}\{S^1[0, n^2] \cap S[0, n^2] \neq \emptyset\} \asymp \begin{cases} 1, & d < 4, \\ (\log n)^{-1}, & d = 4, \\ n^{4-d}, & d \geq 5. \end{cases}$$

These results can be obtained using "second moment" methods. If $J = J_n$ denotes the number of intersections

$$J_n = \sum_{j=0}^{n^2} \sum_{k=0}^{n^2} I\{S^1(j) = S(k)\},$$

then

$$\mathbf{E}[J_n] \asymp n^{4-d},$$

and

$$\mathbf{E}[J_n^2] \asymp \begin{cases} n^{2(4-d)}, & d < 4, \\ \log n, & d = 4, \\ n^{4-d}, & d \geq 5. \end{cases}$$

In fact, for one simple random walk and one loop-erased walk, the probability of intersection is the same as for two simple walks, at least up to multiplicative constant,

$$q_d(n) \asymp \begin{cases} 1, & d < 4, \\ (\log n)^{-1}, & d = 4, \\ n^{4-d}, & d \geq 5. \end{cases}$$

It is not too difficult to see that this must be true for $d = 2$ (and $d = 1$) since geometric considerations require continuous paths conditioned appropriately to intersect. For $d \geq 3$, a different argument is used [6, 13]. The bound for two random walks gives a bound in one direction. For the other direction we consider cut times. A time m is a cut time for $S[0, n^2]$, if

$$S[0, m] \cap S[m + 1, n^2] = \emptyset.$$

Let

$$H_n = \{S(m) : m \text{ cut time for } S[0, n^2]\}.$$

Then

$$H_n \subset L[S(0), \ldots, S(n^2)],$$

and hence it suffices to show

$$\mathbf{P}\{S^1[0,n^2]\cap H_n \neq \emptyset\} \asymp \begin{cases} 1, & d=3, \\ (\log n)^{-1}, & d=4, \\ n^{4-d}, & d\geq 5. \end{cases}$$

Again, second moment methods can be used, but the estimates require good estimates on the probability of being a cut time. Let

$$R_n = \sum_{k=0}^{n^2} I\{S(k) \in S^1[0,n^2];\ k \text{ cut time for } S[0,n^2]\}.$$

It can be shown for $d \geq 3$, that if $n^2/4 \leq j < k \leq 3n^2/4$,

$$\mathbf{P}\{S(j), S(k) \in S^1[0,n^2];\ j,k \text{ cut times}\} \asymp n^{2-d}(k-j)^{(2-d)/2} r(n^2) r(k-j),$$

where $r(m)$ denotes the probability that the paths of two simple random walks of m steps starting at adjacent points do not intersect. We know [11, 12, 14] that

$$r(m) \asymp \begin{cases} m^{-\xi/2}, & d=3, \\ (\log m)^{-1/2}, & d=4, \\ 1, & d\geq 5. \end{cases}$$

Here $\xi = \xi_3$ is the intersection exponent for simple random walk in $\mathbb{Z}^3$ which is known to be strictly less than 1. (Or, in other words the dimension of the set of cut points in three dimensions is $2-\xi > 1$.) By summing we see that

$$\mathbf{E}[R_n] \asymp \begin{cases} n^{2-\xi}, & d=3, \\ (\log n)^{-1/2}, & d=4, \\ n^{4-d}, & d\geq 5, \end{cases}$$

and

$$\mathbf{E}[R_n^2] \asymp \begin{cases} n^{2(2-\xi)}, & d=3, \\ 1, & d=4, \\ n^{4-d}, & d\geq 5. \end{cases}$$

Since

$$\mathbf{P}\{R_n > 0\} \geq \frac{(\mathbf{E}[R_n])^2}{\mathbf{E}[R_n^2]},$$

the lower bound on the probability can be deduced.

12.5 Two Exponent Equalities

There are two exact exponent equalities that can be derived for loop-erased walk. We will refer to them as the *third moment estimate* and the *triple point estimate*.

These equalities are most easily derived for walks with a geometric killing rate, so we will consider such walks in this section; however, it is not difficult to derive similar results for walks of a given number of steps or walks stopped upon reaching a certain distance. We let $S^1, S^2, \ldots$ be independent random walks in $\mathbb{Z}^d$ $(d = 2, 3, 4)$, and we let $\rho^1, \rho^2, \ldots$ be independent geometric random variables, independent of $S^1, S^2, \ldots$, with killing rate λ. That is, the distribution of ρ^i is given by

$$\mathbf{P}\{\rho^i = j\} = (1-\lambda)^j \lambda, \quad j = 0, 1, 2, \ldots$$

Let

$$G(\lambda) = \sum_{x \in \mathbb{Z}^d} \mathbf{P}^{0,x}\{L(S^5[0, \rho^5]) \cap S^6[0, \rho^6] \neq \emptyset\}. \tag{12.5.1}$$

Using the results of the previous section and the fact that a typical path with killing rate λ goes on the order of λ^{-1} steps, it is not difficult to show that

$$G(\lambda) \asymp \lambda^{-d/2} \kappa,$$

where

$$\kappa = \kappa_d(\lambda) = \begin{cases} 1, & d = 2, 3, \\ \log(1/\lambda)^{-1}, & d = 4. \end{cases}$$

Assume S^1, S^2, S^3, S^4 start at the origin, and consider the events

$$V(i,j) = V(i,j,\lambda) = \{S^i(0, \rho^i] \cap L(S^j[0, \rho^j]) = \emptyset\},$$

$$\tilde{V}(i,j) = \tilde{V}(i,j,\lambda) = \{S^i(0, \rho^i] \cap L(S^j[0, \rho^j]) \subset \{0\}\}.$$

The first equality in the next proposition is the third moment estimate; we will discuss its consequences in the next section. The second is the triple point estimate. It is so named because it is related to triple points for uniform spanning trees (see [3]).

Proposition 12.5.1 *For any $\lambda \in (0, 1)$, if all the random walks start at the origin,*

$$\mathbf{P}[V(2,1) \cap V(3,1) \cap \tilde{V}(4,1)] = \lambda^2 G(\lambda) \asymp \lambda^{(4-d)/2} \kappa. \tag{12.5.2}$$

$$\mathbf{P}[V(2,1) \cap V(3,1) \cap V(3,2)] = \lambda^2 G(\lambda) \asymp \lambda^{(4-d)/2} \kappa. \tag{12.5.3}$$

It is not too difficult to show that the proposition implies as well that

$$\mathbf{P}[V(2,1) \cap V(3,1) \cap V(4,1)] \asymp \lambda^{(4-d)/2} \kappa,$$

$$\mathbf{P}[\tilde{V}(2,1) \cap \tilde{V}(3,1) \cap \tilde{V}(4,1)] \asymp \lambda^{(4-d)/2} \kappa.$$

The proof of the proposition comes from transformations of paths. We will give the appropriate transformations here, but we will leave it to the reader to verify that they give the equalities above. In both cases, consider the event

$$\{L(S^5[0, \rho^5]) \cap S^6[0, \rho^6] \neq \emptyset\}.$$

For the third moment estimate (12.5.2), consider the first intersection time where first is defined on the time scale of S^5. More precisely, let

$$\begin{aligned}\sigma_1 &= \inf\{t : S^5(t) \text{ is not erased; } S^5(t) \in S^6[0, \rho^6]\} \\ &= \inf\{t : L(S^5[0,t]) \cap S^5[t+1, \rho^5] = \emptyset;\ S^5(t) \in S^6[0, \rho^6]\},\end{aligned}$$

$$\eta_1 = \inf\{t : S^6(t) = S^5(\sigma_1)\}.$$

Let

$$\begin{aligned} S^1(t) &= S^5(\sigma_1 - t) - S^5(\sigma_1), \quad 0 \le t \le \sigma_1, \\ S^2(t) &= S^5(\sigma_1 + t) - S^5(\sigma_1), \quad 0 \le t \le \rho_5 - \sigma_1, \\ S^3(t) &= S^6(\eta_1 - t) - S^6(\eta_1), \quad 0 \le t \le \eta_1, \\ S^4(t) &= S^6(\eta_1 + t) - S^6(\eta_1), \quad 0 \le t \le \rho_6 - \eta_1. \end{aligned}$$

Then (12.5.2) can be established by examining this transformation. One needs to use the fact that "reverse loop-erasing" is the same as loop-erasing. The extra λ^2 comes from the fact that we have made four walks out of two and hence have to "kill" two extra walks.

For the triple point estimate (12.5.3), consider instead the first time where first is chosen on the time scale of S^6. Let

$$\eta_2 = \inf\{t : S^6(t) \in L(S^5[0, \rho^5])\},$$

$$\sigma_2 = \sup\{t : S^5(t) = S^6(\eta_2)\},$$

and define the random walks

$$\begin{aligned} S^1(t) &= S^5(\sigma_2 - t) - S^5(\sigma_2), \quad 0 \le t \le \sigma_2, \\ S^2(t) &= S^5(\sigma_2 + t) - S^5(\sigma_2), \quad 0 \le t \le \rho^5 - \sigma_2, \\ S^3(t) &= S^6(\eta_2 - t) - S^6(\eta_2), \quad 0 \le t \le \eta_2, \\ S^4(t) &= S^6(\eta_2 + t) - S^6(\eta_2), \quad 0 \le t \le \rho^6 - \eta_2. \end{aligned}$$

Examination of this transformation gives (12.5.3).

12.6 Bounds on the Growth Exponent

Let S, S^1, S^2, S^3 be independent simple random walks starting at the origin. Let

$$L_n = L[S(0), \ldots, S(n^2)],$$

and define the random variables,

$$Z_n = \mathbf{P}\{S^1(0, n^2] \cap L_n = \emptyset \mid L_n\},$$

$$Y_n = \mathbf{P}\{S^i(0, n^2] \cap L_n = \emptyset,\ i = 1, 2, 3 \mid L_n\}.$$

Note that

$$\mathbf{E}[Z_n] = p_d(n^2),$$

where $p_d(\cdot)$ is as defined in Section 12.3, and

$$\mathbf{E}[Y_n] = \mathbf{E}[Z_n^3].$$

It is not difficult to derive from (12.5.2) a corresponding result for fixed length walks. We get

$$\mathbf{E}[Y_n] = \mathbf{E}[Z_n^3] \asymp \begin{cases} n^{d-4}, & d < 4, \\ (\log n)^{-1}, & d = 4, \\ 1, & d \geq 5. \end{cases}$$

As before, we will refer to this as the third moment estimate for loop-erased walks. In this section, we use this exact exponent equality to give some bounds on the growth exponent.

Hölder's inequality tells us that

$$\mathbf{E}[Z_n] \leq \mathbf{E}[Z_n^3]^{1/3}.$$

If the situation were "mean field", then we would expect that

$$\mathbf{E}[Z_n] \approx \mathbf{E}[Z_n^3]^{1/3}.$$

Mean field behavior is expected at the critical dimension $d = 4$. In fact, it has been shown [13], using the ideas of slowly recurrent sets for simple random walk, that

$$\mathbf{E}[Z_n] \asymp \mathbf{E}[Z_n^3]^{1/3} \asymp (\log n)^{-1/3}, \quad d = 4.$$

Below the critical dimension, we do not expect mean field behavior. The third moment estimate and Hölder's inequality imply

$$\alpha_d \geq \frac{4-d}{3}, \quad d = 2, 3;$$

in fact, we would expect (but it has not been proved for $d = 3$) that

$$\alpha_d > \frac{4-d}{3}, \quad d = 2, 3.$$

Recall that it is conjectured that $\alpha_2 = 3/4$.

The recent work of Kenyon [8] establishes the value $\alpha_2 = 3/4$. Here we will sketch the proof of a weaker result, $\alpha_2 < 1$. The discrete Beurling projection theorem, first proved by Kesten [9], tells us that there is an absolute constant $c > 0$ such that if ω is any simple random walk path in $\mathbb{Z}^2$ connecting the origin with the circle of radius n, then the probability a random walk starting adjacent to the origin reaches distance n without hitting ω is bounded above by $cn^{-1/2}$. The maximum probability is obtained (at least up to multiplicative constant) by a half-line. In our notation, this says that $Z_n \leq n^{-1/2}$ (actually this is not quite right, since some atypical loop-erased paths will not actually have radius of order n, but it is essentially true). Using this we get

$$\mathbf{E}[Z_n^3] \le cn^{-1}\mathbf{E}[Z_n], \quad d = 2.$$

(We are varying the value of the constant c from line to line.) This gives the estimate

$$\alpha_2 \le 1,$$

which we have already seen is a trivial estimate. However, loop-erased paths are more crooked than a half-line. It can be shown [16] that, except with extremely small probability, $Z_n \le cn^{-(1/2)-\epsilon}$ for some ϵ. From this we derive the estimate

$$\alpha_2 < 1.$$

There is an analogous problem for random walk intersections. Let

$$\tilde{Z}_n = \mathbf{P}\{S[0,n^2] \cap S^1[1,n^2] = \emptyset \mid S[0,n^2]\}.$$

An argument similar to the one used to prove the third point estimate or the triple point estimate can be used to show that

$$\mathbf{E}[\tilde{Z}_n^2] \asymp \begin{cases} n^{d-4}, & d < 4, \\ (\log n)^{-1}, & d = 4, \\ 1, & d \ge 5. \end{cases}$$

For the critical dimension $d = 4$, the mean-field theory holds [11, 12] and

$$\mathbf{E}[\tilde{Z}_n] \asymp (\log n)^{-1/2}.$$

For $d < 4$, it has been proved [4, 15] that the mean-field theory does not hold: i.e.,

$$\mathbf{E}[\tilde{Z}_n] \asymp n^{-\xi_d},$$

where

$$\frac{4-d}{2} < \xi_d < 4 - d.$$

The left-hand side would be the "mean field" prediction for ξ_d given the value of $\mathbf{E}[\tilde{Z}_n^2]$.

12.7 Wilson's Algorithm

Let S be an irreducible Markov chain on a finite set

$$X = \{y_0, y_1, \ldots, y_k\}$$

with transition probabilities $p(\cdot,\cdot)$. In this section we describe an algorithm, due to Wilson [21, 23], to choose a spanning tree of X. (Here we consider X as an undirected graph with edge (x, y) in the graph if and only if $p(x, y)+p(y, x) > 0$). Start the Markov chain at y_1 and continue until it reaches y_0. Perform loop-erasure on this path and include all the edges of this loop-erased path in the tree. If we have a spanning tree, we stop; otherwise, we choose the vertex y_j of smallest index which has not been added to the tree. Start the Markov chain at this vertex and continue until it reaches a vertex that has already been included in the tree. Perform loop erasure on this path and add these edges to the tree. Continue until we have a completed tree.

It is straightforward to determine the probability that a certain tree $\mathcal{T}$ is chosen in this procedure. Let $\mathcal{T}$ be an undirected spanning tree, and make $\mathcal{T}$ a directed spanning tree by putting arrows on each edge pointing towards y_0, i.e., if $\{x, y\}$ is an edge in the tree, put the arrow pointing from x to y if every path from x to y_0 in the tree must go through y. If we start with the ordering of the vertices

$$\{y_0, y_1, y_2, \ldots, y_k\},$$

we get a new ordering (depending on $\mathcal{T}$),

$$\{z_0, z_1, z_2, \ldots, z_k\},$$

where $z_0 = y_0$; then the vertices in the unique path from y_1 to y_0 are listed; then the vertex of smallest index not already chosen (using the original ordering) is chosen, and then the vertices of its shortest path to the vertices already chosen; etc. One can check by repeated application of (12.2.2) that the probability of $\mathcal{T}$ being chosen is given by

$$f(z_1, \ldots, z_k; \{z_0\}) \prod_{i=1}^{k} p(z_i, z_i^*),$$

where x^* denotes the vertex on $\mathcal{T}$ that is adjacent to x and on the unique self-avoiding path from x to z_0 in $\mathcal{T}$. Here f is as defined in (12.2.3). In Section 12.2, we saw that f is independent of the ordering of $\{z_1, \ldots, z_k\}$, so we can just as well use the original ordering to give

$$\mathbf{P}\{\mathcal{T} \text{ chosen}\} = f(y_1, \ldots, y_k; \{y_0\}) \prod_{i=1}^{k} p(y_i, y_i^*). \tag{12.7.1}$$

Suppose the chain is reversible with respect to its invariant probability ϕ, i.e.,

$$\phi(x)p(x, y) = \phi(y)p(y, x), \quad x, y \in X,$$

and let $\alpha(x, y)$ be the symmetric function,

$$\alpha(x, y) = \phi(x)p(x, y) = \phi(y)\phi(y, x).$$

Let

$$F(y_0) = \frac{f(y_1, \ldots, y_k; \{y_0\})}{\prod_{i=1}^{k} \phi(y_i)},$$

and

$$\nu(\mathcal{T}) = \prod_{e \in T} \alpha(e).$$

Here the product is over all edges in $e \in \mathcal{T}$ and α is considered in a natural way as a function on edges. Note that $\nu(\mathcal{T})$ is a function of the undirected tree, and hence is independent of which vertex we choose to be y_0. We can rewrite (12.7.1) as

$$\mathbf{P}\{\mathcal{T} \text{ chosen}\} = F(y_0)\nu(\mathcal{T}).$$

We could have chosen some other vertex than y_0, say y, to be the initial vertex in our algorithm. In this case we could get

$$\mathbf{P}\{\mathcal{T} \text{ chosen}\} = F(y)\nu(\mathcal{T}).$$

But we know

$$\sum_{\mathcal{T}} \mathbf{P}\{\mathcal{T} \text{ chosen}\} = 1,$$

regardless of which vertex we choose for the first vertex. Hence $F(y)$ must be a constant C independent of y. We have shown that, regardless of how the vertices are ordered,

$$\mathbf{P}\{\mathcal{T} \text{ chosen}\} = C\ \nu(\mathcal{T}).$$

Consider the case of simple random walk on a graph. In this case, $p(x, y) = 1/d(x)$ if x and y are adjacent, where $d(x)$ is the degree of the vertex x. The invariant probability ϕ is proportional to d, and hence $\alpha(e)$ is constant for every edge e. This implies that $\nu(\mathcal{T})$ is the same for all $\mathcal{T}$ and hence every tree has an equally likely chance to be chosen! In other words, this algorithm chooses a spanning tree at random from the uniform distribution on the set of all spanning trees.

12.8 LERW and Conformal Invariance

Exact predictions have been made for the exponents for loop-erased walk in two dimensions using the methods of nonrigorous conformal field theory. These methods use the analogy between LERW and uniform spanning trees and the Q-state Potts model as $Q \to 0$. It may be difficult to make these arguments rigorous, but it is certainly worth investigating what role conformal invariance might have in understanding LERW rigorously. I will not be rigorous or even precise in this section, but rather give some thoughts for future work in the area.

The first thing that we would like is a notion of "loop-erased Brownian motion." It has been established that subsequential limits exist for loop-erased walk [1, 2]; however, there should be some kind of unique limit process. Ideally (and this may be too much to ask), we would like a function that takes any piece of a Brownian path $B[0, t]$ and assigns to it the loop-erased path $L(B[0, t])$ in such a way that:

- $L(B[0, t])$ is a self-avoiding path starting at B_0 and ending at B_t, with

$$L(B[0, t]) \subset B[0, t].$$

Moreover, $L(B[0,t])$ can be determined from $B[0,t]$ and perhaps some independent randomness.

- If $s < t$, then

$$L(B[0,t]) \subset L(B[0,s]) \cup B[s,t],$$

and $L(B[0,t])$ can be determined knowing $L(B[0,s])$ and $B[s,t]$ and some independent randomness.
- The loop-erasing procedure is in some sense conformally invariant in two dimensions.

So, the first open question in both two and three dimensions is: does such a loop erasing scheme exist? If so is it unique, or at least can one identify some such process that is the limit of loop-erased random walk?

Another approach is to try to construct a process with the properties of loop-erased Brownian motion in a method other than erasing loops from Brownian motion. Schramm [22] recently has constructed a process in two dimensions which appears to be a good candidate for the scaling limit of loop-erased walk in two dimensions. In fact, the best approach to produce a process as above might well be to construct the candidate for the loop-erased process and then "add loops" to it to produce a Brownian motion.

For the rest of this section we let $d = 2$, and we will assume that such a loop-erased Brownian motion exists. In fact, we will assume that loop-erased Brownian excursions exist on the annulus

$$\mathcal{A}_n = \{z \in \mathbb{C} : 1 < |z| < e^n\},$$

i.e., Brownian motions starting at $|z| = 1$ and conditioned to stay in the annulus until they leave it at $|z| = e^n$. Assume that we have k independent excursions starting at $z_1 = e^{i\theta_1}, \ldots, z_k = e^{i\theta_k}$. We let ω^j be the path of the jth excursion and we let $L(\omega^j)$ be its loop erasure. The k-point exponent, ζ_k, is defined by saying that the probability that

$$L(\omega^j) \cap \omega^l = \emptyset, \quad 1 \le j < l \le k, \tag{12.8.1}$$

is about $e^{-n\zeta_k}$. Duplantier [5] used methods of nonrigorous conformal field theory to give the conjecture

$$\zeta_k = \frac{k^2 - 1}{4}.$$

Note that $\zeta_3 = 2$, the triple point estimate, has been established (at least the loop-erased random walk analogue of the conjecture). There is also a corresponding conjecture for loop-erased walks restricted to stay in the part of the annulus in the upper half-plane

$$\mathcal{A}_n^+ = \{z \in \mathcal{A}_n : \Im(z) > 0\}.$$

We define the half-plane k-point exponent, $\tilde{\zeta}_k$, by assuming that $z_1, \ldots, z_k \in \mathcal{A}_n^+$ and saying that the probability that both (12.8.1) and

$$\omega^j \subset \mathcal{A}_n^+, \quad j = 1, \ldots, k, \tag{12.8.2}$$

occur is about $e^{-n\tilde{\zeta}_k}$. Duplantier also gave the conjecture

$$\tilde{\zeta}_k = \frac{k^2 + k}{2}.$$

Since the probability that (12.8.2) occurs is about e^{-n}, we could say that the probability that (12.8.1) occurs given (12.8.2) occurs is about $e^{-n\hat{\zeta}_k}$, where $\hat{\zeta}_k = \tilde{\zeta}_k - k$. (When half-plane exponents are given, one has to be careful whether one is talking about the equivalent of $\tilde{\zeta}_k$ or $\hat{\zeta}_k$.)

Define the following random variables, which are functions of ω^1,

$$Z_n(\omega^1) = \mathbf{P}\{\omega^2 \cap L(\omega^1) = \emptyset \mid \omega^1\},$$

$$\tilde{Z}_n(\omega^1) = \mathbf{P}\{\omega^2 \cap L(\omega^1) = \emptyset, \omega^2 \subset \mathcal{A}_n^+ \mid \omega^1\}.$$

Then the exponents $\xi(\lambda), \tilde{\xi}(\lambda)$ are defined by the relations

$$\mathbf{E}[Z_n^\lambda] \approx e^{-n\xi(\lambda)},$$

$$\mathbf{E}[\tilde{Z}_n^\lambda; \omega^1 \subset \mathcal{A}_n^+] \approx e^{-n\tilde{\xi}(\lambda)}.$$

Note that $\xi(1) = \zeta_2, \tilde{\xi}(1) = \tilde{\zeta}_2$. The random variable Z_n has an interesting interpretation through conformal invariance. Consider the region

$$\mathcal{A}_n \setminus \omega^1.$$

If ω_1 is self-avoiding, this is a simply connected region (if by some chance ω_1 is not self-avoiding we consider the simply connected "outer" region). This region can be mapped conformally onto

$$\mathcal{A}^+_{N(\omega^1)}$$

in a way so that the circle of radius 1 is mapped to the half circle of radius 1 and the circle of radius e^n is mapped to the half circle of radius $e^{N(\omega^1)}$. The value $N(\omega^1)$ depends on the path, but a little analysis will show that

$$N(\omega^1) \approx -\log Z_n(\omega^1).$$

If we now consider the k-point exponent ζ_k, focus on one walker ω^1, and do this conformal map, we come to the conformal mapping conjecture

$$\zeta_k = \xi(\tilde{\zeta}_{k-1}). \tag{12.8.3}$$

Similarly,

$$\tilde{\zeta}_k = \tilde{\xi}(\tilde{\zeta}_{k-1}). \tag{12.8.4}$$

Using these relations and the Duplantier conjectures, we can give conjectures for ξ and $\tilde{\xi}$,

$$\xi(\lambda) = \frac{\lambda}{2} + \frac{1}{8}[\sqrt{8\lambda + 1} - 1],$$

$$\tilde{\xi}(\lambda) = \lambda + 1 + \frac{1}{2}[\sqrt{8\lambda + 1} - 1].$$

Note that $\xi(3) = 2$; this is the third moment estimate, and, as we have seen, this is rigorous (or at least the analogous random walk statement is rigorous). While (12.8.3) and (12.8.4) are only conjectures, the analogous results for Brownian motion intersection exponents can be made rigorous [17].

Acknowledgment: This research was supported by the National Science Foundation.

REFERENCES

[1] Aizenman, M. and Burchard, A. (1998). Hölder regularity and dimension bounds for random curves, preprint.

[2] Aizenman, M., Burchard, A., Newman, C., and Wilson, D. (1998). Scaling limits for minimal and random spanning trees in two dimensions, preprint.

[3] Benjamini, I. (1998). Large scale degrees and the number of spanning clusters for the uniform spanning tree, preprint.

[4] Burdzy, K. and Lawler, G. (1990). Non-intersection exponents for random walk and Brownian motion. Part II: Estimates and applications to a random fractal. *Annals of Prob.* **18**, 981–1009.

[5] Duplantier, B. (1992). Loop-erased self-avoiding walks in $2D$, *Physica A* **191**, 516–522.

[6] Fargason, C. (1998). The percolation dimension of Brownian motion in three dimensions, Ph.D. dissertation, Duke University.

[7] Guttmann, A. and Bursill, R. (1990). Critical exponent for the loop-erased self-avoiding walk by Monte Carlo methods. *J. Stat. Phys.* **59**, 1–9.

[8] Kenyon, R. (1998). The asymptotic distribution of the discrete Laplacian, preprint.

[9] Kesten, H. (1987). Hitting probabilities of random walks on $\mathbb{Z}^d$, *Stoc. Proc. and Appl.* **25**, 165–184.

[10] Lawler, G. (1980). A self-avoiding walk. *Duke Math J.* **47**, 655–694.

[11] Lawler, G. (1991). *Intersections of Random Walks.* Birkhäuser, Boston.

[12] Lawler, G. (1992). Escape probabilities for slowly recurrent sets, *Prob. Theor. Rel. Fields* **94**, 91–117.

[13] Lawler, G. (1995). The logarithmic correction for loop-erased walk in four dimensions, Proceedings of the Conference in Honor of Jean-Pierre Kahane (Orsay, 1993), special issue of *J. Fourier Anal. Appl.*, 347–362.

[14] Lawler, G. (1996). Cut points for simple random walk, *Electronic Journal of Probability* **1**, #13.

[15] Lawler, G. (1998) Strict concavity of the intersection exponent for Brownian motion in two and three dimensions, *Math. Phys. Electron. J.* **4**, paper no. 5.

[16] Lawler, G. (1998). A lower bound on the growth exponent for loop-erased random walk in two dimensions, preprint.
[17] Lawler, G. and Werner, W. (1998). Intersection exponents for planar Brownian motion, preprint.
[18] Madras, N. and Slade, G. (1993). *The Self-Avoiding Walk*. Birkhäuser, Boston.
[19] Majumdar, S.N. (1992). Exact fractal dimension of the loop-erased self-avoiding random walk in two dimensions, *Phys. Rev. Letters* **68**, 2329–2331.
[20] Pemantle, R. (1991). Choosing a spanning tree for the integer lattice uniformly, *Annals of Prob.* **19**, 1559–1574.
[21] Propp, J. and D. Wilson (1998). How to get a perfectly random sample from a generic Markov chain and generate a random spanning tree of a directed graph, *Journal of Algorithms* **27**, 170–217.
[22] Schramm, O., in preparation.
[23] D. Wilson (1996). Generating random spanning trees more quickly than the cover time. *Proc. Twenty-Eighth Annual ACM Symposium on Theory of Computing*, 293–303.

Department of Mathematics
Duke University
Durham, NC 27708
jose@math.duke.edu

13

Dominance of the Sum over the Maximum and Some New Classes of Stochastic Compactness

P.S. Griffin and R.A. Maller

ABSTRACT Let S_n be a random walk. Two known results are that S_n is in the domain of attraction of the normal distribution if and only if S_n dominates its increment of largest modulus, $|X_n^{(1)}|$, in that $|S_n -$ median $(S_n)|/|X_n^{(1)}|$ diverges in probability to ∞ (due to Lévy); whereas the subsequential limit random variables of the normed sum S_n/B_n are all normal (possibly degenerate, but not degenerate at 0) if and only if $|S_n|/|X_n^{(1)}|$ diverges in probability to ∞ (due to Kesten and Maller). These represent examples of a much wider class of results connecting various kinds of weak convergence behaviour of S_n, and the question of when the random walk dominates its largest increment. We prove some extensions of the above results and relate them to various generalisations of the Feller class of stochastically compact distributions.

Keywords: Dominance of the sum, random walks, stochastic compactness, domain of attraction, domain of partial attraction, trimmed sums.

AMS Subject Classifications: Primary 60F05, 60J15, 60E07; Secondary 62G30.

13.1 Introduction

In the 1950's Lévy showed that the summand of maximum modulus, $X_n^{(1)}$, of a random walk S_n is asymptotically negligible in probability by comparison with $S_n - m_n$, where m_n is any median of S_n, if and only if S_n is in the domain of attraction of the normal distribution (Lévy [8, pp. 334–336]). Since then, many authors have investigated connections like this between the asymptotic normality, or, more generally, the weak convergence behaviour of S_n, and the way it dominates its largest increments. From Kesten and Maller [7, Theorem 3.1], for example, we have that:

$$\frac{|S_n|}{|X_n^{(1)}|} \xrightarrow{P} \infty \tag{13.1.1}$$

if and only if *there is a nonstochastic norming sequence $B_n \to \infty$ such that each sequence $n_i \to \infty$ of integers contains a subsequence $n_i' \to \infty$ for which $S_{n_i'}/B_{n_i'}$*

converges in distribution to a normal random variable, which may be degenerate, but if so, is not degenerate at 0. We will describe this weak convergence behaviour by saying that $S_n \in D'(N)$. By contrast Lévy's result is that

$$\frac{|S_n - m_n|}{|X_n^{(1)}|} \xrightarrow{P} \infty \tag{13.1.2}$$

if and only if $S_n \in D(N)$, the domain of attraction of the normal; i.e., (13.1.2) holds if and only if *there exist nonstochastic centering and norming sequences A_n and B_n such that* $(S_n - A_n)/B_n \Longrightarrow N(0, 1)$. Here "$\Longrightarrow$" denotes convergence in distribution and $N(0, 1)$ is a standard normal random variable.

Thus dominance conditions can be closely related to weak convergence conditions, which in turn can be characterised by analytical conditions on the distribution of the increments of S_n. In the present paper we continue an exploration of results of this kind. One focus, motivated by the results described above, will be on the role that centering plays in questions of dominance of S_n over $X_n^{(1)}$, and how this is reflected in the corresponding weak convergence equivalences. Thus we study the dichotomy induced by centering: at 0, as in (13.1.1), or at a median (or possibly at a more general sequence of nonstochastic centering constants), as in (13.1.2). We will consider not only random walks but also "trimmed sums", in which extremes are removed from the sample before the increments are summed.

In order to describe our results we need some notation. Let $X_1, X_2, \ldots, X_n$ be i.i.d. with distribution F, and let the random walk be

$$S_n = X_1 + X_2 + \cdots + X_n. \tag{13.1.3}$$

Let $X_n^{(n)}, \ldots, X_n^{(1)}$ denote $X_1, \ldots, X_n$ arranged in increasing order of modulus. (For most of the results, ties may be broken in any way, but see the comments in Section 13.3 below.) For each $r = 1, 2, \ldots$, define the *modulus – trimmed sum*

$$^{(r)}\tilde{S}_n = S_n - X_n^{(1)} - \cdots - X_n^{(r)} \tag{13.1.4}$$

(with $^{(0)}\tilde{S}_n = S_n$). We will need the following functionals of F, defined on $[0, \infty)$:

$$H(x) = P(|X| > x)) = 1 - F(x) + F(-x-); \tag{13.1.5}$$

$$\nu(x) = E(XI(|X| \le x)) = \int_{[-x,x]} y\,dF(y); \tag{13.1.6}$$

and

$$V(x) = E(X^2 I(|X| \le x)) = \int_{[-x,x]} y^2 dF(y). \tag{13.1.7}$$

Throughout the paper we assume $H(x) > 0$ for all $x > 0$.

13.2 Dominance of the Maximum by the Sum

Our first theorem extends Lévy's result by showing that $S_n \in D(N)$ if and only if (13.1.2) holds with m_n replaced by δ_n, for *every* non-stochastic centering sequence δ_n, and gives a generalization in which this is only required to hold for a subsequence $n_i \to \infty$. This provides a version of Lévy's result for S_n in the *domain of partial attraction of the normal distribution*, which we denote by $D_P(N)$, writing $S_n \in D_P(N)$ *if there are nonstochastic sequences* $n_i \to \infty$, A_{n_i} *and* B_{n_i}, *such that* $(S_{n_i} - A_{n_i})/B_{n_i} \Longrightarrow N(0,1)$. Theorem 13.2.1 also concerns the class SC of *stochastically compact distributions*, introduced by Feller [1] as an extension of the concept of the attraction of the sum S_n to a stable random variable. We say that $S_n \in SC$ *if there exist nonstochastic sequences* A_n *and* B_n, *such that every sequence* $n_i \to \infty$ *of integers contains a subsequence* $n_i' \to \infty$ *for which* $(S_{n_i'} - A_{n_i'})/B_{n_i'}$ *converges in distribution to a nondegenerate random variable.* Feller gave an analytic condition ((13.2.5a) below) for SC. We need also the following notion, introduced in Griffin and Maller [4]: *a sequence* Z_n *of random variables has, asymptotically, no mass at* 0, denoted by $Z_n \in NM_0$, if

$$\lim_{\varepsilon \downarrow 0} \limsup_{n\to\infty} P\{|Z_n| \le \varepsilon\} = 0. \tag{13.2.1}$$

Theorem 13.2.1 *For each* $r = 0, 1, 2, \ldots$, $s = 1, 2, 3, \ldots$, *and* $t = 1, 2, 3, \ldots$, *the following are equivalent:*

$${}^{(r)}\tilde{S}_n \in SC, \quad \textit{resp.,} \quad {}^{(r)}\tilde{S}_n \in D(N); \tag{13.2.2a, b}$$

for every nonstochastic sequence δ_n,

$$\frac{|{}^{(s)}\tilde{S}_n - \delta_n|}{|X_n^{(s)}|} \in NM_0, \quad \textit{resp.,} \quad \frac{|{}^{(s)}\tilde{S}_n - \delta_n|}{|X_n^{(s)}|} \xrightarrow{P} \infty; \tag{13.2.3a, b}$$

$$\frac{|{}^{(t)}\tilde{S}_n - \tilde{m}_n^{(t)}|}{|X_n^{(t)}|} \in NM_0, \quad \textit{resp.,} \quad \frac{|{}^{(t)}\tilde{S}_n - \tilde{m}_n^{(t)}|}{|X_n^{(t)}|} \xrightarrow{P} \infty; \tag{13.2.4a, b}$$

$$\liminf_{x\to\infty} \frac{V(x)}{x^2 H(x)} > 0, \quad \textit{resp.,} \quad \lim_{x\to\infty} \frac{V(x)}{x^2 H(x)} = \infty. \tag{13.2.5a, b}$$

Furthermore, (13.2.2b)–(13.2.5b) *remain equivalent if the convergence is along a subsequence* $n_i \to \infty$, *rather than along* $n \to \infty$, *provided* (13.2.2b) *is replaced by* ${}^{(r)}\tilde{S}_n \in D_P(N)$ *and* (13.2.5b) *is replaced by* $\limsup_{x\to\infty} V(x)/x^2H(x) = \infty$.

In Theorem 13.2.1, $\tilde{m}_n^{(t)}$ is any median of ${}^{(t)}\tilde{S}_n$, $t = 1, 2, \ldots$, i.e., any nonstochastic sequence satisfying

$$P\{{}^{(t)}\tilde{S}_n \ge \tilde{m}_n^{(t)}\} \wedge P\{{}^{(t)}\tilde{S}_n \le \tilde{m}_n^{(t)}\} \ge \frac{1}{2}. \tag{13.2.6}$$

It is not hard to show that Theorem 13.2.1 remains true if we replace $\tilde{m}_n^{(t)}$ by $\tilde{m}_n^{(0)} = m_n$, a median of S_n, or by any other sequence satisfying (13.2.6) with $\frac{1}{2}$

replaced by some α in (0, 1). The index of trimming is also irrelevant in that if, for example, any of (13.2.2a, b)–(13.2.4a, b) (or, the version for a subsequence n_i) holds for some $r \geq 0$, $s \geq 1$, or $t \geq 1$, then each of them holds for all such r, s, t, being equivalent to an analytic condition which does not depend on r, s, or t. When $r = 0$, the equivalence of (13.2.2b) and (13.2.5b) is Lévy's classic criterion for $D(N)$, and (13.2.4b) (with $\tilde{m}_n^{(t)}$ replaced by m_n, and $t = 1$) is his formulation for dominance of S_n over $|X_n^{(1)}|$. The condition $\limsup_{x\to\infty} V(x)/x^2H(x) = \infty$ is Lévy's analytical criterion for $D_P(N)$ (see Lévy [8, Theorem 36,3, p. 113]). Lévy does not mention an equivalence of $D_P(N)$ with a condition like (13.2.4b).

Theorem 13.2.1 deals with dominance of the trimmed sum ${}^{(r)}\tilde{S}_n$ over an order statistic $X_n^{(r)}$, after centering at an arbitrary sequence δ_n, or at a median. One could ask for versions of these results for a *specified* nonstochastic sequence δ_n, and in particular, versions for $\delta_n = 0$ are interesting. For this case some results are known. Kesten and Maller [7, Theorem 3.1], and Griffin and Maller [3] proved the equivalence of (13.1.1), or, more generally,

$$\frac{|{}^{(r)}\tilde{S}_n|}{|X_n^{(r)}|} \xrightarrow{P} \infty, \tag{13.2.7}$$

$r = 1, 2, \ldots$, with $S_n \in D'(N)$, and with the following analytic condition: one of

$$\lim_{x\to\infty} \frac{|\nu(x)|}{xH(x)} = \infty \quad \text{or} \quad \lim_{x\to\infty} \frac{V(x)}{x^2H(x) + x|\nu(x)|} = \infty \tag{13.2.8a, b}$$

occurs. These two conditions can in fact be combined into one, namely

$$\lim_{x\to\infty} \frac{x|\nu(x)| + V(x)}{x^2H(x)} = \infty \tag{13.2.8c}$$

(see Griffin and Maller [3]). The components (13.2.8a) and (13.2.8b) have the following intuitive interpretation. (13.2.8a) characterises the *relative stability of* S_n; i.e., that *there exists a nonstochastic sequence* $B_n \to \infty$ *such that* $S_n/B_n \xrightarrow{P} 1$ *or* $S_n/B_n \xrightarrow{P} -1$ (written, $S_n \in RS$), while (13.2.8b) characterises the *domain of attraction of the normal without centering*, written $S_n \in D_0(N)$, where we say that $S_n \in D_0(N)$ *if* $S_n/B_n \Longrightarrow N(0, 1)$, *for some nonstochastic sequence* $B_n \to \infty$. It is easily shown that (13.2.8b) also characterises ${}^{(r)}\tilde{S}_n \in D_0(N), r \geq 1$. We furthermore have the following equivalence. Let Δ_a denote the distribution degenerate at a, $a \in [-\infty, \infty]$. Griffin and Maller [3, Theorem 2.2] showed that $S_n \in D_0(N)$ if and only if, for any $r \geq 1$,

$$\frac{{}^{(r)}\tilde{S}_n}{|X_n^{(r)}|} \stackrel{v}{\Longrightarrow} \frac{1}{2}\Delta_{-\infty} + \frac{1}{2}\Delta_{\infty}. \tag{13.2.9}$$

Here $\stackrel{v}{\Longrightarrow}$ denotes vague convergence (Jain and Orey [5] give a discussion of this) and the right hand side of (13.2.9) should be interpreted as the distribution function equal to 1/2 for all x, i.e., which assigns mass 1/2 to each of $+\infty$ and $-\infty$. Griffin and Maller [3] express the dichotomy, that $D'(N)$ is equivalent to (13.2.8a) or

(13.2.8b), as

$$D'(N) = RS \cup D_0(N),$$

a *disjoint* union. So we have a very satisfying picture of how (13.2.7) occurs.

In order to generalise some of these concepts, Griffin and Maller [4] introduced an extension of Feller's class SC, in the spirit of Kesten and Maller's [7] extension $D'(N)$ of Lévy's class $D(N)$. Say that $S_n \in SC'$ *if there is a nonstochastic sequence B_n such that each sequence of integers $n_i \to \infty$ contains a subsequence $n'_i \to \infty$ for which $S_{n'_i}/B_{n'_i}$ converges in distribution to a proper random variable, which may be degenerate, but if so, is not degenerate at* 0. They proved that SC' is characterised by

$$\liminf_{x\to\infty} \frac{x|\nu(x)| + V(x)}{x^2 H(x)} > 0. \tag{13.2.10}$$

Now in view of (13.2.7) and (13.2.8c), it is natural also to ask if this is also equivalent to

$$\frac{|^{(r)}\tilde{S}_n|}{|X_n^{(r)}|} \in NM_0, \tag{13.2.11}$$

but, as shown by Griffin and Maller [4, Example 2.2], this is not the case. To characterise (13.2.11) we need another version of compactness which we will call SC''. Say that $S_n \in SC''$ *if there is a nonstochastic sequence $B_n \to \infty$ such that each sequence of integers $n_i \to \infty$ contains a subsequence $n'_i \to \infty$ for which $S_{n'_i}/B_{n'_i} \Longrightarrow Z$, where Z is a random variable with no mass at* 0; i.e., $P(Z = 0) = 0$. (Z may be degenerate.) We then have:

Theorem 13.2.2 *For any integers $r = 1, 2, 3, \ldots$ and $s = 0, 1, 2, \ldots$, we have $|^{(r)}\tilde{S}_n|/|X_n^{(r)}| \in NM_0$ if and only if $^{(s)}\tilde{S}_n \in SC''$.*

Griffin and Maller [4] introduced a further class SC_0, the class of distributions which are "stochastically compact without centering constants". We say that $S_n \in SC_0$ if *there is a nonstochastic sequence $B_n \to \infty$ for which each sequence of integers $n_i \to \infty$ contains a subsequence $n'_i \to \infty$ such that $S_{n'_i}/B_{n'_i} \Longrightarrow Z$, where Z is a nondegenerate random variable.* Griffin and Maller [4] showed: $S_n \in SC_0$ if and only if

$$\liminf_{x\to\infty} \frac{V(x)}{x^2 H(x) + x|\nu(x)|} > 0. \tag{13.2.12}$$

Next we ask if there is an analogue for SC_0 of the result (13.2.9), which shows how the mass of $^{(r)}\tilde{S}_n/|X_n^{(r)}|$ "splits" at $\pm\infty$ when $S_n \in D_0(N)$. We prove:

Theorem 13.2.3 $S_n \in SC_0$ *if and only if, for each $r = 0, 1, 2, \ldots$, $^{(r)}\tilde{S}_n \in SC''$, and, in addition, $P\{|Y| = \infty\} > 0$ implies $P\{Y = +\infty\} \wedge P\{Y = -\infty\} > 0$ for any random variable Y which is a vague limit of $^{(r)}\tilde{S}_n/|X_n^{(r)}|$. This remains true if we replace $S_n \in SC_0$ by $^{(s)}\tilde{S}_n \in SC_0$, for any $s = 1, 2, \ldots$.*

The condition on Y in Theorem 13.2.3 can alternatively be expressed by saying that any subsequential (improper) limit Y of ${}^{(r)}\tilde{S}_n/|X_n^{(r)}|$ which has mass at $+\infty$ (i.e., $P\{Y > x\} \geq \delta$ for some $\delta > 0$ for all $x > 0$) also has mass at $-\infty$, and similarly with $+\infty$ and $-\infty$ interchanged.

13.3 Preliminaries to the Proofs

We will need some notation and preliminary lemmas. First we set up an integral representation, commonly used in studies of this kind, for the distribution of the trimmed sum ${}^{(r)}\tilde{S}_n$. This involves a specific method for the breaking of possible ties among $X_n^{(i)}$, $1 \leq i \leq n$, and so we cannot break ties in an arbitrary way. Any of our proofs which require the representation ((13.3.17) below), are thus dependent on this particular tie-breaking scheme. We will index the possible ties among $|X_k|$, $1 \leq k \leq n$ by their order of occurrence. Thus let

$$m_n(j) = \#\{i : |X_i| > |X_j|, \text{ or } |X_i| = |X_j| \text{ and } i < j, 1 \leq i, j \leq n\} \quad (13.3.1)$$

and then define

$$X_n^{(r)} = X_j \quad \text{if} \quad m_n(j) = r. \quad (13.3.2)$$

Whatever scheme is adopted, let $\tilde{I}(n, r)$ denote the indices of the random variables $X_n^{(r+1)}, \ldots, X_n^{(n)}$.

Recall that $H(x) = P\{|X| > x\}$ is a right-continuous function on $[0, \infty)$. It will be convenient to set $H(0-) = 1$. Let

$$\hat{H}(u) = \begin{cases} \sup\{x \geq 0 : H(x) > u\} & \text{if } 0 \leq u < H(0) \\ 0 & \text{if } u \geq H(0). \end{cases}$$

Thus for $0 < u < H(0)$

$$H(\hat{H}(u)) \leq u \leq H(\hat{H}(u)-).$$

The following construction can be found in Mori [10, pp. 508–509]. By the Radon-Nikodym Theorem there exist two Borel functions g^+ and g^-, defined on $(0, \infty)$, such that for all $x > 0$

$$P\{X > x\} = \int_{(x,\infty)} g^+(y)dh(y) \text{ and } P\{X < -x\} = \int_{(x,\infty)} g^-(y)dh(y),$$

where $h(y) = 1 - H(y) = P\{|X| \leq y\}$. We may also assume $0 \leq g^\pm \leq 1$ and $g^+ + g^- \equiv 1$. We will use U to denote a random variable uniformly distributed on $[0, 1]$. Let η satisfy

$$P\{\eta = \pm 1 | U = u\} = g^\pm(\hat{H}(u)), \qquad 0 < u < 1.$$

Then $\eta\hat{H}(U) \stackrel{D}{=} X$. Let (η_i, U_i) be i.i.d. random vectors distributed as (η, U). Let $U_{n,1} \leq \cdots \leq U_{n,n}$ be the order statistics of $U_1, \ldots, U_n$. This ordering is unique and the inequalities are strict except on a null set. Now let $Y_n^{(i)} = \eta_k \hat{H}(U_k)$ if $U_{n,i} = U_k$. Then $(Y_n^{(1)}, \ldots, Y_n^{(n)}) \stackrel{D}{=} (X_n^{(1)}, \ldots, X_n^{(n)})$ and

$$^{(r)}\tilde{S}_n \stackrel{D}{=} Y_n^{(r+1)} + \cdots + Y_n^{(n)}. \tag{13.3.3}$$

Thus we may assume the X_i are given by $\eta_i \hat{H}(U_i)$ and we will not distinguish between $^{(r)}\tilde{S}_n$ (as given by (13.1.4)) and $Y_n^{(r+1)} + \cdots + Y_n^{(n)}$.

For $0 < u < 1$, let $Z(u)$ have the distribution of $\eta \hat{H}(U)$ conditioned by $U \geq u$, i.e.,

$$P\{Z(u) \in B\} = \frac{P\{\eta \hat{H}(U) \in B, U \geq u\}}{1-u}, \tag{13.3.4}$$

for any Borel set B. In particular,

$$P\{|Z(u)| > x\} = \left(\frac{H(x) - u}{1-u}\right) \vee 0. \tag{13.3.5}$$

If $Z_i(u)$ is a sequence of i.i.d. random variables with distribution $Z(u)$, we define

$$J_m(u) = \sum_{i=1}^{m} Z_i(u), \tag{13.3.6}$$

for any $m \geq 1$. For $y > 0$ such that $H(y) < 1$, let $X(y)$ have as distribution the distribution of X conditioned by $|X| \leq y$, i.e., for any Borel set B,

$$P\{X(y) \in B\} = \frac{P\{X \in B, |X| \leq y\}}{1 - H(y)}. \tag{13.3.7}$$

Observe that

$$P\{|X(y)| > x\} = \frac{H(x) - H(y)}{1 - H(y)} \vee 0, \tag{13.3.8}$$

$$E(X(y)) = \frac{\nu(y)}{1 - H(y)}, \quad \text{and} \quad E(X(y))^2 = \frac{V(y)}{1 - H(y)}. \tag{13.3.9}$$

If $X_i(y)$ is a sequence of i.i.d. random variables with distribution the same as that of $X(y)$, let

$$S_m(y) = \sum_{i=1}^{m} X_i(y), \ m = 1, 2, \ldots. \tag{13.3.10}$$

In addition to conditioned variables we will also need to consider truncated ones. So for $y > 0$ let

$$T_n(y) = \sum_{i=1}^{n} X_i I(|X_i| \leq y). \tag{13.3.11}$$

Our interest in conditioned variables comes from the following representation which is easy to prove and will be used frequently: for any Borel set B

$$P\{S_n \in B, |X_n^{(1)}| \leq y\} = P\{S_n(y) \in B\} P\{|X_n^{(1)}| \leq y\}. \tag{13.3.12}$$

We will always use $a_n(\lambda)$ to denote the sequence defined for $\lambda > 0$ by

$$a_n(\lambda) = \sup\{x : H(x) > \frac{1}{\lambda n}\}. \tag{13.3.13}$$

This satisfies, for $\lambda > 1/n$,

$$nH(a_n(\lambda)) \le \frac{1}{\lambda} \le nH(a_n(\lambda)-). \tag{13.3.14}$$

We will write $X(a_n(\lambda))$ and $S_m(a_n(\lambda))$ as $X(\lambda)$ and $S_m(\lambda)$ respectively. This should not cause confusion with $X(x)$ and $S_m(x)$ since we will always use Greek letters in this context. Define $u_n(\lambda)$ for $\lambda > 0$ by

$$u_n(\lambda) = 0 \vee \frac{1}{\lambda n} \wedge 1. \tag{13.3.15}$$

Then, recalling (13.3.13),

$$a_n(\lambda) = \hat{H}(u_n(\lambda)). \tag{13.3.16}$$

For notational convenience we will write $Z(\lambda)$ for $Z(u_n(\lambda))$ and similarly $J_{n-r}(\lambda)$ for $J_{n-r}(u_n(\lambda))$. This should not cause confusion with $Z(u)$ and $J_{n-r}(u)$ since we will always use Greek letters in this setting.

The connection with the previous notation is that $X(x) \stackrel{D}{=} Z(H(x))$. If H is continuous then also $Z(u) \stackrel{D}{=} X(\hat{H}(u))$. However in general this is not the case, and is the reason we need to resort to the representation of X as $\eta\hat{H}(U)$.

We write $U(u)$ for the distribution of U conditioned by $U \ge u$. Thus $U(u)$ is uniform on $(u, 1)$. In keeping with the above, we write $U(\lambda)$ for $U(u_n(\lambda))$. If $U_i(u)$ are i.i.d. with distribution $U(u)$, the order statistics of $U_1(u), \dots, U_m(u)$ will be denoted by $U_{m,1}(u) \le \cdots \le U_{m,m}(u)$. Similarly for $U_{m,1}(\lambda), \dots, U_{m,m}(\lambda)$.

The reason for our interest in these conditioned variables is the following integral representation: for any bounded Borel function φ,

$$\begin{aligned} E\phi({}^{(r)}\tilde{S}_n, U_{n,r+1}, U_{n,r}) &= \int_0^1 E\phi(J_{n-r}(u), U_{n-r,1}(u), u)dP\{U_{n,r} \le u\} \\ &= \int_0^\infty E\phi(J_{n-r}(\lambda), U_{n-r,1}(\lambda), u_n(\lambda))d\nu_{n,r}(\lambda) \end{aligned} \tag{13.3.17}$$

where

$$\nu_{n,r}(\lambda) = P\{U_{n,r} \le u_n(\lambda)\} = P\{|X_n^{(r)}| \le a_n(\lambda)\}.$$

It is well known that $\nu_{n,r}$ converges in the total variation norm to ν_r as $n \to \infty$, where

$$\nu_r(\lambda) = e^{-1/\lambda}\sum_{j=r}^{\infty}\frac{\lambda^{-j}}{j!} = 1 - e^{-1/\lambda}\sum_{j=0}^{r-1}\frac{\lambda^{-j}}{j!}. \tag{13.3.18}$$

This means for example that we can write, as $n \to \infty$,

$$E\phi({}^{(r)}\tilde{S}_n, U_{n,r+1}, U_{n,r}) = \int_0^\infty E\phi(J_{n-r}(\lambda), U_{n-r,1}(\lambda), u_n(\lambda))d\nu_r(\lambda) + o(1). \tag{13.3.19}$$

As a particular instance of (13.3.17), we have for any $r \geq 1$ and $\alpha > 0$

$$\begin{aligned} &P\{{}^{(r)}\tilde{S}_n \in B,\ U_{n,r} < u_n(\alpha) \leq U_{n,r+1}\} \\ &= \int_\alpha^\infty P\{J_{n-r}(\lambda) \in B,\ U_{n-r,1}(\lambda) \geq u_n(\alpha)\}d\nu_{n,r}(\lambda) \\ &= \int_\alpha^\infty P\{J_{n-r}(\alpha) \in B\}P\{U_{n-r,1}(\lambda) \geq u_n(\alpha)\}d\nu_{n,r}(\lambda) \\ &= P\{J_{n-r}(\alpha) \in B\}P\{U_{n,r} < u_n(\alpha) \leq U_{n,r+1}\}. \end{aligned} \tag{13.3.20}$$

The second equality uses an obvious analogue for conditioned random variables of

$$P\{S_m \in B, U_{m,1} \geq u\} = P\{J_m(u) \in B\}P\{U_{m,1} \geq u\}, \tag{13.3.21}$$

which in turn is an analogue of (13.3.12).

Next we need to recall some properties relating to the weak convergence of normed and centered sums. Recall that A_n and B_n always denote nonstochastic centering and norming sequences, respectively, while $n_i \to \infty$ and $n'_i \to \infty$ are sequences of integers. Tightness of a sequence R_n of random variables is equivalent to the property that each sequence $n_i \to \infty$ contains a subsequence $n'_i \to \infty$ for which $R_{n'_i}$ converges in distribution to a (proper) random variable, possibly degenerate. Necessary and sufficient for tightness of $(S_n - A_n)/B_n$, for some A_n, are

$$\limsup_{n\to\infty} \frac{nV(B_n)}{B_n^2} < \infty \quad \text{and} \quad \lim_{x\to\infty} \limsup_{n\to\infty} nH(xB_n) = 0. \tag{13.3.22}$$

(See Jain and Orey [6] or Feller [1, Lemma, p. 309]). It follows from (13.3.22) that $B_n \to \infty$ as $n \to \infty$. Furthermore, convergence in distribution of $(S_{n_i} - A_{n_i})/B_{n_i}$ to a proper (infinitely divisible) random variable implies

$$n_i H(xB_{n_i}) \to T(x) \quad (i \to \infty) \tag{13.3.23}$$

for $x > 0$, where $T(x)$ is the tail of a Lévy measure; thus $T(x)$ is decreasing with $T(+\infty) = 0$ (Gnedenko and Kolmogorov [2, pp. 116 and 84]).

The notation $a_n \asymp b_n$ will mean that the ratio of a_n and b_n is bounded away from 0 and ∞, while $a_n \sim b_n$ if $a_n/b_n \to 1$ as $n \to \infty$. We often omit $n \to \infty$ or $x \to \infty$ etc., if it is obvious. If $S_n \in SC$ for a specific norming sequence B_n, write $S_n \in SC(B_n)$, and similarly for SC' and SC''. Random variables obtained as limits of some sequence of random variables along a subsequence will be referred to as "subsequential limits" of the original sequence. We will also need:

Lemma 13.3.1 *Let $r = 0, 1, 2, \ldots$. We have $S_n \in SC(B_n)$ if and only if ${}^{(r)}\tilde{S}_n \in SC(B_n)$. If $S_n \in SC(B_n)$ and Z_r is any subsequential limit of $({}^{(r)}\tilde{S}_n - A_n)/B_n$, then Z_r is a continuous random variable.*

PROOF. The first statement follows from Mori [10, Theorem 1], so let $S_n \in SC(B_n)$ and let Z_r be a subsequential limit of $({}^{(r)}\tilde{S}_n - A_n)/B_n$. When $r = 0$ we know that Z_r is continuous by Pruitt [12] or Maller [9, p. 274], so keep $r \geq 1$. If $z \in (-\infty, +\infty)$ and $\varepsilon > 0$ we have by (13.3.21) with $u = u_n(\lambda)$ that

$$P\left\{\left|\frac{S_n - A_{n+r}}{B_{n+r}} - z\right| \leq \varepsilon\right\} \geq P\left\{\left|\frac{S_n - A_{n+r}}{B_{n+r}} - z\right| \leq \varepsilon, U_{n,1} \geq u_n(\lambda)\right\}$$
$$= P\left\{\left|\frac{J_n(\lambda) - A_{n+r}}{B_{n+r}} - z\right| \leq \varepsilon\right\} P\left\{U_{n,1} \geq u_n(\lambda)\right\}. \tag{13.3.24}$$

We also have

$$P\left\{U_{n,1} \geq u_n(\lambda)\right\} = (1 - u_n(\lambda))^n = (1 - \frac{1}{\lambda n})^n \to e^{-1/\lambda}.$$

Now we claim that, for all $\lambda > 0$,

$$\lim_{\varepsilon \to 0} \limsup_{n \to \infty} P\left\{\left|\frac{J_n(\lambda) - A_{n+r}}{B_{n+r}} - z\right| \leq \varepsilon\right\} = 0. \tag{13.3.25}$$

If (13.3.25) fails for some $\lambda > 0$ then by (13.3.24) we can find $\varepsilon_i \downarrow 0$ and $n_i \to \infty$ such that

$$\liminf_{i \to \infty} P\left\{\left|\frac{S_{n_i} - A_{n_i+r}}{B_{n_i+r}} - z\right| \leq \varepsilon_i\right\} > 0. \tag{13.3.26}$$

Taking a further subsequence of n_i if necessary we can assume $(S_{n_i} - A_{n_i})/B_{n_i} \Longrightarrow Z$, a continuous random variable. Also by Gnedenko and Kolmogorov [2, p. 117], $A_n = n\nu(B_n) + O(B_n)$, so

$$\begin{aligned} A_{n+r} - A_n &= n(\nu(B_{n+r}) - \nu(B_n)) + r\nu(B_{n+r}) + O(B_{n+r}) + O(B_n) \\ &= n \int_{B_n < |x| \leq B_{n+r}} x dF(x) + O(B_{n+r}) + O(B_n) \\ &= O(nH(B_n))B_{n+r} + O(B_{n+r}) + O(B_n) = O(B_{n+r}) + O(B_n) \end{aligned}$$

since it follows easily from (13.3.22) that $nH(B_n) = 0(1)$. Also, $\nu(B_n) = O(B_n)$, trivially. By Feller [1, Theorem, p. 387], $B_{n+r} \asymp B_n$, so we have, taking a further subsequence if necessary,

$$\frac{S_{n_i} - A_{n_i+r}}{B_{n_i+r}} \Longrightarrow aZ + b$$

for some constants $a > 0$ and b. But since Z is continuous, $aZ + b$ is continuous, and we have a contradiction with (13.3.26). Thus indeed (13.3.25) holds. Now by (13.3.19), as $n \to \infty$,

$$P\left\{\left|\frac{{}^{(r)}\tilde{S}_n - A_n}{B_n} - z\right| \leq \varepsilon\right\} = \int_0^\infty P\left\{\left|\frac{J_{n-r}(\lambda) - A_n}{B_n} - z\right| \leq \varepsilon\right\} d\nu_r(\lambda) + o(1).$$

We now apply a version of Fatou's lemma which allows us to take a limsup under an integral if the integrand is dominated by an integrable function (e.g.,

Rudin [13, p. 247]). Thus

$$\limsup_{n\to\infty} P\left\{\left|\frac{{}^{(r)}\tilde{S}_n - A_n}{B_n} - z\right| \le \varepsilon\right\}$$
$$\le \int_0^\infty \limsup_{n\to\infty} P\left\{\left|\frac{J_{n-r}(\lambda) - A_n}{B_n} - z\right| \le \varepsilon\right\} d\nu_r(\lambda). \quad (13.3.27)$$

Take the limit as $\varepsilon \to 0$ of both sides, then take the limit under the integral on the right by another application of the dominated version of Fatou's lemma. By (13.3.25) this will give 0 on the right hand side, so we have proved that

$$\lim_{\varepsilon\to 0}\limsup_{n\to\infty} P\left\{\left|\frac{{}^{(r)}\tilde{S}_n - A_n}{B_n} - z\right| \le \varepsilon\right\} = 0. \quad (13.3.28)$$

It follows that any subsequential limit of $({}^{(r)}\tilde{S}_n - A_n)/B_n$ is continuous. □

13.4 Proof of Theorem 13.2.1

Let (13.2.2a) hold, and we will show (13.2.3a). If (13.2.3a) fails we can find a nonstochastic sequence δ_n and integers $n_i \to \infty$ such that

$$\liminf_{i\to\infty} P\{|{}^{(s)}\tilde{S}_{n_i} - \delta_{n_i}| \le x_i |X_{n_i}^{(s)}|\} \ge \eta > 0 \quad (13.4.1)$$

for some $x_i \downarrow 0$ and $\eta > 0$, where $s \ge 1$. (See (13.2.1).) Now if (13.2.2a) holds it holds with r replaced by 0, by Lemma 13.3.1, and hence with r replaced s. So we can take a further subsequence of n_i and nonstochastic A_n and B_n such that

$$\frac{{}^{(s)}\tilde{S}_{n_i} - A_{n_i}}{B_{n_i}} \Longrightarrow Z_s \quad (13.4.2)$$

for some continuous random variable Z_s, and also

$$\frac{\delta_{n_i} - A_{n_i}}{B_{n_i}} \to C \text{ for some constant } C \text{ in } [-\infty, \infty]. \quad (13.4.3)$$

Suppose first that $|C| < \infty$. Then

$$\frac{{}^{(s)}\tilde{S}_{n_i} - \delta_{n_i}}{B_{n_i}} \Longrightarrow Z_s - C. \quad (13.4.4)$$

Fix $x > 0$ and choose i so large that $x_i < x$. Write

$$P\{|{}^{(s)}\tilde{S}_{n_i} - \delta_{n_i}| \le x_i|X_{n_i}^{(s)}|\}$$
$$\le P\{|{}^{(s)}\tilde{S}_{n_i} - \delta_{n_i}| \le x|X_{n_i}^{(s)}|,\ |X_{n_i}^{(s)}| \le x^{-1/2}B_{n_i}\} + P\{|X_{n_i}^{(s)}| > x^{-1/2}B_{n_i}\}$$
$$\le P\{|{}^{(s)}\tilde{S}_{n_i} - \delta_{n_i}| \le x^{\frac{1}{2}}B_{n_i}\} + P\{|X_{n_i}^{(1)}| > x^{-\frac{1}{2}}B_{n_i}\} \quad (13.4.5)$$
$$\le P\{|Z_s - C| \le x^{\frac{1}{2}}\} + o(1) + n_i H(x^{-\frac{1}{2}}B_{n_i}), \quad (13.4.6)$$

where (13.4.6) follows from (13.4.5) by (13.4.4). Since $S_n \in SC(B_n)$, we can by (13.3.23) take a further subsequence of n_i if necessary so that $n_i H(x^{-\frac{1}{2}} B_{n_i}) \to T(x^{-\frac{1}{2}})$ for some Lévy tail function T. Thus (13.4.6) gives

$$\limsup_{i\to\infty} P\{|{}^{(s)}\tilde{S}_{n_i} - \delta_{n_i}| \le x_i |X_{n_i}^{(s)}|\} \le P\{|Z_s - C| \le x^{\frac{1}{2}}\} + T(x^{-\frac{1}{2}}).$$

But the right hand side of this converges to 0 as $x \to 0$, since, by Lemma 13.3.1, Z is continuous at C, while $T(+\infty) = 0$. This contradicts (13.4.1).

Next suppose that $|C| = \infty$. Then (13.4.2) and (13.4.3) show that

$$\frac{|{}^{(s)}\tilde{S}_{n_i} - \delta_{n_i}|}{B_{n_i}} \ge \left| \frac{|{}^{(s)}\tilde{S}_{n_i} - A_{n_i}|}{B_{n_i}} - \frac{|\delta_{n_i} - A_{n_i}|}{B_{n_i}} \right| \xrightarrow{P} \infty$$

and we see from (13.4.5) that

$$\limsup_{i\to\infty} P\{|{}^{(s)}\tilde{S}_{n_i} - \delta_{n_i}| \le x |X_{n_i}^{(s)}|\} \le T(x^{-\frac{1}{2}})$$

so again we get a contradiction with (13.4.1) as $x \to 0$. Thus (13.2.2a) implies (13.2.3a).

Next we show that (13.2.2b) implies (13.2.3b). Let ${}^{(r)}\tilde{S}_n \in D(N)$ and suppose $|{}^{(s)}\tilde{S}_n - \delta_n|/|X_n^{(s)}|$ does not diverge in probability to ∞ for some δ_n and some $s \ge 1$. Then there are integers $n_i \to \infty$ and a constant $x_0 > 0$ such that

$$\liminf_{i\to\infty} P\{|{}^{(s)}\tilde{S}_{n_i} - \delta_{n_i}| \le x_0 |X_{n_i}^{(s)}|\} \ge \eta > 0. \tag{13.4.7}$$

By Mori [10, Theorem 3], we can replace r by s and by 0 to deduce from (13.2.2b) that (13.4.2) holds, with Z_s and Z_0 replaced by $N(0, 1)$, for some A_n, B_n, and we can also assume that (13.4.3) holds again. If $|C| < \infty$ use an argument like that used to obtain (13.4.6) (with x_i replaced by x_0 and $x^{-\frac{1}{2}}$ replaced by ε/x_0) to get

$$\limsup_{i\to\infty} P\{|{}^{(s)}\tilde{S}_{n_i} - \delta_{n_i}| \le x_0 |X_{n_i}^{(s)}|\} \le P\{|N(-C, 1)| \le \varepsilon\} + T(\varepsilon/x_0), \tag{13.4.8}$$

where $\varepsilon > 0$ is arbitrary and $T(\lambda)$ is the limit of $n_i H(\lambda B_{n_i})$. But for convergence to normality, $T(\lambda) \equiv 0$, so letting $\varepsilon \to 0$ in (13.4.8) gives a contradiction with (13.4.7), since of course $N(-C, 1)$ is a continuous random variable. On the other hand if $|C| = \infty$ then as before $|{}^{(s)}\tilde{S}_{n_i} - \delta_{n_i}|/B_{n_i} \xrightarrow{P} \infty$, and we obtain a contradiction with (4.7) via

$$\begin{aligned} P\{|{}^{(s)}\tilde{S}_{n_i} - \delta_{n_i}| \le x_0 |X_{n_i}^{(s)}|\} &\le P\{|{}^{(s)}\tilde{S}_{n_i} - \delta_{n_i}| \le x_0 B_{n_i}\} + P\{|X_{n_i}^{(s)}| > B_{n_i}\} \\ &\le o(1) + n_i H(B_{n_i}) \to 0. \end{aligned}$$

This proves that (13.2.2b) implies (13.2.3b).

The proof that (13.2.2b) implies (13.2.3b) along a subsequence is practically the same. If ${}^{(r)}\tilde{S}_n \in D_P(N)$ then ${}^{(s)}\tilde{S}_n \in D_P(N)$ by Mori [10] and so again (13.4.2) holds with Z_s replaced by $N(0, 1)$, for the given n_i, and some A_{n_i}, B_{n_i}. If (13.2.3b) fails for this subsequence n_i, we have (13.4.7). This is impossible exactly as before, proving that (13.2.2b) implies (13.2.3b).

Now note that the implications (13.2.3a) ⇒ (13.2.4a) and (13.2.3b) ⇒ (13.2.4b) are trivial, as is the implication (13.2.3b) ⇒ (13.2.4b) along a subsequence. So we concentrate next on showing that (13.2.4a) ⇒ (13.2.5a). Let (13.2.4a) hold and suppose (13.2.5a) fails, so there is a sequence $x_j \to \infty$ such that

$$\frac{V(x_j)}{x_j^2 H(x_j)} \to 0 \quad (j \to \infty). \tag{13.4.9}$$

Fix $\eta > 0$ so small that $e^{-\eta} > 3/4$. For j large we can define integers $n_j \to \infty$ by

$$n_j = \max\{n : nH(x_j) \le \eta\}, \tag{13.4.10}$$

so that $n_j H(x_j) \le \eta < (n_j + 1)H(x_j)$, and $n_j H(x_j) \to \eta$. (13.4.9) then shows that $n_j V(x_j)/x_j^2 \to 0$. Fix $\varepsilon \in (0, 1]$ and suppose $|X_n^{(t+1)}| \le \varepsilon x_j$. Then

$$\begin{aligned} {}^{(t)}\tilde{S}_n &= \sum_{i \in \tilde{I}(n,t)} X_i = \sum_{i \in \tilde{I}(n,t)} X_i I(|X_i| \le \varepsilon x_j) \\ &= \sum_{i=1}^{n} X_i I(|X_i| \le \varepsilon x_j) - \sum_{i=1}^{t} X_n^{(i)} I(|X_n^{(i)}| \le \varepsilon x_j) \end{aligned}$$

so that (see (13.3.11) for T_n)

$$|{}^{(t)}\tilde{S}_n - T_n(\varepsilon x_j)| \le t\varepsilon x_j \quad (\text{on } \{|X_n^{(t+1)}| \le \varepsilon x_j\}). \tag{13.4.11}$$

Since $E(T_n(\varepsilon x_j)) = n\nu(\varepsilon x_j)$ it follows from Chebychev's inequality that

$$\begin{aligned} &P\{|{}^{(t)}\tilde{S}_{n_j} - n_j\nu(\varepsilon x_j)| \le (t+1)\varepsilon x_j\} \\ &\quad \ge P\{|T_{n_j}(\varepsilon x_j) - n_j\nu(\varepsilon x_j)| \le \varepsilon x_j, |X_{n_j}^{(t+1)}| \le \varepsilon x_j\} \\ &\quad \ge 1 - \frac{n_j V(\varepsilon x_j)}{\varepsilon^2 x_j^2} - P\{|X_{n_j}^{(t+1)}| > \varepsilon x_j\}. \end{aligned} \tag{13.4.12}$$

Since $\varepsilon \le 1$ we have $V(\varepsilon x_j) \le V(x_j)$ so $n_j V(\varepsilon x_j)/x_j^2 \to 0$ as $j \to \infty$. Also

$$\frac{n_j V(x_j)}{x_j^2} = \frac{n_j V(\varepsilon x_j)}{x_j^2} + \frac{n_j \int_{\varepsilon x_j < |y| \le x_j} y^2 dF(y)}{x_j^2} \ge \varepsilon^2 n_j [H(\varepsilon x_j) - H(x_j)]$$

and thus

$$n_j H(\varepsilon x_j) \le n_j H(x_j) + \frac{n_j V(x_j)}{\varepsilon^2 x_j^2} = \eta + o(1).$$

Also, because $\varepsilon \le 1, n_j H(\varepsilon x_j) \ge n_j H(x_j) = \eta + o(1)$, so $n_j H(\varepsilon x_j) \to \eta$. This means that

$$P\{|X_{n_j}^{(t+1)}| \le \varepsilon x_j\} \ge P\{|X_{n_j}^{(1)}| \le \varepsilon x_j\} = (1 - H(\varepsilon x_j))^{n_j} \to e^{-\eta}. \tag{13.4.13}$$

By our choice of η this exceeds 3/4 so (13.4.12) shows that

$$P\{|{}^{(t)}\tilde{S}_{n_j} - n_j\nu(\varepsilon x_j)| \le (t+1)\varepsilon x_j\} > \frac{1}{2} \tag{13.4.14}$$

if j is large enough. Now any interval of probability exceeding $\frac{1}{2}$ containing ${}^{(t)}\tilde{S}_{n_j}$ must contain any median of ${}^{(t)}\tilde{S}_{n_j}$. In other words,

$$|\tilde{m}_{n_j}^{(t)} - n_j\nu(\varepsilon x_j)| \le (t+1)\varepsilon x_j \tag{13.4.15}$$

if j is large enough. Next, by (13.4.11) and (13.4.15)

$$\begin{aligned} P\{|{}^{(t)}\tilde{S}_{n_j} - \tilde{m}_{n_j}^{(t)}| &\le (2t+2)\varepsilon|X_{n_j}^{(t)}|\} \\ &\ge P\{|{}^{(t)}\tilde{S}_{n_j} - n_j\nu(\varepsilon x_j)| \le (t+1)\varepsilon x_j, |X_{n_j}^{(t+1)}| \le \varepsilon x_j \le x_j < |X_{n_j}^{(t)}|\} \\ &\ge P\{|T_{n_j}(\varepsilon x_j) - n_j\nu(\varepsilon x_j)| \le \varepsilon x_j\} - P\{|X_{n_j}^{(t+1)}| \le \varepsilon x_j \le x_j < |X_{n_j}^{(t)}|\}^c. \end{aligned} \tag{13.4.16}$$

But since $n_j H(\varepsilon x_j) \to \eta$ and $n_j H(x_j) \to \eta$ we have

$$\begin{aligned} P\{|X_{n_j}^{(t+1)}| &\le \varepsilon x_j \le x_j < |X_{n_j}^{(t)}|\} \\ &= P\{ \text{exactly } t \text{ of the } |X_i| \text{ exceed } x_j, 1 \le i \le n_j, \\ &\qquad \text{and } n_j - t \text{ of the } |X_i| \text{ are smaller than } \varepsilon x_j\} \\ &= \binom{n_j}{t} H^t(x_j)[1 - H(\varepsilon x_j)]^{n_j - t} \to e^{-\eta}\frac{\eta^t}{t!}. \end{aligned} \tag{13.4.17}$$

From (13.4.16) (and Chebychev's inequality, as in (13.4.12)) we now deduce

$$P\{|{}^{(t)}\tilde{S}_{n_j} - \tilde{m}_{n_j}^{(t)}| \le (2t+2)\varepsilon|X_{n_j}^{(t)}|\} \ge 1 - o(1) - (1 - e^{-\eta}\frac{\eta^t}{t!}) = e^{-\eta}\frac{\eta^t}{t!} - o(1). \tag{13.4.18}$$

However the assumption (13.2.4a) means

$$\limsup_{n\to\infty} P\{|{}^{(t)}\tilde{S}_n - \tilde{m}_n^{(t)}| \le \varepsilon|X_n^{(t)}|\} \to 0 \quad (\varepsilon \to 0),$$

giving a contradiction with (13.4.18). Thus we have proved (13.2.5a).

The remaining proofs are similar. Let (13.2.4b) hold and suppose (13.2.5b) fails, so

$$\frac{V(x_j)}{x_j^2 H(x_j)} \le C \tag{13.4.19}$$

for some $x_j \to \infty$ and $C < \infty$. Again choose $\eta > 0$ so small that $e^{-\eta} > 3/4$, and define n_j by (13.4.10), so that $n_j H(x_j) \to \eta$ again. We then have

$$\frac{n_j V(x_j)}{x_j^2} \le C\eta. \tag{13.4.20}$$

Keep $|X_n^{(t+1)}| \le x_j$ and argue as in (13.4.11)–(13.4.13) (setting $\varepsilon = 1$) to get $|{}^{(t)}\tilde{S}_n - T_n(x_j)| \le t x_j$, and, for any $x > 0$,

$$\begin{aligned} P\{|{}^{(t)}\tilde{S}_{n_i} - n_j\nu(x_j)| &\le (x+t)x_j\} \\ &\ge P\{|T_{n_j}(x_j) - n_j\nu(x_j)| \le x x_j, |X_{n_j}^{(t+1)}| \le x_j\} \end{aligned}$$

$$\geq 1 - \frac{n_j V(x_j)}{x^2 x_j^2} - P\{|X_{n_j}^{(t+1)}| > x_j\}$$

$$\geq 1 - \frac{C\eta}{x^2} - (1 - e^{-\eta}) + o(1) \geq \frac{3}{4} - \frac{C\eta}{x^2} + o(1).$$

Choosing $x^2 = 8C\eta$, this expression exceeds 1/2, so (13.4.15) takes the form, here,

$$|\tilde{m}_{n_j}^{(t)} - n_j \nu(x_j)| \leq (t + \sqrt{8C\eta})x_j,$$

for j large enough. Apply a similar argument as in (13.4.16) and (13.4.17) to get

$$P\{|{}^{(t)}\tilde{S}_{n_j} - \tilde{m}_{n_j}^{(t)}| \leq (x + 2t + \sqrt{8C\eta})|X_{n_j}^{(t)}|\}$$
$$\geq P\{|T_{n_j}(x_j) - n_j \nu(x_j)| \leq x x_j\} - P\{|X_{n_j}^{(t+1)}| \leq x_j < |X_{n_j}^{(t)}|\}^c$$
$$\geq 1 - \frac{n_j V(x_j)}{x^2 x_j^2} - \left(1 - e^{-\eta}\frac{\eta^t}{t!}\right) + o(1) \geq e^{-\eta}\frac{\eta^t}{t!} - \frac{C\eta}{x^2} + o(1) \quad (13.4.21)$$

for $x > 0$. If x is large enough the last expression is positive, whereas the left hand side of (13.4.21) converges to 0 for any $x > 0$ if (13.2.4b) holds. Thus we have a contradiction and (13.2.5b) must hold.

Finally, suppose that (13.2.4b) holds along a subsequence of integers, n_i, say, but that $\limsup_{x\to\infty} V(x)/x^2 H(x)$ is finite, so for all x large enough

$$\frac{V(x)}{x^2 H(x)} \leq C \quad (13.4.22)$$

for some $C < \infty$. Again choose $\eta > 0$ so small that $e^{-\eta} > 3/4$, and for the specified n_i define $x_i \to \infty$ by

$$x_i = \inf\{x : H(x) \leq \frac{\eta}{n_i}\}. \quad (13.4.23)$$

Then $n_i H(x_i) \leq \eta \leq n_i H(x_i-)$. From (13.4.22) it follows that for large x

$$C \geq \frac{V(x)}{x^2 H(x)} = \frac{V(x-) + x^2(H(x-) - H(x))}{x^2 H(x)} \geq \frac{H(x-)}{H(x)} - 1.$$

Thus for large i

$$n_i H(x_i) \geq \frac{n_i H(x_i-)}{1+C} \geq \frac{\eta}{1+C},$$

so by taking a further subsequence if necessary we can assume $n_i H(x_i) \to \eta' \in [\eta/(1+C), \eta]$. An argument analogous to that which gave (13.4.21) now leads to a contradiction. Thus $\limsup V(x)/x^2 H(x) = \infty$.

To complete the proof of Theorem 13.2.1 it only remains to note that (13.2.5a) implies (13.2.2a) with $r = 0$ by Feller [1], hence for all r by Lemma 13.3.1, while (13.2.5b) implies (13.2.2b) and $\limsup V(x)/x^2 H(x) = \infty$ implies ${}^{(r)}\tilde{S}_n \in D_P(N)$, for $r = 0$, by Lévy [8, Theorem 36,3, p. 113], and for $r \geq 1$ by Mori [10, Theorem 3]. □

13.5 Proof of Theorem 13.2.2

Suppose $|{}^{(r)}\tilde{S}_n|/|X_n^{(r)}| \in NM_0$ for some $r = 1, 2, \ldots$. We first show that, for all $\lambda > 0$,

$$\lim_{\varepsilon\downarrow 0}\limsup_{n\to\infty} P\{|J_{n-r}(\lambda)| \le \varepsilon a_n(\lambda)\} = 0. \tag{13.5.1}$$

This is proved as follows. Fix $\lambda > 0$ and $\varepsilon > 0$ and take $n > r$. Then by (13.3.20)

$$\begin{aligned} P\{|{}^{(r)}\tilde{S}_n| \le \varepsilon|X_n^{(r)}|\} &\ge P\{|{}^{(r)}\tilde{S}_n| \le \varepsilon\hat{H}(U_{n,r}), U_{n,r} < u_n(\lambda) \le U_{n,r+1}\} \\ &\ge P\{|{}^{(r)}\tilde{S}_n| \le \varepsilon a_n(\lambda), U_{n,r} < u_n(\lambda) \le U_{n,r+1}\} \\ &= P\{|J_{n-r}(\lambda)| \le \varepsilon a_n(\lambda)\}P\{U_{n,r} < u_n(\lambda) \le U_{n,r+1}\}. \end{aligned} \tag{13.5.2}$$

Since $|{}^{(r)}\tilde{S}_n|/|X_n^{(r)}| \in NM_0$ the left hand side of (13.5.2) converges to 0 as $n \to \infty$ then $\varepsilon \downarrow 0$, consequently the same is true for the right hand side of (13.5.2). But since

$$P\{U_{n,r} < u_n(\lambda) \le U_{n,r+1}\} \to \frac{e^{-1/\lambda}(1/\lambda)^r}{r!} \quad \text{(by (13.3.18))} \tag{13.5.3}$$

this forces (13.5.1) to be true.

Next we prove that (13.2.10) holds. This proof is similar to that of (13.2.4a) implies (13.2.5a) in Theorem 13.2.1, so we merely sketch it. Suppose (13.2.10) fails, so

$$\frac{|\nu(x_j)|}{x_j H(x_j)} + \frac{V(x_j)}{x_j^2 H(x_j)} \to 0 \ (j \to \infty) \tag{13.5.4}$$

for some $x_j \to \infty$. Define

$$n_j = \sup\{n : nH(x_j) \le 1\},$$

so we have $n_j H(x_j) \to 1$. By (13.5.4) this means that $n_j|\nu(x_j)|/x_j \to 0$ so, given $\varepsilon > 0$, we may choose j large enough for $n_j|\nu(x_j)| < \varepsilon x_j$. Then by (13.4.11)

$$\begin{aligned} P\{|{}^{(r)}\tilde{S}_{n_j}| \le (r+1)\varepsilon|X_{n_j}^{(r)}|\} &\ge P\{|{}^{(r)}\tilde{S}_{n_j}| \le (r+1)\varepsilon x_j,\ |X_{n_j}^{(r+1)}| \le x_j < |X_{n_j}^{(r)}|\} \\ &\ge P\{|T_{n_j}(x_j)| \le \varepsilon x_j\} - P\{|X_{n_j}^{(r+1)}| \le x_j < |X_{n_j}^{(r)}|\}^c, \end{aligned}$$

while by Chebychev's inequality

$$\begin{aligned} P\{|T_{n_j}(x_j)| \le \varepsilon x_j\} &\ge P\{|T_{n_j}(x_j) - E(T_{n_j}(x_j))| \le \varepsilon x_j - n_j|\nu(x_j)|\} \\ &\ge 1 - \frac{n_j V(x_j)}{(\varepsilon x_j - n_j|\nu(x_j)|)^2} = 1 - \frac{n_j V(x_j)}{\varepsilon^2 x_j^2(1+o(1))}. \end{aligned}$$

Furthermore, since $n_j H(x_j) \to 1$, (13.4.17) gives

$$P\{|X_{n_j}^{(r+1)}| \le x_j < |X_{n_j}^{(r)}|\} \to e^{-1}/r!$$

Since $n_j V(x_j)/x_j^2 \to 0$ by (13.5.4), these show that

$$\liminf_{j\to\infty} P\{|{}^{(r)}\tilde{S}_{n_j}| \le (r+1)\varepsilon|X_{n_j}^{(r)}|\} \ge 1-(1-e^{-1}/r!) = e^{-1}/r!$$

But this is impossible if $|{}^{(r)}\tilde{S}_n|/|X_n^{(r)}| \in NM_0$, proving (13.2.10).

Now we will show that ${}^{(s)}\tilde{S}_n \in SC''$. (13.2.10) and Griffin and Maller [4, Theorem 2.1] give $S_n \in SC'(D_n)$, where D_n is the norming sequence defined in Griffin and Maller [4, (3.18)]. Take $n_i \to \infty$ such that $S_{n_i}/D_{n_i} \Longrightarrow Z$. By Mori [10, Theorem 1], this implies ${}^{(s)}\tilde{S}_{n_i}/D_{n_i}$ is tight; also, no further subsequential limit of ${}^{(s)}\tilde{S}_{n_i}/D_{n_i}$ can be degenerate at 0, otherwise $S_{n_i}/D_{n_i} \xrightarrow{P} 0$ through a subsequence (see the comment on Mori [10, p. 511]), contradicting $S_n \in SC'(D_n)$. Consequently ${}^{(s)}\tilde{S}_n \in SC'(D_n)$. To prove, further, that ${}^{(s)}\tilde{S}_n \in SC''$, we assume the contrary, that is that

$$\lim_{\varepsilon\to 0}\liminf_{j\to\infty} P\{|{}^{(s)}\tilde{S}_{n_j}| \le \varepsilon D_{n_j}\} > 0 \tag{13.5.5}$$

for some $n_j \to \infty$. We will contradict this by induction on s. Thus first consider the case $s = 0$. Choose a further subsequence, if necessary so that

$$\frac{a_{n_j}(\lambda)}{D_{n_j}} \to \rho > 0 \ \text{ for some } \lambda > 0 \tag{13.5.6}$$

or

$$\frac{a_{n_j}(\lambda)}{D_{n_j}} \to 0 \ \text{ for all } \lambda > 0. \tag{13.5.7}$$

Suppose first that (13.5.6) holds. Then since $a_n(\lambda)$ increases in λ (see (13.3.16))

$$\liminf_{j\to\infty} \frac{a_{n_j}(\alpha)}{D_{n_j}} \ge \rho \ \text{ for all } \alpha \ge \lambda.$$

Thus if $\alpha > \lambda$, $\varepsilon > 0$, and j is sufficiently large, we have by (13.3.21)

$$\begin{aligned} P\{|S_{n_j}| \ge \varepsilon D_{n_j}\} &\ge P\{|S_{n_j}| \ge 2\varepsilon a_{n_j}(\alpha)/\rho\} \\ &\ge P\{|J_{n_j}(\alpha)| \ge 2\varepsilon a_{n_j}(\alpha)/\rho\} P\{U_{n_j,1} \ge u_{n_j}(\alpha)\}. \end{aligned}$$

By (13.5.1), the first factor in the last expression converges to 1 as $j \to \infty$ then $\varepsilon \downarrow 0$. Since, by (13.3.18), $P\{U_{n,1} \ge u_n(\alpha)\} \to e^{-1/\alpha}$, we have that

$$\lim_{\varepsilon\to 0}\liminf_{j\to\infty} P\{|S_{n_j}| \ge \varepsilon D_{n_j}\} \ge e^{-1/\alpha}.$$

But when $\alpha \to \infty$ the right hand side of this converges to 1, contradicting (13.5.5) (with $s = 0$). Alternatively, (13.5.7) holds, in which case

$$\limsup_{j\to\infty} n_j H(\varepsilon D_{n_j}) \le \limsup_{j\to\infty} n_j H(a_{n_j}(\lambda)) \le 1/\lambda$$

for any $\varepsilon > 0$ and $\lambda > 0$. Letting $\lambda \to \infty$ we obtain for each $\varepsilon > 0$

$$\lim_{j\to\infty} n_j H(\varepsilon D_{n_j}) = 0. \tag{13.5.8}$$

However, since $S_n \in SC'(D_n)$, we can take a further subsequence of n_j if necessary so that $S_{n_j}/D_{n_j} \Longrightarrow Z$. (13.5.8) shows that the tail function $T(\varepsilon)$ of the Lévy canonical measure corresponding to Z satisfies $T(\varepsilon) \equiv 0$. Thus Z is a normal random variable, possibly degenerate but not degenerate at 0. Hence again (13.5.5) does not hold with $s = 0$, completing the proof that $S_n \in SC''(D_n)$.

Now assume that we have proved ${}^{(s-1)}\tilde{S}_n \in SC''(D_n)$, $s \ge 1$. Suppose there is a sequence n_j for which (13.5.5) holds. Let

$$\lambda = \inf\{\alpha : \limsup_{j\to\infty} \frac{a_{n_j}(\alpha)}{D_{n_j}} > 0\} \in [0, \infty].$$

If $\lambda = \infty$, then (13.5.7) holds and hence as before (13.5.8) holds. Thus

$$P\{|{}^{(s)}\tilde{S}_{n_j} - S_{n_j}| > s\varepsilon D_{n_j}\} \le P\{|X_{n_j}^{(1)}| > \varepsilon D_{n_j}\} \le n_j H(\varepsilon D_{n_j}) \to 0.$$

This means that if (13.5.5) holds as stated, then it also holds with $s = 0$. But this contradicts $S_n \in SC''(D_n)$. Thus we may assume $\lambda < \infty$. Then for each $\delta > 0$, there is a $\rho > 0$ and a subsequence m_j of n_j so that

$$\liminf_{j\to\infty} \frac{a_{m_j}(\alpha)}{D_{m_j}} \ge \rho \ \text{ if } \alpha > \lambda + \delta \tag{13.5.9}$$

and

$$\limsup_{j\to\infty} \frac{a_{m_j}(\alpha)}{D_{m_j}} = 0 \ \text{ if } \alpha < \lambda. \tag{13.5.10}$$

If $\lambda = 0$, (13.5.10) is taken to be vacuous. Thus, using (13.3.17)–(13.3.19),

$$\begin{aligned} P\{|{}^{(s)}\tilde{S}_{m_j}| \le \varepsilon D_{m_j}\} \le & \int_{\lambda+\delta}^{\infty} P\{|J_{m_j - s}(\alpha)| \le 2\varepsilon\rho^{-1} a_{m_j}(\alpha)\} d\nu_s(\alpha) + o(1) \\ & + P\{|{}^{(s)}\tilde{S}_{m_j}| \le \varepsilon D_{m_j},\ U_{m_j,s} > u_{m_j}((\lambda-\delta)\vee 0)\} \\ & + P\{u_{m_j}(\lambda+\delta) \le U_{m_j,s} \le u_{m_j}((\lambda-\delta)\vee 0)\} \\ = & \, \mathrm{I} + \mathrm{II} + \mathrm{III}, \ \text{ say.} \end{aligned} \tag{13.5.11}$$

If $\lambda = 0$, then $\mathrm{II} = 0$. If $\lambda > 0$ we will only consider $\delta < \lambda$. Now by (13.5.1) and the version of Fatou's lemma used in Lemma 13.3.1 it follows that $\mathrm{I} \to 0$ as $j \to \infty$, then $\varepsilon \downarrow 0$. For II when $\lambda > 0$, on $\{U_{m_j,s} > u_{m_j}(\lambda - \delta)\}$ we have $|X_{m_j}^{(s)}| \le a_{m_j}(\lambda - \delta) = o(D_{m_j})$. Thus as $j \to \infty$ then $\varepsilon \downarrow 0$,

$$\mathrm{II} \le P\{|{}^{(s-1)}\tilde{S}_{m_j}| \le (\varepsilon + o(1)) D_{m_j}\} \to 0$$

by the induction hypothesis. Finally for III, it follows from (13.3.18) that if $\lambda > 0$

$$\mathrm{III} \to e^{-1/(\lambda-\delta)} \sum_{j=0}^{s-1} \frac{(\lambda-\delta)^{-j}}{j!} - e^{-1/(\lambda+\delta)} \sum_{j=0}^{s-1} \frac{(\lambda+\delta)^{-j}}{j!},$$

while if $\lambda = 0$,

$$\text{III} \to e^{-1/\delta} \sum_{j=0}^{s-1} \frac{\delta^{-j}}{j!}.$$

In either case the limit approaches 0 as $\delta \downarrow 0$. Thus by (13.5.11) we have that (13.5.5) cannot hold, completing the proof that ${}^{(s)}\tilde{S}_n \in SC''(D_n)$.

It remains to show that ${}^{(s)}\tilde{S}_n \in SC''$ implies $|{}^{(r)}\tilde{S}_n|/|X_n^{(r)}| \in NM_0$. Assume ${}^{(s)}\tilde{S}_n \in SC''$, so ${}^{(s)}\tilde{S}_n \in SC'$, and by Griffin and Maller [4, Theorem 2.1],

$$\lim_{\lambda\to\infty} \limsup_{x\to\infty} \frac{x^2 H(\lambda x)}{x|\nu(x)| + V(x)} = 0.$$

The same result tells us that $n(D_n|\nu(D_n)| + V(D_n)) \asymp D_n^2$, so we deduce that $\lim_{\lambda\to\infty} \limsup_{n\to\infty} nH(\lambda D_n) = 0$. Since by the definition of $a_n(\lambda)$, $H(a_n(\lambda)-) \geq 1/(\lambda n)$, this forces

$$\liminf_{n\to\infty} \frac{D_n}{a_n(\lambda)} > 2C \tag{13.5.12}$$

for some $C = C(\lambda) > 0$ for all $\lambda > 0$. Now fix $\lambda > 0$. Then by (13.3.20)

$$\begin{aligned} P\{|{}^{(s)}\tilde{S}_n| \leq \varepsilon D_n\} &\geq P\{|{}^{(s)}\tilde{S}_n| \leq C\varepsilon a_n(\lambda), U_{n,s} < u_n(\lambda) \leq U_{n,s+1}\} \\ &= P\{|J_{n-s}(\lambda)| \leq C\varepsilon a_n(\lambda)\} P\{U_{n,s} < u_n(\lambda) \leq U_{n,s+1}\}. \end{aligned}$$

Since ${}^{(s)}\tilde{S}_n \in SC''$ the left hand side of this converges to 0 as $n \to \infty$ then $\varepsilon \downarrow 0$, so, recalling (13.5.3), we see that (13.5.1) holds with r replaced by s. But

$$\begin{aligned} P\{|J_{n-r}(\lambda) - J_{n-s}(\lambda)| > |r-s|\varepsilon a_n(\lambda)\} &\leq |r-s| P\{|Z(\lambda)| > \varepsilon a_n(\lambda)\} \\ &\leq \frac{|r-s| H(\varepsilon a_n(\lambda))}{(1 - 1/(\lambda n))} \to 0 \end{aligned} \tag{13.5.13}$$

as $n \to \infty$. Thus (13.5.1) holds as stated.

It remains to show that (13.5.1) implies ${}^{(r)}\tilde{S}_n/X_n^{(r)} \in NM_0$. But this follows from the integral representation (13.3.19) and Fatou's lemma applied as before. □

13.6 Proof of Theorem 13.2.3

Let $S_n \in SC_0$. Then $S_n \in SC''$ follows from Griffin and Maller [4, Corollary 2.4]. Fix $r \geq 1$ and suppose that Y is a vague limit of ${}^{(r)}\tilde{S}_n/|X_n^{(r)}|$, and that Y has positive mass at $+\infty$. Then for some deterministic sequence $n_i \to \infty$ and some $\delta > 0$,

$$\liminf_{n_i\to\infty} P\{{}^{(r)}\tilde{S}_{n_i} > x|X_{n_i}^{(r)}|\} \geq \delta \tag{13.6.1}$$

for all $x > 0$. We will show that Y has positive mass at $-\infty$, also.

Since $S_n \in SC_0$ we may take a further subsequence of n_i, if necessary, so that $S_{n_i}/B_{n_i} \Rightarrow Z$, where B_n is a deterministic sequence and Z is a proper, nondegenerate, infinitely divisible r.v. We need some preliminary results and notation. Denote the Lévy components of Z by σ^2, $N(x)$, $M(-x)$ and γ, so that we can write the characteristic function, $f(t)$, say, of Z in the form (cf. Petrov [11, Eq. 2.12, pp. 30–31]):

$$f(t) = \exp\left(i\gamma t - \frac{1}{2}\sigma^2 t^2 + \int_{(0,\infty)} \left(e^{itx} - 1 - \frac{itx}{1+x^2}\right)(-dN(x)) + \int_{(-\infty,0]} \left(e^{itx} - 1 - \frac{itx}{1+x^2}\right) dM(x)\right). \quad (13.6.2)$$

(Our $N(x) = -L(x)$, $x > 0$, and $M(x) = L(x)$, $x < 0$, in Petrov's [11, p. 31] notation, and Petrov's notation for $G(\cdot)$ corresponds to

$$dG(y) = \begin{cases} \dfrac{y^2}{1+y^2} dL(y) = \dfrac{-y^2}{1+y^2} dN(y), & \text{for } y > 0, \\ \dfrac{y^2}{1+y^2} dL(y) = \dfrac{y^2}{1+y^2} dM(y), & \text{for } y < 0, \end{cases}$$

in our notation; our σ^2 equals $G(0+) - G(0-)$.) The functions $N(x)$, $M(-x)$ and the function $T(x)$ defined by

$$T(x) = N(x) + M(-x) \quad (13.6.3)$$

are nonincreasing on $x > 0$ and satisfy $N(\infty) = M(-\infty) = T(\infty) = 0$, and

$$-\int_{(0,1]} y^2 dT(y) < \infty. \quad (13.6.4)$$

Now for $S_{n_i}/B_{n_i} \Rightarrow Z$ we must have, by Petrov [11, Theorem 6, pp. 77–78], for $x > 0$, x a continuity point of $N(\cdot)$ and $M(-\cdot)$,

$$\lim_{n_i\to\infty} n_i(1 - F(xB_{n_i})) = N(x), \qquad \lim_{n_i\to\infty} n_i F(-xB_{n_i}) = M(-x), \quad (13.6.5)$$

$$\lim_{\varepsilon\downarrow 0} \liminf_{n_i\to\infty} \frac{n_i\left(V(\varepsilon B_{n_i}) - \nu^2(\varepsilon B_{n_i})\right)}{B_{n_i}^2} = \lim_{\varepsilon\downarrow 0} \limsup_{n_i\to\infty} \frac{n_i\left(V(\varepsilon B_{n_i}) - \nu^2(\varepsilon B_{n_i})\right)}{B_{n_i}^2} = \sigma^2, \quad (13.6.6)$$

and

$$\lim_{n_i\to\infty} \frac{n_i \nu(xB_{n_i})}{B_{n_i}} = \gamma - \int_{(0,x)} \frac{y^3}{1+y^2} dN(y) + \int_{(-x,0)} \frac{y^3}{1+y^2} dM(y) + \int_{[x,\infty)} \frac{y}{1+y^2} dN(y) - \int_{(-\infty,-x]} \frac{y}{1+y^2} dM(y). \quad (13.6.7)$$

Note that, under (13.6.7), (13.6.6) is equivalent to

$$\lim_{\varepsilon\downarrow 0}\liminf_{n_i\to\infty}\frac{n_i V(\varepsilon B_{n_i})}{B_{n_i}^2}=\lim_{\varepsilon\downarrow 0}\limsup_{n_i\to\infty}\frac{n_i V(\varepsilon B_{n_i})}{B_{n_i}^2}=\sigma^2. \tag{13.6.8}$$

Taking a further subsequence of n_i if necessary, we can by Helley's Theorem assume that $n_i V(xB_{n_i})/B_{n_i}^2$ has a limit as $n_i\to\infty$ for each $x>0$, and it follows then from (13.6.5) and (13.6.8) that we must have, for $x>0$,

$$\lim_{n_i\to\infty}\frac{n_i V(xB_{n_i})}{B_{n_i}^2}=\sigma^2-\int_{(0,x]}y^2dT(y). \tag{13.6.9}$$

Now we begin the proof that Y has positive mass at $-\infty$. Note that (13.6.5) implies, for $x>0$ and $s=1,2,\dots$,

$$\begin{aligned}P\{|X_{n_i}^{(s)}|\le xB_{n_i}\}&=\sum_{j=0}^{s-1}\binom{n_i}{j}H^j(xB_{n_i})(1-H(xB_{n_i}))^{n_i-j}\\&\to\sum_{j=0}^{s-1}\frac{(T(x))^j}{j!}e^{-T(x)},\end{aligned} \tag{13.6.10}$$

as $n_i\to\infty$, at points of continuity of $N(\cdot)$ and $M(-\cdot)$. Since $T(\infty)=0$, the right hand side of (13.6.10) is a proper c.d.f., hence in particular $|X_{n_i}^{(1)}|/B_{n_i}$ is tight, and so ${}^{(r)}\tilde{S}_{n_i}/B_{n_i}$ is tight.

Our next step is to show that $T(0+)<\infty$. Suppose that $T(0+)=\infty$. Then, using (13.6.10), we have for $x>0$, $\varepsilon>0$, as $n_i\to\infty$,

$$\begin{aligned}P\{{}^{(r)}\tilde{S}_{n_i}>x|X_{n_i}^{(r)}|\}&\le P\{{}^{(r)}\tilde{S}_{n_i}>x\varepsilon B_{n_i}\}+P\{|X_{n_i}^{(r)}|\le\varepsilon B_{n_i}\}\\&\le P\{{}^{(r)}\tilde{S}_{n_i}>x\varepsilon B_{n_i}\}+e^{-T(\varepsilon)}\sum_{j=0}^{r-1}(T(\varepsilon))^j/j!+o(1).\end{aligned} \tag{13.6.11}$$

Since $T(0+)=\infty$ we may choose $\varepsilon=\varepsilon(\delta)$ so small that the second term on the right hand side of (13.6.11) does not exceed $\delta/2$. For this ε, the first term on the right hand side of (13.6.11) converges to 0 as $n_i\to\infty$ then $x\to\infty$ through points of continuity of $N(\cdot)$ and $M(-\cdot)$, since ${}^{(r)}\tilde{S}_{n_i}/B_{n_i}$ is tight. This gives a contradiction with (13.6.1), so we deduce that $T(0+)<\infty$.

Next we show that $\sigma^2>0$. Suppose $\sigma^2=0$. If, in addition, $T(0+)=0$, then Z is degenerate, which is not possible by the definition of SC_0. Thus $T(0+)>0$ in this case. By (13.2.12) we have, for some $c_0>0$, all $x>0$, and for n_i large enough,

$$x^2 n_i H(xB_{n_i})+\frac{n_i x|\nu(xB_{n_i})|}{B_{n_i}}\le\frac{c_0 n_i V(xB_{n_i})}{B_{n_i}^2}. \tag{13.6.12}$$

Choose x_0 so small that $T(x)>0$ for $0<x\le x_0$, as is possible since $T(0+)>0$. Then letting $n_i\to\infty$ in (13.6.12) and using (13.6.9) gives, for $0<x\le x_0$,

$$x^2T(x)\le c_0\Big(\sigma^2-\int_{(0,x]}y^2dT(y)\Big)=-c_0\int_{(0,x]}y^2dT(y)\qquad(\text{since }\sigma^2=0)$$

$$= c_0\left(\int_0^x 2yT(y)dy - x^2T(x)\right) = 2c_0x^2\int_0^1 y(T(xy) - T(x))dy.$$

Consequently, for $0 < x < x_0$,

$$\int_0^1 y\left(\frac{T(xy)}{T(x)} - 1\right) \geq \frac{1}{2c_0}. \tag{13.6.13}$$

Now $T(xy)/T(x) - 1 \leq T(0+)/T(x_0)$ for $0 < y \leq 1$ and $0 < x < x_0$, and the bound here is finite since $T(0+) < \infty$ and $T(x_0) > 0$. Also $T(xy)/T(x) - 1 \to T(0+)/T(0+) - 1 = 0$ as $x \to 0+$. So by dominated convergence the integral in (13.6.13) converges to 0 as $x \to 0+$, a contradiction. Thus $\sigma^2 > 0$.

We thus have, for the remainder of the proof, that $T(0+) < \infty$ and $\sigma^2 > 0$. We will use the following inequality, valid for $x > r$, $r \geq 1$, $\varepsilon > 0$:

$$\begin{aligned} P\{{}^{(r)}\tilde{S}_n &\leq -(x-r)|X_n^{(r)}|\} \geq P\{S_n + r|X_n^{(1)}| \leq -(x-r)|X_n^{(1)}|\} \\ &= P\{S_n \leq -x|X_n^{(1)}|\} \geq P\{S_n \leq -\varepsilon B_n, |X_n^{(1)}| \leq \varepsilon B_n/x\} \\ &= P\{S_n(\varepsilon B_n/x) \leq -\varepsilon B_n\}P\{|X_n^{(1)}| \leq \varepsilon B_n/x\}. \end{aligned} \tag{13.6.14}$$

In the last step we used (13.3.12), $S_n(y)$ being as in (13.3.10). Now we need:

Lemma 13.6.1 *Suppose n_i (integers) and B_{n_i} are nonstochastic sequences such that $S_{n_i}/B_{n_i} \Rightarrow Z$, where Z is a proper, nondegenerate, infinitely divisible random variable with Lévy components σ^2, $N(x)$, $M(-x)$ and γ, so that* (13.6.5)–(13.6.8) *hold for n_i and B_{n_i}. Suppose also that $T(0+) = N(0+) + M(0-) < \infty$. Then for each $y > 0$, y being a continuity point of $N(\cdot)$ and $M(-\cdot)$,*

$$\frac{S_{n_i}(yB_{n_i})}{B_{n_i}} \Rightarrow Z_y \text{ as } n_i \to \infty, \tag{13.6.15}$$

where Z_y is an infinitely divisible r.v. with Lévy components $\sigma_y^2 = \sigma^2$, $\gamma_y = \gamma$, and, for $z > 0$,

$$N_y(z) = (N(z) - N(y))I(z < y), \ M_y(-z) = (M(-z) - M(-y))I(z < y). \tag{13.6.16}$$

Proof. Fix $y > 0$, y a continuity point of $N(\cdot)$ and $M(-\cdot)$. Using (13.3.10), we can regard

$$\frac{S_n(yB_n)}{B_n} = \sum_{j=1}^n \frac{X_j(yB_n)}{B_n} := \sum_{j=1}^n X_{n,j} \tag{13.6.17}$$

as the row sum of a triangular array of independent r.v.'s

$$X_{n,j} = X_{n,j}(y) := X_j(yB_n)/B_n, \quad 1 \leq j \leq n. \tag{13.6.18}$$

These are uniformly asymptotically negligible or "infinitely small" (Petrov [11, p. 63]), since for $z > 0$

$$\begin{aligned} \max_{1\leq j\leq n} P\{|X_{n,j}| > z\} &= \max_{1\leq j\leq n} P\{|X_j(yB_n)| > zB_n\} \\ &\leq P\{\max_{1\leq j\leq n} |X_j(yB_n)| > zB_n\} = 0, \quad \text{once } z \geq y. \end{aligned} \tag{13.6.19}$$

Now we will restrict ourselves to the subsequence n_i. Keep n_i so large that $P\{|X| \le yB_{n_i}\} > 0$. For $z > 0$, z a continuity point of $N(\cdot)$, we have

$$\begin{aligned}\sum_{j=1}^{n_i} P\{X_{n_i,j} > z\} &= \frac{n_i P\{X > zB_{n_i}, |X| \le yB_{n_i}\}}{P\{|X| \le yB_{n_i}\}} \\ &= \frac{n_i P\{zB_{n_i} < X \le yB_{n_i}\} I(z < y)}{P\{|X| \le yB_{n_i}\}} \\ &\to (N(z) - N(y))I(z < y) = N_y(z) \qquad (13.6.20)\end{aligned}$$

by (13.6.5) and (13.6.16). Similarly, for $z > 0$, z a continuity point of $M(-\cdot)$,

$$\begin{aligned}\sum_{j=1}^{n_i} P\{X_{n_i,j} < -z\} &= \frac{n_i P\{X < -zB_{n_i}, |X| \le yB_{n_i}\}}{P\{|X| \le yB_{n_i}\}} \\ &\to (M(-z) - M(y))I(z < y) = M_y(z). \qquad (13.6.21)\end{aligned}$$

Define, for $z > 0$,

$$\begin{aligned}T_y(z) = N_y(z) + M_y(-z) &= (N(z) + M(-z) - N(y) - M(-y))I(z < y) \\ &= (T(z) - T(y))I(z < y), \qquad (13.6.22)\end{aligned}$$

where $T(x)$ is as in (13.6.3). Next we have, for $z > 0$,

$$\begin{aligned}&\sum_{j=1}^{n_i} E\left(X_{n_i,j}^2 I(|X_{n_i,j}| \le z)\right) \\ &= \sum_{j=1}^{n_i} \left(2\int_0^z uP\{|X_{n_i,j}| > u\}du - z^2 P\{|X_{n_i,j}| > z\}\right) \\ &= \frac{n_i \left(2\int_0^z uP\{|X| > uB_{n_i}, |X| \le yB_{n_i}\}du - z^2 P\{|X| > zB_{n_i}, |X| \le yB_{n_i}\}\right)}{P\{|X| \le yB_{n_i}\}} \\ &= \begin{cases} \dfrac{2n_i \int_0^z uP\{uB_{n_i} < |X| \le zB_{n_i}\}du}{P\{|X| \le yB_{n_i}\}}, & \text{if } z < y \\ \dfrac{2n_i \int_0^y uP\{uB_{n_i} < |X| \le yB_{n_i}\}du}{P\{|X| \le yB_{n_i}\}}, & \text{if } z \ge y. \end{cases}\end{aligned}$$

It is easy to see that the last expression equals

$$\begin{cases} \dfrac{n_i V(zB_{n_i})}{B_{n_i}^2 P\{|X| \le yB_{n_i}\}}, & \text{if } z < y \\ \dfrac{n_i V(yB_{n_i})}{B_{n_i}^2 P\{|X| \le yB_{n_i}\}}, & \text{if } z \ge y \end{cases}$$

which is the same as

$$\frac{n_i V((z \wedge y)B_{n_i})}{B_{n_i}^2 P\{|X| \le yB_{n_i}\}}. \qquad (13.6.23)$$

As $n_i \to \infty$ the last expression converges to

$$\sigma^2 - \int_{(0,z\wedge y]} u^2 dT(u) \ \big(\text{by } (13.6.9)\big) \ = \sigma^2 - \int_{(0,z]} u^2 dT_y(u) \ \big(\text{by } (13.6.22)\big). \tag{13.6.24}$$

Again keep $z > 0$. We also need

$$\begin{aligned}\sum_{j=1}^{n_i} E\left(X_{n_i,j} I(0 \le X_{n_i,j} \le z)\right) &= \sum_{j=1}^{n_i}\left(\int_0^z P\left\{X_{n_i,j} > u\right\} du - zP\left\{X_{x_{i,j}} > z\right\}\right) \\ &= \frac{n_i \int_{[0,z\wedge y]} u dF(u)}{B_{n_i} P\{|X| \le yB_{n_i}\}},\end{aligned} \tag{13.6.25}$$

by similar calculations as led to (13.6.23). Likewise,

$$\sum_{j=1}^{n_i} E\big(X_{n_i,j} I(-z \le X_{n_i,j} < 0)\big) = \frac{n_i \int_{[-z\wedge y,0)} u dF(u)}{B_{n_i} P\{|X| \le yB_{n_i}\}}. \tag{13.6.26}$$

Adding (13.6.25) and (13.6.26) gives

$$\sum_{j=1}^{n_i} E\left(X_{n_i,j} I(|X_{n_i,j}| \le z)\right) = \frac{n_i \nu\left((z\wedge y)B_{n_i}\right)}{B_{n_i} P\{|X| \le yB_{n_i}\}}. \tag{13.6.27}$$

Now we need the limit of the right hand side of (13.6.27). Recall that we have $T(0+) = N(0+) + M(0-) < \infty$. Consequently we can let $x \downarrow 0$ in (13.6.7) to get

$$\lim_{\varepsilon\to 0+} \lim_{n_i\to\infty} \frac{n_i \nu(\varepsilon B_{n_i})}{B_{n_i}} = \gamma + \int_{(0,\infty)} \frac{y}{1+y^2} dN(y) - \int_{(-\infty,0)} \frac{y}{1+y^2} dM(y),$$

where the right hand side is finite. Call the right hand side ν_0. Substituting for ν_0 in (6.7) we then see that, for $x > 0$,

$$\lim_{n_i\to\infty} \frac{n_i \nu(xB_{n_i})}{B_{n_i}} = \nu_0 + \int_{(0,x)} u\,(-dN(u)) + \int_{(-x,0)} u dM(u),$$

and this can be used in place of (13.6.7) (along with (13.6.5) and (13.6.8)) to characterise the convergence $S_{n_i}/B_{n_i} \Rightarrow Z$, when $T(0+) < \infty$. But from (13.6.27) we then obtain, for all $z > 0$,

$$\begin{aligned}\sum_{j=1}^{n_i} E\left(X_{n_i,j} I(|X_{n_i,j}| \le z)\right) &\to \nu_0 + \int_{(0,z\wedge y)} u\,(-dN(u)) + \int_{(-z\wedge y,0)} u dM(u) \\ &= \nu_0 + \int_{(0,z)} u\left(-dN_y(u)\right) + \int_{(-z,0)} u dM_y(u),\end{aligned} \tag{13.6.28}$$

as required (along with (13.6.20), (13.6.21), and (13.6.29) below) to characterise the convergence $S_{n_i}(y)/B_{n_i} \Rightarrow Z_y$, when $T_y(0+) < \infty$.

We need one more condition. Again keep $z > 0$. Then, as in (13.6.27),

$$\sum_{j=1}^{n_i} \left(E(X_{n_i,j} I(|X_{n_i,j}| \le z)\right)^2 = \frac{n_i \nu^2\left((z \wedge y)B_{n_i}\right)}{B_{n_i}^2 P\{|X| \le yB_{n_i}\}} = o\left(\frac{n_i|\nu_i((z \wedge y)B_{n_i})|}{B_{n_i}}\right)$$

and this is $o(1)$ as $n_i \to \infty$. Putting this together with (13.6.24) gives, for all $z > 0$,

$$\sum_{j=1}^{n_i} \left\{ E\left(X_{n_i,j}^2 I(|X_{n_i,j}| \le z)\right) - E^2\left(X_{n_i,j} I(|X_{n_i,j}| \le z)\right)\right\}$$

$$\to \sigma^2 - \int_{(0,z]} u^2 dT_y(u). \tag{13.6.29}$$

(13.6.20), (13.6.21), (13.6.28), and (13.6.29) show that $\sum_{j=1}^{n_i} X_{n_i,j}$ converges in distribution to an infinitely divisible r.v. Z_y whose Lévy components are as stated. □

Now we can complete this part of the proof. Going back to (13.6.14), and using Lemma 13.6.1 and (13.6.10) we get

$$\liminf_{n_i \to \infty} P\left\{ {}^{(r)}\tilde{S}_{n_i} \le -(x-r)|X_{n_i}^{(r)}| \right\} \ge P\left\{Z_{\varepsilon/x} \le -\varepsilon\right\} e^{-T(\varepsilon/x)}. \tag{13.6.30}$$

This holds for $\varepsilon > 0$, $x > r$, ε/x a point of continuity of $N(\cdot)$ and $M(-\cdot)$. Now as $x \to \infty$, $N_{\varepsilon/x}(z) \to 0$ and $M_{\varepsilon/x}(z) \to 0$ for each $\varepsilon > 0$, $z > 0$, so if we let $x \to \infty$ through a sequence of integers we have by the convergence lemma for infinitely divisible distributions (Petrov [11, Lemma 3, p. 29]) that $Z_{\varepsilon/x} \Rightarrow Z_0$, where Z_0 has Lévy components σ^2, γ, and $N(\cdot) = M(-\cdot) = 0$. In other words, $Z_0 \sim N(\gamma, \sigma^2)$. Letting $x \to \infty$ in such a way, then, we deduce from (13.6.30) that

$$\lim_{x\to\infty} \liminf_{n_i \to \infty} P\left\{ {}^{(r)}\tilde{S}_{n_i} \le -(x-r)|X_{n_i}^{(r)}| \right\} \ge P\left\{N(\gamma, \sigma^2) \le -\varepsilon\right\} e^{-T(0+)}.$$

Since $T(0+) < \infty$ and $\sigma^2 > 0$, when $\varepsilon \downarrow 0$ the right hand side of this converges to a positive constant, showing that Y indeed has positive mass at $-\infty$.

If, instead, we start with a Y having positive mass at $-\infty$, some obvious modifications of the above arguments show that Y also has positive mass at $+\infty$.

For the converse, suppose $S_n \in SC''$, and if any subsequential vague limit Y has mass at ∞ then it also has mass at $-\infty$. We must show $S_n \in SC_0$. Now $S_n \in SC''$ implies $S_n \in SC'$, thus (13.2.10) holds, and we wish to show (13.2.12). Suppose $x|\nu(x)| \le C_1 V(x)$ for some $C_1 < \infty$ and all large x. Then from (13.2.10), $x^2 H(x) \le C_2 V(x)$ for some $C_2 < \infty$ and all large x, so (13.2.12) holds. Alternatively, there is a sequence $x_i \to \infty$ with

$$\frac{V(x_i)}{x_i|\nu(x_i)|} \to 0. \tag{13.6.31}$$

If in addition $x_i^2 H(x_i) \le C_3 V(x_i)$ for some $C_3 < \infty$, then by (13.6.31)

$$\frac{\nu^2(x_i)}{H(x_i)V(x_i)} \to \infty \quad (i \to \infty) \tag{13.6.32}$$

so by Kesten and Maller [7, Lemma 4.5], ${}^{(r)}\tilde{S}_{n_i}/B_{n_i} \xrightarrow{P} \pm 1$ for some $n_i \to \infty$, $B_{n_i} \to \infty$. This entails $X_{n_i}^{(r)}/B_{n_i} \xrightarrow{P} 0$, so ${}^{(r)}\tilde{S}_{n_i}/|X_{n_i}^{(r)}| \xrightarrow{P} \pm\infty$. But this contradicts our assumption that any vague limit of ${}^{(r)}\tilde{S}_{n_i}/|X_{n_i}^{(r)}|$ has mass at both $+\infty$ and $-\infty$ (if it has mass at either). Consequently we must have, taking a further subsequence if necessary,

$$\frac{x_i^2 H(x_i)}{V(x_i)} \to \infty. \tag{13.6.33}$$

If in addition $x_i H(x_i) \le C_4 |\nu(x_i)|$, then (13.6.31) shows that (13.6.32) holds again, and this is impossible. Consequently we have

$$\frac{x_i H(x_i)}{|\nu(x_i)|} \to \infty. \tag{13.6.34}$$

Fix $\eta > 0$ and choose $n_i = n_i(\eta)$ so that $n_i H(x_i) \le \eta < (n_i + 1)H(x_i)$. This together with (13.6.33) and (13.6.34) then gives

$$n_i H(x_i) \to \eta, \quad \frac{n_i|\nu(x_i)|}{x_i} \to 0, \quad \text{and} \quad \frac{n_i V(x_i)}{x_i^2} \to 0.$$

But then by (13.4.11)

$$\begin{aligned} P\{|{}^{(r)}\tilde{S}_{n_i}| \le (r+1)\varepsilon|X_{n_i}^{(r)}|\} &\ge P\{|{}^{(r)}\tilde{S}_{n_i}| \le (r+1)\varepsilon x_i,\ |X_{n_i}^{(r+1)}| \le x_i < |X_{n_i}^{(r)}|\} \\ &\ge P\{|T_{n_i}(x_i)| \le \varepsilon x_i\} - P\{|X_{n_i}^{(r+1)}| \le x_i < |X_{n_i}^{(r)}|\}^c. \end{aligned} \tag{13.6.35}$$

Since $n_i \nu(x_i) = o(x_i)$ we obtain, for i large enough, by Chebychev's inequality,

$$P\{|T_{n_i}(x_i)| \le \varepsilon x_i\} \ge P\{|T_{n_i}(x_i) - n_i \nu(x_i)| \le \varepsilon x_i/2\} \ge 1 - \frac{4 n_i V(x_i)}{\varepsilon^2 x_i^2} = 1 - o(1)$$

while, as in (13.4.17),

$$P\{|X_{n_i}^{(r+1)}| \le x_i < |X_{n_i}^{(r)}|\} \to e^{-\eta}\eta^r/r! \quad (i \to \infty).$$

But then (13.6.35) shows that $|{}^{(r)}\tilde{S}_{n_i}|/|X_{n_i}^{(r)}|$ has mass at 0 in the limit as $n_i \to \infty$, which by Theorem 13.2.2 contradicts $S_n \in SC''$.

Finally, $S_n \in SC_0$ if and only if ${}^{(s)}\tilde{S}_n \in SC_0$, for any $s = 1, 2, \ldots$, follows easily from Mori [10, Theorem 1]. □

Acknowledgments: We are grateful to Brenda Churchill for typing help. The research of the first author was supported by the NSF through a grant to Syracuse University.

REFERENCES

[1] Feller, W. On regular variation and local limit theorems. *Proceedings of the Fifth Berkeley Symposium on Mathematical Statistics and Probability* II, Part I (1965–66). University of California Press, Berkeley and Los Angeles, 1967, 373–388.

[2] Gnedenko, B.V. and Kolmogorov, A.N. *Limit Distributions for Sums of Independent Random Variables*, 2nd Edition. Addison–Wesley, Reading, MA, 1968.

[3] Griffin, P.S. and Maller, R.A. On the rate of growth of the overshoot and the maximum partial sum. *Adv. Appl. Prob.* **30** (1998), 181–196.

[4] Griffin, P.S. and Maller, R.A. On compactness properties of the exit position of a random walk from an interval. *Proc. London Math. Soc.*, 1999.

[5] Jain, N.C. and Orey, S. Vague convergence of sums of independent random variables. *Israel J. Math.* **33** (1979), 317–348.

[6] Jain, N.C. and Orey, S. Domains of partial attraction and tightness conditions. *Ann. Prob.* **8** (1980), 584 599.

[7] Kesten, H. and Maller, R.A. Infinite limits and infinite limit points of random walks and trimmed sums. *Ann. Prob.* **22** (1994), 1473–1513.

[8] Lévy, P. *Théorie de l'Addition des Variables Aléatoires*, 2nd Edition. Gauthier–Villars, Paris, 1954.

[9] Maller, R.A. Some properties of stochastic compactness. *J. Australian Math. Soc.* **30** (1981), 264–277.

[10] Mori, T. On the limit distributions of lightly trimmed sums. *Math. Proc. Cambridge Philos. Soc.* **96** (1984), 507–516.

[11] Petrov, V.V. *Sums of Independent Random Variables*. Springer-Verlag, Berlin, 1975.

[12] Pruitt, W.E. The class of limit laws for stochastically compact normed sums. *Ann. Prob.* **11** (1983), 962–969.

[13] Rudin, W. *Principles of Mathematical Analysis*, 2nd Edition. McGraw–Hill, New York, 1964.

P.S. Griffin
Department of Mathematics
Syracuse University
Syracuse, NY 13244-1150

R.A. Maller
Department of Mathematics
The University of Western Australia
Nedlands 6907
Western Australia

14

Stability and Heavy Traffic Limits for Queueing Networks

Maury Bramson

ABSTRACT We discuss here two topics of recent interest in queueing theory. The first is the question of when strictly subcritical queueing networks are stable. Namely, given a network whose stations all serve customers more quickly than the long-term rate at which customers visit the system, when is the underlying Markov process positive recurrent? The other topic is the existence of heavy traffic limits for queueing networks. That is, when does a sequence of networks, under diffusive scaling, converge to a reflecting Brownian motion?

Keywords: Queueing networks, stability, positive Harris recurrence, fluid models, heavy traffic limits.

AMS Subject Classification: 60K25.

14.1 Introduction

The purpose of this article is to present two topics of current activity in queueing theory. Both topics deal with the long-term behavior for queueing networks, that is, systems of queues, where customers enter the network and move from one station to the next, until leaving the network. The behavior for such systems is substantially more complex than that of a single queue, and the general theory is only partially understood. This includes the topic of when a queueing network is stable, that is, its underlying Markov process is positive recurrent. Contrary to what one might expect, it is not sufficient for all stations, when busy, to serve customers more quickly than the long-term rate at which they visit the system. This behavior may give rise to unpleasant surprises in actual practice. Sharp conditions for stability are presently known only in a restricted number of cases.

A continuing topic in queueing theory is the study of heavy traffic limits. In addition to their applicability as approximations of heavily-loaded queueing networks, such limits provide interesting settings for the invariance principle. Again, examples with unanticipated behavior exist, which adds to the complexity of the general theory.

Here, we present a short discussion on the background and present status of these topics. The choice of subjects reflects, to be sure, a degree of subjectivity on the author's part; a broader spectrum is available through the references. It is hoped that this article will provide interesting reading for the queueing nonspecialist.

The outline of the paper is as follows. In Section 14.2, we introduce the topics of stability and heavy traffic limits in the setting of the elementary M/M/1 queue. In Section 14.3, we provide the basic terminology for queueing networks. Section 14.4 presents some well behaved classical examples. In Section 14.5, some of the basic examples of instability for strictly subcritical networks are given. Fluid models are important tools for the demonstration of stability of queueing networks. Fluid models are introduced in Section 14.6, and applications are given in Section 14.7. The topic of global stability is discussed in Section 14.8. There, one studies the parameter ranges where all disciplines, i.e., service rules, are stable. Section 14.9 provides a brief background of heavy traffic limits. Section 14.10 discusses the role of state space collapse in recent work on heavy traffic limits, and its connections with fluid models.

14.2 The M/M/1 Queue

The M/M/1 queue is the most basic example of a queueing network. It is familiar to the general probability community, and is simple to discuss. We therefore begin with a summary of some of its basic properties to motivate the general case.

The setup consists of a server at a workstation, and customers who line up at the server until they are served, one by one. They then leave the system. These customers are assumed to arrive at the server according to a Poisson process with intensity 1; equivalently, the interarrival times of succeeding customers are given by independent rate-1 exponential random variables. The service times of customers are given by independent rate-μ exponential random variables. We are interested in the behavior of $Z(t)$, the number of customers in the queue at time t.

The process $Z(\cdot)$ can be interpreted in several ways. It is a continuous time Markov chain, and is also a birth and death process on $\{0, 1, 2, \ldots\}$. Consequently, when an equilibrium measure exists, the process is reversible. Letting $m = 1/\mu$ denote the mean time for serving a customer, it is easy to show from the accompanying detailed balance equations that, when $m < 1$, $Z(\cdot)$ is positive recurrent with the geometric equilibrium measure

$$\pi_m(n) = (1-m)m^n, \quad n = 0, 1, 2, \ldots. \tag{14.2.1}$$

The mean of π_m is $m(1-m)^{-1}$, which blows up as $m \uparrow 1$.

One can also interpret $Z(\cdot)$ as a continuous time biased simple random walk, which is reflected at 0. This viewpoint is useful when one sets $\hat{Z}^r(t) = Z^r(r^2 t)/r$, for a sequence of such processes $Z^r(\cdot)$. When

$$r(m^r - 1) \to \theta \quad \text{as } r \to \infty, \tag{14.2.2}$$

for some θ, $\hat{Z}^r(\cdot)$ converges weakly to a reflecting Brownian motion with drift θ

and variance 2. Such a limit is an example of a *heavy traffic limit.* We postpone the formal definitions involved in the limiting procedure until Section 14.9.

14.3 Queueing Networks

The M/M/1 queue admits natural generalizations in a number of directions. It is unnecessary to assume that the service and interarrival distributions are exponential, assumptions we now drop. Also, the single queue discussed above can be extended to a finite system of queues, or a *queueing network*, where customers, upon leaving a queue, line up at another queue or leave the system. Here, "customers" can be envisioned as people circulating about through a bureaucratic maze, products of some sort of complex manufacturing process with multiple steps, or, as tasks that need to be performed by a computer or communication system. Depending on a customer's previous history, we may also wish to prescribe different service distributions at a queue, or routing to the next queue. (For instance, semiconductor components are fabricated using the same machines repeatedly at different stages of the procedure, with the service rule depending on the individual step. More mundanely, patients at the receptionist's desk at a doctor's office will follow different rules depending on whether they are checking in or out.) While making such distinctions, we will wish to maintain the Markov nature of the system. To do this, we assign one or more *classes* or *buffers* to each queue or *station*, and assume that all customers in such a class satisfy the same random service and routing behavior. We will say that the queueing network is *multiclass* when at least one station has more than one class; otherwise, the network is *single class*. It will be assumed that the server at each station is *non-idling*, that is, remains busy as long as there are customers present. Stations are assumed to have *infinite capacity*, with customers never being turned away.

We now introduce terminology appropriate for queueing networks. Stations are labelled $j = 1, \ldots, J$ and classes by $k, \ell = 1, \ldots K$. We use $\mathcal{C}(j)$ to denote the set of classes belonging to a station j, and $s(k)$ to denote the station to which class k belongs; when j and k appear together, we implicitly set $j = s(k)$.

The triples $(E(\cdot), V(\cdot), \Phi(\cdot))$ of "input data" are employed to construct individual realizations of the queueing network. The *external arrival process* $E(t) = \{E_k(t),\ k = 1, \ldots, K\}$, $t \geq 0$, counts the number of arrivals at each class from outside the network by time t. For $k \in \mathcal{A}$, the subset of classes where external arrivals are allowed, we define

$$E_k(t) = \max\{n : U_k(n) \leq t\}, \tag{14.3.1}$$

where $U_k(n) = \sum_{i=1}^{n} u_k(i)$. The *external interarrival times* $\{u_k(i),\ i = 2, 3, \ldots\}$ are assumed here to be i.i.d. positive random variables with means $1/\alpha_k$, $\alpha_k > 0$; the term $u_k(1) > 0$ is arbitrary and independent of these times, and is the *residual time* of the customer initially waiting to enter the network. The vector $\alpha = \{\alpha_k,\ k = 1, \ldots, K\}$ is referred to as the *arrival rate*, where we set $\alpha_k = 0$ for $k \notin \mathcal{A}$.

The *cumulative service time process* $V(n) = \{V_k(n_k),\ k = 1, \ldots, K\}$, $n = (n_1, \ldots, n_K)$ with $n_k = 1, 2, \ldots$, is given by $V_k(n_k) = \sum_{i=1}^{n_k} v_k(i)$, where $v_k(i)$ is the *service time* of the ith customer of class k. One assumes that $\{v_k(i),\ i = 2, 3, \ldots\}$ are i.i.d. positive random variables with finite means $m_k > 0$; the residual time of the customer initially being served, $v_k(1)$, $v_k(1) > 0$, is arbitrary and independent of these times. Denote by M the diagonal matrix with m_k, $k = 1, \ldots, K$, as its entries, and write $\mu_k = 1/m_k$ for the rate of service at the class k.

The *routing process* $\Phi(n) = \{\Phi^k(n),\ k = 1, \ldots, K\}$, $n = 1, 2, \ldots$, is given by $\Phi^k(n) = \sum_{i=1}^{n} \phi^k(i)$. Here, $\phi^k(i) = \{\phi^k_\ell(i),\ \ell = 1, \ldots, K\}$, $i = 1, 2, \ldots$, are i.i.d. random vectors, at most one of whose components is 1, the others being 0; the nonzero component indicates the class to which the ith customer served at k is routed, with $\phi^k(i) = 0$ indicating a departure from the system. Denote by $P = \{P_{k\ell},\ k, \ell = 1, \ldots, K\}$ the *mean transition matrix* corresponding to $\phi^k(\cdot)$, $k = 1, \ldots, K$, i.e., $P_{k\ell}$ is the probability a customer departing from class k goes to class ℓ. In many interesting cases, the routing will be deterministic, with all customers entering the system at the same class, and moving along a given route, until they exit from the system. Such networks are referred to as *re-entrant lines*. We will consider here only *open* networks, that is, the matrix

$$Q \stackrel{\text{def.}}{=} (I - P^t)^{-1} = I + P^t + (P^t)^2 + \cdots \tag{14.3.2}$$

is finite. ("^{t}" denotes the transpose.) This means that customers at any class are capable of ultimately leaving the network. We also assume throughout that the sequences defining $E(\cdot)$, $V(\cdot)$ and $\Phi(\cdot)$ each have independent components (indexed by k), and that the different sequences, except for the residual times, are independent.

To investigate open multiclass queueing networks, one employs the solutions λ_ℓ, $\ell = 1, \ldots, K$, of the *traffic equations*

$$\lambda_\ell = \alpha_\ell + \sum_{k=1}^{K} \lambda_k P_{k\ell}, \tag{14.3.3}$$

or equivalently, in vector form, of $\lambda = \alpha + P^t\lambda$. (All vectors in this article are to be interpreted as column vectors.) Solving (14.3.3), one obtains $\lambda = Q\alpha$. The term λ_k is referred to as the *total arrival rate* for class k; to avoid degeneracies, we assume that $\lambda_k > 0$ for all k. Employing m and λ, one defines the *traffic intensity* ρ_j for the jth server as

$$\rho_j = \sum_{k \in C(j)} m_k \lambda_k, \tag{14.3.4}$$

with ρ being the corresponding vector. The station j is said to be *subcritical* when $\rho_j \leq 1$, *strictly subcritical* when $\rho_j < 1$, and *critical* when $\rho_j = 1$. When all stations of a network share these properties, we refer to the network as subcritical, strictly subcritical, and critical, respectively; we abbreviate these conditions by writing $\rho \leq \mathbf{1}$, $\rho < \mathbf{1}$ and $\rho = \mathbf{1}$, where $\mathbf{1} = (1, \ldots, 1)$.

For re-entrant lines whose customers are assumed to enter the network at rate 1, $\lambda_k \equiv 1$, and so $\rho_j = \sum_{k \in C(j)} m_k$. In particular, $\rho = m$ for the M/M/1 queue. For re-entrant lines, it is not difficult to show that when a station j is strictly subcritical, it will process work faster than it is arriving, on the average, and so will repeatedly become empty. This behavior also holds for general routing, where ρ_j is given by (14.3.4). One can think of λ_k as the long-term rate at which customers arrive at a class k, assuming the network is "in equilibrium." (Such an equilibrium need not exist for a particular network, e.g., when $\rho_j > 1$ for some j.) In this setting, ρ_j is the long-term rate at which work arrives at the station j, and is thus the fraction of time the server is busy. In general, ρ_j provides an upper bound.

Associated with each queueing network is a *discipline*, which specifies the order in which customers receive service. Except in Section 14.4, we consider only *head-of-the-line* (HL) disciplines, where only the first (or oldest) customer in each class at a station may receive service. For multiclass networks, the proportion of service to be devoted to each class needs to be specified. Well-known disciplines are first-in first-out (FIFO), where the first customer at a station receives all of the service irrespective of its class, and static priority disciplines, where classes are assigned a strict ranking and customers of higher ranked classes are always served before customers of lower ranked classes. (The static priority disciplines here are assumed to be *preemptive resume*. That is, customers of higher rank interrupt the service of customers of lower rank; when the service of these higher ranked customers is completed, service of the lower ranked customers continues where it left off.) In the setting of re-entrant lines, examples of static priority disciplines are first-buffer-first-served (FBFS) and last-buffer-first-served (LBFS), where customers at the earlier, respectively latter, classes have priority. When the queueing network is single class with exponential service and interarrival times, it is referred to as a *Jackson network*; when this restriction on the service and interarrival times is removed, it is called a *generalized Jackson network.*

Once a discipline is given, the triple $(E(\cdot), V(\cdot), \Phi(\cdot))$ and the initial data uniquely specify the evolution of the queueing network along each realization. This defines an underlying Markov process. Depending on the discipline, and the service and interarrival distributions, the appropriate state space may vary in complexity. For instance, for static priority disciplines, one only needs to keep track of the number of customers in each class, whereas for FIFO networks, the order of arrival of customers at the class must be given. When all of the service and interarrival times are exponentially distributed, residual times can be ignored, and the state space of the Markov process is discrete. For general service and interarrival times, one needs to know the residual service time of the first customer in each class and the residual interarrival time at each $k \in \mathcal{A}$. This requires one to use Markov chain theory for general state spaces, and concepts such as Harris recurrence. When the underlying Markov process of a queueing network is positive Harris recurrent, we say that the queueing network is *stable*. Although the description of the state space can be a bit of a notational burden, standard results, such as the strong Markov property, can be verified. (See Dai (1995) for background and additional references on the application of Markov chain theory to this setting.)

14.4 FIFO Networks of Kelly Type and Some Other Examples

We recall from Section 14.2 that it is simple to analyze the behavior of the Markov chain $Z(\cdot)$ for an M/M/1 queue. In particular, its equilibrium measure, when $m < 1$, is given by (14.2.1). Such an explicit representation cannot be expected in general for queueing networks, but analogs exist for certain families. An important family exhibiting this behavior is the FIFO queueing networks with exponential service and interarrival times, where the mean service times for all classes at the same station are the same, i.e., $m_k = m_\ell$ for $s(k) = s(\ell)$. Networks satisfying this latter condition are said to be of *Kelly type*.

Suppose that such a FIFO network is strictly subcritical, i.e., $\rho < \mathbf{1}$. Then, the queueing network is stable. Moreover, its equilibrium distribution factors, and can be described in terms of the geometric distributions $\pi_{\rho_j}(\cdot)$, where $\pi_m(\cdot)$ is given in (14.2.1). The states at different stations are independent, and the probability there are n_j customers at station j is $\pi_{\rho_j}(n_j)$. Conditioned on n_j, the number of customers at each class $k \in \mathcal{C}(j)$ is given by a multinomial distribution with frequencies $m_k\lambda_k/\rho_j$, and the different possible orders of customers are equally likely. (Since Jackson networks are a special case of these networks, the above behavior, in a simplified format, holds for them as well.)

The equilibria for FIFO networks of Kelly type and several other networks, when $\rho < \mathbf{1}$, are given in Baskett *et al.* (1975) and Kelly (1975, 1979). These networks are examples of *quasi-reversible* Markov chains. Their equilibria satisfy *partial balance equations*, which are weaker versions of the detailed balance equations satisfied by reversible processes. Employing these equations, one can show that the equilibria factor into independent states at different stations. As before, the number of customers at each $k \in \mathcal{C}(j)$, for given n_j, is a multinomial distribution with frequencies $m_k\lambda_k/\rho_j$. The distribution of n_j will vary with the discipline, however.

Additional examples of such networks are given by the *processor sharing*, *last-come-first-served*, and *infinite server* disciplines. For the processor sharing discipline, all customers at a station, at a given time, are assumed to receive service at equal rates; this continues for an individual customer until it attains its random service requirements, at which point it is routed to the next class. Last-come-first-served, as the name suggests, is the discipline where the last customer to enter a station immediately receives all of the service being allocated there. The infinite server discipline assumes the presence of a reserve of an infinite number of servers at the station, so that a customer begins service as soon as it arrives at the station. None of these three disciplines is HL, since customers other than the first in each class receive service. Kelly (1979) includes these three examples under the heading of *symmetric queues*. In this setting, the assumptions that the service times are exponential and that $m_k = m_\ell$ for $s(k) = s(\ell)$ are not needed for the equilibrium to factor as above, in contrast to the situation for FIFO networks. (The assumption that the interarrival times are exponential cannot be dropped, however.)

14.5 Unstable Strictly Subcritical Networks

As we saw in Section 14.2, the M/M/1 queue is stable exactly when $m < 1$. For FIFO networks of Kelly type with exponential service and interarrival times, and for the other networks in Section 14.4, the corresponding condition for stability is that $\rho < \mathbf{1}$. It is tempting to try to generalize this to all networks. It is not difficult to see that $\rho < \mathbf{1}$ is a necessary condition for stability. But, is this condition sufficient, modulo mild conditions on the service and interarrival distributions (in order to avoid periodicity at the individual stations)?

For a long while, this was generally believed to be the case. In Lu-Kumar (1991) and Rybko-Stolyar (1992), examples were given of strictly subcritical, but unstable, queueing networks. (These examples were motivated by related examples of unstable *clearing policies* in the less well known paper, Kumar-Seidman (1990). There, all service at a station is concentrated on a single class, until the class empties.) Because of the specialized nature of the static priority discipline in both cases, these examples were initially treated insufficiently seriously. Since then, examples of unstable FIFO networks have been given in Bramson (1994a) and Seidman (1994), and it is now realized that the phenomenon of unstable strictly subcritical networks is not exceptional.

These examples are easier to analyze in a deterministic setting, where customers are replaced by a continuous mass that flows through the network according to rules that are analogous to those for the original network. (Such solutions are examples of fluid models, which are discussed in the next section.) We use this setting to present a variant of the Lu-Kumar example.

This network is a re-entrant line consisting of two stations with two classes each. Write (1, 1), (1, 2), (2, 1) and (2, 2) for the different classes, with the first number corresponding to the station, and the second number corresponding to the class at the station. The discipline is static priority, with the higher priority class at each station being designated by the lower number. The routing is

$$\to (1,2) \to (2,1) \to (2,2) \to (1,1) \to, \qquad (14.5.1)$$

with $\alpha_{(1,2)} = 1$, all other $\alpha_k = 0$, $m_{(2,1)} = m_{(1,1)} = 2/3$ and $m_{(1,2)} = m_{(2,2)} = 1/6$. That is, after entering the system at rate 1, customer mass first flows through a quick class, followed by slow, quick and slow classes, at which point it exits from the system. The slow classes have priority over the quick classes. The system is strictly subcritical, with $\rho_1 = \rho_2 = 5/6$.

We provide a sketch showing why the network is unstable, with the reader being invited to fill in the details. Let $Z_k(t)$ denote the amount of mass at the class k at time t, and let $Z(t)$ be the corresponding vector. Starting with $Z_{(1,2)}(0) = 1$ and $Z_{(2,1)}(0) = Z_{(2,2)}(0) = Z_{(1,1)}(0) = 0$, it is not difficult to follow the evolution of the system. Service at (1, 2) immediately begins and sends mass to (2, 1), at which service also immediately begins. Service continues there until all of the mass, together with the mass arriving in the meantime into the system, has been served at (2, 1). This occurs at $t = 2$, with $Z_{(2,2)}(2) = 3$ and $Z_{(1,2)}(2) = Z_{(2,1)}(2) = Z_{(1,1)}(2) = 0$. Since (2, 2) has low priority, no service there has taken place by then. At $t = 2$, service begins at (2, 2) and sends mass to (1, 1), at which service also begins. This continues until $t = 4$, by which time all of the mass, previously at (2, 2), has been served at (1, 1). Over $t \in (2, 4]$, all 2 units of new mass entering the system is obliged to wait at (1, 2). Thus, $Z_{(1,2)}(4) = 2$ and $Z_{(2,1)}(4) = Z_{(2,2)}(4) = Z_{(1,1)}(4) = 0$, which shows that $Z(4) = 2Z(0)$. Iterating in this manner, one sees that

$$\liminf_{t\to\infty} |Z(t)|/t = 1/2, \tag{14.5.2}$$

where $|\cdot|$ denotes the sum of the coordinates. Similar behavior holds if one instead assumes that $m_{(1,2)}$ and $m_{(2,2)}$ are each in $[0, 1/3)$, and if one considers the discrete stochastic analog (for instance, with exponentially distributed service times and arrivals) along individual sample paths.

The unstable strictly subcritical FIFO example in Bramson (1994a) is a two station re-entrant line, with routing

$$\to (1,2) \to (2,1) \to (2,2) \to \cdots \to (2,L) \to (1,1) \to . \tag{14.5.3}$$

One assumes the interarrival and service times are exponential with $\alpha_{(1,2)} = 1$ (with all other $\alpha_k = 0$), $m_{(2,1)} = m_{(1,1)} = c$ and $m_{(1,2)} = m_{(2,\ell)} = \delta$ for $\ell = 2, \ldots, L$, and

$$c \in [399/400, 1), \quad c^L \leq 1/50, \quad \delta \leq (1-c)/50L^2. \tag{14.5.4}$$

For instance, one can set $c = 399/400$, $L = 1600$ and $\delta = 10^{-11}$. (More careful estimation allows one to considerably improve the bounds in (14.5.4). For $L = 4$, simulations show unstable strictly subcritical behavior for appropriate choices of c and δ.) The labelling of classes emphasizes the similarity between (14.5.3) and the route given in (14.5.1). Note that in both cases, the first station has two classes, which are positioned at the beginning and at the end of the route, whereas in (14.5.3), the low priority quick class in the second station is replaced by $L-1$ quick classes. (Of course, there are no priorities among classes in (14.5.3).) The almost sure limiting behavior of $|Z(t)|/t$ along sample paths is analogous to (14.5.2), with the growth taking place in "cycles".

Various questions naturally come up with the two preceding examples in mind. For instance, since both examples consist only of 2 stations, they can only be unstable if $\max_j \rho_j \geq 1/2$. One can ask whether, by adding more stations, one can construct unstable networks with $\max_j \rho_j$ as small as desired.

This is not difficult to see in the continuous deterministic setting for an appropriate static priority network. For instance, assume that a re-entrant line consisting of J stations, each with two classes, has route

$$\begin{aligned} &\to (1,2) \to (2,1) \to (2,2) \to (3,1) \to \cdots \\ &\qquad\qquad \to (J,1) \to (J,2) \to (1,1) \to \end{aligned} \tag{14.5.5}$$

where, as in (14.5.1), the lower class number corresponds to the higher priority class. Assume that $\alpha_{(1,2)} = 1$, with $m_{(j,1)} = 2m_{(j,2)} = \gamma < 1/2$ for $j = 1, \ldots, J$. Then, $\rho_j = 3\gamma/2 < 1$. Also, assume that $Z_{(1,2)}(0) = 1$ and $Z_k(0) = 0$ for

$k \neq (1,2)$. Then, one can apply the same reasoning repeatedly, over $j = 2, \ldots, J$, as was used for the network in (14.5.1), to show that at time $t_0 = [(1+\gamma)/(1-\gamma)^{J-1}] - 1$, $Z_{(1,2)}(t_0) = \gamma/(1-\gamma)^{J-1}$ and $Z_k(t_0) = 0$ for $k \neq (1,2)$. So, $Z(t_0) = [\gamma/(1-\gamma)^{J-1}]Z(0)$. Iteration of this shows that $\liminf_{t\to\infty} |Z(t)|/t > 0$ for any given $\gamma \in (0, 1/2)$, if J is large enough. So, as claimed, $\max_j \rho_j$ can be chosen as small as desired. As in (14.5.1), it is easy to generalize $m_{(j,2)}$ ($m_{(j,2)} < m_{(j,1)}(1 - m_{(j,1)})$ suffices); one can also replace the network by its stochastic analog.

Substantially more work is required if one wishes to transfer the setting to that of FIFO networks. Because of the possible complications arising due to some customers moving through the network faster than the main body of customers, it is simpler to modify the network by allowing two types of customers, which progress along their almost identical routes. This is done in Bramson (1994b). The networks employed there have the following interesting property. When the mean service times are altered by increasing them to $m_k = \gamma$ for all k, the model becomes a FIFO network of Kelly type; for small γ, it is strictly subcritical. So, from Section 14.4, the altered network is stable. This shows that it is possible to stabilize an unstable network by slowing down service for the different classes. Another example of this phenomenon is given in Dumas (1997).

There is, in fact, a general technique for stabilizing unstable strictly subcritical networks, which is related to the *leaky buckets* procedure employed in the engineering literature to reduce the *burstiness* (unevenness of flow of customers) throughout a system. One inserts *regulators* (single class stations) between all classes, which have exponential service times and are strictly subcritical, but with the traffic intensity ρ_j in each case greater than those of the original stations. This, in effect, insulates classes at the original stations from other classes which can rapidly discharge customers, and so prevents any buildup there. When these classes are omitted from the network, one is left with the inserted stations, which together form a strictly subcritical Jackson network, which we know is stable. Using these ideas one can show the entire network is stable (see Humes (1994) and Bramson (1998a)).

14.6 Fluid Models

Fluid models are now the main tool for showing the stability of queueing networks. They allow one, in essence, to replace a queueing network with its continuous deterministic analog, consisting of mass flowing through the system. It is typically a considerably easier problem to show stability in this deterministic setting. Under mild conditions, the stability of the original queueing network will then follow.

The basic idea is as follows. The evolution of a queueing network can be measured by that of certain random vectors, such as $A(t)$, $D(t)$, $T(t)$, $W(t)$, $Y(t)$ and $Z(t)$. The vectors $A(t)$ and $D(t)$ denote the number of arrivals, respectively, the number of departures, by time t, $T(t)$ is the cumulative service time up until

t, and $Z(t)$ is the number of customers at time t. These four quantities are all *class vectors*, with components, indexed by $k = 1, \ldots, K$, corresponding to the individual classes. The vectors $W(t)$ and $Y(t)$ are both *station vectors*, with $W(t)$ being the *immediate workload* (the time required to serve customers currently at $j = 1, \ldots, J$), and $Y(t)$ is the cumulative idletime. Typically, not all of these quantities are needed, and the choice depends on the particular setting. We will denote the corresponding n-tuple by $\mathfrak{X}(t)$. In the above setting,

$$\mathfrak{X}(t) = (A(t), D(t), T(t), W(t), Y(t), Z(t)).$$

One connects these quantities together by *queueing network equations*, which include

$$A(t) = E(t) + \sum_k \Phi^k(D_k(t)), \tag{14.6.1}$$

$$Z(t) = Z(0) + A(t) - D(t), \tag{14.6.2}$$

$$W(t) = CV(A(t) + Z(0)) - CT(t), \tag{14.6.3}$$

$$CT(t) + Y(t) = \mathbf{1}t, \tag{14.6.4}$$

$$Y_j(t) \text{ can only increase when } W_j(t) = 0, \; j = 1, \ldots, J, \tag{14.6.5}$$

for all $t \geq 0$. Here, C is the *constituency matrix*,

$$\begin{aligned} C_{jk} &= 1 \quad \text{if } k \in C(j) \\ &= 0 \quad \text{otherwise,} \end{aligned}$$

and $\mathbf{1}$ denotes the J-vector of all 1's. Recall that the external arrivals $E(t)$, cumulative service times $V(n)$ and routing matrices $\Phi(n)$ were defined in Section 14.3. These equations do not say anything about the discipline of the network. HL queueing networks, in addition, satisfy

$$V(D(t)) \leq T(t) < V(D(t) + \mathbf{1}), \tag{14.6.6}$$

where the inequalities are componentwise, and $\mathbf{1}$ denotes the K-vector of all 1's. Additional equation(s) still need to be given which correspond to the particular discipline. For example, for the FIFO discipline, one employs

$$D_k(t + W_j(t)) = Z_k(0) + A_k(t), \quad k = 1, \ldots, K, \tag{14.6.7}$$

for all $t \geq 0$.

For our purposes here, the exact nature of the equations (14.6.1)–(14.6.7) is not too important. One should think of there as being enough equations to determine the evolution of the queueing network. We will use these equations in conjunction with their deterministic analogs, which are obtained by replacing $E(t)$, $V(n)$ and $\Phi(n)$ by their means. In place of (14.6.1)–(14.6.7), one obtains

$$\bar{A}(t) = \alpha t + P^t \bar{D}(t), \tag{14.6.8}$$

$$\bar{Z}(t) = \bar{Z}(0) + \bar{A}(t) - \bar{D}(t), \tag{14.6.9}$$

$$\bar{W}(t) = CM(\bar{A}(t) + \bar{Z}(0)) - C\bar{T}(t), \tag{14.6.10}$$

$$C\bar{T}(t) + \bar{Y}(t) = \mathbf{1}t, \tag{14.6.11}$$

$$\bar{Y}_j(t) \text{ can only increase when } \bar{W}_j(t) = 0, \quad j = 1, \ldots, J, \tag{14.6.12}$$

$$\bar{T}(t) = M\bar{D}(t), \tag{14.6.13}$$

$$\bar{D}_k(t + \bar{W}_j(t)) = \bar{Z}_k(0) + \bar{A}_k(t), \quad k = 1, \ldots, K, \tag{14.6.14}$$

for all $t \geq 0$. Equations such as (14.6.8)–(14.6.14) are known as *fluid model equations*; their solutions

$$\bar{\mathfrak{X}}(t) = (\bar{A}(t), \bar{D}(t), \bar{T}(t), \bar{W}(t), \bar{Y}(t), \bar{Z}(t))$$

are referred to as *fluid model solutions*. (The equation (14.6.14), of course, will change with the discipline.) We restrict our attention to solutions with continuous and nonnegative components, where $\bar{A}(t)$, $\bar{D}(t)$, $\bar{T}(t)$ and $\bar{Y}(t)$ are nondecreasing in t. We remark here that fluid model solutions will, in general, not be unique, even when the discipline and all other relevant information is included in the corresponding equations. This occurs, for example, with the Lu-Kumar example in (14.5.1), when one specifies that $\bar{Z}(0) = 0$: both $\bar{Z}(t) \equiv 0$ and a solution corresponding to (14.5.2) exist.

Solutions of the fluid model equations (14.6.8)–(14.6.14) are connected with solutions of the queueing network equations (14.6.1)–(14.6.7) by the corresponding *fluid limits*. A fluid limit is any limit of $\mathfrak{X}(\cdot)$ under hydrodynamic scaling, that is, where the weight of individual customers and time are scaled proportionally. (We avoid the technical details here.) Fluid limits are solutions of the fluid model equations; solution of the latter therefore gives information about the original queueing network. The fluid model is said to be *stable* if, for some $\delta > 0$ and all solutions of the fluid model equations, $\bar{Z}(t) = 0$ for $t \geq \delta|\bar{Z}(0)|$. (As earlier, $|\cdot|$ denotes the sum of the coordinates.) Since the solutions of a fluid model correspond to a queueing network with the randomness removed, stability of the fluid model says that, in essence, the total number of customers in the queueing network has a net negative drift. For instance, the fluid model which corresponds to a strictly subcritical M/M/1 queue is clearly stable, with derivative $\bar{Z}'(t) = 1 - \mu < 0$ as long as $\bar{Z}(t) > 0$.

Using elementary properties of Markov processes on general state spaces, one can show that, under mild assumptions on the service and interarrival times:

A queueing network is stable whenever the corresponding fluid model is stable. (14.6.15)

By applying (14.6.15), one can demonstrate the stability of a queueing network by instead studying systems of fluid model equations such as (14.6.8)–(14.6.14). In particular, the distributions of the service and interarrival times are not needed in this setting. This enables one, for example, to very simply demonstrate the stability of strictly subcritical generalized Jackson networks, whereas a direct argument is quite tedious.

Fluid limits and fluid models were first employed in Rybko-Stolyar (1992), where they served to provide motivation for the stability of the corresponding queueing networks. They were systematized in Dai (1995), which is the standard reference. ((14.6.15), in its general setting, is taken from Theorem 4.2 of Dai (1995).) Stolyar (1994) has related results in the setting of exponential service and interarrival times, and also has a somewhat weaker requirement on fluid models than the stability assumed here.

14.7 Application of Fluid Models

Fluid models are valuable for showing the stability of strictly subcritical queueing networks under a variety of standard disciplines. Besides generalized Jackson networks, the use of fluid models gives easy proofs, for example, of the stability of FBFS and LBFS re-entrant lines. We know, however, from the Lu-Kumar example, that not all strictly subcritical static priority networks are stable; a general theory for static priority networks remains to be worked out.

We also know that not all strictly subcritical FIFO networks are stable. What happens if one restricts oneself to FIFO networks of Kelly type? From Section 14.3, we know that when the service and interarrival distributions are all exponential, then not only is the queueing network stable, but the equilibrium distribution can be written down explicitly. Although one cannot expect an explicit representation in general, it is reasonable to hope that stability will still always hold.

This is, in fact, the case. In Bramson (1996), the entropy function

$$\mathcal{H}(t) = \sum_k \int_t^{t+\bar{W}_j(t)} h_k(\bar{D}_k'(r))dr, \quad t \geq 0, \tag{14.7.1}$$

with $h_k(x) = x\log(x/\lambda_k)$, $x \geq 0$, is examined. For strictly subcritical FIFO fluid models of Kelly type, it is shown that $\mathcal{H}'(t) \leq -b < 0$ until $\bar{Z}(t) = 0$. The stability of the fluid model follows quickly from this and (14.7.1). By (14.6.15), the stability of the corresponding FIFO networks of Kelly type follows.

Another entropy function works for the fluid models of *head-of-the-line proportional processor sharing* (HLPPS) queueing networks. For this discipline, the fraction of time spent serving a class present at a station is proportional to the quantity of the class there, with all of the service going into the first customer of each class. This differs from the processor sharing networks mentioned in Section 14.4, where the service directed to a class is instead equally spread among all of the customers present. When the service times are all exponentially distributed, the two models have the same queue length process $Z(\cdot)$, and so the stability of one implies the stability of the other. Since HLPPS networks are HL, (14.6.15) applies. To show the stability of the corresponding fluid models, Bramson (1997) employs the entropy function

$$\mathcal{H}(t) = \sum_{k} m_k Z_j^{\Sigma}(t) h_k(\bar{D}_k'(t)), \quad t \geq 0, \tag{14.7.2}$$

where $Z_j^{\Sigma}(t) = \sum_{k \in C(j)} Z_k(t)$ and $h_k(x) = x \log(x/\lambda_k)$.

After having seen a number of examples where stability of fluid models is employed to show stability of the corresponding queueing networks, it is natural to ask about the opposite direction. Certainly, the converse of (14.6.15) should "usually" be true. Little is known in this direction, however, with the only positive results (Dai (1996), Meyn (1995)) stating that when the fluid limits all have uniformly positive drift, then the queueing network is unstable. (For subcritical fluid models, $Z(t) \equiv 0$ will always produce a solution.) Also, examples of stable static priority networks with unstable fluid models do exist (Bramson (1999)). So, at this point it is unclear to what extent the stability of a queueing network can be reduced to the analogous problem for its fluid model.

14.8 Global Stability

Rather than working with a particular queueing network having a given discipline, one can instead allow the discipline to vary over some appropriate family, such as all HL disciplines. One can then inquire whether all such networks are stable for the given choices of the interarrival and service distributions; under mild conditions, stability should only depend on the means α and m. The *global stability region* (GSR) of a queueing network is defined to be the domain (α, m) where this stability holds. An interesting problem is to determine how this region varies with certain network quantities, such as the traffic intensity ρ.

As discussed in Section 14.6, fluid models are a natural tool with which to investigate the stability of a particular network; the same is true for global stability. The GSR of a fluid model can be defined analogously to that of the corresponding queueing network, as the intersection of the stability regions over all disciplines. This can also be phrased as being the stability region of the fluid model obtained by removing any discipline-specifying equation (e.g., (14.6.14)) from the accompanying fluid model equations.

Precise conditions can be given on the global stability of fluid models with two stations. In addition to the obvious stipulation that $\rho < \mathbf{1}$, added requirements are given by *virtual stations* and *push starts*.

Virtual stations, for a given discipline, consist of sets of classes from different stations which cannot simultaneously be served after an initial period of time, which depends on the initial state. The activity of one such class serves to delay the arrival of customers at the other classes. For example, the high priority classes

(1, 1) and (2, 1) of the Lu-Kumar network in (14.5.1) form a virtual station. This virtual station introduces the additional stability condition

$$m_{(1,1)} + m_{(2,1)} < 1$$

for the network.

A push start occurs when continual service at a class restricts the service available for other classes at the same station. For example, a push start occurs in the six class static priority re-entrant line

$$\rightarrow (1,1) \rightarrow (2,1) \rightarrow (2,3) \rightarrow (1,2) \rightarrow (1,3) \rightarrow (2,2) \rightarrow, \qquad (14.8.1)$$

where, as before, the lower the second index, the higher the priority. The point here is that the fractions of service time devoted to the classes (1, 1) and (2, 1) are always at least $m_{(1,1)}$ and $m_{(2,1)}$, which leaves only the service times $1 - m_{(1,1)}$ and $1 - m_{(2,1)}$ for all of the other classes at their respective stations. The four class network obtained by deleting the first two classes in (14.8.1) is equivalent to the network in (14.5.1) after scaling m. So, (1, 2) and (2, 2) together form a virtual station as above. Now, because of the service at $m_{(1,1)}$ and $m_{(2,1)}$, the condition

$$\frac{m_{(1,2)}}{1 - m_{(1,1)}} + \frac{m_{(2,2)}}{1 - m_{(2,1)}} < 1 \qquad (14.8.2)$$

needs to hold for stability.

Virtual stations can be employed in the context of either queueing networks or fluid models. Push starts, as phrased here, can only be used for fluid models, since one needs the service at the relevant classes to be uninterrupted. In Dai-Vande Vate (1999), it is shown that the conditions given by strict subcriticality, virtual stations and push starts identify the precise GSR of fluid models which have two stations and are re-entrant lines. (Presumably, this remains valid for two station networks with general routing.) Moreover, whenever (α, m) is not in the global stability region, there is an unstable static priority fluid model with these parameters. Using (14.6.15), one also obtains lower bounds on the GSR of the corresponding queueing networks. Although complete results are not yet known, the GSRs of the queueing networks and the fluid models should be the same, except possibly on the boundary. We also note that, in Bertsimas-Gamarnik-Tsitsiklis (1996), a linear program is given which determines whether any fixed pair (α, m) is in the GSR of a fluid model.

The GSR for a two station re-entrant line is always monotone—that is, by decreasing (α, m), one remains in the region. This is no longer the case for three or more stations, as is shown in Dai-Hasenbein-Vande Vate (1999). Moreover, the GSR is no longer given by the intersection of the stability regions of static priority fluid models. In this more general setting, the linear program in Bertsimas-Gamarnik-Tsitsiklis (1996) gives a nonsharp lower bound on the GSR. No general theory has been worked out.

14.9 Heavy Traffic Limits

An important approach to understanding the behavior of queueing networks consists of their approximation, through diffusive scaling, by reflecting Brownian motion in the positive orthant. The equilibria of such systems, although not typically explicitly solvable, are much easier to estimate using analytic techniques. When a system is heavily loaded, such an equilibrium, one would anticipate, should be a close approximation of that of the original network. In addition to such practical applications, the study of queueing networks under this scaling provides examples of the invariance principle in a variety of settings.

At the end of Section 14.2, the scaling $\hat{Z}^r(t) = Z^r(r^2 t)/r$ was discussed for the queue lengths $Z^r(t)$ of a sequence of M/M/1 queues. Under the limiting behavior of m^r in (14.2.2), $\hat{Z}^r(\cdot)$ converges weakly to a reflecting Brownian motion. One expects that, under appropriate restrictions, analogous behavior should hold for sequences of queueing networks. In this setting, the diffusively scaled immediate workload $\hat{W}^r(\cdot) = (\hat{W}^r_j(\cdot))_{j=1,\dots,J}$, rather than $\hat{Z}^r(\cdot) = (\hat{Z}^r_k(\cdot))_{k=1,\dots,K}$, is the proper quantity to consider, since random fluctuations will occur at the station level rather than at the class level. For M/M/1 queues, the two formats are equivalent up to a constant.

The functions $\hat{W}^r(\cdot)$ take values in the space of J-dimensional right continuous functions with left limits, which is equipped with the usual Skorokhod topology. We employ the traffic intensities ρ^r, and, instead of (14.2.2), assume that m^r converges and that

$$r(\rho^r_j - 1) \to \gamma_j \quad \text{as } r \to \infty, \tag{14.9.1}$$

for $j = 1, \dots, J$ and constants $\gamma_j \in \mathbb{R}$. (γ will denote the corresponding vector.) The goal is to show that

$$\hat{W}^r(\cdot) \Rightarrow W(\cdot) \quad \text{as } r \to \infty \tag{14.9.2}$$

for a continuous process $W(\cdot)$ taking values in the J-dimensional positive orthant, which has a semimartingale decomposition of the form

$$W(t) = X(t) + RY(t). \tag{14.9.3}$$

("$\Rightarrow$" denotes weak convergence.) Here, $X(\cdot)$ is a J-dimensional Brownian motion with constant drift $\theta = (\theta_j)_{j=1,\dots,J}$, $\theta = R\gamma$, and a nondegenerate covariance matrix (depending on $(E^r(\cdot), V^r(\cdot), \Phi^r(\cdot))$), where $X(t) - X(0) - \theta t$, $t \geq 0$, is a martingale with respect to the filtration defined by $W(\cdot)$ and $Y(\cdot)$. The linear map $R : \mathbb{R}^J \to \mathbb{R}^J$ is called the reflection matrix. The J-dimensional process $Y(\cdot)$, with $Y(0) = 0$, is continuous and nondecreasing, and $Y_j(\cdot)$ can only increase when $W_j(\cdot)$ is zero.

Such a $W(\cdot)$ is known as a *semimartingale reflecting Brownian motion* (SRBM). One should think of (14.9.3) as saying that $W(\cdot)$ evolves as a Brownian motion when all of its coordinates are positive, and that when the process hits one of the zero faces, it is reflected back into the positive orthant according to the matrix R. This matrix depends on C, M, P and the discipline of the queueing networks under consideration. The random vector $Y(t)$ is a multiple of the local time of $W(t)$ on the faces. The limit in (14.9.2) is known as a *heavy traffic limit*.

The derivation of heavy traffic limits has been an ongoing project over the last three decades, with the conditions on the networks under examination becoming successively less restrictive. Two important criteria in this regard are whether the network is single class or multiclass, and whether or not it allows *feedback* (i.e., the output from a station can eventually become part of its input). The original works on single class networks without feedback are Iglehart-Whitt (1970a, 1970b); single class networks with feedback (i.e., generalized Jackson networks) are analyzed in Reiman (1984a). Static priority and FIFO multiclass networks without feedback are considered in Peterson (1991); static priority networks, with a single station, had been investigated in Whitt (1971). The representation (14.9.3) for the heavy traffic limit, which has been fundamental for much of this work, was first given in Harrison (1978). A more substantial summary of these and related results is given in Williams (1996).

General approximation schemes were presented in Harrison-Nguyen (1990, 1993) for FIFO multiclass networks. As is the case for stability of queueing networks, the anticipated results do not hold in the most general setting; in Dai-Wang (1993), it was first shown that the approximations proposed in Harrison-Nguyen (1990) are not always valid. The theory for multiclass networks with feedback is presently incomplete. A summary of recent work utilizing state space collapse for general disciplines is given in the next section. FBFS static priority disciplines are analyzed in Chen-Zhang (1996).

Heavy traffic limits, as in (14.9.2), implicitly assume the existence of SRBMs satisfying (14.9.3), for appropriate R. One may interpret (14.9.3) in either the sense of strong or weak solutions, with the latter being valid for a wider choice of R. For a strong solution, one requires, for each given Brownian motion $X(\cdot)$, a pair $(W(\cdot), Y(\cdot))$ which is adapted to $X(\cdot)$ so that the properties below (14.9.3) hold. For a weak solution, it suffices for a triple, as in (14.9.3), to exist on some probability space (where $(W(\cdot), Y(\cdot))$ is not required to be adapted to $X(\cdot)$). In the context of sequences of single class queueing networks, it is enough to examine strong solutions, as was done in Harrison-Reiman (1981).

For sequences of multiclass networks, one, in general, requires weak solutions. As was shown in Reiman-Williams (1988) and Taylor-Williams (1993), a necessary and sufficient condition for the existence of a weak solution to (14.9.3) is that R be *completely-S*. For this, we recall that a *principal submatrix* of the $J \times J$ matrix R is any square matrix obtained from R by deleting all rows and columns from R with indices in some subset of $\{1, \ldots, J\}$. The matrix R is, by definition, completely-S when, for each principal submatrix $\widetilde{R}$ of R, there is an $\widetilde{x} \geq 0$ so that $\widetilde{R}\widetilde{x} > 0$. This says that, at each point x on the boundary of the positive orthant, there is a positive linear combination of the columns of R, corresponding to the zero coordinates of x, which points into the orthant. In the above Dai-Wang reference, it is shown that the reflection matrix corresponding to a particular sequence of FIFO queueing networks is not completely-S, and so (14.9.3) cannot hold. A good survey of SRBMs is given in Williams (1995).

14.10 State Space Collapse

Implicit in the formulation of (14.9.2) is the assumption that the states of the corresponding networks are, for large r, essentially given by $\hat{W}^r(t)$ at time t. More detailed information about the system, such as the queue lengths corresponding to individual classes, or the order of customers at a station, should not be necessary to study the evolution of the limit $W(t)$. This type of behavior has been phrased in a variety of settings as *state space collapse*; the typical requirement is that

$$\|\hat{Z}^r(\cdot) - \Delta \hat{W}^r(\cdot)\|_T \to 0 \quad \text{in probability} \tag{14.10.1}$$

as $r \to \infty$. Here, Δ is an appropriate linear map from $\mathbb{R}^J$ to $\mathbb{R}^K$ depending on the service discipline, and $\|\cdot\|_T$ is the uniform norm over $[0, T]$. The term state space collapse has been used since Reiman (1984b), with related ideas going back to Whitt (1971). It has sometimes been employed in settings besides that given by the diffusive scaling " ^ ".

Assuming (14.10.1), the appendix of Harrison-Williams (1992) gives a brief discussion on how one might proceed to show (14.9.2)–(14.9.3) for multiclass networks. In Williams (1998), a rigorous derivation of (14.9.2)–(14.9.3) is given in the HL setting. There, the requirement (14.10.1) is reduced to the weaker assumption of *multiplicative state space collapse* (MSSC). That is, for each $T > 0$,

$$\frac{\|\hat{Z}^r(\cdot) - \Delta \hat{W}^r(\cdot)\|_T}{\max(\|\hat{W}^r(\cdot)\|_T, 1)} \to 0 \quad \text{in probability} \tag{14.10.2}$$

as $r \to \infty$.

Williams (1998) also requires that the means α^r, M^r and P^r of the queueing networks satisfy

$$\alpha^r \to \alpha, \quad M^r \to M, \quad P^r \to P \quad \text{as } r \to \infty \tag{14.10.3}$$

for some (α, M, P), and that (14.9.1) hold. (This last condition also implies that the traffic intensity ρ corresponding to (α, M, P) satisfies $\rho = \mathbf{1}$.) The vector $\theta = R\gamma$ gives the drift of the limiting SRBM, where $R = (CMQ\Delta)^{-1}$; the matrix R needs to be completely-$\mathcal{S}$. The variances of the interarrival and service times are assumed to have limits, from which the covariance matrix of the SRBM is defined. In Williams (1998), it is assumed that $\hat{W}^r(0) \Rightarrow W(0)$ for some $W(0)$. To simplify matters here, we replace this by $\hat{W}^r(0) = 0$ for all r, i.e., the system is assumed to be initially empty. In addition, certain mild regularity conditions on the distributions of the incremental random variables underlying the triples $(E^r(\cdot), V^r(\cdot), \Phi^r(\cdot))$ are needed.

The above assumptions, for the most part, concern only the asymptotic nature of $(E^r(\cdot), V^r(\cdot), \Phi^r(\cdot))$, and can routinely be checked. The exceptions are MSSC, and verification that R exists and is completely-$\mathcal{S}$, which depend on the discipline under consideration. These conditions on R are deterministic, whereas MSSC involves the limiting behavior of random sequences. It is desirable to replace MSSC with a simpler condition.

In Bramson (1998b), it is shown that MSSC, for HL queueing networks, follows from the convergence of the queue length $\bar{Z}(t)$ of the corresponding fluid model. To avoid complications here, we restrict discussion to several familiar disciplines. Consider, for example, sequences of static priority networks satisfying (14.10.3), with $\rho = \mathbf{1}$, and certain mild regularity conditions on the random variables underlying $(E^r(\cdot), V^r(\cdot), \Phi^r(\cdot))$. Moreover, assume that, for all $|\bar{Z}(0)| \leq 1$,

$$|\bar{Z}(t) - \bar{Z}(\infty)| \leq H(t) \tag{14.10.4}$$

holds for a fixed function $H(t)$, with $H(t) \to 0$ as $t \to \infty$, and for appropriate $\bar{Z}(\infty)$ (depending on $\bar{Z}(0)$) of the form

$$\bar{Z}(\infty) = \Delta \bar{W} \quad \text{for some } \bar{W} \in \mathbb{R}^J. \tag{14.10.5}$$

MSSC then holds whenever the initial data of the queueing networks satisfy $\hat{Z}^r(0) = 0$. The main work in showing that (14.9.2)–(14.9.3) hold for a particular sequence of static priority networks thus reduces to verifying (14.10.4), and that the reflection matrix R exists and is completely-$\mathcal{S}$. Of course, (14.9.2)–(14.9.3) will not hold for all disciplines; the Lu-Kumar network in (14.5.1), for instance, easily leads to a counterexample. The general question of when (14.10.4) is satisfied remains to be investigated. Some particular cases, with the corresponding heavy traffic limits, are worked out in Bramson-Dai (1999).

One can also analyze the behavior of FIFO networks of Kelly type and HLPPS networks by investigating their corresponding fluid models. One can apply the entropy functions in (14.7.1) and (14.7.2), for critical FIFO fluid models of Kelly type and critical HLPPS fluid models, to show that the analog of (14.10.4) is always satisfied, where $H(t)$ is replaced by $B_1 e^{-B_2 t}$, for appropriate B_1 and $B_2 > 0$. As before, we assume that (14.10.3) is satisfied, with $\rho = \mathbf{1}$, that $\hat{Z}^r(0) = 0$ for all r, and that the same regularity conditions hold for the random variables underlying $(E^r(\cdot), V^r(\cdot), \Phi^r(\cdot))$. MSSC then follows for FIFO networks of Kelly type and for HLPPS networks. Moreover, it is known that the R matrix for sequences of both families of networks is always completely-$\mathcal{S}$ (Dai-Harrison (1993) and Williams (1998)). Consequently, by the above result from Williams (1998), the heavy traffic limits in (14.9.2)–(14.9.3) follow for FIFO networks of Kelly type and HLPPS networks, under the conditions given below (14.10.3).

The basic motivation for why MSSC follows from (14.10.4) can be explained by comparing the diffusive and hydrodynamic limits obtained by scaling time and the queue length vectors. The hydrodynamic scaling associated with fluid limits, and hence with fluid models, gives the behavior of a queueing network on an infinitesimal stretch of time relative to the Brownian motion limits in (14.9.2). So, on the time scale given by the diffusive limit, (14.10.4) says that the limit of the normalized queue length vectors collapses instantaneously to a state of the form given in (14.10.5). Consequently, if the limit begins at such a state, for instance $Z(0) = 0$, it must always remain at such states. This leads to (14.10.2).

Acknowledgments: The author thanks Jim Dai and Ruth Williams for helpful comments on an earlier draft. He also thanks his thesis advisor, Harry Kesten, for his share in making Cornell University an inspiring place for the author and other young probabilists in the late 1970's. This research was partially supported by NSF Grant DMS-96-26196.

REFERENCES

[1] Baskett, F., Chandy, K.M., Muntz, R.R., and Palacios, F.G. (1975). Open, closed and mixed networks of queues with different classes of customers. *J. ACM* **22** 248–260.

[2] Bertsimas, D., Gamarnik, D., and Tsitsiklis, J.N. (1996). Stability conditions for multiclass fluid queueing networks. *IEEE Trans. Autom. Control* **41** 1618–1631. Correction (1997): **42** 128.

[3] Bramson, M. (1994a). Instability of FIFO queueing networks. *Ann. Appl. Probab.* **4** 414–431.

[4] Bramson, M. (1994b). Instability of FIFO queueing networks with quick service times. *Ann. Appl. Probab.* **4** 693–718.

[5] Bramson, M. (1996). Convergence to equilibria for fluid models of FIFO queueing networks. *Queueing Systems* **22** 5–45.

[6] Bramson, M. (1997). Convergence to equilibria for fluid models of head-of-the-line proportional processor sharing queueing networks. *Queueing Systems* **23** 1–26.

[7] Bramson, M. (1998a). Stability of two families of queueing networks and a discussion of fluid limits. *Queueing Systems* **28** 7–31.

[8] Bramson, M. (1998b). State space collapse with application to heavy traffic limits for multiclass queueing networks. *Queueing Systems* **30** 89–148.

[9] Bramson, M. (1999). A stable queueing network with unstable fluid model. *Ann. Appl. Probab.*, to appear.

[10] Bramson, M. and Dai, J. (1999). Heavy traffic limits for some queueing networks, in preparation.

[11] Chen, H. and Zhang, H. (1996). Diffusion approximations for re-entrant lines with a first-buffer-first-served priority discipline. *Queueing Systems* **23** 177–195.

[12] Dai, J. (1995). On positive Harris recurrence of multiclass queueing networks: a unified approach via fluid models. *Ann. Appl. Probab.* **5** 49–77.

[13] Dai, J. (1996). A fluid-limit criterion for instability of multiclass queueing networks. *Ann. Appl. Probab.* **6** 751–757.

[14] Dai, J. and Harrison, J.M. (1993). The QNET method for two-moment analysis of closed manufacturing systems. *Ann. Appl. Probab.* **3** 968–1012.

[15] Dai, J., Hasenbein, J., and Vande Vate, J. (1999). Stability of a three-station fluid network. *Queueing Systems*, to appear.

[16] Dai, J. and Vande Vate, J. (1999). The stability of two-station multi-type fluid networks. *Operations Research*, to appear.

[17] Dai, J. and Wang, Y. (1993). Nonexistence of Brownian models for certain multiclass queueing networks. *Queueing Systems* **13** 41–46.

[18] Dumas, V. (1997). A multiclass network with non-linear, non-convex, non-monotone stability conditions. *Queueing Systems* **25** 1–43.

[19] Harrison, J.M. (1978). The diffusion approximation for tandem queues in heavy traffic. *Adv. Appl. Probab.* **10** 886-905.

[20] Harrison, J.M. and Nguyen, V. (1990). The QNET method for two-moment analysis of open queueing networks. *Queueing Systems* **6** 1–32.

[21] Harrison, J.M. and Nguyen, V. (1993). Brownian models of multiclass queueing networks: current status and open problems. *Queueing Systems* **13** 5–40.

[22] Harrison, J.M. and Reiman, M.I. (1981). Reflected Brownian motion on an orthant. *Ann. Probab.* **9** 302–308.

[23] Harrison, J.M. and Williams, R.J. (1992). Brownian models of feedforward queueing networks: quasireversibility and product form solutions. *Ann. Appl. Probab.* **2** 263–293.

[24] Humes, C. (1994). A regulator stabilization technique: Kumar-Seidman revisited. *IEEE Trans. Automat. Control* **39** 191–196.

[25] Iglehart, D.L. and Whitt, W. (1970a). Multiple channel queues in heavy traffic I. *Adv. Appl. Probab.* **2** 150–177.

[26] Iglehart, D.L. and Whitt, W. (1970b). Multiple channel queues in heavy traffic II. *Adv. Appl. Probab.* **2** 355–364.

[27] Kelly, F.P. (1975). Networks of queues with customers of different types. *J. Appl. Probab.* **12** 542–554.

[28] Kelly, F.P. (1979). *Reversibility and Stochastic Networks*. Wiley, New York.

[29] Kumar, P.R. and Seidman, T.I. (1990). Dynamic instabilities and stabilization methods in distributed real-time scheduling of manufacturing systems. *IEEE Trans. Automat. Control* **35** 289–298.

[30] Lu, S.H. and Kumar, P.R. (1991). Distributed scheduling based on due dates and buffer priorities. *IEEE Trans. Autom. Control* **36** 1406–1416.

[31] Meyn, S. (1995). Transience of multiclass queueing networks via fluid limit models. *Ann. Appl. Probab.* **5** 946–957.

[32] Peterson, W.P. (1991). Diffusion approximations for networks of queues with multiple customer types. *Math. Oper. Res.* **9** 90–118.

[33] Reiman, M.I. (1984a). Open queueing networks in heavy traffic. *Math. Oper. Res.* **9** 441–458.

[34] Reiman, M.I. (1984b). Some diffusion approximations with state space collapse. *Proceedings International Seminar on Modeling and Performance Evaluation Methodology*, Lecture Notes in Control and Informational Sciences, F. Baccelli and G. Fayolle (eds.), Springer-Verlag, New York, 209–240.

[35] Reiman, M.I. and Williams, R.J. (1988). A boundary property of semimartingale reflecting Brownian motions. *Probab. Theory Related Fields* **77** 87–97. (Correction (1989) **80** 633.)

[36] Rybko, S. and Stolyar, A. (1992). Ergodicity of stochastic processes that describe the functioning of open queueing networks. *Problems Inform. Trans.* **28** 3–26 (in Russian).

[37] Seidman, T.I. (1994). "First come, first served" can be unstable! *IEEE Trans. Automat. Control* **39** 2166–2171.

[38] Stolyar, A. (1994). On the stability of multiclass queueing networks. *Proc. 2nd Conf. on Telecommunication Systems - Modeling and Analysis*, Nashville 1020–1028.

[39] Taylor, L.M. and Williams, R.J. (1993). Existence and uniqueness of semimartingale reflecting Brownian motions in an orthant. *Probab. Theory Related Fields* **96** 283–317.

[40] Whitt, W. (1971). Weak convergence theorems for priority queues: preemptive-resume discipline. *J. Appl. Probab.* **8** 74–94.

[41] Williams, R.J. (1995). Semimartingale reflecting Brownian motions in the orthant. *Stochastic Networks*, IMA Volumes in Mathematics and its Applications, Vol. **71**. Springer-Verlag, New York, 125–138.

[42] Williams, R.J. (1996). On the approximation of queueing networks in heavy traffic. *Stochastic Networks, Theory and Applications*, Royal Statistical Society Lecture Note Series, F.P. Kelly, S. Zachary, I. Ziedlins (eds.), Clarendon Press, Oxford, 35–56.

[43] Williams, R.J. (1998). Diffusion approximations for open multiclass queueing networks: sufficient conditions involving state space collapse. *Queueing Systems* **30** 27–88.

School of Mathematics
University of Minnesota
Minneapolis, MN 55455
bramson@math.umn.edu

15

Rescaled Particle Systems Converging to Super-Brownian Motion

Ted Cox, Richard Durrett, and Edwin A. Perkins

ABSTRACT Super-Brownian motion was originally constructed as a scaling limit of branching random walk. Here we describe recent results which show that, in two or more dimensions, it is also the limit of long range contact processes and long, short, and medium range voter models.

Keywords: Voter model, contact process, super-Brownian motion, measure-valued diffusion, stochastic spatial models.

AMS Subject Classifications: Primary 60K35, 60G67; Secondary 60F05, 60J80.

Super-Brownian motion and its close relatives have arisen in a variety of different contexts in the last few years. In this article we focus on their appearance as limits of rescaled interacting particle systems and, in particular, on the results in Mueller and Tribe (1995), Durrett and Perkins (1998), and Cox, Durrett and Perkins (1998). Rather than providing complete proofs, which at present are still rather lengthy in some cases, we will focus on explaining why these theorems hold.

These results are related in spirit, if not methodology, to recent work of Derbez and Slade (1998) on the convergence of "sufficiently spread out" rescaled lattice trees to Integrated Super-Excursion (ISE) in more than 8 dimensions, and to ongoing work of Derbez, van der Hofstad and Slade on convergence of sufficiently spread out oriented percolation to super-Brownian motion in more than 4 spatial dimensions. These results are described by Slade (1999) elsewhere in this volume. Super-Brownian motion arises as the limit of rescaled branching random walk and so is "trivial" in the mathematical physics terminology. In this same parlance one expects rescaled interacting systems to converge to this trivial limit above a critical dimension and at the critical dimension, perhaps with logarithmic corrections. For the particle systems we will consider in detail, this dimension is two, the critical dimension for recurrence of simple random walk or Brownian motion. We note that for the contact process this critical dimension (as opposed to $d = 4$ as in the above work on critical oriented percolation) arises because of our long range scaling in this setting.

In order to describe super-Brownian motion we start with

Example 15.1 (Branching Random Walk) The state space for our particles is $S = \varepsilon\mathbf{Z}^d$ where $\varepsilon > 0$. Branching random walk is described by giving a displacement distribution $p : S \to [0, 1]$, a birth rate B and a death rate D. Each particle dies at rate D and gives birth with rate B. A particle born at x is sent to a neighbouring site y with probability $p(y - x)$. The dynamics for each particle operate independently. The $\mathbf{Z}_+^S$-valued process $\xi_t^0(x) =$ the number of particles at x at time t is our branching random walk.

Super-Brownian motion is constructed from branching random walk for critical (or near critical) branching by rescaling space and time in a manner similar to that used to build Brownian motion from ordinary random walk. To define this process, choose a variance parameter $\sigma^2 > 0$, a drift parameter $\theta \in (-\infty, \infty)$, and a branching rate $b > 0$. We will take limits using a set-up that is much more elaborate than necessary. However in doing so we will introduce notation and assumptions that we will use throughout the rest of the paper.

Let $B_N = Nb + \theta$ and $D_N = Nb$. We are speeding up time by a factor of N, so we scale space by defining

$$\varepsilon_N = 1/M_N\sqrt{N} \quad \text{and} \quad S_N = \varepsilon_N \mathbf{Z}^d.$$

To explain the extra factor of M_N, think of scaling the displacement distribution by its standard deviation. In symbols, we let $W_N \in \mathbf{Z}^d/M_N$ and let the displacement distribution p_N on S_N be the law of $W_N/\sqrt{N}$. This double scaling is not only convenient mathematically but introduces another modeling parameter: the number M_N^d represents the potential local population size, the number of sites in a cube with sides equal to the standard deviation of the displacement distribution.

In this paper we will be interested in two cases.

(FK) Fixed Kernel $M_N \equiv 1$. The distribution of W_N does not depend on N, is symmetric, irreducible, has finite support, and covariance

$$\sum_x x^i x^j p(x) = \sigma^2 \delta_{ij}.$$

(UD) Uniform dispersal W_N is uniform on $\mathbf{Z}^d/M_N \cap ([-1, 1]^d \setminus \{0\})$. In this case the limiting distribution as $M_N \to \infty$ is uniform on $[-1, 1]^d$, so for the purposes of stating results like Theorem 15.1 we set $\sigma^2 = 1/3$.

Let ξ_t^N denote the rescaled branching random walk with these parameters, i.e., it is a branching random walk in which each particle at x gives birth to a new particle at $x + W_N/\sqrt{N}$ with rate $Nb + \theta$ and dies with rate Nb, for a "total effective branching rate" of $2Nb + \theta$. Using this we define the measure valued process

$$X_t^N = \frac{1}{N'} \sum_{x \in S_N} \xi_t^N(x)\, \delta_x. \tag{15.1}$$

To see what value of N' to choose, assume for the moment that $\theta = 0$ so that the total population size $\sum_x \xi_t^N(x)$ is a critical Galton-Watson branching process. If we start with a single particle then the probability of survival to time t will approach 0 at rate C/Nt as $N \to \infty$. See, e.g., Section I.10 of Harris (1963). To compensate for this we will therefore let $N' = N$ and start with $O(N)$ particles at time 0. This will also be the correct normalization for our general value of θ, since the extra birth rate θ has only a moderate effect over bounded time intervals.

We consider $X_t^N(\omega)$ as an element of the space $M_F(\mathbf{R}^d)$ of finite measures on $\mathbf{R}^d$ equipped with the topology of weak convergence, denoted by $\to$. The process X_t^N has sample paths in Ω_D, the space of cadlag $M_F(\mathbf{R}^d)$-valued paths with the Skorohod topology. Let Ω_C be the corresponding space of continuous paths with the topology of uniform convergence on compact sets. If μ is a measure on $\mathbf{R}^d$ and ϕ is a real-valued function on $\mathbf{R}^d$, $\langle \mu, \phi \rangle$ or $\mu(\phi)$ denotes the integral of ϕ with respect to μ. We use $\Rightarrow$ to denote weak convergence of probability measures, usually on the space Ω_D. $C_0^\infty(\mathbf{R}^d)$ is the space of infinitely differentiable functions on $\mathbf{R}^d$ with compact support.

Before giving the natural definition of super-Brownian motion as a limit of the above branching random walks, we give a less intuitive but technically more convenient one.

Definition An adapted a.s. continuous $M_F(\mathbf{R}^d)$-valued process $(X_t, t \geq 0)$ on a complete filtered probability space $(\Omega, \mathcal{F}, \mathcal{F}_t, P)$, is an $(\mathcal{F}_t)$-super-Brownian motion with branching rate $\gamma > 0$, diffusion coefficient $\sigma^2 > 0$ and drift $\theta \in (-\infty, \infty)$ starting at $X_0 \in M_F(\mathbf{R}^d)$ if and only if it solves the following martingale problem:

$(MP)_{X_0}^{\gamma,\sigma^2,\theta}$: For all $\phi \in C_0^\infty(\mathbf{R}^d)$,

$$M_t(\phi) = X_t(\phi) - X_0(\phi) - \int_0^t X_s(\sigma^2 \Delta\phi/2 + \theta\phi)\, ds$$

is a continuous $(\mathcal{F}_t)$-martingale with $M_0(\phi) = 0$, and

$$\langle M(\phi) \rangle_t = \int_0^t X_s(\gamma\phi^2) ds.$$

The existence of such a process and the fact that its law on Ω_C is unique are well-known. Theorems 7.2.2, Lemma 7.2.1 and Theorem 6.1.3 of Dawson (1993) show the above martingale problem is well-posed. See the Appendix of Cox, Durrett and Perkins (1998) if you are concerned about the smaller class of test functions used here.

The following construction of super-Brownian motion gives one a better understanding of it, and allows one to easily understand the elements of the above martingale problem. Recall our rescaled branching random walk measures defined in (15.1) were denoted by X_t^N.

Theorem 15.1 (Branching Random Walk) *Assume* (FK) *or* (UD). *If* $X_0^N \to X_0$ *in* $M_F(\mathbf{R}^d)$ *then* $P(X^N \in \cdot) \Rightarrow P_{X_0}^{2b,b\sigma^2,\theta}$ *in* Ω_D.

There is an extensive literature on super-Brownian motion and superprocesses in general. Introductions to this subject may be found in Chapter 9 of Ethier and Kurtz (1986) (where one may find slight perturbations of Theorem 15.1), Dawson (1993), Dawson and Perkins (1999) or Le Gall (1995). Much is known about the qualitative behavior of super-Brownian motion. In one spatial dimension, it will have a jointly continuous density $u(t, x)$ that satisfies the stochastic partial differential equation

$$\frac{\partial u}{\partial t} = \frac{\sigma^2}{2}\frac{\partial^2 u}{\partial x^2} + \theta u + \sqrt{\gamma u}\,\dot{W}. \tag{15.2}$$

See Konno and Shiga (1988) or Reimers (1989). Here $\dot{W}$ is a space-time white noise. In higher dimensions, super-Brownian motion $X_t(\cdot)$ is a singular measure for all $t > 0$ a.s. (see Perkins (1989)). This rich and detailed theory is made possible by the independence of particles in the underlying branching random walk ξ_t^N, which persists in the limiting super-Brownian motion as a superposition property. Super-Brownian motion starting from $\mu + \nu$ is the sum of independent processes starting from μ and ν.

This independence property leads to a rather explicit expression for the Laplace transform of super-Brownian motion which is crucial in the proof of uniqueness of solutions to the martingale problem posed above. On the other hand, for more realistic population models, it is natural to include interactions between individuals. For example, one would like to introduce branching, diffusion and drift parameters that depend on the individual's location and the configuration of the entire population. Particularly natural are "local interactions" in which these parameters depend on the population in the immediate vicinity of an individual.

For interactive drift terms, Dawson's Girsanov theorem (Theorem 7.2.2 of Dawson (1993)) gives a general approach to the study of these models. This approach shows that under the appropriate boundedness or integrability conditions on the predictable drift $\theta(t, x, \omega)$, the law of the resulting measure valued diffusion is absolutely continuous with respect to that of the corresponding super-Brownian motion with $\theta = 0$, and gives a formula for this Radon-Nikodym derivative. For interactions which affect the spatial motion, a pair of strong stochastic equations was introduced in Perkins (1992, 1995). These equations describe: (i) how a typical particle obeys an Itô equation with a spatial drift vector and diffusion matrix depending on the rest of the population, and (ii) how these particles constitute the entire population. Donnelly and Kurtz (1998) incorporated these ideas into their exchangeable particle representation and were able to refine the uniqueness results for the historical martingale problem in Perkins (1995) to the actual measure valued martingale problem (see Kurtz (1998)). The situation for interactive branching is still not well-understood. A restrictive set of such interactions in which the branching rate involves some time averaging is handled in Perkins (1995), but there are still some basic uniqueness problems, even when the state space is finite and the interactive super-Brownian motion is the solution of a finite-dimensional degenerate stochastic differential equation (sde).

All of these problems seem to be compounded when the interaction is local, particularly above one dimension when the measures are typically singular. In this setting, Dawson's Girsanov approach for drifts can break down–an example in Evans and Perkins (1994) (described briefly in Example 15.4 below) shows that the resulting process may not be absolutely continuous with respect to ordinary super-Brownian motion. For spatial motions, Adler and Tribe (1997) extend the strong equation approach to a class of singular drift vectors. Robert Adler has defined an interesting model in which individuals move toward directions of higher concentration. Intuitively their drift is $c\nabla u$, where $c > 0$ and u is the (non-existent) density of the process in two dimensions. Simulations suggest that the random measure should display a collapse phase transition (reduction to point masses) for c sufficiently large.

The situation for local interactions in the branching rates is again not at all understood. In one spatial dimension, however, Mytnik (1998) has used an approximate duality argument to analyze solutions of (15.2) in which $\sqrt{u}$ is replace with u^α for $1/2 < \alpha \leq 1$. This corresponds to branching at a rate given by the $(\alpha - 1/2)$th power of the local density. In Dawson and Perkins (1998) and ongoing work of Dawson, Etheridge, Fleischmann, Jie, Mytnik and Perkins, a mutually catalytic branching model is studied in which there are two types, and the branching rate of each type is proportional to the local density of the other. In one dimension the solutions are given by a system of stochastic partial differential equations (spde's):

$$\frac{\partial u_i}{\partial t} = \frac{\sigma^2}{2}\frac{\partial^2 u_i}{\partial x^2} + \sqrt{u_1 u_2}\,\dot{W}_i, \quad i = 1, 2,$$

where $\dot{W}_1$ and $\dot{W}_2$ are a pair of independent white noises. In two dimensions the solutions still exist (although at present the proof can only be done for $\sigma^2 > 1.3$) and have densities. However, instead of the above system of spde's, they are constructed as the solution of a martingale problem in which the square function of the martingale part of $X_t^i(\phi) = \int u_i(t,x)\phi(x)dx$ is given by

$$L(X^1, X^2)_t(\phi) = \int_0^t \int\int \delta_0(x_1 - x_2)\phi(x_1)\, X_s^1(dx_1)\, X_s^2(dx_2)\, ds.$$

This is the collision local time of X^1 and X^2 introduced in Barlow, Evans and Perkins (1991), where a precise definition can be found. It can be used to describe local interactions between a pair of populations (see Evans and Perkins (1998) and Example 15.4 below).

In the particle system limits studied here, one encounters local interactions in the branching rates for the voter model, and in the drift for the contact process. The reason we will, for the most part, avoid some of the difficulties described above is that in the limit there is an asymptotic independence between spatially separated locations. This leads to a mean field simplification, and hence constant drift and branching rates in the limit. With this in mind, we now describe these two classical interacting particle systems as branching random walks with density dependent birth rates. We will continue to use $\xi_t(x)$ as the process of interest and X_t^N as the associated measure valued process.

Example 15.2 (Voter Model) In this case, $\xi_t(x) \in \{0, 1\}$ gives the opinion of the voter at $x \in S$. The voter model is described by giving a symmetric displacement distribution $p : S \to [0, 1]$ and a birth rate B. At rate $B > 0$, the voter at x chooses a neighbor y with probability $p(y - x)$, and convinces the voter at y to adopt its opinion (i.e., $\xi_t(y)$ changes to $\xi_t(x)$). The dynamics for different sites are independent.

To compare the voter model with branching random walk, we change our point of view and consider sites with $\xi_t(x) = 1$ as occupied and $\xi_t(x) = 0$ as unoccupied. With this interpretation, a particle at x at time t produces an offspring at y with rate $Bp(y-x)1(\xi_t(y) = 0)$, and dies with rate $Bv(t,x)$, where

$$v(t,x) = \sum_y p(y-x)1(\xi_t(y) = 0) \tag{15.3}$$

is the density of vacant sites near x at time t. Since the total birth rate is also $Bv(t,x)$, we may view the voter model as a critical branching random walk with a state dependent branching rate proportional to the local density of 0's.

Using the set-up of Example 15.1, but taking $N' = N$ in (15.1), we define a measure valued process by

$$X_t^N = \frac{1}{N} \sum_{x \in S_N} \xi_t^N(x)\delta_x.$$

The next result is Theorem 1.1 of Cox, Durrett and Perkins (1998).

Theorem 15.2 (Long Range Voter Model) *Suppose* (UD) *with*

$$\begin{aligned} M_N/\sqrt{N} &\to \infty && \text{in } d = 1,\\ M_N/\sqrt{\log N} &\to \infty && \text{in } d = 2,\\ M_N &\to \infty && \text{in } d \geq 3. \end{aligned}$$

If $X_0^N \to X_0$ *in* $M_F(\mathbf{R}^d)$ *and then* $P(X^N \in \cdot) \Rightarrow P_{X_0}^{2,1/3,0}$ *in* Ω_D.

To explain why this should be true, we need to recall the duality between the voter model and coalescing random walks. Let ζ_t^x, $x \in S_N$, be random walks that (i) take steps according to the displacement distribution p at rate $2B$, (ii) move independently before they hit, and (iii) stick together after they hit. It is well know that (see Griffeath (1978) or Liggett (1985)) that the following duality relationship holds,

$$P(\xi_t(x) = 1 \text{ for all } x \in A) = P(\xi_0(\zeta_t^x) = 1 \text{ for all } x \in A). \tag{15.4}$$

In the setting of Theorem 15.2, the number of 1's is $O(N)$ while the number of available sites is $(M_N\sqrt{N})^d \gg N$. Hence there is an asymmetry between types in which 1's are much rarer than 0's. The fact that the overall density of 1's is small does not necessarily mean the above local density of 0's near x satisfying $\xi_t(x) = 1$ must be close to 1, as there are local correlations: a 1 tends to be close to other 1's. It is intuitively clear, however, that if $M_N \to \infty$ quickly enough, then the above local density of 0's will indeed approach 1, and the voter model will behave like a branching random walk. By comparison with Theorem 15.1, we would expect to get super-Brownian motion with branching rate 2 in the limit.

To explain the reasons behind the choice of M_N, let S_t^x and S_t^y be independent random walks that individually have the same distribution as ζ_t^x and ζ_t^y. Standard random walk calculations show that the number of close encounters that occur by time 1 (i.e., jumps that occur when the two particles are within distance $1/\sqrt{N}$ on the rescaled lattice $\mathbf{Z}^d/M_N\sqrt{N})$ is

$$\approx \begin{cases} C\sqrt{N} & \text{in } d = 1, \\ C\log N & \text{in } d = 2, \\ C & \text{in } d \geq 3. \end{cases} \tag{15.5}$$

Here $\approx$ means that for large C, the right hand side is an upper bound for the left hand side for large N, while for small C, it is a lower bound. From (15.5) we see that for our choice of M_N, the probability of a collision between S_t^x and S_t^y tends to 0 as $N \to \infty$.

From the long range case we turn now to the case of a fixed displacement distribution. The next result is Theorem 1.2 of Cox, Durrett and Perkins (1998).

Theorem 15.3 (Short Range Voter Model) *Suppose* (FK) *and $d \geq 2$. Let $M_N = 1$ and let*

$$N' = \begin{cases} N/\log N & \text{in } d = 2, \\ N & \text{in } d \geq 3. \end{cases}$$

If $X_0^N \to X_0$ in $M_F(\mathbf{R}^d)$, then $P(X^N \in \cdot) \Rightarrow P_{X_0}^{2b,\sigma^2,0}$ in Ω_D, where

$$b = \begin{cases} 2\pi\sigma^2 & \text{in } d = 2, \\ \gamma_e, & \text{in } d \geq 3. \end{cases}$$

Here γ_e is the "escape probability," i.e., the probability that the random walk with jump distribution p never returns to its starting point.

To explain this, we first consider the nearest neighbor case $p(x) = 1/2d$ for $|x| = 1$ in $d \geq 3$. The effective branching rate at an occupied site x at time t is $2Nv_N(t,x)$ where $v_N(t,x)$ is the local density of vacant sites as defined in (15.3). In Theorem 15.2, $v_N(t,x) \approx 1$ at most occupied sites. In the new setting, $v_N(t,x)$ will take on all of the values $m/2d, 0 \leq m \leq 2d$, with positive probability. However, using duality, one can argue that $E(v_N(t,x)|\xi_t(x) = 1) \approx \gamma_e$. To see this observe that if the dual random walks ζ_t^x and ζ_t^y coalesce and $\xi_t(x) = 1$ then $\xi_t(y) = 1$, while if the walk from y does not hit the one from x it is very likely to be a zero due to the low initial density.

The emergence of a constant branching rate in the limit is to due to a local averaging of the $v_N(t,x)$, which we call a *"mean field simplification."* To see the intuition behind this, note that the transience of random walks in $d \geq 3$ implies that collisions will happen quickly or not at all. This means that the number of vacant neighbors $v_N(t,x)$ and $v_N(t,x')$ for two occupied sites x and x' separated by a large multiple of $1/\sqrt{N}$ on the scaled lattice $\mathbf{Z}^d/\sqrt{N}$ will be almost independent. A second moment calculation then shows that the total branching rate in an area that contains a large number of 1's will be roughly γ_e times the number of 1's there.

In $d = 2$, the recurrence of random walk will make the local density of 0's near a 1 asymptotically 0. Thus if we let $N' = N$, we will get super-Brownian motion with branching rate 0, i.e., deterministic heat flow (see the remark at the end of Section 2 in Presutti and Spohn (1983) or Theorem 2 in Cox and Durrett (1995)). To get a random limit, we note that the density of 0's around a 1 is $O(1/\log N)$, so we can counteract the slow down in the rate of change by increasing our branching rate by a factor of $\log N$, or what is the same, by decreasing the number of initial particles by the same factor of $\log N$. The rescaled local densities of vacant sites, $(\log N)v_N(t, x)$, will again be random with a nontrivial distribution, so to prove the result we have to establish that the mean field simplification occurs.

Finally, we consider the case $d = 1$ which is not covered by Theorem 15.3. If we restrict our attention to the nearest neighbor case, take $M_N = 1$, $N' = N^{1/2}$, and then use the reasoning that led to Theorem 2 of Cox and Griffeath (1986), we see that X_t^N defined above converges to a measure valued process where the density at any positive time is 1 or 0 on alternating intervals of random length, with the endpoints of these intervals undergoing annihilating Brownian motions.

In $d \geq 3$, Theorems 15.2 and 15.3 cover all the possibilities. In $d = 2$ the next result, which is Theorem 1.3 in Cox, Durrett, and Perkins (1998) covers the middle ground.

Theorem 15.4 (Two-Dimensional Medium Range Voter Model) *Suppose* (UD) *with* $M_N^2/\log N \to \rho \in [0, \infty)$, *and let* $N' = NM_N^2/\log N$. *If* $X_0^N \to X_0$ *in* $M_F(\mathbf{R}^2)$, *then* $P(X_t^N \in \cdot) \Rightarrow P_{X_0}^{2\gamma,1/3,0}$, *where* $\gamma = 1/[\rho + (3/2\pi)]$.

Finally consider (UD) in $d = 1$ with $M_N^2/N \to 1$. In this case $S_N = \mathbf{Z}/M_N\sqrt{N} \sim \mathbf{Z}/N$, so the densities of 0's and 1's are comparable. The total effective branching rate at x is $2Nv_N(t, x)$ as always. Let $u_N(t, x) = 1 - v_N(t, x) \in [0, 1]$ for $x \in S_N$ and interpolate linearly to get a continuous function on the rescaled lattice. Our choice of M_N shows that for $x \in S_N$

$$u_N(t, x) = \frac{NX_t^N([x - \frac{1}{\sqrt{N}}, x + \frac{1}{\sqrt{N}}] - \{x\})}{2M_N} \sim \frac{X_t^N([x - \frac{1}{\sqrt{N}}, x + \frac{1}{\sqrt{N}}] - \{x\})}{2/\sqrt{N}}.$$

Assuming, for the moment, that X^N has a weak limit X, this limit will necessarily have a density u bounded by 1. Comparing the above branching rate, $N2(1 - u_N(t, x))$, with that in Theorem 15.1 and anticipating weak convergence of u_N to u, we expect that the effective branching rate for the limit at x and time t should be $2(1 - u(t, x))$. Recalling in (15.2) that γ is the constant branching rate, we deduce that u should solve

$$\frac{\partial u}{\partial t} = \frac{1}{6}\frac{\partial^2 u}{\partial x^2} + \sqrt{2(1-u)u}\,\dot{W}. \tag{15.6}$$

This result was proved by Mueller and Tribe (1995) whose result we now state. Let C_b^+ denote the space of bounded continuous non-negative functions $\mathbf{R}$ with the topology of uniform convergence on compact sets.

Theorem 15.5 (One-Dimensional Medium Range Voter Model) *Assume* (UD) *in $d = 1$ with $M_N^2/N \to 1$ and let $N' = N$. If $u_N(0,\cdot) \to u(0,\cdot)$ in C_b^+ then $X_t^N \Rightarrow X_t$ in Ω_D, where $X_t(dx) = u(t,x)\,dx$ and the jointly continuous density u satisfies* (15.6).

Mueller and Tribe actually prove a stronger result giving the anticipated weak convergence of the approximate densities $u_N(t,x)$ to the solution of (15.6) in an appropriate space of continuous functions with sub-exponential growth in x. Uniqueness in law of solutions to (15.6) holds by a duality argument (see Shiga (1988)). This spde originally arose in populations genetics to model the frequency of a genotype in a continuous stepping stone model. See, e.g., Shiga (1980ab), (1987), Shiga and Uchiyama (1986), or Cox and Greven (1994).

Example 15.3 (Contact Process) In this case, $\xi_t(x) = 1$ indicates an occupied site and $\xi_t(x) = 0$ a vacant one. The contact process is described by giving a symmetric displacement distribution $p : S \to [0,1]$, a birth rate B and a death rate D. For each site x in S, with rate D a particle at x dies, and with rate B a particle at x produces an offspring which is sent to the neighbouring site y with probability $p(y-x)$. If y is vacant it becomes occupied. If y is occupied nothing happens. The mechanisms at distinct sites are assumed to operate independently.

Our goal in this section is to define a measure valued process using the recipe in (15.1) and investigate the convergence as $N \to \infty$. To do this we consider a rescaled long range contact process with uniform displacement (UD), birth rate $B_N = N + \theta$, and death rate $D_N = N$. The only difference between ξ_t^N and the corresponding branching random walk with the same birth and death rates, which we now call $\xi_t^{0,N}$, is that offspring born onto occupied sites are immediately killed for ξ_t^N. Thus, if $\xi_0^N \le \xi_0^{N,0}$, we will have $\xi_t^N \le \xi_t^{0,N}$ for all $t \ge 0$. ($\xi \le \xi'$ means $\xi(x) \le \xi'(x)$ for all x.) This interference reduces the birth rate from x at time t so that the effective drift parameter at $x \in S_N$ at time t is now the random quantity

$$\theta(t,x) = \theta - (N+\theta)\cdot u_N(t,x), \tag{15.7}$$

where $u_N(t,x) = 1 - v_N(t,x)$ is the local density of occupied sites in the lattice. It is clear from (15.7) that we must choose M_N so that this local density is $O(1/N)$.

Well known results for branching random walk imply that the mean number of particles in the neighborhood of a tagged occupied site (i.e., within distance $1/\sqrt{N}$ on the rescaled lattice $Z^d/M_N\sqrt{N})$ is

$$\approx \begin{cases} C\sqrt{N} & \text{in } d = 1, \\ C\log N & \text{in } d = 2, \\ C & \text{in } d \ge 3. \end{cases} \tag{15.8}$$

Here $\approx$ means the same as it did in (15.5), because what is being estimated is the same. To see (15.8), condition on the number of generations back, n, the neighborhood particle branched off from the family tree of the tagged particle and sum over n. From (15.8) it follows that we have to take the number of neighbors

$$M_N^d = \begin{cases} N^{3/2} & \text{in } d = 1, \\ N \log N & \text{in } d = 2, \\ N & \text{in } d \geq 3. \end{cases} \tag{15.9}$$

We first discuss the one-dimensional case. Recalling that super-Brownian motion has a density, we see that the bound $\xi^N \leq \xi^{0,N}$ and Theorem 15.1 show that any weak limit point of $\{X^N\}$ must also have a density. Note that the above choice of M_N implies that for $x \in S_N$

$$Nu_N(t,x) = \frac{X_t^N([x - \frac{1}{\sqrt{N}}, x + \frac{1}{\sqrt{N}}] - \{x\})}{N^{-2}2M_N} = \frac{X_t^N([x - \frac{1}{\sqrt{N}}, x + \frac{1}{\sqrt{N}}] - \{x\})}{2N^{-1/2}},$$

which, if there is any justice, should converge weakly to the density, $u(t,x)$, of the limiting measure-valued process X_t. Recalling (15.2), the fact that our limiting branching rate should now be 2 (from Theorem 15.1 with $b = 1$), and our effective drift at x at time t, $\theta(t,x)$ from (15.7), we expect that our limiting density u should solve the stochastic pde

$$\frac{\partial u}{\partial t} = \frac{1}{6}\frac{\partial^2 u}{\partial x^2} + (\theta - u)u + \sqrt{2u}\dot{W}. \tag{15.10}$$

Convergence to this limiting equation was conjectured by one of us (D) and proved by Mueller and Tribe (1995). Here is their result.

Theorem 15.6 (One-Dimensional Medium Range Contact Process) *Assume $d = 1$, $M_N = N^{3/2}$, and let $N' = N$. If $u_N(0,\cdot) \to u(0,\cdot)$ in C_b^+ as $N \to \infty$, then $X^N \to X$ in Ω_D, where $X_t(dx) = u(t,x)dx$ and u is the unique (in law) jointly continuous solution of* (15.10).

To bring out the analogy with the voter model results, we call this the medium range limit. In the "long range case" $M_N/N^{3/2} \to \infty$, the limit will be super-Brownian motion, i.e., the middle term in (15.10) below will be just θu. Uniqueness in law of the solution to (15.10) follows from Dawson's Girsanov theorem, described earlier (Dawson (1993) and Evans and Perkins (1994)). As for the voter model, Mueller and Tribe (1995) actually prove a stronger result than given in Theorem 15.5, giving weak convergence of the approximate densities u_N, extended linearly to **R**.

To describe the higher-dimensional cases, let $S_n = x_1 + \cdots + x_n$ be a random walk in $\mathbf{R}^d$ with steps $\{x_i\}$ which are i.i.d. uniform random variables on $[-1,1]^d$. Let

$$\kappa_d = \begin{cases} 3/2\pi & \text{in } d = 2, \\ \displaystyle\sum_{n=1}^{\infty} 2^{-d} P(S_n \in [-1,1]^d) & \text{in } d \geq 3. \end{cases} \tag{15.11}$$

The next result is proved in Durrett and Perkins (1998).

Theorem 15.7 (Higher-Dimensional Medium Range Contact Process) *Assume $d \geq 2$, M_N is as in (15.8), and the initial measures satisfy $X_0^N \to X_0$ in $M_F(\mathbf{R}^d)$, where X_0 is an atomless measure. Then $P(X^N \in \cdot) \Rightarrow P_{X_0}^{2,1/3,\theta-\kappa_d}$ in Ω_D.*

Comparing this result with the effective drift $\theta(t,x)$ given by (15.7), we see that the result reduces to showing that

$$Nu_N(t,x) \approx \kappa_d \quad \text{when } \xi_t(x) = 1 \tag{15.12}$$

in an appropriate sense. As in the short and medium range voter model, $Nu_N(t,x)$ will be random and have a nontrivial distribution, so we again have to show that a mean field simplification occurs. For $d \geq 3$, the transience of the continuous time random walk means that only particles which have branched off from the family tree of the particle at (t,x) in the last time interval of length $O(1/N)$ will contribute to the local density $u_N(t,x)$. This means that the number of occupied neighbors $Nu_N(t,x)$ and $Nu_N(t,x')$ at sites x and x' separated by a large multiple of $1/\sqrt{N}$ (on the rescaled lattice $\mathbf{Z}^d/M_N\sqrt{N}$) will be almost independent.

In $d = 2$ things are more delicate, but using the calculus fact

$$\int_1^{Nt} \frac{ds}{s} = \log(Nt),$$

we see that it is enough to include only those particles which have branched off at times $t \leq 1/\log N$. This means that the number of occupied neighbors $Nu_N(t,x)$ and $Nu_N(t,x')$ at sites x and x' separated by a large multiple of $1/\sqrt{\log N}$ (on the rescaled lattice $\mathbf{Z}^d/M_N\sqrt{N}$) will be almost independent. In contrast, for $d = 1$, the strong recurrence of random walks means that $Nu^N(t,x)$ will receive significant contributions even from unrelated particles. This results in the density being continuous on the rescaled lattice and gives rise to the $-u^2$ term in (15.10).

This explicit calculation of the limiting drift in Theorem 15.7 allows us to obtain some new information about the underlying contact processes. Return now to the simple contact processes of Example 15.3 with a uniform dispersal distribution (UD). Since we are looking at one fixed process we will set $N = 1$, i.e., use the rescaled lattice $\mathbf{Z}^d/M$, and let the displacement distribution $p = p_M$ be the uniform distribution on $\mathbf{Z}^d/M \cap ([-1,1]^d \setminus \{0\})$. To follow the traditional notation of the contact process, we will set the death rate $D = 1$ and the birth rate $B = \beta$.

Harris (1974) showed there is a critical birth rate $1 < \beta_c(M) < \infty$ such that if $\xi_0 = \delta_0$, then for $\beta > \beta_c(M)$, we have $P(\xi_t \neq 0 \text{ for all } t > 0) > 0$, while for $\beta < \beta_c(M)$ this probability is 0. $\beta_c(M) - 1$ represents the additional birth rate needed to compensate for the deaths of those offspring that land on occupied sites. This latter effect should become negligible as M gets large, and so $\beta_c(M)$ should approach 1 as $M \to \infty$. Bramson, Durrett and Swindle (1989) showed that

$$\beta_c(M) - 1 \approx \begin{cases} C/M^{2/3} & \text{in } d = 1, \\ C \log M/M^2 & \text{in } d = 2, \\ C/M^d & \text{in } d \geq 3, \end{cases} \tag{15.13}$$

where $\approx$ means the right-hand side is a lower, respectively, upper, bound for the left-hand side if M is large enough and C is sufficiently small positive, respectively, large. Theorem 15.7 was used in Durrett and Perkins (1998) to refine this result as follows:

Theorem 15.8 *If $d \geq 2$, then*

$$\beta_c(M) - 1 \sim \begin{cases} \kappa_2 \log(M^2)/M^2 & \text{in } d = 2, \\ \kappa_d/M^d & \text{in } d \geq 3, \end{cases} \tag{15.14}$$

where $\sim$ means the ratio approaches one as $M \to \infty$.

To see how this follows from Theorem 15.7, note first that $\theta = \kappa_d$ is critical for survival of the limiting super-Brownian motion in that

$$\lim_{t\to\infty} P^{2,1/3,\theta-\kappa_d}_{X_0}(X_t \neq 0) > 0 \quad \text{if and only if } \theta > \kappa_d.$$

If we ignore a few details then we can derive Theorem 15.8 from Theorem 15.7 by interchanging limits to conclude:

$$\lim_{N\to\infty} \lim_{t\to\infty} P(X_t^N \neq 0) = \lim_{t\to\infty} \lim_{N\to\infty} P(X_t^N \neq 0).$$

To justify the last step, we use the block construction methodology as explained in Durrett (1992a, 1995): we find an event in a finite space time box for super-Brownian motion that guarantees a positive probability of survival, and then use our weak convergence result to prove the existence of a suitable block event for the long range contact process.

In $d = 1$, Mueller and Tribe (1994) have shown that the limiting spde in Theorem 15.6 has a critical value, θ_c, below which there is a.s. extinction, and above which there is long term survival with positive probability. In view of this, it is natural to formulate

Conjecture 15.1 *In $d = 1$ as $M \to \infty$, $\beta_c(M) - 1 \sim \theta_c/M^{2/3}$.*

To prove this seems difficult. Our proof of Theorem 15.8 makes crucial use of the fact that if $\theta > \theta_c$, we can identify a suitable block event that guarantees survival for the limiting process. Bezuidenhout and Grimmett (1990) have shown that block events exist for supercritical contact processes, but extending their result to the limiting spde seems difficult.

Example 15.4 (Multitype Contact Process) We now generalize the set-up of the previous example so that $\xi_t(x) \in \{0, 1, 2\}$, where 1 and 2 denote two types of particles such that type i particles die at rate $D_i = N$ and give birth at rate $B_i = N + \theta_i$. Births occur uniformly on $[-M, M]^d \setminus \{0\}$, subject to the exclusion rule that births can only occur onto empty sites. Based on work of Mueller and Tribe, it is natural to expect that in one dimension we will get convergence to a system of spde's

$$
\begin{aligned}
\frac{\partial u_1}{\partial t} &= \frac{1}{6}\frac{\partial^2 u_1}{\partial x^2} + (\theta_1 - u_1 - u_2)u_1 + \sqrt{2u_1}\,\dot{W}_1 \\
\frac{\partial u_2}{\partial t} &= \frac{1}{6}\frac{\partial^2 u_2}{\partial x^2} + (\theta_2 - u_1 - u_2)u_2 + \sqrt{2u_2}\,\dot{W}_2
\end{aligned}
\tag{15.15}
$$

where $\dot{W}_1$ and $\dot{W}_2$ are independent.

The two components of the spde have a nontrivial interaction through their drift terms. This disappears in dimensions $d \geq 2$. Calculations in Durrett and Perkins (1998) show that collisions between unrelated individuals can be ignored in a single type contact process. (See Lemma 5.1.) From this observation it follows that (at least in the special case $\theta_1 = \theta_2$) the limit for the two type process will be a pair of INDEPENDENT super-Brownian motions.

To get a nontrivial limit then we must increase the interaction between the species. Following the rules in the colicin system discussed by Durrett and Levin (1996) where type 1 releases a chemical toxic for 2's into the environment, we change the death rate for type 2 from a constant to $1+\gamma f_1$, where f_1 is the fraction of neighbors in state 1. If γ remains bounded then again nothing happens, so we must let γ tend to ∞ with N.

To describe our conjectured limit, we recall the measure valued diffusions studied in Section 4 of Evans and Perkins (1994). (X^1, X^2) are a pair of interacting super-Brownian motions for which collisions between types are bad for the health of members of the X^2 population. They satisfy the following martingale problem (recall the description of collision local time given earlier) with $\lambda_1 = 0$ and λ_2 a positive constant

For all $\phi_i \in C_0^\infty(\mathbf{R}^d)$, $M_0^i(\phi_i) = 0$,

$$
M_t^i(\phi_i) = X_t^i(\phi_i) - X_0^i(\phi_i) + \lambda_i L(X^1, X^2)_t(\phi_i) - \int_0^t X_s^i(\sigma^2 \Delta\phi_i/2 + \theta_i\phi_i)\,ds
$$

is a continuous $\mathcal{F}_t$*-martingale, and* $\langle M^i(\phi_i), M^j(\phi_j)\rangle_t = \delta_{i,j}\int_0^t X_s^i(\phi_i^2)ds$.

Evans and Perkins (1994) study this martingale problem with $\theta_i = 0$ but the basic existence, non-existence and uniqueness results extend easily to the above by a version of Dawson's Girsanov theorem—the processes with drifts θ_i are absolutely continuous with respect to the processes with no such drifts. Solutions do not exist in $d \geq 4$. Assuming some mild regularity conditions on the initial measures, solutions exist in $d \leq 3$ and their law is unique.

Our preliminary calculations suggest that the colicin system when rescaled converges to the interacting measure valued diffusions described above. Specifically, we believe

Conjecture 15.2 *If* $\gamma = \log N$ *in* $d = 2$ *and* $\gamma = N^{1/2}$ *in* $d = 3$*, then we get convergence to the unique solution of the above martingale problem for an appropriate choice of* λ.

To explain the guesses for γ, and why there is no nontrivial limit in dimensions $d \geq 4$ (as is suggested by the non-existence of solutions to the above martingale problem), consider the case $d = 3$. In N generations population 1 will spread over an area $N^{3/2}$ (on $\mathbf{Z}^d/M_N$) and have about N members. Assuming the local density per unit area (again on $\mathbf{Z}^d/M_N$) is bounded this means that a fraction $N^{-1/2}$ of the unit cubes on $\mathbf{Z}^d/M_N$ have some mass. An independent copy of the population will then have $N \cdot N^{-1/2}$ occupied unit cubes at any one time that feel the presence of the 1's. Interchanging space and time we conclude that each individual feels the presence of the other type about $N^{-1/2}$ of the time, so taking $\gamma = N^{1/2}$ will bring this effect to a positive level.

In dimensions $d \geq 4$ the fraction of boxes that contain mass is $N/N^{d/2}$ so the amount of contact is $N^2/N^{d/2}$, which stays bounded in $d = 4$ and tends to 0 for $d > 4$. In these dimensions then one cannot push hard enough on the intersection to have an effect. In $d = 2$ the number of occupied boxes is $N/\log N$, with a typical box having $\log N$ members. This may sound like γ bounded, but one must remember that in $d = 2$ the local neighborhood size is $M^2 = N \log N$, introducing an extra factor of $\log N$ in the denominator of the observed density of the other type.

Of course one could modify the above particle system so that each species has a toxic effect on the other. One expects an analogous limit theorem to be true but with the competing species model studied in Evans and Perkins (1998) and Mytnik (1998) as the limit. Here the processes satisfy the same martingale problem as above but now there collision local time drifts in both types. This seems to make the analysis of the limit much more complicated.

Acknowledgments: The research of the first author was supported in part by NSF Grant DMS-96-26675 and by an NSERC Collaborative Grant. The research of the second author was supported in part by NSF Grant DMS-96-26201 and by an NSERC Collaborative Grant. The research of the third author was supported in part by an NSERC Research grant and an NSERC Collaborative Grant.

REFERENCES

Adler, R. and Tribe, R. (1997) Uniqueness for a historical SDE with a singular drift. *Preprint.*

Barlow, M.T., Evans, S.N., and Perkins, E. (1991) Collision local times and measure-valued processes. *Ann. Prob.* **43**, 897–938.

Bramson, M., Durrett, R., and Swindle, G. (1989) Statistical mechanics of crab grass. *Ann. Prob.* **17**, 444–481.

Cox, J.T., and Durrett, R. (1995) Hybrid zones and voter model interfaces. *Bernoulli* **1**, 343–370.

Cox, J.T., Durrett, R., and Perkins, E. (1998) Rescaled voter models converge to super-Brownian motion. Preprint.

Cox, J.T. and Greven, A. (1994) Ergodic theorems for infinite systems of locally interacting diffusions. *Ann. Prob.* **22**, 833–853.

Cox, J.T. and Griffeath, D. (1986) Diffusive clustering in the two dimensional voter model. *Ann. Prob.* **14**, 347–370.

Dawson, D.A. (1978) Geostochastic calculus. *Can. J. Statistics* **6**, 143–168.

Dawson, D.A. (1993) Measure-valued Markov processes. Pages 1–260 in *Ecole d'été de Probabilités de St. Flour, XXI.* Lecture Notes in Mathematics 1541, Springer-Verlag, New York.

Dawson and Perkins (1998) Long-time behaviour and co-existence in a mutually catalytic branching model. *Ann. Prob.* **26**, 1088–1138.

Dawson and Perkins (1999) Measure-valued processes and renormalization of branching particle systems. Pages 45–106 in *Stochastic partial differential equations: Six perspectives.* Edited by R. Carmona and B. Rozovsky. AMS Math Surveys and Monographs **64**.

Derbez, E. and Slade, G. (1998) The scaling limit of lattice trees in high dimensions. *Comm. Math. Phys.* **193**, 69–104.

Donnelly and Kurtz (1998) Particle representations for measure-valued populations models. *Ann. Prob.*, to appear.

Durrett, R. (1988) *Lecture Notes on Particle Systems and Percolation.* Wadsworth Pub. Co., Belmont, CA.

Durrett, R. (1992a) A new method for proving the existence of phase transitions. Pages 141–170 in *Spatial Stochastic Processes.* Edited by K.S. Alexander and J.C. Watkins. Birkhäuser, Boston.

Durrett, R. (1992b) The contact process: 1974-89. Pages 1–18 in *Mathematics of Random Media* Edited by W. Kohler and B. White. American Math. Society, Providence, RI.

Durrett, R. (1995) Ten lectures on particle systems. Pages 97–201 in *Ecole d'été de Probabilités de St. Flour, XXIII.* Lecture Notes in Mathematics 1608, Springer-Verlag, New York.

Durrett, R. and Levin, S.A. (1997) Allelopathy in spatially distributed populations. *J. Theor. Biol.*, **185**, 165–172.

Durrett, R. and Perkins, E. (1998) Rescaled contact processes converge to super-Brownian motion in two or more dimensions. *Prob. Th. Rel. Fields.*, to appear.

Ethier, S. and Kurtz, T. (1986) *Markov Processes: Characterization and Convergence.* John Wiley, New York.

Evans, S.N. and Perkins, E. (1994) Measure-valued branching diffusions with singular interactions. *Can. J. Math.* **46**, 120–168.

Evans, S.N. and Perkins, E. (1998) Collision local times, historical stochastic calculus and competing superprocesses. *Elect. J. Prob.* **3**.

Griffeath, D. (1978) *Additive and Cancellative Interacting Particle Systems.* Lecture Notes in Mathematics 724. Springer-Verlag, New York.

Harris, T. (1963) *Branching Processes.* Springer-Verlag, New York.

Jacod, J. and Shiryaev, A.N. (1987) *Limit Theorems for Stochastic Processes.* Springer-Verlag, Berlin.

Konno, N. and Shiga, T. (1988) Stochastic differential equations for some measure-valued diffusions. *Prob. Theory Rel. Fields* **79**, 201–225.

Kurtz, T. (1998) Martingale problems for conditional distributions of Markov processes. *Elect. J. Prob.* **3**.

LeGall, J.-F. (1995) Mouvement brownien, processus du branchement et superprocesses. Unpublished course notes.

Liggett, T. (1985) *Interacting Particle Systems.* Springer-Verlag, New York.

Mueller, C. and Tribe, R. (1994) A phase transition for a stochastic PDE related to the contact process. *Prob. Theory Rel. Fields* **100**, 131–156.

Mueller, C. and Tribe, R. (1995) Stochastic PDE's arising from the long range contact process and long range voter model. *Prob. Theory Rel. Fields* **102**, 519–546.

Mytnik, L. (1998) Weak uniqueness for the heat equation with noise. *Ann. Prob.* **26**, 968–984.

Mytnik, L. (1998) Uniqueness for a competing species model. *Can. J. Math.*, to appear.

Perkins, E. (1989) The Hausdorff measure of the closed support of super-Brownian motion. *Ann. Inst. Henri Poincaré Prob. et Stat.* **25**, 205–224.

Perkins, E. (1992) Measure-valued branching diffusions with spatial interactions. *Prob. Th. Rel. Fields* **94**, 189–245.

Perkins, E. (1995) On the martingale problem for interactive measure-valued branching diffusions. *Memoirs of the American Math. Soc.* **115** no. 549.

Presutti, E. and Spohn, H. (1983) Hydrodynamics for the voter model. *Ann. Prob.* **4**, 867–875.

Reimers, M. (1989) One dimensional stochastic partial differential equations and the branching measure diffusion. *Prob. Theory Rel. Fields* **81**, 319–340.

Shiga, T. (1980a) An interacting system in population genetics. *J. Math. Kyoto U.* **20**, 213–242.

Shiga, T. (1980b) An interacting system in population genetics, II. *J. Math. Kyoto U.* **20**, 723–732.

Shiga, T. (1987) A certain class of infinite-dimensional diffusion processes arising in population genetics. *J. Math. Soc. Japan.* **39**, 17–25.

Shiga, T. (1988) Stepping stone models in population genetics and population dynamics. In *Stochastic Proceses in Physics and Engineering.* Edited by S. Albeverio et al.

Shiga, T. and Uchiyama, K. (1986) Stationary states and their stability of the stepping stone model involving mutation and selection. *Prob. Th. Rel. Fields.* **73**, 87–117.

Slade, G. (1999) Lattice trees, percolation and super-Brownian motion. In this volume.

Walsh, J. (1986) An introduction to stochastic partial differential equations. Pages 265–439 in *Ecole d'été de Probabilités de St. Flour, XIV* Lecture Notes in Mathematics 1180, Springer-Verlag, New York.

Ted Cox
Department of Mathematics
Syracuse University
Syracuse, NY 13244

Richard Durrett
Department of Mathematics
Cornell University
Ithaca, NY 14853

Edwin A. Perkins
Department of Mathematics
The University of British Columbia
Vancouver, BC V6T 1Z2
Canada

16

The Hausdorff Measure of the Range of Super-Brownian Motion

Jean-François Le Gall

ABSTRACT In dimension $d \geq 4$, the total occupation measure of super-Brownian motion coincides with the restriction of a Hausdorff measure to the range of the process. A similar result holds for the random measure called ISE.

Keywords: Super-Brownian motion, range, Brownian snake, Hausdorff measure, occupation measure.

AMS Subject Classifications: Primary 60G57; Secondary 60J80.

16.1 Introduction

The main goal of this work is to give an exact Hausdorff measure function for the range of super-Brownian motion, or equivalently of the Brownian snake, in dimension $d \geq 4$. More precisely, we prove that the total occupation measure of super-Brownian motion coincides with the restriction to the range of a certain Hausdorff measure. A slightly less precise result has been obtained by Dawson, Iscoe and Perkins [3] in dimension $d \geq 5$. As an easy corollary of our main result, we get that the random measure known as ISE (Integrated Super-Brownian Excursion measure) is the restriction to its closed support of a Hausdorff measure. A motivation for the last result comes from the recent work of Derbez and Slade [6] and Hara and Slade [8] showing that ISE arises in limit theorems for models of statistical mechanics.

Super-Brownian motion is the continuous Markov process $X = (X_t, t \geq 0; \mathbf{P}_\mu, \mu \in M_f(\mathbb{R}^d))$ with values in the space $M_f(\mathbb{R}^d)$ of all finite measures on $\mathbb{R}^d$, whose transition kernels can be characterized as follows. Let Q_t denote the transition semigroup of d-dimensional Brownian motion. Then, for any bounded continuous nonnegative function φ on $\mathbb{R}^d$,

$$\mathbf{E}_\mu[\exp -\langle X_t, \varphi\rangle] = \exp -\langle \mu, V_t\varphi\rangle,$$

where the function $(V_t\varphi(x), t \geq 0, x \in \mathbb{R}^d)$ is the unique nonnegative solution of the integral equation

$$V_t\varphi(x) + 2\gamma \int_0^t Q_{t-s}((V_s\varphi)^2)(x)\, ds = Q_t\varphi(x).$$

Here $\gamma > 0$ is a positive constant determining the branching rate, and the multiplicative factor 2 is for convenience.

The range of X is then the subset of $\mathbb{R}^d$ defined by

$$\mathbf{R} = \bigcup_{\varepsilon>0} \Big(\overline{\bigcup_{t\geq\varepsilon} \operatorname{supp} X_t} \Big),$$

where supp X_t stands for the closed support of X_t, and $\bar{A}$ denotes the closure of A. The total occupation measure of X is the (a.s. finite) random measure $\mathbf{Z}$ supported on $\mathbf{R}$ which is defined by

$$\mathbf{Z} = \int_0^\infty dt\, X_t.$$

The range $\mathbf{R}$ has positive Lebesgue measure (and $\mathbf{Z}$ is absolutely continuous with respect to Lebesgue measure) if and only if $d \leq 3$ (see [3] and [16]). From now on, we assume that $d \geq 4$.

If h is a monotone increasing function from $[0, \infty)$ into $[0, \infty)$, we denote by $h-m$ the Hausdorff measure associated with h. We let $\mathcal{B}(\mathbb{R}^d)$ be the Borel σ-field of $\mathbb{R}^d$.

Theorem 16.1.1 *Set*

$$h(r) = \begin{cases} r^4 \log \frac{1}{r} \log\log\log \frac{1}{r} & \text{if } d = 4, \\ r^4 \log\log \frac{1}{r} & \text{if } d \geq 5. \end{cases}$$

There exists a positive constant K_d depending only on d, such that a.s. for every $A \in \mathcal{B}(\mathbb{R}^d)$,

$$h - m(A \cap \mathbf{R}) = K_d\, \gamma^{-1}\, \mathbf{Z}(A).$$

In dimension $d \geq 5$, Dawson, Iscoe and Perkins [3, Theorem 1.4] proved that the ratio $h-m(A\cap\mathbf{R})/\mathbf{Z}(A)$ is bounded above and below by positive constants, which is only slightly weaker than the previous statement. Therefore, the main novelty of Theorem 16.1.1 is the case $d = 4$, where the form of the function h had also been conjectured in [3]. We present below a complete derivation including the case $d \geq 5$. Our proof in this case is different from the one in [3] and maybe simpler. We believe that our approach to the case $d \geq 5$ is also useful for pedagogical reasons, to understand the critical dimension $d = 4$, which is significantly more difficult.

Our method of proof relies on a systematic use of the path-valued process called the Brownian snake [9, 10] which is closely related to super-Brownian motion. The strong Markov property of the Brownian snake provides a powerful tool that is not available in traditional approaches to super-Brownian motion. This idea was exploited in [13] to give an exact Hausdorff measure function for the support of X_t in the critical dimension $d = 2$. Serlet [15] used an approach very similar to the present work to give precise information on the Hausdorff measure of multiple points of super-Brownian motion.

Let us come to our result concerning ISE. Informally, ISE is the random measure $\mathbf{Z}^*$ whose law is the distribution of $\mathbf{Z}$ under the excursion measure of super-Brownian motion from 0, conditionally on $\mathbf{Z}(\mathbb{R}^d) = 1$. A precise definition in terms of the Brownian snake is given in Section 16.2 below. We denote by $\mathbf{R}^*$ the closed support of $\mathbf{Z}^*$.

Corollary 16.1.1 *The conclusion of Theorem* 16.1.1 *still holds if* $\mathbf{Z}$ *and* $\mathbf{R}$ *are replaced by* $\mathbf{Z}^*$ *and* $\mathbf{R}^*$ *respectively, with the same constant* K_d.

The fact that Corollary 16.1.1 can be deduced from Theorem 16.1.1 was already pointed out in [5, Theorem 4.9].

The paper is organized as follows. In Section 16.2, we recall the definition and some basic properties of the Brownian snake, which play an important role in our proofs. Section 16.3 gives precise estimates on the total occupation measure of small balls. In Section 16.4, we obtain lower and upper bounds for the Hausdorff measure of the range in dimension $d \geq 5$. The corresponding results in the critical dimension $d = 4$ are derived in Section 16.5, which contains the most difficult technical part of the paper. Finally, the main results are proved in Section 16.6.

16.2 The Brownian Snake

In this section, we briefly recall the definition and a few properties of the Brownian snake, which will be used in the subsequent proofs.

We denote by $\mathcal{W}$ the set of all stopped paths in $\mathbb{R}^d$. An element w of $\mathcal{W}$ is a continuous mapping $\mathrm{w} : [0, \zeta] \to \mathbb{R}^d$, where $\zeta = \zeta_{\mathrm{w}}$, the lifetime of w, can be any nonnegative real number. We write $\hat{\mathrm{w}} = \mathrm{w}(\zeta)$ for the endpoint of w. The distance on $\mathcal{W}$ is defined by $d(\mathrm{w}, \mathrm{w}') = \sup_{t\geq 0} |\mathrm{w}(t \wedge \zeta) - \mathrm{w}'(t \wedge \zeta')| + |\zeta - \zeta'|$.

Let us fix a point $x \in \mathbb{R}^d$ and write $\mathcal{W}_x$ for the subset of $\mathcal{W}$ consisting of those paths w for which $\mathrm{w}(0) = x$. The trivial path in $\mathcal{W}_x$ with $\zeta_{\mathrm{w}} = 0$ is identified with the point x of $\mathbb{R}^d$.

The Brownian snake with initial point x is the continuous strong Markov process $(W_s, s \geq 0)$ in $\mathcal{W}_x$ whose distribution is characterized by the following properties:

(a) The "lifetime process" $\zeta_s = \zeta_{W_s}$ is a reflecting Brownian motion in $\mathbb{R}_+$.
(b) Conditionally on $(\zeta_s, s \geq 0)$, the distribution of $(W_s, s \geq 0)$ is that of an inhomogeneous Markov process whose transition kernels are described as follows: For every $s < s'$,

- $W_{s'}(t) = W_s(t)$ for every $t \leq m(s, s') := \inf_{[s,s']} \zeta_r$.
- $(W_{s'}(m(s, s')+t) - W_{s'}(m(s, s')), 0 \leq t \leq \zeta_{s'} - m(s, s'))$ is a Brownian motion in $\mathbb{R}^d$ independent of W_s.

Informally, W_s should be seen as a Brownian path in $\mathbb{R}^d$ with random lifetime ζ_s evolving like reflecting Brownian motion. When ζ_s decreases, the path erases itself. When ζ_s increases the path is extended by adding "little pieces" of Brownian motion at its tip.

We may and will assume that the process W is defined on the canonical space $C(\mathbb{R}_+, \mathcal{W})$ of all continuous mappings from $\mathbb{R}_+$ into $\mathcal{W}$. We then denote by $\mathbb{P}_{\mathrm{w}}$ the law of W started at w. We also denote by $\mathbb{P}^*_{\mathrm{w}}$ the law under $\mathbb{P}_{\mathrm{w}}$ of $(W_{s\wedge\sigma}, s \geq 0)$, where $\sigma = \inf\{s > 0, \zeta_s = 0\}$.

It is obvious that x is a regular recurrent point for W. The associated excursion measure is denoted by $\mathbb{N}_x$. The law of $(\zeta_s, s \geq 0)$ under $\mathbb{N}_x$ is the Itô measure of positive excursions of linear Brownian motion, and σ represents the length or duration of the excursion. Note that $\mathbb{N}_x$ is an infinite measure on $C(\mathbb{R}_+, \mathcal{W})$ and that $W_s = x$ for every $s \geq \sigma$, $\mathbb{N}_x$ a.e. We normalize $\mathbb{N}_x$ so that $\mathbb{N}_x[\sup_{s\geq 0} \zeta_s > \varepsilon] = (2\varepsilon)^{-1}$ for every $\varepsilon > 0$. Note the useful scaling property of $\mathbb{N}_0$: If $\tilde{W}_s(t) = \lambda^{-1} W_{\lambda^4 s}(\lambda^2 t)$, for $0 \leq t \leq \tilde{\zeta}_s = \lambda^{-2}\zeta_{\lambda^4 s}$, the law of $(\tilde{W}_s, s \geq 0)$ under $\mathbb{N}_0$ is $\lambda^{-2}\mathbb{N}_0$.

The strong Markov property holds under $\mathbb{N}_x$ in the following form: If $(\mathcal{F}_t)$ is the canonical filtration on $C(\mathbb{R}_+, \mathcal{W})$ and T is a stopping time of the filtration $(\mathcal{F}_t)$ such that $T > 0$, $\mathbb{N}_x$ a.e., then the law of $(W_{T+s}, s \geq 0)$ under $\mathbb{N}_x$ conditionally on $\mathcal{F}_T$, is $\mathbb{P}^*_{W_T}$ (see [10] for a more precise statement).

The range of the Brownian snake (under the excursion measure $\mathbb{N}_x$ or under the probability measure $\mathbb{P}^*_{\mathrm{w}}$) is the compact subset of $\mathbb{R}^d$ defined by

$$\mathcal{R} = \{\hat{W}_s\, ; 0 \leq s \leq \sigma\}.$$

The occupation measure $\mathcal{Z}$ is the random measure supported on $\mathcal{R}$ defined by

$$\mathcal{Z}(A) = \int_0^\sigma ds\, 1_A(\hat{W}_s) \tag{16.2.1}$$

for every $A \in \mathcal{B}(\mathbb{R}^d)$. It will be important to have formulas for the moments of $\mathcal{Z}$ under $\mathbb{N}_x$. We denote by G the Green function of Brownian motion in $\mathbb{R}^d$:

$$G(y, z) = G(z - y) = \frac{\Gamma(\frac{d}{2} - 1)}{2\pi^{d/2}}\, |z - y|^{2-d}.$$

Then, for every $A \in \mathcal{B}(\mathbb{R}^d)$,

$$\mathbb{N}_x[\mathcal{Z}(A)] = \int_A dy\, G(x, y) \tag{16.2.2}$$

and, for every integer $p \geq 2$,

$$\mathbb{N}_x[\mathcal{Z}(A)^p] = 2\sum_{j=1}^{p-1} \binom{p}{j} \int_{\mathbb{R}^d} dz\, G(x, z)\, \mathbb{N}_z[\mathcal{Z}(A)^j]\, \mathbb{N}_z[\mathcal{Z}(A)^{p-j}]. \tag{16.2.3}$$

An easy way to derive these formulas is to use the relationship between the Brownian snake and super-Brownian motion [9] to see that for every $\lambda > 0$,

$$u_\lambda(x) := \mathbb{N}_x[1 - \exp -\lambda\mathcal{Z}(A)] = -\log \mathbf{E}_{\delta_x}[\exp -\lambda\mathbf{Z}(A)].$$

By a standard result of the theory of superprocesses, u_λ solves the integral equation

$$u_\lambda(x) + 2\int_{\mathbb{R}^d} dy\, G(x, y) u_\lambda^2(y) = \lambda \int_A dy\, G(x, y),$$

from which it is a simple matter to obtain (16.2.2) and (16.2.3).

For every $y \in \mathbb{R}^d$ and $r > 0$, denote by $B(y, r)$ the open ball of radius r centered at y. Fix $K > 0$. For technical reasons, it will be useful to consider the Brownian snake "stopped at the boundary of the ball $B(0, K)$." This means that the underlying spatial motion becomes Brownian motion stopped when it exits the ball $B(0, K)$. Equivalently, we may replace W_s by $W_s^K(t) := W_s(t \wedge \tau_K(W_s))$, where $\tau_K(W_s)$ is the (possibly infinite) first exit time of W_s from $B(0, K)$. We can then introduce the occupation measure $\mathcal{Z}^K$ of W^K (replace $\hat{W}_s$ by $\hat{W}_s^K = W_s(\zeta_s \wedge \tau_K(W_s))$ in (16.2.1)). Formulas (16.2.2) and (16.2.3) remain valid for $\mathcal{Z}^K$ provided A is contained in the ball $B(0, K)$ and we replace G by the Green function G_K of the ball $B(0, K)$.

In the next lemma, we collect some well known bounds on hitting probabilities of balls. Both assertions of this lemma are consequences of [3, Theorem 3.2] and its proof (see the bounds (3.3.2) and (3.3.3)).

Lemma 16.2.1 (i) *There exists a constant $c = c(d)$ such that for every $x \in \mathbb{R}^d$ with $|x| \geq 1$ and every $\varepsilon \in (0, 1/2)$,*

$$\mathbb{N}_x[\mathcal{R} \cap B(0,\varepsilon) \neq \emptyset] \leq \begin{cases} c\,\varepsilon^{4-d} & \text{if } d \geq 5, \\ c\left(\log \frac{1}{\varepsilon}\right)^{-1} & \text{if } d = 4. \end{cases}$$

(ii) *For every $\delta > 0$, set $\mathcal{R}_\delta = \{\hat{W}_s; 0 \leq s \leq \sigma, \zeta_s \geq \delta\}$. There exists a constant $c' = c'(d, \delta)$ such that for every $x \in \mathbb{R}^d$ and $\varepsilon \in (0, 1/2)$,*

$$\mathbb{N}_x[\mathcal{R}_\delta \cap B(0,\varepsilon) \neq \emptyset] \leq \begin{cases} c'\,\varepsilon^{4-d} & \text{if } d \geq 5, \\ c'\left(\log \frac{1}{\varepsilon}\right)^{-1} & \text{if } d = 4. \end{cases}$$

A trivial consequence of the lemma is the polarity of points in dimension $d \geq 4$: $\mathbb{N}_x[0 \in \mathcal{R}] = 0$ if $x \neq 0$.

Let us fix a path $\mathrm{w} \in \mathcal{W}_x$. Then, $\mathbb{P}^*_{\mathrm{w}}$ a.s., for every $s \in [0, \sigma)$ the path W_s coincides with $W_0 = \mathrm{w}$ up to time $m(0, s) > 0$. We want to describe more precisely the structure of the "tree of paths" $(W_s, 0 \leq s \leq \sigma)$. To this end, let $(\alpha_i, \beta_i)_{i \in I}$ be the excursion intervals away from 0 of the process $(\zeta_s - m(0, s), s \in [0, \sigma])$. For every $i \in I$, define $\omega^i \in C(\mathbb{R}_+, \mathcal{W})$ by the formulas: For $s \geq 0$,

$$\omega_s^i(t) = W_{(\alpha_i+s)\wedge\beta_i}(\zeta_{\alpha_i} + t), \quad 0 \leq t \leq \zeta_s^i := \zeta_{(\alpha_i+s)\wedge\beta_i} - \zeta_{\alpha_i}.$$

Theorem 16.2.1 ([10]) *Under $\mathbb{P}^*_{\mathrm{w}}$, the point measure*

$$\mathcal{N}(dtd\omega) := \sum_{i \in I} \delta_{\zeta_{\alpha_i}, \omega^i}(dtd\omega)$$

is a Poisson point measure with intensity $2 \cdot 1_{[0,\zeta_{\mathrm{w}}]}(t)\, dt\, \mathbb{N}_{\mathrm{w}(t)}(d\omega)$.

We will use Theorem 16.2.1 in connection with the following remark. Since the complement in $[0, \sigma]$ of the union of the intervals $[\alpha_i, \beta_i]$ has zero Lebesgue measure, we have $\mathbb{P}^*_{\mathrm{w}}$ a.s.,

$$\mathcal{Z} = \sum_{i \in I} \mathcal{Z}(\omega^i) = \int \mathcal{N}(dt d\omega)\, \mathcal{Z}(\omega).$$

In our study of the critical case $d = 4$ in Section 16.5, we shall need two other results concerning the Brownian snake. The first one is a Palm measure formula for the occupation measure $\mathcal{Z}$. For every $a \geq 0$, we denote by P^a_x the law of Brownian motion in $\mathbb{R}^d$ started at x and stopped at time a. This law can be viewed as a probability measure on the set $\mathcal{W}^a_x = \{\mathrm{w} \in \mathcal{W}_x;\ \zeta_{\mathrm{w}} = a\}$.

Proposition 16.2.1 *Let F be a nonnegative measurable function on $\mathbb{R}^d \times M_f(\mathbb{R}^d)$. Then,*

$$\mathbb{N}_x\Big[\int \mathcal{Z}(dy)\, F(y, \mathcal{Z})\Big]$$
$$= \int_0^\infty da \int P^a_x(d\mathrm{w})\, E^{(\mathrm{w})}\Big[F\Big(\mathrm{w}(a), \int \mathcal{N}(dt d\omega)\mathcal{Z}(\omega)\Big)\Big],$$

where for every $\mathrm{w} \in \mathcal{W}^a_x$, the probability measure $P^{(\mathrm{w})}$ is defined on an auxiliary probability space and such that, under $P^{(\mathrm{w})}$, $\mathcal{N}(dt d\omega)$ is a Poisson point measure with intensity $4 \cdot 1_{[0,\zeta_{\mathrm{w}}]}(t)\, dt\, \mathbb{N}_{\mathrm{w}(t)}(d\omega)$.

This proposition is a simple consequence of Theorem 16.2.1: See [10, Proposition 4.1] for a very similar proof. Alternatively, one can also use analytic arguments along the lines of the proof of [4, Theorem 4.1.1].

Finally, we will need a form of the special Markov property for the Brownian snake. We let D be a domain in $\mathbb{R}^d$ and $x \in D$. For every $\mathrm{w} \in \mathcal{W}_x$, we set $\tau(\mathrm{w}) = \inf\{t \in [0, \zeta_{\mathrm{w}}], \mathrm{w}(t) \notin D\}$ with the usual convention $\inf \emptyset = \infty$. Then, $\mathbb{N}_x$ a.e., the limit

$$\langle X^D, \varphi \rangle := \lim_{\varepsilon \to 0} \frac{1}{\varepsilon} \int_0^\sigma ds\, \varphi(W_s(\tau(W_s)))\, 1_{\{\tau(W_s) < \zeta_s < \tau(W_s) + \varepsilon\}}$$

exists for every bounded continuous function φ on ∂D, and defines a random measure X^D on ∂D called the exit measure from D (see [10] or [12]). Roughly speaking, the special Markov property states that, conditionally on the exit measure X^D, the behavior of the paths W_s after their exit time from D is independent of what happened before. To make this precise, we need some more notation.

For every $s \geq 0$, we set

$$\eta^D_s = \inf\{r \geq 0, \int_0^r du\, 1_{\{\zeta_u \leq \tau(W_u)\}} > s\}$$

and we denote by $\mathcal{E}^D$ the σ-field generated by the collection $(W_{\eta^D_s}, s \geq 0)$ (this represents the information "before the first exit time from D"). The random variable X^D is measurable with respect to $\mathcal{E}^D$.

Next, we introduce the excursions of the Brownian snake outside D. We let (a_j, b_j), $j \in J$ be the connected components of the open set $\{s \geq 0, \tau(W_s) < \zeta_s\}$. It is easy to see that, for every $j \in J$, the paths W_s, $s \in (a_j, b_j)$ coincide up to their first exit time from D, which is equal to $\zeta_{a_j} = \zeta_{b_j}$. We thus define $W^j \in C(\mathbb{R}_+, \mathcal{W})$ by the formulas

$$W_s^j(t) = W_{(a_j+s)\wedge b_j}(\zeta_{a_j} + t), \quad 0 \le t \le \zeta_s^j := \zeta_{(a_j+s)\wedge b_j} - \zeta_{a_j}.$$

Theorem 16.2.2 ([12]) *Conditionally on $\mathcal{E}^D$, the point measure $\sum_{j\in J} \delta_{W^j}$ is a Poisson point measure with intensity $\int X^D(dy)\,\mathbb{N}_y[\cdot]$.*

We conclude this section with the Brownian snake definition of ISE. Informally, the law of ISE is the distribution of $\mathcal{Z}$ under $\mathbb{N}_0[\cdot \mid \sigma = 1]$. To precisely justify the conditioning, we introduce the continuous process $(W_s^*, 0 \le s \le 1)$ with values in $\mathcal{W}_0$, such that

(a′) The lifetime process $\zeta_s^* = \zeta_{W_s^*}$ is a normalized Brownian excursion (positive Brownian excursion conditioned to have duration 1).
(b′) Exactly as (b) in the beginning of this section, replacing W and ζ by W^* and ζ^* respectively.

Then ISE is the random probability measure $\mathbf{Z}^*$ on $\mathbb{R}^d$ defined by

$$\mathbf{Z}^*(A) = \int_0^1 ds\, 1_A(\hat{W}_s^*).$$

The equivalence of this presentation of ISE with other definitions follows from Aldous' coding of the continuum random tree by a normalized Brownian excursion [1] (our definition differs by a scaling factor 2, but we will ignore this).

16.3 Estimates for the Occupation Measure of Balls

Our first goal is to obtain good estimates for the "probability" under $\mathbb{N}_x$ that the total occupation measure of a small ball $B(y, \varepsilon)$ takes unusually large values. By translation invariance, we may restrict our attention to a ball centered at the origin. We start by deriving precise upper and lower bounds for $\mathcal{Z}(B(0, \varepsilon))$.

Proposition 16.3.1 *Suppose that $d \ge 5$. There exist two positive constants C_1, C_2 depending only on d, such that, for every $p \ge 1$, $x \in \mathbb{R}^d$, $\varepsilon \in (0, 1]$,*

$$C_1^p\, p!(G(x)\wedge\varepsilon^{2-d})\varepsilon^{d+4(p-1)} \le \mathbb{N}_x[\mathcal{Z}(B(0,\varepsilon))^p] \le C_2^p\, p!(G(x)\wedge\varepsilon^{2-d})\varepsilon^{d+4(p-1)}.$$

PROOF. We first prove the upper bound. The case $p = 1$ follows easily from formula (16.2.2). We then argue by induction on p, using (16.2.3) and the following elementary lemma, whose easy proof is left to the reader.

Lemma 16.3.1 *There exist two positive constants C_3, C_4 such that, for every $x \in \mathbb{R}^d$, $\varepsilon \in (0, 1]$,*

$$C_3(G(x)\wedge\varepsilon^{2-d})\varepsilon^{4-d} \le \int dz\, G(z-x)\,(G(z)\wedge\varepsilon^{2-d})^2 \le C_4(G(x)\wedge\varepsilon^{2-d})\varepsilon^{4-d}.$$

We set $a_1 = 1$ and for every $j \geq 2$,

$$a_j = (2j-3) \times (2j-5) \times \cdots \times 3 \times 1.$$

From the equality $1 - \sqrt{1-x} = \sum_{j=1}^{\infty} (2^{-j} a_j / j!) x^j$ (for $|x| < 1$) one gets the combinatorial identity: For $p \geq 2$,

$$2\, a_p = \sum_{j=1}^{p-1} \binom{p}{j} a_j a_{p-j}. \qquad (16.3.1)$$

We then prove by induction on j that for every $j \geq 1$, $x \in \mathbb{R}^d$ and $\varepsilon \in (0,1]$,

$$\mathbb{N}_x[\mathcal{Z}(B(0,\varepsilon))^j] \leq \eta(\eta')^{j-1} a_j (G(x) \wedge \varepsilon^{2-d}) \varepsilon^{d+4(j-1)}. \qquad (16.3.2)$$

Here $\eta > 0$ is chosen so that this inequality holds for $j = 1$ (we already noticed that this is possible), and $\eta' = 4\eta C_4$, where C_4 is the constant in Lemma 16.3.1.

Suppose that (16.3.2) holds for $j = 1, \ldots, p-1$. Then, by (16.2.3),

$$\begin{aligned}
&\mathbb{N}_x[\mathcal{Z}(B(0,\varepsilon))^p)] \\
&\leq 2\eta^2(\eta')^{p-2} \varepsilon^{2d+4(p-2)} \sum_{j=1}^{p-1} \binom{p}{j} a_j a_{p-j} \int_{\mathbb{R}^d} dz\, G(x,z)\, (G(z) \wedge \varepsilon^{2-d})^2 \\
&\leq 4\eta^2(\eta')^{p-2} a_p\, C_4 (G(x) \wedge \varepsilon^{2-d}) \varepsilon^{d+4(p-1)},
\end{aligned}$$

by (16.3.1) and Lemma 16.3.1. This completes the proof of (16.3.2). The upper bound of the proposition follows, since $a_p \leq 2^p p!$. The proof of the lower bound is similar (in fact easier) using now the lower bound of Lemma 16.3.1. □

We now state and prove the analogue of Proposition 16.3.1 in the critical dimension $d = 4$. For technical reasons, we fix $K > 1$ and we deal with the total occupation measure $\mathcal{Z}^K$ of the snake "stopped at the boundary of $B(0,K)$" as defined in Section 16.2. We will use the easy bound

$$G_K(x,y) \geq C_5\, G(x,y) \qquad (16.3.3)$$

which is valid for $x, y \in B(0, K/2)$ with a constant $C_5 > 0$.

Proposition 16.3.2 *Let $d = 4$. There exist two positive constants $C_6 = C_6(K)$, $C_7 = C_7(K)$ such that, for every $p \geq 1$, $x \in B(0, K/2)$, $\varepsilon \in (0, 1/2]$,*

$$\begin{aligned}
&C_6^p\, p! (G(x) \wedge \varepsilon^{-2}) \varepsilon^{4p} \Big(1 + \log_{|} \frac{|x|}{\varepsilon}\Big)^{p-1} \\
&\qquad \leq \mathbb{N}_x[\mathcal{Z}^K(B(0,\varepsilon))^p] \leq C_7^p\, p! (G(x) \wedge \varepsilon^{-2}) \varepsilon^{4p} \Big(\log \frac{1}{\varepsilon}\Big)^{p-1}.
\end{aligned}$$

Proof. The case $p = 1$ is straightforward from the first moment formula for $\mathcal{Z}^K$ applying (16.3.3) for the lower bound. We then proceed by induction using the next lemma, whose proof is again left to the reader.

Lemma 16.3.2 (i) *There exists a constant* $C_8 = C_8(K)$ *such that, for every* $x \in B(0, K)$, $\varepsilon \in (0, 1/2]$,

$$\int_{B(0,K)} dz\, G(z-x)\,(G(z) \wedge \varepsilon^{-2})^2 \le C_8 (G(x) \wedge \varepsilon^{-2}) \log \frac{1}{\varepsilon}. \tag{16.3.4}$$

(ii) *There exists a positive constant* $C_9 = C_9(K)$ *such that, for every* $x \in B(0, K/2)$, $\varepsilon \in (0, 1/2]$ *and every integer* $p \ge 1$,

$$\int_{B(0,K/2)} dz\, G(z-x)\,(G(z) \wedge \varepsilon^{-2})^2 \Big(1 + \log_+ \frac{|z|}{\varepsilon}\Big)^{p-1}$$
$$\ge \frac{C_9}{p} (G(x) \wedge \varepsilon^{-2}) \Big(1 + \log_+ \frac{|x|}{\varepsilon}\Big)^p. \tag{16.3.5}$$

Let us complete the proof of Proposition 16.3.2. We first establish the lower bound. We argue by induction. Let $p \ge 2$. Assume that, for $j = 1, \ldots, p-1$, $x \in B(0, K/2)$, $\varepsilon \in (0, 1/2)$, we have

$$\mathbb{N}_x[\mathcal{Z}^K(B(0,\varepsilon))^j] \ge \gamma (\gamma')^{j-1}\, j!(G(x) \wedge \varepsilon^{-2})\, \varepsilon^{4j} \Big(1 + \log_+ \frac{|x|}{\varepsilon}\Big)^{j-1} \tag{16.3.6}$$

where $\gamma > 0$ is chosen so that this inequality holds for $j = 1$, and $\gamma' = 2\gamma C_5 C_9$. We have then, using (16.3.3), (16.2.3) (with $\mathcal{Z}$ replaced by $\mathcal{Z}^K$ and G by G_K) and (16.3.6),

$$\begin{aligned}
&\mathbb{N}_x[\mathcal{Z}^K(B(0,\varepsilon))^p] \\
&\ge 2\gamma^2 (\gamma')^{p-2} \varepsilon^{4p} \sum_{j=1}^{p-1} \binom{p}{j} j!(p-j)! \\
&\qquad \int_{B(0,K)} dz\, G_K(x,z)(G(z) \wedge \varepsilon^{-2})^2 \Big(1 + \log_+ \frac{|z|}{\varepsilon}\Big)^{p-2} \\
&\ge 2C_5\, \gamma^2\, (\gamma')^{p-2} \varepsilon^{4p}\, (p-1)\, p! \\
&\qquad \int_{B(0,K/2)} dz\, G(z-x)(G(z) \wedge \varepsilon^{-2})^2 \Big(1 + \log_+ \frac{|z|}{\varepsilon}\Big)^{p-2} \\
&\ge \gamma (\gamma')^{p-1} p!\, \varepsilon^{4p} (G(x) \wedge \varepsilon^{-2}) \Big(1 + \log_+ \frac{|x|}{\varepsilon}\Big)^{p-1},
\end{aligned}$$

by (16.3.5) and our choice of γ'. We obtain that (16.3.6) also holds for $j = p$, which completes the proof of the lower bound.

For the upper bound, we use the same sequence (a_j) as in the proof of Proposition 16.3.1. We show by induction that, for every $j \ge 1$, $x \in B(0, K)$, $\varepsilon \in (0, 1/2]$,

$$\mathbb{N}_x[\mathcal{Z}^K(B(0,\varepsilon))^j] \le \eta (\eta')^{j-1}\, a_j (G(x) \wedge \varepsilon^{-2})\, \varepsilon^{4j} \Big(\log \frac{1}{\varepsilon}\Big)^{j-1},$$

where $\eta > 0$ is chosen so that this bound holds for $j = 1$, and $\eta' = 4\eta C_8$. The argument is exactly the same as in the proof of Proposition 16.3.1, using the bound (16.3.4). □

Corollary 16.3.1 (i) ($d \geq 5$) *For every $A > 1$, there exist two positive constants C_{10}, C_{11} such that, for every $x \in B(0, A)$, $\lambda > 0$, $\varepsilon \in (0, 1]$,*

$$\mathbb{N}_x[\mathcal{Z}(B(0,\varepsilon)) \geq \lambda\varepsilon^4] \geq C_{10}\,\varepsilon^{d-4}\,\exp(-C_{11}\lambda).$$

(ii) ($d = 4$) *For every $A > 1$, there exist two positive constants C_{12}, C_{13} such that, for every $x \in B(0, A)$, $\varepsilon \in (0, 1/2]$, $\lambda > 0$,*

$$\mathbb{N}_x\Big[\mathcal{Z}(B(0,\varepsilon)) \geq \lambda\,\varepsilon^4 \log\frac{1}{\varepsilon}\Big] \geq C_{12}\Big(\log\frac{1}{\varepsilon}\Big)^{-1}\,\exp(-C_{13}\lambda).$$

Proof. (i) Clearly, we may restrict our attention to the case $\lambda > 1$. We first assume that $|x| \geq 2$. Then, for every integer $p \geq 1$ and every $r > 0$,

$$\mathbb{N}_x[\mathcal{Z}(B(0,\varepsilon))^p] \leq r^p\,\mathbb{N}_x[\mathcal{Z}(B(0,\varepsilon)) > 0] + \mathbb{N}_x[\mathcal{Z}(B(0,\varepsilon))^p\,1_{\{\mathcal{Z}(B(0,\varepsilon))\geq r\}}].$$

The Cauchy-Schwarz inequality then gives the lower bound

$$(\mathbb{N}_x[\mathcal{Z}(B(0,\varepsilon)) \geq r])^{1/2} \geq \frac{\mathbb{N}_x[\mathcal{Z}(B(0,\varepsilon))^p] - r^p\,\mathbb{N}_x[\mathcal{Z}(B(0,\varepsilon)) > 0]}{(\mathbb{N}_x[\mathcal{Z}(B(0,\varepsilon))^{2p}])^{1/2}}.$$

From Lemma 16.2.1, we know that, for $|x| \geq 2$, $\varepsilon \in (0, 1]$,

$$\mathbb{N}_x[\mathcal{Z}(B(0,\varepsilon)) > 0] \leq C_{14}\,\varepsilon^{d-4}.$$

for some constant C_{14} independent of x, ε.

Then, by Proposition 16.3.1, we have for $2 \leq |x| \leq A$

$$(\mathbb{N}_x[\mathcal{Z}(B(0,\varepsilon)) \geq r])^{1/2} \geq \frac{c\,C_1^p\,p!\varepsilon^{d+4(p-1)} - C_{14}\varepsilon^{d-4}r^p}{c'C_2^p((2p)!)^{1/2}\varepsilon^{(d+4(2p-1))/2}},$$

where the positive constants c, c' only depend on A. Taking $r = \lambda\,\varepsilon^4$ gives

$$(\mathbb{N}_x[\mathcal{Z}(B(0,\varepsilon)) \geq \lambda\,\varepsilon^4])^{1/2} \geq \frac{c\,C_1^p\,p! - C_{14}\lambda^p}{c'C_2^p((2p)!)^{1/2}}\,\varepsilon^{(d-4)/2}.$$

Then, by choosing $p = [M\lambda]$, where $M = M(A)$ is taken sufficiently large, we arrive at the desired estimate.

It remains to remove the restriction $|x| \geq 2$. To this end, we apply the strong Markov property of the Brownian snake at $T = \inf\{s > 0, \zeta_s = 1\}$. Note that, for $|x| \leq 2$,

$$\mathbb{N}_x[T < \infty,\ 2 \leq |W_T(t)| \leq 4, \forall t \in [1/2, 1]]$$

is bounded below by a positive constant α. This is so because, conditionally on the event $\{T < \infty\}$, $(W_T(t), 0 \leq t \leq 1)$ is distributed as a Brownian path starting at x. Then, by applying Theorem 16.2.1 under $\mathbb{P}^*_{W_T}$, we get that, for $|x| \leq 2$,

$$\begin{aligned}\mathbb{N}_x[\mathcal{Z}(B(0,\varepsilon)) \geq \lambda\varepsilon^4] &\geq \mathbb{N}_x\Big[T < \infty,\ 2 \leq |W_T(t)| \leq 4, \forall t \in [1/2, 1],\\ &\qquad \times\Big(1 - \exp -2\int_{1/2}^1 dt\,\mathbb{N}_{W_T(t)}[\mathcal{Z}(B(0,\varepsilon)) \geq \lambda\varepsilon^4]\Big)\Big]\\ &\geq \alpha\,(1 - \exp(-\inf_{2\leq|y|\leq 4}\mathbb{N}_y[\mathcal{Z}(B(0,\varepsilon)) \geq \lambda\varepsilon^4])).\end{aligned}$$

Therefore, the case $|x| \leq 2$ also follows from the first part of the proof.

(ii) We first observe that it suffices to prove the bound of the proposition when $\mathcal{Z}$ is replaced by $\mathcal{Z}^K$, where $K = 2A$. In fact the distribution of $\mathcal{Z}^K(B(0,\varepsilon))$ is obviously dominated by the distribution of $\mathcal{Z}(B(0,\varepsilon))$. From Lemma 16.2.1, we know that there exists a constant C_{15} such that, for $1 \leq |x| \leq A$ and $\varepsilon \in (0, 1/2]$,

$$\mathbb{N}_x[\mathcal{Z}^K(B(0,\varepsilon)) > 0] \leq C_{15}\Big(\log\frac{1}{\varepsilon}\Big)^{-1}$$

(this bound holds with $\mathcal{Z}$ instead of $\mathcal{Z}^K$). Then, by arguing as in the proof of (i) and using Proposition 16.3.2 instead of Proposition 16.3.1, we arrive at the bound

$$(\mathbb{N}_x[\mathcal{Z}^K(B(0,\varepsilon)) \geq \lambda\,\varepsilon^4 \log\frac{1}{\varepsilon}])^{1/2} \geq \frac{c\,C_6^p\,p! - C_{15}\lambda^p}{c' C_7^p((2p)!)^{1/2}}\Big(\log\frac{1}{\varepsilon}\Big)^{-1/2},$$

valid for $1 \leq |x| \leq A$, $\lambda > 1$, $\varepsilon \in (0, 1/2]$, with positive constants c, c' depending only on A. We can again choose $p = [M\lambda]$, with $M = M(A)$ sufficiently large, to get the desired lower bound when $|x| \geq 1$. This last restriction is then removed by the same reasoning as in the proof of Corollary 16.3.1. □

16.4 The Case $d \geq 5$

Throughout this section, we suppose that $d \geq 5$.

Proposition 16.4.1 *There exists a constant $C_{16} = C_{16}(d)$ such that $\mathbb{N}_0$ a.e.*

$$\limsup_{r\to 0} \frac{\mathcal{Z}(B(y,r))}{h(r)} \leq C_{16}, \quad \mathcal{Z}(dy) \text{ a.e.} \tag{16.4.1}$$

In consequence, there exists a positive constant $M_1 = M_1(d)$ such that $\mathbb{N}_0$ a.e. for every Borel subset A of $\mathbb{R}^d$

$$h - m(\mathcal{R} \cap A) \geq M_1\,\mathcal{Z}(A).$$

PROOF. The second statement follows from the first one by using the density theorems of Rogers and Taylor (see, e.g., Perkins [14, Theorem 1.4]). To prove (16.4.1), we first check that there exists a constant C_{17} such that, for every $K > 0$, for every integer $p \geq 0$ and every $r \in (0, 1/d]$,

$$\mathbb{N}_0\Big[\int_{B(0,K)} \mathcal{Z}(dy)\,\mathcal{Z}(B(y,r))^p\Big] \leq c_K\,C_{17}^p\,p!\,r^{4p}, \tag{16.4.2}$$

for some constant c_K depending on K. The bound (16.4.2) follows easily from Proposition 16.3.1. First notice that

$$\int_{B(0,K)} \mathcal{Z}(dy)\,\mathcal{Z}(B(y,r))^p \leq \sum_{a\in\mathbb{Z}^d, |ra|\leq K+1} \mathcal{Z}(B(ra, (\sqrt{d}+1)r))^{p+1}.$$

Then, by Proposition 16.3.1,

$$\begin{aligned}
\mathbb{N}_0\Big[\int_{B(0,K)} \mathcal{Z}(dy)\, \mathcal{Z}(B(y,r))^p\Big]
&\leq \sum_{a\in\mathbb{Z}^d,|ra|\leq K+1} C_2^{p+1}(p+1)!\,(dr)^{d+4p}\,(G(ra)\wedge r^{2-d})\\
&\leq d^{d+4p} C_2^{p+1}(p+1)!\, r^{4p} \sum_{a\in\mathbb{Z}^d,|ra|\leq K+1} (G(ra)\wedge r^{2-d})\, r^d\\
&\leq c_K\,(d^4C_2)^{p+1}(p+1)!\, r^{4p}
\end{aligned}$$

which gives the bound (16.4.2) with $C_{17} = 2d^4C_2$. From (16.4.2), we see that, for every $K > 0$ there exists a constant c'_K such that, for every $r \in (0, 1/d]$,

$$\mathbb{N}_0\Big[\int_{B(0,K)} \mathcal{Z}(dy)\, \exp\Big(\frac{\mathcal{Z}(B(y,r))}{2C_{17}r^4}\Big)\Big] \leq c'_K.$$

Hence, taking $r = 2^{-k}$, we have for every $\alpha > 0$,

$$\begin{aligned}
\mathbb{N}_0\Big[\int_{B(0,K)} \mathcal{Z}(dy)\, 1\{\mathcal{Z}(B(y,2^{-k})) \geq \alpha\, h(2^{-k})\}\Big]\\
\leq c'_K\, \exp(-(2C_{17})^{-1}\alpha \log\log 2^k).
\end{aligned}$$

By choosing $\alpha = 4C_{17}$, we get a convergent series, and therefore

$$\mathbb{N}_0\Big[\int_{B(0,K)} \mathcal{Z}(dy) \sum_{k=0}^{\infty} 1\{\mathcal{Z}(B(y,2^{-k})) \geq \alpha\, h(2^{-k})\}\Big] < \infty$$

which implies

$$\mathbb{N}_0 \text{ a.e., } \mathcal{Z}(dy) \text{ a.e. on } B(0,K),\ \limsup_{k\to\infty} \frac{\mathcal{Z}(B(y,r_k))}{h(r_k)} \leq \alpha,$$

and also, by a monotonicity argument,

$$\mathbb{N}_0 \text{ a.e., } \mathcal{Z}(dy) \text{ a.e. on } B(0,K),\ \limsup_{r\to 0} \frac{\mathcal{Z}(B(y,r))}{h(r)} \leq C_{16},$$

with $C_{16} = 16\alpha = 128d^4C_2$. Since this holds for every $K > 0$, with a constant C_{16} independent of K, the proof of Proposition 16.4.1 is complete. □

We now turn to the upper bound corresponding to Proposition 16.4.1.

Proposition 16.4.2 *There exists a constant $M_2 = M_2(d)$ such that, $\mathbb{N}_0$ a.e. for every Borel subset A of $\mathbb{R}^d$,*

$$h - m(\mathcal{R}\cap A) \leq M_2\, \mathcal{Z}(A). \tag{16.4.3}$$

PROOF. The proof of Proposition 16.4.2 relies on the following basic ideas. By the density theorems of Rogers and Taylor, it is enough to bound the Hausdorff measure of the subset of $\mathcal{R}$ consisting of all points z such that the total occupation measure of balls centered at z is unusually small. To this end, we will apply the strong Markov property of the Brownian snake at the first time when a path W_s comes "close to z", in order to use the estimates of Section 16.3. We will need a result of [11], showing that the behavior of the paths W_s near their endpoints can be controlled uniformly.

We fix $\delta > 0$. Recall that

$$\mathcal{R}_\delta = \{z = \hat{W}_s,\ s \geq 0,\ \zeta_s \geq \delta\} \subset \mathcal{R}.$$

It is enough to prove that (16.4.3) holds when $\mathcal{R}$ is replaced by $\mathcal{R}_\delta$, with a constant M_2 independent of δ.

For every $c > 0$, introduce the set of "good points"

$$\mathcal{G}(c) = \left\{z \in \mathcal{R}_\delta,\ \limsup_{r \to 0} \frac{\mathcal{Z}(B(z,r))}{h(r)} \geq c\right\}$$

and set $\mathcal{B}(c) = \mathcal{R}_\delta \backslash \mathcal{G}(c)$. From [14, Theorem 1.4], we know that there exists a universal constant c_1 such that, for every Borel subset A of $\mathbb{R}^d$,

$$h - m(A \cap \mathcal{G}(c)) \leq \frac{c_1}{c}\, \mathcal{Z}(A).$$

Therefore, in order to prove Proposition 16.4.2, it will be enough to check that one can choose $c > 0$ small enough (independently of δ) so that

$$h - m(\mathcal{B}(c)) = 0, \quad \mathbb{N}_0 \text{ a.e.} \tag{16.4.4}$$

Set $r_n = 2^{-n}$ and for every integer $N \geq 1$ set

$$\mathcal{B}_N(c) = \{z \in \mathcal{R}_\delta \cap B(0,N);\ \mathcal{Z}(B(z, d\, r_n)) \leq c\, h(r_n),\ \forall n \geq N\}.$$

Then,

$$\mathcal{B}(c) \subset \bigcup_{N \geq 1} \mathcal{B}_N(d^4 c).$$

Therefore, (16.4.4) will follow if we can check that for c small enough we have $\mathbb{N}_0$ a.e. for every N

$$h - m(\mathcal{B}_N(c)) = 0. \tag{16.4.5}$$

The proof of (16.4.5) uses explicit coverings of the set $\mathcal{B}_N(c)$. We fix $N \geq 2$ such that $r_N \leq \delta$, and, for $n \geq 1$, we denote by $\mathcal{C}_n^N$ the collection of all closed cubes of sidelength r_n whose vertices belong to $r_n\mathbb{Z}^d$, and which are contained in $B(0, 2N)$. Let us consider one such cube $\Gamma \in \mathcal{C}_n^N$, with $n > N$. We aim to bound $\mathbb{N}_0[\Gamma \cap \mathcal{B}_N(c) \neq \emptyset]$. Let

$$T = T_\Gamma = \inf\{s \geq 0,\ \hat{W}_s \in \Gamma,\ \zeta_s \geq \delta\}$$

so that T is a stopping time, which is finite on the event $\{\Gamma \cap \mathcal{B}_N(c) \neq \emptyset\}$.

We then need to distinguish two cases, according as the path W_T is well-behaved or not. Let n_0 be the smallest integer such that $2^{-n_0} \leq \delta$. For every $m \geq n_0$, $A > 0$

and every $\mathrm{w} \in \mathcal{W}$ such that $\zeta = \zeta_{\mathrm{w}} \geq \delta$, we set

$$F^A_{n_0,m}(\mathrm{w}) = \frac{1}{m}\mathrm{Card}\{k \in \{n_0, \dots, m\};\\ |\mathrm{w}(t) - \hat{\mathrm{w}}| < A2^{-k/2},\ \forall t \in [\zeta - 2^{-k}, \zeta]\}.$$

As a consequence of the main result of [11], we can fix $A > 0$, independently of δ, so that $\mathbb{N}_0$ a.e. there exists a (random) integer $m_0 > n_0$ such that, for every $s \geq 0$ with $\zeta_s \geq \delta$,

$$F^A_{n_0,m}(W_s) > \frac{7}{8}, \quad \forall m \geq m_0. \tag{16.4.6}$$

We first bound the quantity

$$\mathbb{N}_0[\Gamma \cap \mathcal{B}_N(c) \neq \emptyset,\ F^A_{n_0,m}(W_T) > \frac{7}{8},\ \forall m \geq N]$$

(note that $N \geq n_0$, since $r_N \leq \delta$). If $\Gamma \cap \mathcal{B}_N(c) \neq \emptyset$, then there exists a point $y \in \Gamma$ such that

$$\mathcal{Z}(B(y, d\,r_k)) \leq c\,h(r_k), \quad \forall k \geq N.$$

Note that, if z_Γ denotes the center of Γ, the ball $B(z_\Gamma, r_k)$ is contained in the ball $B(y, d\,r_k)$, for every $k \leq n$. Hence, we have also

$$\mathcal{Z}(B(z_\Gamma, r_k)) \leq c\,h(r_k), \quad \forall k \in [N, n].$$

Then, denote by $\mathcal{Z}^T$ the occupation measure of $\hat{W}$ over $[T, \infty)$:

$$\langle \mathcal{Z}^T, \varphi \rangle = \int_T^\infty ds\, \varphi(\hat{W}_s).$$

We have a fortiori, on the event $\{\Gamma \cap \mathcal{B}_N(c) \neq \emptyset\}$,

$$\mathcal{Z}^T(B(z_\Gamma, r_k)) \leq c\,h(r_k), \quad \forall k \in [N, n].$$

As a consequence of the previous remarks, we have

$$\begin{aligned}
&\mathbb{N}_0[\Gamma \cap \mathcal{B}_N(c) \neq \emptyset,\ F^A_{n_0,m}(W_T) > \frac{7}{8},\ \forall m \geq N]\\
&\leq \mathbb{N}_0[T < \infty;\ F^A_{n_0,m}(W_T) > \frac{7}{8},\ \forall m \geq N;\\
&\qquad \mathcal{Z}^T(B(z_\Gamma, r_k)) \leq c\,h(r_k),\ \forall k \in [N, n]]\\
&= \mathbb{N}_0[T < \infty;\ F^A_{n_0,m}(W_T) > \frac{7}{8},\ \forall m \geq N;\\
&\qquad \mathbb{P}^*_{W_T}[\mathcal{Z}(B(z_\Gamma, r_k)) \leq c\,h(r_k),\ \forall k \in [N, n]]],
\end{aligned} \tag{16.4.7}$$

using the strong Markov property of the Brownian snake at time T. At this point, we need a lemma.

Lemma 16.4.1 *We may choose $c \in (0, 1]$ (independently of δ) and a constant $C_{18} > 0$ so that for every $N \geq n_0 \vee 2$, $n \geq 2N$ and for every $\mathrm{w} \in \mathcal{W}_0$ such that*

$\zeta_{\mathrm{w}} \geq 2^{-n_0}$, $|\hat{\mathrm{w}} - z_\Gamma| \leq r_n$ *and* $F^A_{n_0,m}(\mathrm{w}) > 7/8$, $\forall m \geq N$, *we have*

$$\mathbb{P}^*_{\mathrm{w}}[\mathcal{Z}(B(z_\Gamma, r_k)) \leq c\, h(r_k),\ \forall k \in [N, n]] \leq \exp(-C_{18}\, n^{1/2}).$$

PROOF. Let $\mathrm{w} \in \mathcal{W}_x$ and write $\zeta = \zeta_{\mathrm{w}}$ for simplicity. By Theorem 16.2.1, we have $\mathbb{P}^*_{\mathrm{w}}$ a.s.

$$\mathcal{Z} = \int \mathcal{Z}(\omega)\, \mathcal{N}(dt d\omega),$$

where $\mathcal{N}$ is a Poisson measure on $\mathbb{R}_+ \times C_0(\mathbb{R}_+, \mathcal{W})$ with intensity

$$2 \cdot 1_{[0,\zeta]}(t) dt\, \mathbb{N}_{\mathrm{w}(t)}(d\omega).$$

For $k \in [N, n]$, set

$$\mathcal{Z}^{(k)} = \int 1_{[\zeta - r_{k-1}^2, \zeta - r_k^2)}(t)\, \mathcal{Z}(\omega)\, \mathcal{N}(dt d\omega).$$

By properties of Poisson measures, the random measures $\mathcal{Z}^{(k)}$ are independent under $\mathbb{P}^*_{\mathrm{w}}$. We have for every $c > 0$

$$\begin{aligned}
&\mathbb{P}^*_{\mathrm{w}}[\mathcal{Z}^{(k)}(B(z_\Gamma, r_k)) > c\, h(r_k)] \\
&\geq 1 - \exp\Big(-2 \int_{\zeta - 2^{-2(k-1)}}^{\zeta - 2^{-2k}} dt\, \mathbb{N}_{\mathrm{w}(t)}[\mathcal{Z}(B(z_\Gamma, r_k)) > c\, h(r_k)]\Big) \\
&= 1 - \exp\Big(-2^{2k+1} \int_{\zeta - 2^{-2(k-1)}}^{\zeta - 2^{-2k}} dt\, \mathbb{N}_0\Big[\mathcal{Z}\Big(B\Big(\frac{z_\Gamma - \mathrm{w}(t)}{2^{-k}}, 1\Big)\Big) > c\, \frac{h(r_k)}{2^{-4k}}\Big]\Big),
\end{aligned}$$

using translation invariance and scaling in the last equality. Let $\mathcal{L}$ be the set of all integers $k \in [N, n]$ such that $\mathrm{w}([\zeta - 2^{-2(k-1)}, \zeta]) \subset B(\hat{\mathrm{w}}, A2^{-(k-1)})$. Note that $B(\hat{\mathrm{w}}, A2^{-(k-1)}) \subset B(z_\Gamma, (A+1)2^{-(k-1)})$ for $k \leq n$. Observe that Card $\mathcal{L} \geq n/4$ because $F^A_{n_0,2n}(\mathrm{w}) > 7/8$ and $n \geq 2N$. If $k \in \mathcal{L}$, the previous inequality and Corollary 16.3.1 (i) (with $\varepsilon = 1$) give

$$\mathbb{P}^*_{\mathrm{w}}[\mathcal{Z}^{(k)}(B(z_\Gamma, r_k)) > c\, h(r_k)] \geq 1 - \exp(-C_{19}\, k^{-cC_{20}}),$$

for some constants C_{19}, C_{20} depending only on A.

Using the independence of the random measures $\mathcal{Z}^{(k)}$, we finally get

$$\begin{aligned}
&\mathbb{P}^*_{\mathrm{w}}[\mathcal{Z}(B(z_\Gamma, r_k)) \leq c\, h(r_k),\ \forall k \in [N, n]] \\
&\leq \prod_{k=N}^{n} \Big(1 - \mathbb{P}^*_{\mathrm{w}}[\mathcal{Z}^{(k)}(B(z_\Gamma, r_k)) > c\, h(r_k)]\Big) \\
&\leq \prod_{k \in \mathcal{L}} \exp(-C_{19}\, k^{-cC_{20}}) \\
&\leq \exp(-\frac{1}{4} C_{19}\, n^{1-cC_{20}}),
\end{aligned}$$

and the desired result follows by taking c such that $cC_{20} < 1/2$. □

From (16.4.7) and Lemma 16.4.1, we get

$$\mathbb{N}_0[\Gamma \cap \mathcal{B}_N(c) \neq \emptyset;\ F^A_{n_0,m}(W_T) > \frac{7}{8},\ \forall m \geq N] \leq \mathbb{N}_0[T < \infty] \exp(-C_{18}\, n^{1/2}).$$

By Lemma 16.2.1 (ii), there exists a constant c_δ independent of n and Γ such that

$$\mathbb{N}_0[T < \infty] \leq c_\delta\, r_n^{d-4}.$$

By summing over $\Gamma \in \mathcal{C}_n^N$, we obtain

$$\mathbb{N}_0\Big[\sum_{\Gamma \in \mathcal{C}_n^N} 1\{\Gamma \cap \mathcal{B}_N(c) \neq \emptyset\}\, 1\{F^A_{n_0,m}(W_T) > \frac{7}{8},\ \forall m \geq N\}\Big]$$
$$\leq c_\delta\, (4N r_n^{-1})^d\, r_n^{d-4} \exp(-C_8\, n^{1/2}).$$

We multiply by $h(r_n)$ each side of the previous inequality and sum over n to get a convergent series. Therefore, $\mathbb{N}_0$ a.e.

$$\lim_{n\to\infty} h(r_n) \sum_{\Gamma \in \mathcal{C}_n^N} 1\{\Gamma \cap \mathcal{B}_N(c) \neq \emptyset\}\, 1\{F^A_{n_0,m}(W_{T_\Gamma}) > \frac{7}{8},\ \forall m \geq N\} = 0, \tag{16.4.8}$$

Notice that (16.4.8) holds for every $N \geq n_0 \vee 2$. On the other hand, (16.4.6) shows that $\mathbb{N}_0$ a.e. for every N sufficiently large,

$$\sum_{\Gamma \in \mathcal{C}_n^N} 1\{\Gamma \cap \mathcal{B}_N(c) \neq \emptyset\}\, 1\{F^A_{n_0,m}(W_{T_\Gamma}) \leq \frac{7}{8},\ \text{for some } m \geq N\} = 0. \tag{16.4.9}$$

Combining (16.4.8) and (16.4.9) gives $\mathbb{N}_0$ a.e. for N large

$$\lim_{n\to\infty} h(r_n) \sum_{\Gamma \in \mathcal{C}_n^N} 1\{\Gamma \cap \mathcal{B}_N(c) \neq \emptyset\} = 0,$$

and (16.4.5) follows by the definition of Hausdorff measures. This completes the proof of Proposition 16.4.2. □

16.5 The Critical Case

Throughout this section, we suppose that $d = 4$ and we aim at proving analogues of the results of the previous section. It turns out that the proof of the result analogous to Proposition 16.4.2 is essentially similar to the case $d \geq 5$, whereas the lower bound on the Hausdorff measure is much harder. Therefore we start by deriving the upper bound.

Proposition 16.5.1 *There exists a constant M_2 such that, $\mathbb{N}_0$ a.e. for every Borel subset A of $\mathbb{R}^4$,*

$$h - m(\mathcal{R} \cap A) \leq M_2\, \mathcal{Z}(A). \tag{16.5.1}$$

PROOF. The proof follows the same outline as for Proposition 16.4.2 but is a little more complicated essentially because we have to use a different sequence (r_n). We will carefully explain those parts of the arguments that differ from the proof of Proposition 16.4.2. As in that proof, we first fix $\delta > 0$ and observe that it is enough to prove that (16.5.1) holds when $\mathcal{R}$ is replaced by $\mathcal{R}_\delta$, provided that the constant M_2 does not depend on δ. We then introduce the sets $\mathcal{G}(c)$, $\mathcal{B}(c)$ as previously. The definition of $\mathcal{B}_N(c)$ is also the same as before, except that the sequence (r_n) is now defined by

$$r_n = 2^{-2^n}.$$

Again, it suffices to prove that, for a suitable value of c, $h - m(\mathcal{B}_N(c)) = 0$, for N sufficiently large, $\mathbb{N}_0$ a.e.

We define the sets $\mathcal{C}_n^N$ as previously (but of course with the new choice of r_n). We then fix $N \geq 3$ such that $r_N \leq \delta$ and, if $\Gamma \in \mathcal{C}_n^N$, for some $n > N$, we look for a suitable bound on

$$\mathbb{N}_0[\Gamma \cap \mathcal{B}_N(c) \neq \emptyset,\ F^A_{n_0,m}(W_T) > \frac{7}{8},\ \forall m \geq 2^N].$$

Here $A > 0$ has been chosen as in the proof of Proposition 16.4.2 (so that (16.4.6) holds) and similarly,

$$T = T_\Gamma = \inf\{s \geq 0,\ \hat{W}_s \in \Gamma,\ \zeta_s \geq \delta\}.$$

By the same arguments used to derive (16.4.7), we have

$$\begin{aligned}
&\mathbb{N}_0[\Gamma \cap \mathcal{B}_N(c) \neq \emptyset,\ F^A_{n_0,m}(W_T) > \frac{7}{8},\ \forall m \geq 2^N] \qquad (16.5.2)\\
&= \mathbb{N}_0[T < \infty;\ F^A_{n_0,m}(W_T) > \frac{7}{8},\ \forall m \geq 2^N;\\
&\qquad \mathbb{P}^*_{W_T}[\mathcal{Z}(B(z_\Gamma, r_k)) \leq c\, h(r_k),\ \forall k \in [N,n]]\,],
\end{aligned}$$

so that we can use the following analogue of Lemma 16.4.1.

Lemma 16.5.1 *We may choose $c > 0$ (independently of δ) and a constant $C_{21} > 0$ so that, for every $N \geq n_0 \vee 3$, $n > N$ and for every $\mathrm{w} \in \mathcal{W}_0$ such that $\zeta_\mathrm{w} \geq 2^{-n_0}$, $|\hat{\mathrm{w}} - z_\Gamma| \leq r_n$ and $F^A_{n_0,m}(\mathrm{w}) > 7/8$, $\forall m \geq 2^N$, we have*

$$\mathbb{P}^*_\mathrm{w}[\mathcal{Z}(B(z_\Gamma, r_k)) \leq c\, h(r_k),\ \forall k \in [N,n]] \leq \exp(-C_{21}\,(n-N)^{1/2}).$$

PROOF. The argument here is significantly different from the proof of Lemma 16.4.1. We start similarly, writing under $\mathbb{P}^*_\mathrm{w}$

$$\mathcal{Z} = \int \mathcal{Z}(\omega)\,\mathcal{N}(dt\,d\omega),$$

and defining for $k \in [N,n]$

$$\mathcal{Z}^{(k)} = \int 1_{[\zeta - r^2_{k-1}, \zeta - r^2_k)}(t)\ \mathcal{Z}(\omega)\,\mathcal{N}(dt\,d\omega).$$

However, we also introduce, for every $k \in [N, n]$ and $2^k < l \leq 2^{k+1}$,

$$\mathcal{Z}_l^{(k)} = \int 1_{[\zeta-2^{-(l-1)},\zeta-2^{-l})}(t)\, \mathcal{Z}(\omega)\, \mathcal{N}(dt d\omega),$$

in such a way that $\mathcal{Z}^{(k)} = \sum_l \mathcal{Z}_l^{(k)}$. The random measures $\mathcal{Z}_l^{(k)}$ are independent under $\mathbb{P}_{\mathrm{w}}^*$.

Then,

$$\begin{aligned}
&\mathbb{P}_{\mathrm{w}}^*[\mathcal{Z}^{(k)}(B(z_\Gamma, r_k)) \leq c\, h(r_k)] \\
&\leq \mathbb{P}_{\mathrm{w}}^*\Bigg[\bigcap_{2^k<l\leq\frac{3}{2}2^k} \Big\{\mathcal{Z}_l^{(k)}(B(z_\Gamma, r_k)) \leq c\, h(r_k)\Big\}\Bigg] \\
&= \prod_{2^k<l\leq\frac{3}{2}2^k} \Big(1 - \mathbb{P}_{\mathrm{w}}^*\Big[\mathcal{Z}_l^{(k)}(B(z_\Gamma, r_k)) > c\, h(r_k)\Big]\Big). \qquad (16.5.3)
\end{aligned}$$

Using the exponential formula for Poisson measures, we have

$$\begin{aligned}
&\mathbb{P}_{\mathrm{w}}^*[\mathcal{Z}_l^{(k)}(B(z_\Gamma, r_k)) > c\, h(r_k)] \\
&\geq 1 - \exp\Big(- 2\int_{\zeta-2^{-(l-1)}}^{\zeta-2^{-l}} dt\, \mathbb{N}_{\mathrm{w}(t)}\Big[\mathcal{Z}(B(z_\Gamma, r_k)) > c\, h(r_k)\Big]\Big) \\
&= 1 - \exp\Big(- 2^{l+1}\int_{\zeta-2^{-(l-1)}}^{\zeta-2^{-l}} dt \\
&\qquad \times \mathbb{N}_0\Big[\mathcal{Z}(B(2^{l/2}(z_\Gamma - \mathrm{w}(t)), 2^{l/2}r_k)) > c2^{2l}h(r_k)\Big]\Big) \qquad (16.5.4)
\end{aligned}$$

using translation invariance and the scaling properties of the measure $\mathbb{N}_0$.

For $k \in [N, n]$, denote by $\mathcal{L}_k$ the set of all integers l such that $2^k < l \leq \frac{3}{2}2^k$ and

$$\mathrm{w}([\zeta - 2^{-(l-1)}, \zeta]) \subset B(\hat{\mathrm{w}}, A2^{-(l-1)/2}) \subset B(z_\Gamma, (A+1)2^{-(l-1)/2}).$$

Our assumption $F_{n_0,m}^A(\mathrm{w}) > \frac{7}{8}$, for $m = \frac{3}{2}2^k$, guarantees that $\mathrm{Card}\,\mathcal{L}_k \geq 2^k/4$. Moreover, if $l \in \mathcal{L}_k$ and $t \in [\zeta - 2^{-(l-1)}, \zeta - 2^{-l}]$, we have

$$|2^{l/2}(z_\Gamma - \mathrm{w}(t))| \leq (A+1)2^{1/2},$$

so that, by Corollary 16.3.1 (ii),

$$\begin{aligned}
&\mathbb{N}_0\Big[\mathcal{Z}(B(2^{l/2}(z_\Gamma - \mathrm{w}(t)), 2^{l/2}r_k)) > c2^{2l}h(r_k)\Big] \\
&\geq C_{12}|\log 2^{l/2}r_k|^{-1}\exp\Big(- cC_{13}|\log 2^{l/2}r_k|^{-1}r_k^{-4}h(r_k)\Big) \geq C_{22}\, 2^{-k}\, k^{-cC_{23}},
\end{aligned}$$

for some positive constants C_{22}, C_{23} independent of $c > 0$. Here we used the conditions $2^k < l \leq \frac{3}{2}2^k$ and $k \geq N \geq 3$. From (16.5.4), we have thus, for $k \in [N, n]$ and $l \in \mathcal{L}_k$,

$$\mathbb{P}_{\mathrm{w}}^*[\mathcal{Z}_l^{(k)}(B(z_\Gamma, r_k)) > c\, h(r_k)] \geq 1 - \exp(-2C_{22}\, 2^{-k}\, k^{-cC_{23}}).$$

By considering only the values $l \in \mathcal{L}_k$ in the product of (16.5.3), we arrive at

$$\mathbb{P}^*_{\mathrm{w}}[\mathcal{Z}^{(k)}(B(z_\Gamma, r_k)) \leq c\,h(r_k)] \leq (\exp(-2C_{22}\,2^{-k}\,k^{-cC_{23}}))^{\operatorname{Card}\mathcal{L}_k}$$
$$\leq \exp(-\frac{C_{22}}{2}k^{-cC_{23}}).$$

We take c so small that $cC_{23} < 1/2$. Then,

$$\begin{aligned}\mathbb{P}^*_{\mathrm{w}}[\mathcal{Z}(B(z_\Gamma, r_k)) &\leq c\,h(r_k),\ \forall k \in [N, n]]\\ &\leq \prod_{k=N}^{n} \mathbb{P}^*_{\mathrm{w}}[\mathcal{Z}^{(k)}(B(z_\Gamma, r_k)) \leq c\,h(r_k)]\\ &\leq \exp\Big(-\frac{C_{22}}{2}\sum_{k=N}^{n} k^{-cC_{23}}\Big),\end{aligned}$$

which gives the bound of Lemma 16.5.1. □

The end of the proof of Proposition 16.5.1 is now very similar to the proof of Proposition 16.4.2. We fix c as given in Lemma 16.5.1. From (16.5.2) and Lemma 16.5.1, we have

$$\begin{aligned}\mathbb{N}_0[\Gamma \cap \mathcal{B}_N(c) \neq \emptyset,\ F^A_{n_0,m}(W_T) > \frac{7}{8},\ \forall m \geq 2^N]&\\ \leq \mathbb{N}_0[T_\Gamma < \infty]\exp(-C_{21}(n-N)^{1/2}).& \qquad (16.5.5)\end{aligned}$$

Lemma 16.2.1 (ii) gives the bound

$$\mathbb{N}_0[T_\Gamma < \infty] \leq C_{24}\,|\log r_n|^{-1},$$

with a constant C_{24} depending only on δ. By summing over $\Gamma \in \mathcal{C}^N_n$ and then multiplying by $h(r_n)$, we obtain $\mathbb{N}_0$ a.e.

$$\lim_{n\to\infty} h(r_n) \sum_{\Gamma\in\mathcal{C}^N_n} 1\{\Gamma \cap \mathcal{B}_N(c) \neq \emptyset\}\,1\{F^A_{n_0,m}(W_{T_\Gamma}) > \frac{7}{8},\ \forall m \geq 2^N\} = 0,$$

and the estimate $h - m(\mathcal{B}_N(c)) = 0$ (for N large, $\mathbb{N}_0$ a.e.) follows by using (16.4.9), which holds in any dimension. □

We now propose to prove the lower bound corresponding to Proposition 16.4.1. The proof is significantly more difficult. It will rely on the Palm formula for $\mathcal{Z}$ (Proposition 16.2.1) and the special Markov property (Theorem 16.2.2).

Proposition 16.5.2 *There exists a constant C_{25} such that, $\mathbb{N}_0$ a.e.*

$$\limsup_{r\to 0} \frac{\mathcal{Z}(B(y,r))}{h(r)} \leq C_{25}, \quad \mathcal{Z}(dy)\ a.e. \qquad (16.5.6)$$

In consequence, there exists a constant $M_1 > 0$ such that, $\mathbb{N}_0$ a.e. for every Borel subset A of $\mathbb{R}^4$,

$$h - m(\mathcal{R} \cap A) \geq M_1 \mathcal{Z}(A).$$

PROOF. As for Proposition 16.4.1, we need only prove the first part of the proposition. We start from the Palm formula for $\mathcal{Z}$ (Proposition 16.2.1), and we also write $\sigma_y(\mathcal{Z})$ for the image of $\mathcal{Z}$ under the mapping $z \longrightarrow z - y$. Then, from Proposition 16.2.1 and an easy time-reversal argument, we have

$$\mathbb{N}_0\Big[\int \mathcal{Z}(dy)\, F(\sigma_y(\mathcal{Z}))\Big] = \int_0^\infty da \int P_0^a(d\mathrm{w})\, E^{(\mathrm{w})}\Big[F\Big(\int \mathcal{N}(dtd\omega)\, \mathcal{Z}(\omega)\Big)\Big], \tag{16.5.7}$$

where $\mathcal{N}(dtd\omega)$ is under $P^{(\mathrm{w})}$ a Poisson point measure on $\mathbb{R}_+ \times C_0(\mathbb{R}_+, \mathcal{W})$ with intensity

$$4 \cdot 1_{[0,\zeta_\mathrm{w}]}(t)\, dt\, \mathbb{N}_{\mathrm{w}(t)}(d\omega).$$

By applying (16.5.7) to the function

$$F(\nu) = 1\Big\{\limsup_{r\to 0} \frac{\nu(B(0,r))}{h(r)} > c\Big\},$$

with a suitable value of c, we see that the proof of (16.5.6) reduces to checking that, for every $a > 0$, $P_0^a(d\mathrm{w})$ a.s.,

$$\limsup_{r\to 0} \frac{1}{h(r)} \int \mathcal{N}(dtd\omega)\, \mathcal{Z}(\omega)(B(0,r)) \le c\,, \quad P^{(\mathrm{w})} \text{ a.s.} \tag{16.5.8}$$

From now on, we will fix $a > 0$ and $\mathrm{w} \in \mathcal{W}_0$ such that $\zeta_\mathrm{w} = a$. We will prove (16.5.8) for these particular values of a, w. We may and will assume that w satisfy certain properties that hold P_0^a a.s. More precisely, we will assume that $\mathrm{w}(t) \neq 0$ for every $t \in (0, a]$ and furthermore that w satisfies the following two properties:

(H1) There exists a constant $c(\mathrm{w})$ such that, for every $\rho \in (0, 1/2)$,

$$\frac{1}{|\log \rho|} \int_0^a dt\, (G(\mathrm{w}(t)) \wedge \rho^{-2}) \le c(\mathrm{w}).$$

(H2) For every $k \ge 1$, set $\delta_k(\mathrm{w}) = \int_0^a dt\, 1\{|\mathrm{w}(t)| \le 2^{-k}\}$. There exists a constant $c'(\mathrm{w})$ such that, for every $k \ge 1$,

$$\delta_k(\mathrm{w}) \le c'(\mathrm{w})\, 2^{-2k} \log\log(2^k).$$

Both (H1) and (H2) hold outside a set of zero P_0^a-probability. For (H1), one may use Birkhoff's ergodic theorem to verify that

$$\lim_{k\to\infty} \frac{1}{k} \int_{2^{-k}}^a \frac{dt}{|\mathrm{w}(t)|^2} = c < \infty\,, \quad P_0^a(d\mathrm{w}) \text{ a.s.}$$

For (H2), see, e.g., Ciesielski and Taylor [2].

For simplicity, we write

$$\mathcal{Y} = \int \mathcal{N}(dtd\omega)\, \mathcal{Z}(\omega)$$

so that $\mathcal{Y}$ and $\mathcal{Y}_r$ are random measures on $\mathbb{R}^4$, defined on the probability space that supports $P^{(\mathrm{w})}$. Recall the notation $\mathcal{Z}^K$ for the occupation measure of the snake stopped at the boundary of $B(0, K)$. We also set

$$\mathcal{Y}^K = \int \mathcal{N}(dtd\omega)\, \mathcal{Z}^K(\omega), \quad \mathcal{Y}_r^K = \int_{\{t\le r\}} \mathcal{N}(dtd\omega)\, \mathcal{Z}^K(\omega).$$

These definitions make sense provided that $|\mathrm{w}(t)| < K, \forall t \in [0, a]$.

We set $\rho_j = 2^{-j}$ for every $j \geq 1$.

Lemma 16.5.2 *There exists a constant C_{25}, independent of a, w such that*

$$\limsup_{j\to\infty} \frac{1}{h(\rho_{2j})}\mathcal{Y}(B(0, \rho_{2j})) \leq C_{25}\,, \quad P^{(\mathrm{w})}\ a.s. \tag{16.5.9}$$

Proof. Let $K > 0$ be such that $|\mathrm{w}(t)| < K/2, \forall t \in [0, a]$. For every $\lambda > 0$,

$$E^{(\mathrm{w})}\Big[\exp(\lambda\mathcal{Y}^K(B(0, \rho_{2j})))\Big] = \exp\left(4\int_0^u dt\, \mathbb{N}_{\mathrm{w}(t)}\Big[e^{\lambda\mathcal{Z}^K(B(0,\rho_{2j}))} - 1\Big]\right).$$

However, by Proposition 16.3.2, for $x \in B(0, K/2)$,

$$\begin{aligned}\mathbb{N}_x\Big[e^{\lambda\mathcal{Z}^K(B(0,\rho_{2j}))} - 1\Big] &= \sum_{p=1}^{\infty}\frac{\lambda^p}{p!}\mathbb{N}_x[\mathcal{Z}^K(B(0, \rho_{2j}))^p] \\ &\leq (G(x) \wedge \rho_{2j}^{-2})\Big(\log\frac{1}{\rho_{2j}}\Big)^{-1}\sum_{p=1}^{\infty} C_7^p\lambda^p\Big(\rho_{2j}^4 \log\frac{1}{\rho_{2j}}\Big)^p.\end{aligned}$$

By taking $\lambda = \Big(2C_7\rho_{2j}^4 \log 1/\rho_{2j}\Big)^{-1}$, we get

$$\begin{aligned}E^{(\mathrm{w})}\Big[\exp\Big(\frac{\mathcal{Y}^K(B(0, \rho_{2j}))}{2C_7\rho_{2j}^4 \log 1/\rho_{2j}}\Big)\Big] &\leq \exp\Big(\frac{4}{|\log\rho_{2j}|}\int_0^a dt\,(G(\mathrm{w}(t)) \wedge \rho_{2j}^{-2})\Big) \\ &\leq \exp(4c(\mathrm{w})),\end{aligned}$$

using assumption (H1). For every $M > 0$, we have thus

$$P^{(\mathrm{w})}[\mathcal{Y}^K(B(0, \rho_{2j})) \geq Mh(\rho_{2j})] \leq \exp\Big(4c(\mathrm{w}) - \frac{M}{2C_7}\log\log\log\Big(\frac{1}{\rho_{2j}}\Big)\Big)$$

and the Borel-Cantelli lemma immediately gives

$$\limsup_{j\to\infty}\frac{1}{h(\rho_{2j})}\mathcal{Y}^K(B(0, \rho_{2j})) \leq 2C_7\,, \quad P^{(\mathrm{w})}\ \text{a.s.}$$

This is not yet the desired result, because we need to replace $\mathcal{Y}^K$ by $\mathcal{Y}$, and furthermore the constant $C_7 = C_7(K)$ depends on K (see Proposition 16.3.2). However, we can argue as follows.

Let us fix any value of K, say $K = 1$ for definiteness. Clearly, if $r > 0$ is small enough the paths $W_s(\omega), s \geq 0$, for $(t, \omega) \in \operatorname{supp}(1_{[0,r]}(t)\mathcal{N}(dtd\omega))$, will remain in the ball $B(0,1)$. Thus, $P^{(\mathrm{w})}$ a.s. we can choose a rational $r \in (0,a]$ so that $|\mathrm{w}(u)| < 1/2$ for every $u \in [0,r]$ and

$$\mathcal{Y}_r = \mathcal{Y}_r^1.$$

On the other hand, we have also $P^{(\mathrm{w})}$ a.e. for all j large enough,

$$\mathcal{Y}(B(0,\rho_{2j})) = \mathcal{Y}_r(B(0,\rho_{2j})).$$

This is so because $P^{(\mathrm{w})}$ a.s. for every $u \in (0,a]$ the closed support of the measure $\mathcal{Y} - \mathcal{Y}^u$ does not contain 0 (recall that we assume $\mathrm{w}(t) \neq 0$ for every $t \in (0,a]$ and that for every $x \neq 0$, $\mathbb{N}_x[0 \in \mathcal{R}] = 0$). By combining the previous two observations, we obtain that

$$\limsup_{j\to\infty} \frac{1}{h(\rho_{2j})} \mathcal{Y}(B(0,\rho_{2j})) = \limsup_{j\to\infty} \frac{1}{h(\rho_{2j})} \mathcal{Y}_r^1(B(0,\rho_{2j})) \le (2C_7(1))^{-1},$$

where for the last bound we used the first part of the proof applied to the path w stopped at time r. This completes the proof of Lemma 16.5.2. □

Clearly, the estimate (16.5.9) would yield (16.5.8) if we were able to replace ρ_{2j} by ρ_j in (16.5.9). In other words, we need to control the values $\mathcal{Y}(B(0,\rho_k))$ from the information on $\mathcal{Y}(B(0,\rho_{2j}))$ provided by the previous lemma. We will achieve this task by showing that, if $2^{j-1} \le k \le 2^j$ and if $\mathcal{Y}(B(0,\rho_k))$ is large, then $\mathcal{Y}(B(0,\rho_{2j}))$ must also be large with a high probability. The precise implementation of this idea requires introducing the exit measures from the open sets $\mathbb{R}^4 \backslash B(0,\rho)$ and using the special Markov property.

We first give the form of the special Markov property under the measures $P^{(\mathrm{w})}$ that is suitable for our needs. Let us write

$$\mathcal{N}(dtd\omega) = \sum_{i\in I} \delta_{(t_i,\omega^i)}(dtd\omega).$$

We fix $k \ge 1$ and set $I_k = \{i \in I, |\mathrm{w}(t_i)| > \rho_k\}$. For each excursion ω^i, $i \in I_k$, we may consider the exit measure $X^{\Omega^{(k)}}(\omega^i)$ of ω^i from the domain $\Omega^{(k)} = \mathbb{R}^4 \backslash \bar{B}(0,\rho_k)$. We write

$$Y^{(k)} = \sum_{i\in I_k} X^{\Omega^{(k)}}(\omega^i).$$

Then, for every $i \in I_k$, we set $\omega^{i,(k)}(s) = \omega^i(\eta_i^{(k)}(s))$, where

$$\eta_i^{(k)}(s) = \inf\left\{r \ge 0, \int_0^r du\, 1\{\zeta_{\omega^i(u)} \le \tau^{(k)}(\omega^i(u))\} > s\right\},$$

and $\tau^{(k)}$ denotes the first exit time from $\Omega^{(k)}$. We let $\mathcal{E}^{(k)}$ be the σ-field generated by the point measure

$$\sum_{i\in I_k} \delta_{\omega^{i,(k)}}.$$

Note that $Y^{(k)}$ is $\mathcal{E}^{(k)}$-measurable. From our definitions and the fact that the sequence $\Omega^{(k)}$ is increasing, the sequence of σ-fields $\mathcal{E}^{(k)}$ is also increasing.

Finally, for every $i \in I_k$, we denote by ω^i_j, $j \in J_{k,i}$ the excursions of ω^i outside $\Omega^{(k)}$ (cf the definitions preceding Theorem 16.2.2). An application of Theorem 16.2.2 shows that under $P^{(\mathrm{w})}$, conditionally on the σ-field $\mathcal{E}^{(k)}$, the point measure

$$\mathcal{M}_k(d\omega) \equiv \sum_{i\in I_k}\sum_{j\in J_{k,i}} \delta_{\omega^i_j}(d\omega)$$

is a Poisson point measure with intensity $\int Y^{(k)}(dy)\,\mathbb{N}_y(d\omega)$.

Lemma 16.5.3 *There exists a constant C_{26} such that, $P^{(\mathrm{w})}$ a.s.*

$$\limsup_{k\to\infty} \frac{\rho_k^2}{h(\rho_k)} \langle Y^{(k)}, 1\rangle \le C_{26}. \tag{16.5.10}$$

Proof. Let us fix $k \ge 1$ and j such that $2^{j-1} \le k < 2^j$. We will show that if $\langle Y^{(k)}, 1\rangle$ is large then, with a probability bounded from below, $\mathcal{Y}(B(0, \rho_{2j}))$ must also be large. To make this more precise, let $\tilde{\mathcal{Y}}(B(0, \rho_{2j}))$ be the contribution to $\mathcal{Y}(B(0, \rho_{2j}))$ of those excursions ω^i that start ouside $B(0, \rho_k)$:

$$\tilde{\mathcal{Y}}(B(0, \rho_{2j})) = \sum_{i\in I_k} \mathcal{Z}(\omega^i)(B(0, \rho_{2j})) \le \mathcal{Y}(B(0, \rho_{2j})).$$

With the previous notation, we may write

$$\tilde{\mathcal{Y}}(B(0, \rho_{2j})) = \int \mathcal{M}_k(d\omega)\, \mathcal{Z}(\omega)(B(0, \rho_{2j}))$$

and the special Markov property (in the form stated above) allows us to compute the conditional moments of $\tilde{\mathcal{Y}}(B(0, \rho_{2j}))$: If y_ρ denotes any fixed point such that $|y_\rho| = \rho$,

$$\begin{aligned} E^{(\mathrm{w})}\Big[\tilde{\mathcal{Y}}(B(0, \rho_{2j}))\Big|\mathcal{E}^{(k)}\Big] &= \langle Y^{(k)}, 1\rangle \mathbb{N}_{y_{\rho_k}}[\mathcal{Z}(B(0, \rho_{2j}))] \\ &\ge c\langle Y^{(k)}, 1\rangle \rho_k^{-2}\, \rho_{2j}^4, \end{aligned}$$

and

$$\begin{aligned} \mathrm{var}_{P^{(\mathrm{w})}}\Big(\tilde{\mathcal{Y}}(B(0, \rho_{2j}))\Big|\mathcal{E}^{(k)}\Big) &= \langle Y^{(k)}, 1\rangle \mathbb{N}_{y_{\rho_k}}[\mathcal{Z}(B(0, \rho_{2j}))^2] \\ &\le c'\langle Y^{(k)}, 1\rangle \rho_k^{-2}\, \rho_{2j}^8 \log\frac{\rho_k}{\rho_{2j}}. \end{aligned}$$

Here, for $0 < r \le \rho/2$, we used the bounds

$$\mathbb{N}_{y_\rho}[\mathcal{Z}(B(0, r))] \ge c\,\rho^{-2} r^4, \quad \mathbb{N}_{y_\rho}[\mathcal{Z}(B(0, r))^2] \le c'\,\rho^{-2} r^8 \log\frac{\rho}{r},$$

that follow easily from the explicit expressions of the moments of $\mathcal{Z}$.

By Tchebicheff's inequality, we have for $\alpha > 0$

$$P^{(\mathrm{w})}\Big[\Big|\tilde{\mathcal{Y}}(B(0, \rho_{2j})) - E^{(\mathrm{w})}\Big[\tilde{\mathcal{Y}}(B(0, \rho_{2j}))\Big|\mathcal{E}^{(k)}\Big]\Big| \ge \alpha \,\Big|\, \mathcal{E}^{(k)}\Big]$$

$$\le \alpha^{-2} \operatorname{var}_{P^{(w)}} \Big(\tilde{\mathcal{Y}}(B(0, \rho_{2^j})) \Big| \mathcal{E}^{(k)} \Big).$$

Applying this inequality with $\alpha = 2\langle Y^{(k)}, 1\rangle^{1/2} (c' |\log \rho_{2^j}|)^{1/2} \rho_k^{-1} \rho_{2^j}^4$, we arrive at

$$P^{(w)}\Big[\tilde{\mathcal{Y}}(B(0, \rho_{2^j})) \ge$$
$$\Big(c\langle Y^{(k)}, 1\rangle \rho_k^{-1} - 2\langle Y^{(k)}, 1\rangle^{1/2} (c'|\log \rho_{2^j}|)^{1/2} \Big) \rho_k^{-1} \rho_{2^j}^4 \;\Big|\; \mathcal{E}^{(k)} \Big] \ge \frac{3}{4}.$$

Therefore, we can choose two positive constants c'', M, independent of k, so that on the $\mathcal{E}^{(k)}$-measurable event $\{\langle Y^{(k)}, 1\rangle \ge M \rho_k^2 |\log \rho_k|\}$, one has

$$P^{(w)}\Big[\tilde{\mathcal{Y}}(B(0, \rho_{2^j})) \ge c'' \rho_k^{-2} \rho_{2^j}^4 \langle Y^{(k)}, 1\rangle \;\Big|\; \mathcal{E}^{(k)} \Big] \ge \frac{3}{4}. \tag{16.5.11}$$

We claim that (16.5.10) follows from (16.5.9) and (16.5.11), with a constant C_{26} given by $C_{26} = 2C_{25}/c''$. To verify this, let $c''' > 2C_{25}/c''$ and let $(k_n, n \in \mathbb{N})$ be the successive values of k for which $\langle Y^{(k)}, 1\rangle \ge c''' \rho_k^{-2} h(\rho_k)$, with the convention that $k_n = \infty$ if the nth such value does not exist. Since $Y^{(k)}$ is $\mathcal{E}^{(k)}$-measurable, each k_n is a stopping time of the filtration $(\mathcal{E}^{(k)})$. We have to show that $P^{(w)}[k_n < \infty, \forall n] = 0$. When $k_n < \infty$, let j_n be such that $2^{j_n - 1} \le k_n < 2^{j_n}$. Let $\bar{c}$ be such that $c''c'''/2 > \bar{c} > C_{25}$ and let $\mathcal{A}_n$ be the event

$$\mathcal{A}_n = \{k_n < \infty \text{ and } \tilde{\mathcal{Y}}(B(0, \rho_{2^{j_n}})) \ge \bar{c}\, h(\rho_{2^{j_n}})\}.$$

The estimate (16.5.11), and the fact that k_n is a stopping time imply that, for n sufficiently large,

$$P^{(w)}[\mathcal{A}_n] \ge \frac{3}{4} P^{(w)}[k_n < \infty]$$

(note that $\rho_k^{-4} \rho_{2^j}^4 h(\rho_k) \ge (\bar{c}/c''c''') h(\rho_{2^j})$ for j sufficiently large and $2^{j-1} \le k \le 2^j$). It follows that

$$P^{(w)}\Big[\limsup_{n\to\infty} \mathcal{A}_n \Big] \ge \limsup_{n\to\infty} P^{(w)}[\mathcal{A}_n] \ge \frac{3}{4} P^{(w)}[k_n < \infty, \forall n].$$

The left side is zero by (16.5.9) and the proof is complete. □

We now complete the proof of Proposition 16.5.2. We have to check that (16.5.8) holds, or equivalently that

$$\limsup_{k\to\infty} \frac{1}{h(\rho_k)} \mathcal{Y}(B(0, \rho_k)) \le c, \quad P^{(w)} \text{ a.s.} \tag{16.5.12}$$

for a suitable constant c. Following the ideas of the proof of Lemma 16.5.3 we introduce for every $k \ge 1$

$$\tilde{\mathcal{Y}}(B(0, \rho_k)) = \int \mathcal{N}(dt d\omega)\, 1\{|\mathrm{w}(t)| > \rho_k\}\, \mathcal{Z}(\omega)(B(0, \rho_k)),$$

$$\hat{\mathcal{Y}}(B(0, \rho_k)) = \int \mathcal{N}(dt d\omega)\, 1\{|\mathrm{w}(t)| \le \rho_k\}\, \mathcal{Z}(\omega)(B(0, \rho_k)),$$

so that $\mathcal{Y}(B(0,\rho_k)) = \tilde{\mathcal{Y}}(B(0,\rho_k)) + \hat{\mathcal{Y}}(B(0,\rho_k))$. We first verify that

$$\lim_{k\to\infty} \frac{1}{h(\rho_k)} \hat{\mathcal{Y}}(B(0,\rho_k)) = 0\,, \qquad P^{(\mathrm{w})} \text{ a.s.} \tag{16.5.13}$$

This follows from simple moment calculations. Using the notation introduced in (H2), we have

$$\begin{aligned} E^{(\mathrm{w})}[\hat{\mathcal{Y}}(B(0,\rho_k))] &= 4\int_0^a dt\, 1\{|\mathrm{w}(t)| \le \rho_k\}\, \mathbb{N}_{\mathrm{w}(t)}[\mathcal{Z}(B(0,\rho_k))] \\ &\le 4\,\delta_k(\mathrm{w}) \sup_{|y|\le\rho_k} \mathbb{N}_y[\mathcal{Z}(B(0,\rho_k))] \\ &\le 4\,C\delta_k(\mathrm{w})\,\rho_k^2. \end{aligned}$$

On the other hand, we have

$$\begin{aligned} \mathrm{var}_{P^{(\mathrm{w})}}(\hat{\mathcal{Y}}(B(0,\rho_k))) &= 4\int_0^a dt\, 1\{|\mathrm{w}(t)| \le \rho_k\}\, \mathbb{N}_{\mathrm{w}(t)}[\mathcal{Z}(B(0,\rho_k))^2] \\ &\le 4\,C'\delta_k(\mathrm{w})\rho_k^6, \end{aligned}$$

using the easy bound $\sup_{z\in\mathbb{R}^4} \mathbb{N}_z[\mathcal{Z}(B(0,\rho))^2] \le C'\rho^6$ which follows via straightforward calculations from the explicit formula for $\mathbb{N}_z[\mathcal{Z}(B(0,\rho))^2]$ given by (16.2.3). By combining the previous two bounds and using (H2), we get the existence of a finite constant $C''(\mathrm{w})$ such that

$$E^{(\mathrm{w})}[\hat{\mathcal{Y}}(B(0,\rho_k))^2] \le C''(\mathrm{w})\,\rho_k^8\,(\log\log\frac{1}{\rho_k})^2.$$

Then the Tchebicheff inequality gives for every $\varepsilon > 0$

$$\sum_{k=1}^{\infty} P^{(\mathrm{w})}[\hat{\mathcal{Y}}(B(0,\rho_k)) \ge \varepsilon\, h(\rho_k)] < \infty,$$

which implies (16.5.13).

It remains to control $\tilde{\mathcal{Y}}(B(0,\rho_k))$. Let $C_{27} = C_{26}+1$. We will prove the existence of a constant C_{28} such that

$$\sum_{k=1}^{\infty} P^{(\mathrm{w})}[\langle Y^{(k)}, 1\rangle \le C_{27}\rho_k^{-2}h(\rho_k),\ \tilde{\mathcal{Y}}(B(0,\rho_k)) \ge C_{28}\, h(\rho_k)] < \infty. \tag{16.5.14}$$

Together with (16.5.10), (16.5.14) gives

$$\limsup_{k\to\infty} \frac{1}{h(\rho_k)} \tilde{\mathcal{Y}}(B(0,\rho_k)) \le C_{28}\,, \qquad P^{(\mathrm{w})} \text{ a.s.} \tag{16.5.15}$$

and (16.5.12) follows from (16.5.13) and (16.5.15).

To prove (16.5.14), we use the special Markov property as in the proof of Lemma 16.5.3. We have

$$\tilde{\mathcal{Y}}(B(0,\rho_k)) = \int \mathcal{M}_k(d\omega)\, \mathcal{Z}(\omega)(B(0,\rho_k))$$

and this formula allows us to compute conditional moments of $\tilde{\mathcal{Y}}(B(0,\rho_k))$ knowing $\mathcal{E}^{(k)}$. First,

$$E^{(\mathrm{w})}\Big[\tilde{\mathcal{Y}}(B(0,\rho_k))\,\Big|\,\mathcal{E}^{(k)}\Big]=\int Y^{(k)}(dy)\,\mathbb{N}_y[\mathcal{Z}(B(0,\rho_k)]\le C_{29}\langle Y^{(k)},1\rangle\rho_k^2,$$

where the constant C_{29} is chosen so that $\mathbb{N}_y[\mathcal{Z}(B(0,\rho))]\le C_{29}\rho^2$ for every $\rho>0$, $y\in\mathbb{R}^4$. Then, by a well-known formula for Poisson measures,

$$\begin{aligned}
&E^{(\mathrm{w})}\Big[\Big(\tilde{\mathcal{Y}}(B(0,\rho_k))-E^{(\mathrm{w})}\Big[\tilde{\mathcal{Y}}(B(0,\rho_k))\,\Big|\,\mathcal{E}^{(k)}\Big]\Big)^4\,\Big|\,\mathcal{E}^{(k)}\Big]\\
&=3\Big(\int Y^{(k)}(dy)\,\mathbb{N}_y[\mathcal{Z}(B(0,\rho_k)^2]\Big)^2+\int Y^{(k)}(dy)\,\mathbb{N}_y[\mathcal{Z}(B(0,\rho_k)^4]\\
&\le 3c^2\rho_k^{12}\langle Y^{(k)},1\rangle^2+c'\rho_k^{14}\langle Y^{(k)},1\rangle,
\end{aligned}$$

using the bounds $\mathbb{N}_y[\mathcal{Z}(B(0,\rho_k))^2]\le c\,\rho_k^6$, $\mathbb{N}_y[\mathcal{Z}(B(0,\rho_k))^4]\le c'\,\rho_k^{14}$.

We take $C_{28}=C_{27}C_{29}+1$. Then

$$\begin{aligned}
&P^{(\mathrm{w})}[\langle Y^{(k)},1\rangle\le C_{27}\rho_k^{-2}h(\rho_k),\ \tilde{\mathcal{Y}}(B(0,\rho_k))\ge C_{28}\,h(\rho_k)]\\
&\quad\le P^{(\mathrm{w})}\Big[\langle Y^{(k)},1\rangle\le C_{27}\rho_k^{-2}h(\rho_k),\\
&\qquad\qquad \tilde{\mathcal{Y}}(B(0,\rho_k))-E^{(\mathrm{w})}\Big[\tilde{\mathcal{Y}}(B(0,\rho_k))\,\Big|\,\mathcal{E}^{(k)}\Big]\ge h(\rho_k)\Big]\\
&\quad\le E^{(\mathrm{w})}\Big[1\{\langle Y^{(k)},1\rangle\le C_{27}\rho_k^{-2}h(\rho_k)\}\\
&\qquad\times\ h(\rho_k)^{-4}E^{(\mathrm{w})}\Big[\Big(\tilde{\mathcal{Y}}(B(0,\rho_k))-E^{(\mathrm{w})}\Big[\tilde{\mathcal{Y}}(B(0,\rho_k))\,\Big|\,\mathcal{E}^{(k)}\Big]\Big)^4\,\Big|\,\mathcal{E}^{(k)}\Big]\Big]\\
&\quad\le\frac{3c^2C_{27}^2\rho_k^8h(\rho_k)^2+c'C_{27}\rho_k^{12}h(\rho_k)}{h(\rho_k)^4},
\end{aligned}$$

using the previous estimates. Summing over k gives a convergent series. This completes the proof of (16.5.14) and of Proposition 16.5.2. □

16.6 Proof of the Main Results

The results stated in the introduction are straightforward consequences of the following theorem.

Theorem 16.6.1 *There exists a positive constant K_d such that, $\mathbb{N}_0$ a.e. for every Borel subset A of $\mathbb{R}^d$,*

$$h-m(A\cap\mathcal{R})=K_d\,\mathcal{Z}(A).$$

PROOF. By combining Propositions 16.4.2 and 16.4.1 if $d\ge5$, Propositions 16.5.1 and 16.5.2 if $d=4$, we know that there exist two positive constants $M_1=M_1(d)$, $M_2=M_2(d)$ such that, $\mathbb{N}_0$ a.e. for every Borel subset A of $\mathbb{R}^d$,

$$M_1\,\mathcal{Z}(A)\le h-m(A\cap\mathcal{R})\le M_2\,\mathcal{Z}(A).$$

Let us set

$$\theta(A) = h - m(A \cap \mathcal{R}).$$

Then, $\mathbb{N}_0$ a.e. the measure θ is absolutely continuous with respect to $\mathcal{Z}$ and its Radon-Nikodym derivative satisfies

$$0 < M_1 \le \frac{d\theta}{d\mathcal{Z}}(y) \le M_2 < \infty\,, \qquad \mathcal{Z}(dy) \text{ a.e.} \tag{16.6.1}$$

To prove Theorem 16.6.1, we have to check that this Radon-Nikodym derivative is constant, $\mathcal{Z}(dy)$ a.e. To this end, we use the same argument as in [13], relying on a result of geometric measure theory which can be found in [7, Theorems 2.9.5 and 2.9.7]. Let (r_n) be a fixed deterministic sequence of positive numbers decreasing to 0. Then,

$$\frac{d\theta}{d\mathcal{Z}}(y) = \limsup_{n\to\infty} \frac{\theta(B(y, r_n))}{\mathcal{Z}(B(y, r_n))}\,, \qquad \mathcal{Z}(dy) \text{ a.e.} \tag{16.6.2}$$

The statement of Theorem 16.6.1 follows from (16.6.1), (16.6.2) and the next lemma.

Lemma 16.6.1 *There exists a constant $K_d \in [0, \infty]$ such that*

$$\limsup_{n\to\infty} \frac{\theta(B(y, r_n))}{\mathcal{Z}(B(y, r_n))} = K_d\,, \qquad \mathcal{Z}(dy) \textit{ a.e.}$$

Remark Because of (16.6.1) and (16.6.2) we have $M_1 \le K_d \le M_2$, so that K_d is both positive and finite.

PROOF. We again rely on the Palm formula (16.5.7). Recall the notation $\mathcal{Y}$, $\mathcal{Y}_r$ from the previous section and set

$$U = \limsup_{n\to\infty} \frac{h - m(\operatorname{supp}\mathcal{Y} \cap B(0, r_n))}{\mathcal{Y}(B(0, r_n))}\,,$$

so that U is a random variable defined on the probability space that supports the measures $P^{(\mathrm{w})}$, $\mathrm{w} \in \mathcal{W}_0$. We first prove that there exists a constant $K_d \in [0, \infty]$ such that, for every $a > 0$,

$$U = K_d \quad P^{(\mathrm{w})} \text{ a.s., } P_0^a(d\mathrm{w}) \text{ a.s.} \tag{16.6.3}$$

Let us fix $a > 0$ and consider $u \in (0, a)$. Let $\mathrm{w} \in \mathcal{W}_0$ with lifetime a and such that $\mathrm{w}(t) \neq 0$ for every $t > 0$. Then $P^{(\mathrm{w})}$ a.s. the support of $\mathcal{Y} - \mathcal{Y}_u$ does not contain 0 (recall that $\mathbb{N}_y[0 \in \mathcal{R}] = 0$ if $y \neq 0$). Hence, $P^{(\mathrm{w})}$ a.s. for all n sufficiently large we have

$$\mathcal{Y}(B(0, r_n)) = \mathcal{Y}_u(B(0, r_n)),$$
$$\operatorname{supp}\mathcal{Y} \cap B(0, r_n) = \operatorname{supp}\mathcal{Y}_u \cap B(0, r_n).$$

Therefore, we have also

$$U = \limsup_{n\to\infty} \frac{h - m(\operatorname{supp}\mathcal{Y}_u \cap B(0, r_n))}{\mathcal{Y}_u(B(0, r_n))}. \tag{16.6.4}$$

Thus, U is measurable with respect to the σ-field

$$\mathcal{G}_u = \sigma\Big(1_{[0,u]}(t)\,\mathcal{N}(dt d\omega)\Big).$$

By a standard argument, the tail σ-field $\cap_{u>0}\mathcal{G}_u$ is $P^{(\mathrm{w})}$-trivial, so that U must be constant $P^{(\mathrm{w})}$ a.s.,

$$U = K(\mathrm{w}), \quad P^{(\mathrm{w})} \text{ a.s.}$$

where $K(\mathrm{w}) \in [0, \infty]$. Notice that $K(\mathrm{w})$ is a measurable function of w, because the collection $(P^{(\mathrm{w})}, \mathrm{w} \in \mathcal{W}_0)$ can clearly be chosen to be measurable. Next observe that, by (16.6.4) again, if $\mathrm{w}_{(u)}$ denotes the path w stopped at time u,

$$K(\mathrm{w}) = K(\mathrm{w}_{(u)}), \quad P_0^a(d\mathrm{w}) \text{ a.s.} \tag{16.6.5}$$

The standard zero-one law for Brownian motion then implies that $K(\mathrm{w})$ is constant $P_0^a(d\mathrm{w})$ a.s. Furthermore, (16.6.5) shows that this constant does not depend on a. This completes the proof of (16.6.3).

We then apply the Palm formula (16.5.7) to the function

$$F(\nu) = 1\left\{\limsup_{n\to\infty} \frac{h-m(\operatorname{supp}\nu \cap B(0,r_n))}{\nu(B(0,r_n))} \neq K_d\right\}$$

and the statement of Lemma 16.6.1 follows from (16.6.3). □

Proof of Theorem 16.1.1. By a trivial scaling argument, it is enough to treat the case $\gamma = 1$ (if we replace X by λX, for $\lambda > 0$, then $\mathbf{Z}$ is replaced by $\lambda\mathbf{Z}$, and γ by $\lambda\gamma$, whereas $\mathbf{R}$ is unchanged). We then recall the connection between super-Brownian motion and the Brownian snake which is relevant to the proof (see [9]). Let $\mu \in M_f(\mathbb{R}^d)$ and let

$$\mathcal{M}(dx d\omega) = \sum_{i\in I} \delta_{(x_i,\omega_i)}(dx d\omega)$$

be a Poisson measure on $\mathbb{R}^d \times C(\mathbb{R}_+, \mathcal{W})$ with intensity $\mu(dx)\mathbb{N}_x(d\omega)$. Then there exists a super-Brownian motion X, with $\gamma = 1$ and initial value $X_0 = \mu$, whose range $\mathbf{R}$ and total occupation measure $\mathbf{Z}$ satisfy the relations

$$\mathbf{R} \cup \{x_i, i \in I\} = \bigcup_{i\in I} \mathcal{R}(\omega_i), \tag{16.6.6}$$

and

$$\mathbf{Z} = \sum_{i\in I} \mathcal{Z}(\omega_i). \tag{16.6.7}$$

By using Lemma 16.2.1 (or the deeper results of Serlet [15] on multiple points of super-Brownian motion), one easily gets that

$$h-m(\mathcal{R}(\omega_i) \cap \mathcal{R}(\omega_j)) = 0,$$

for every $i \neq j$, a.s. Therefore, the relation (16.6.6) implies that a.s. for every $A \in \mathcal{B}(\mathbb{R}^d)$,

$$h - m(A \cap \mathbf{R}) = \sum_{i \in I} h - m(A \cap \mathcal{R}(\omega_i)).$$

Theorem 16.1.1 follows from this last identity, Theorem 16.6.1 and (16.6.7). □

PROOF OF COROLLARY 16.1.1. Recall the notation σ for the duration of the excursion under $\mathbb{N}_x$. Set $\mathbb{N}_0^{(1)} = \mathbb{N}_0[\cdot \mid \sigma \geq 1]$, and define for every $s \in [0, 1]$ a finite path $W_s^* = (W_s^*(t), 0 \leq t \leq \zeta_s^*)$ by

$$W_s^*(t) = \sigma^{-1/4} W_{\sigma s}(\sigma^{1/2} t), \qquad \zeta_s^* = \sigma^{-1/2} \zeta_{\sigma s}.$$

From the scaling properties of Brownian motion and the Itô excursion measure, one easily verifies that the law of the process $(W_s^*, 0 \leq s \leq 1)$ under $\mathbb{N}_0^{(1)}$ is as described at the end of Section 16.2. As a consequence, the random measure $\mathbf{Z}^*$ defined under $\mathbb{N}_0^{(1)}$ by

$$\mathbf{Z}^*(A) = \int_0^1 ds \, 1_A(\hat{W}_s^*) = \sigma^{-1} \mathcal{Z}(\sigma^{1/4} A)$$

has the distribution of ISE. Furthermore the closed support of $\mathbf{Z}^*$ is

$$\mathbf{R}^* = \{\hat{W}_s^*, 0 \leq s \leq 1\} = \sigma^{-1/4} \mathcal{R}.$$

To complete the proof, use Theorem 16.6.1 and the scaling property of $h - m$ to write $\mathbb{N}_0^{(1)}$ a.s. for every $A \in \mathcal{B}(\mathbb{R}^d)$,

$$h - m(A \cap \mathbf{R}^*) = \sigma^{-1} h - m(\sigma^{1/4} A \cap \mathcal{R}) = K_d \sigma^{-1} \mathcal{Z}(\sigma^{1/4} A) = K_d \mathbf{Z}^*(A) \square$$

REFERENCES

[1] ALDOUS, D.J. (1993) The continuum random tree III. *Ann. Probab.* **21**, 248–289.

[2] CIESIELSKI, Z. AND TAYLOR, S.J. (1962) First passage and sojourn times for Brownian motion in space and the exact Hausdorff measure of the sample path. *Trans. Amer. Math. Soc.* **103**, 434–450.

[3] DAWSON, D.A., ISCOE, I., AND PERKINS, E.A. (1989) Super-Brownian motion: Path properties and hitting probabilities. *Probab. Th. Rel. Fields* **83**, 135–205.

[4] DAWSON, D.A. AND PERKINS, E.A. (1991) Historical processes. *Mem. Amer. Math. Soc.* **454**.

[5] DAWSON, D.A. AND PERKINS, E.A. (1998) Measure-valued processes and renormalization of branching processes. To appear in *Stochastic partial differential equations: Six perspectives*, AMS Math. Surveys and Monographs.

[6] DERBEZ, E. AND SLADE, G. (1998) The scaling limit of lattice trees in high dimensions. *Comm. Math. Phys.* **198**, 69–104.

[7] FEDERER, H. (1969) *Geometric Measure Theory.* Springer-Verlag, Berlin.

[8] HARA, T. AND SLADE, G. (1998) The incipient infinite cluster in high-dimensional percolation. *Electron. Res. Announc. Amer. Math. Soc.* **4**, 48–55.

[9] LE GALL, J.F. (1993) A class of path-valued Markov processes and its applications to superprocesses. *Probab. Th. Rel. Fields* **95**, 25–46.

[10] LE GALL, J.F. (1994) A path-valued Markov process and its connections with partial differential equations. In: *Proc. First European Congress of Mathematics, Vol.II*, pp.185–212. Birkhäuser, Boston.
[11] LE GALL, J.F. (1994) A lemma on super-Brownian motion with some applications. In *Festschrift in Honor of E.B. Dynkin* (M. Freidlin ed.) 237–251. Birkhäuser, Boston.
[12] LE GALL, J.F. (1995) The Brownian snake and solutions of $\Delta u = u^2$ in a domain. *Probab. Th. Rel. Fields* **102**, 393–432.
[13] LE GALL, J.F. AND PERKINS, E.A. (1995) The Hausdorff measure of the support of two-dimensional super-Brownian motion. *Ann. Probab.* **23**, 1719–1747.
[14] PERKINS, E.A. (1988) A space-time property of a class of measure-valued branching diffusions. *Trans. Amer. Math. Soc.* **305**, 743–795.
[15] SERLET, L. (1995) On the Hausdorff measure of multiple points and collision points of super-Brownian motion. *Stoch. Stoch. Reports* **54**, 169–198.
[16] SUGITANI, S. (1989) Some properties for the measure-valued branching diffusion process. *J. Math. Soc. Japan* **41**, 437–462.

DMI
Ecole Normale Supérieure
45, rue d'Ulm
75005 Paris
France
Jean-Francois.LeGall@ens.fr

17

Branching Random Walks on Finite Trees

Thomas M. Liggett

ABSTRACT Consider a branching random walk on the ball of radius N in a homogeneous tree. We obtain precise asymptotics on the critical value and on the extinction time (in critical and subcritical cases) as $N \to \infty$.

Keywords: Branching random walks, growth models on trees, contact processes.

AMS Subject Classification: 60K35.

17.1 Introduction

During the past several years, there has been a significant amount of interest in contact processes and branching random walks on homogeneous and inhomogeneous trees. This interest has been stimulated largely by the fact that these processes usually have two phase transition points, rather than the single one that they have on Z^d. Part of the theory of the contact process on Z^d deals with the way in which the size of the extinction time of the process on a large finite box reflects the critical behavior of the process on the infinite lattice. In this paper, we will consider a branching random walk on a large finite tree, and will obtain some explicit results on the asymptotic behavior of the process as the tree expands to an infinite homogeneous tree. Our technique involves an analysis of the spectral decomposition for the mean number of particles in the system, using the setup of multi-type branching processes. The results we obtain can be used to suggest results that may be true for the contact process on a large finite tree, though one should not expect the results to be as precise in that case. While we do not use their results directly, we should mention an early work on the spectral theory of single type branching processes: Karlin and McGregor (1966).

Harry Kesten has made so many contributions to so many areas of probability theory that it would be hard to write a paper in probability that did not relate in some manner to his work. The present paper connects with his contributions in a number of ways, including (a) the use of branching processes, a field in which Harry made seminal contributions thirty years ago, and (b) the connection with the contact process, the theory of which parallels in many ways percolation theory, which is the subject of Harry's book and the focus of much of his more recent work.

We begin with some definitions. If S is a finite or countable graph and λ is a positive parameter, we can define a branching random walk ζ_t on S in the following way:

1. At any given time, $\zeta_t \in \{0, 1, \dots\}^S$, and $\zeta_t(x)$ is interpreted as the number of particles at $x \in S$.
2. Particles die at rate 1.
3. For each pair of neighbors x, y in S, each particle at x generates a new particle at y at rate λ.

Under mild conditions on S and the initial configuration, this description determines a well defined process. The contact process η_t on S is defined in exactly the same way, except that now $\eta_t \in \{0, 1\}^S$, and when a new particle is generated at a point that is already occupied, the birth is suppressed. The branching random walk has been used in several papers to suggest results that might be proved for the contact process.

As is by now well known, the contact process on the d-dimensional integer lattice Z^d has a single critical point $\lambda_c^{(d)}$, with the property that the process dies out in all conceivable ways if $\lambda \leq \lambda_c^{(d)}$ and survives in a very strong sense if $\lambda > \lambda_c^{(d)}$. Some references on this are Bezuidenhout and Grimmett (1990, 1991), Durrett (1991), and Liggett (1985, 1999). In a series of papers, Durrett et al (1988a, 1988b, 1989) showed that even though the contact process on the finite set $\{0, \dots, N\}$ dies out, the extinction time has quite a different behavior depending on whether the process on Z^1 is subcritical, critical or supercritical. To state their results, let τ_N be the extinction time of the contact process on $\{0, \dots, N\}$ with initial configuration $\eta_0 \equiv 1$. Then there exist constants C_j so that

$$\frac{\tau_N}{\log N} \to C_1$$

in probability if $\lambda < \lambda_c^{(1)}$,

$$\lim_{N\to\infty} P\big(C_2 N \leq \tau_N \leq C_3 N^4\big) = 1$$

if $\lambda = \lambda_c^{(1)}$, and

$$\frac{\log \tau_N}{N} \to C_4$$

in probability if $\lambda > \lambda_c^{(1)}$. Using the results of Bezuidenhout and Grimmett, Chen (1994) has obtained an extension of the first statement, and a partial extension of the third statement for the contact process on $\{0, \dots, N\}^d$ for $d > 1$.

We turn now to results on trees. Let T_d be the homogeneous tree in which each vertex has $d + 1$ neighbors, and denote one of the vertices by O. If $d \geq 2$, both the contact process and the branching random walk have two distinct critical points that we will denote by $\lambda_1^{(d)} < \lambda_2^{(d)}$. (We will use the same notation for the critical points for the two processes, though they are of course not the same for the two processes.) Here are some of the results that are known about these processes. Starting with a single particle at O, the process is eventually $\equiv 0$ with probability one if $\lambda \leq \lambda_1^{(d)}$, the process survives with positive probability, but dies out locally (i.e., each vertex is visited only finitely often) with probability one if $\lambda_1^{(d)} < \lambda \leq \lambda_2^{(d)}$, and the process survives locally with positive probability (in fact, it survives locally whenever it survives at all) if $\lambda > \lambda_2^{(d)}$. For the branching random walk, these critical values can be computed explicitly:

$$\lambda_1^{(d)} = \frac{1}{d+1}, \qquad \lambda_2^{(d)} = \frac{1}{2\sqrt{d}}.$$

These results were proved in a series of papers by a number of authors over the past decade. See Liggett (1997, 1999) for detailed references. Recent results on the contact process and branching random walks on inhomogeneous trees are given in Pemantle and Stacey (1999). Of greatest interest are some that show that the two processes on a given tree can have substantially different behavior.

Let $T_{d,N}$ be the finite subtree of T_d obtained by retaining only those vertices that can be reached from O by a path of length $\leq N$. Stacey (1999) has considered the contact process on $T_{d,N}$ with a singleton initial state for $\lambda > \lambda_2^{(d)}$, and proved that with significant probability, the process survives for a time that is almost doubly exponential in N.

Our primary interest in this paper is the asymptotic behavior of the branching random walk on $T_{d,N}$ as $N \to \infty$. Unlike the contact process on a finite set, the branching random walk on $T_{d,N}$ will survive for sufficiently large λ. This is clear for $\lambda > 1$, for example, since then every particle has birth rate larger than 1, and death rate equal to 1. Let λ_c^N be the critical value for the branching random walk on $T_{d,N}$:

$$\lambda_c^N = \inf\{\lambda > 0 : P(\zeta_t \equiv 0) \not\to 1 \text{ as } t \to \infty\}.$$

It should not be surprising that $\lambda_c^N \to \lambda_2^{(d)}$ as $N \to \infty$. Our first result gives the rate in this convergence:

Theorem 17.1.1 *For the branching random walk on $T_{d,N}$,*

$$\lim_{N\to\infty} N^2\big(2\sqrt{d}\lambda_c^N - 1\big) = \frac{\pi^2}{2}.$$

This theorem is proved in Section 17.2. Next, we ask what the time scale for survival is for the branching random walk ζ_t on $T_{d,N}$. This is only of interest if $\lambda \leq \lambda_2^{(d)} = \frac{1}{2\sqrt{d}}$, since otherwise Theorem 17.1.1 implies that for large N, the process survives forever with positive probability.

Theorem 17.1.2 *Consider the branching random walk on $T_{d,N}$ with initial condition $\zeta_0(x) \equiv 1$. Fix a number $w > 0$. Then for N beyond some point there are times t_N so that*

$$E \sum_{x \in T_{d,N}} \zeta_{t_N}(x) = w \tag{17.1.1}$$

and

$$\lim_{N\to\infty}\left[t_N - N\log d\right] = -\log w + \log\frac{d+1}{d-1}$$

if $\lambda = 0$,

$$\lim_{N\to\infty}\left[t_N\left(1-2\lambda\sqrt{d}\right) - N\log d + \frac{3}{2}\log N\right] = -\log w \qquad (17.1.2)$$

$$+\log\left[\frac{(d+1)d^{\frac{1}{4}}}{2(\sqrt{d}-1)^4\sqrt{\pi}}\right] + \frac{3}{2}\log\left[\frac{1-2\lambda\sqrt{d}}{\lambda\log d}\right]$$

if $0 < \lambda < 1/2\sqrt{d}$, *and*

$$\lim_{N\to\infty}\frac{t_N}{N^3} = \frac{2\log d}{\pi^2}$$

if $\lambda = 1/2\sqrt{d}$.

The $\lambda = 0$ case is elementary, of course, and is included only for comparison purposes. The rest of Theorem 17.1.2 is proved in Section 17.3. Note that if N is large, there is always at least one t_N satisfying (17.1.1), since the mean number of particles is large at time $t = 0$, and tends to 0 as $t \to \infty$ if $\lambda < \lambda_c^N$. One somewhat surprising feature of Theorem 17.1.2 is that nothing unusual seems to happen in this limit at $\lambda = \lambda_1^{(d)}$. In fact, the reason we wanted to give such precise asymptotics in the case $0 < \lambda < 1/2\sqrt{d}$ was to see whether some nonsmoothness at the first critical value appeared in the lower order terms, but we see that it does not.

The fact that the first critical value is not "felt" in this limit may be a consequence of our choice of initial condition, which puts many particles close to the boundary of $T_{d,N}$. It would be of interest to determine the asymptotics of t_N for more sparse initial configurations, such as those that put one particle at each point in a copy of Z^1. In Section 17.3 we carry out part of the analysis that would be required for this. In fact, the results there easily imply an analogue of Theorem 17.1.2 for initial configurations that put m^k particles on sites a distance k from O, provided that $m > \sqrt{d}$. More interesting, though, is the case $m < \sqrt{d}$. In the extreme case in which initially there is a single particle at O (i.e., $m = 0$), the behavior of t_N is clearly different on the two sides of $\lambda_1^{(d)}$. It is bounded if $\lambda < \lambda_1^{(d)}$, and tends to ∞ if $\lambda > \lambda_1^{(d)}$. What happens if $0 < m < \sqrt{d}$?

17.2 Asymptotics of the Critical Value

This section is devoted to the proof of Theorem 17.1.1. It is useful to think of the branching random walk on $T_{d,N}$ as a multi-type branching process, with the type of a particle being its distance from the root O. Thus, there are types $0, 1, \ldots, N$. We will use the notation of Section 7 of Chapter V of Athreya and Ney (1972).

So, let $M_N(t)$ be the $(N+1)\times(N+1)$ matrix whose (j,k) entry gives the mean number of particles of type k at time t produced by a single parent particle of type j at time 0. The generator of this matrix semigroup is denoted by A_N, and in the present case is the tri-diagonal matrix whose (j,k) entry is given by

$$a_{j,k} = \begin{cases} -1 & \text{if } j = k, \\ \lambda & \text{if } j = k+1, \\ \lambda d & \text{if } j = k-1 \geq 1, \\ \lambda(d+1) & \text{if } j = 0, k = 1, \end{cases}$$

where $0 \leq j, k \leq N$. Referring to Athreya and Ney (1972), page 203, we see that the branching process is supercritical if and only if A_N has a positive eigenvalue. Writing $A_N = -I + \lambda C_N$, we see that the eigenvalues of A_N are of the form $-1+\lambda\sigma$, where σ is an eigenvalue of C_N (with the same eigenvector). In particular, if σ is a positive eigenvalue of C_N, the corresponding eigenvalue of A_N is an increasing function of λ. Therefore, λ_c^N is the smallest value of λ so that A_N has zero as an eigenvalue.

Let $f_N(x) = \det(xI - A_N)$ be the characteristic polynomial of A_N. It turns out that f_N can be computed exactly. To do this, expand the determinant that defines f_N along the bottom row to get the recursion

$$f_N(x) = (x+1)f_{N-1}(x) - \lambda^2 d f_{N-2}(x) \tag{17.2.1}$$

for $N \geq 2$, with initial conditions $f_0(x) = x+1$, $f_1(x) = (x+1)^2 - \lambda^2(d+1)$. This is a simple second order difference equation. The solution to (17.2.1) is given by

$$f_N(x) = \lambda^2\big[2d + N(d-1)\big]\left(\frac{x+1}{2}\right)^{N-1} \tag{17.2.2}$$

if $4\lambda^2 d = (x+1)^2$, and if not,

$$f_N(x) = C_1\gamma_1^N + C_2\gamma_2^N, \tag{17.2.3}$$

where

$$\gamma_1 = \frac{x+1-\sqrt{(x+1)^2-4\lambda^2 d}}{2}, \quad \gamma_2 = \frac{x+1+\sqrt{(x+1)^2-4\lambda^2 d}}{2},$$

$$C_1 = \frac{2\lambda^2(1+d)-(x+1)^2}{2\sqrt{(x+1)^2-4\lambda^2 d}} + \frac{x+1}{2},$$

$$C_2 = \frac{-2\lambda^2(1+d)+(x+1)^2}{2\sqrt{(x+1)^2-4\lambda^2 d}} + \frac{x+1}{2},$$

even if the argument of the square root above is negative. Note that (17.2.2) is not zero if $x \geq 0$. Therefore, 0 is an eigenvalue of A_N if and only if $4\lambda^2 d \neq 1$ and

$$\left(\frac{1-\sqrt{1-4\lambda^2 d}}{1+\sqrt{1-4\lambda^2 d}}\right)^N = \frac{2\lambda^2(1+d)-1-\sqrt{1-4\lambda^2 d}}{2\lambda^2(1+d)-1+\sqrt{1-4\lambda^2 d}}. \tag{17.2.4}$$

Making the substitution $u = \sqrt{1-4\lambda^2 d}$ in (17.2.4) when $4\lambda^2 d < 1$, it is clear that the left side of (17.2.4) decreases from 1 to 0 as u increases from 0 to 1. The right side of (17.2.4) can be written as

$$\frac{(1+u)\,[(d-1)+u(d+1)]}{(1-u)\,[(d-1)-u(d+1)]},$$

which is > 1 for $0 < u < (d-1)/(d+1)$, and < 0 for $(d-1)/(d+1) < u < 1$. Therefore, as expected, there is no solution to (17.2.4) for $4\lambda^2 d < 1$. By the comments made above, the critical value is the smallest value of $\lambda > 1/2\sqrt{d}$ that satisfies (17.2.4). Note that in this range, both sides of (17.2.4) are complex numbers of modulus one, and therefore (17.2.4) is just the statement that the arguments of the two sides agree.

Now consider a new change of variables:

$$4\lambda^2 d = 1 + \frac{u^2}{N^2}, \quad u \geq 0 \tag{17.2.5}$$

for each N. Making this substitution in (17.2.4) gives

$$\left(\frac{1-i\frac{u}{N}}{1+i\frac{u}{N}}\right)^N = \frac{(d-1)N^2-(d+1)u^2+2dNui}{(d-1)N^2-(d+1)u^2-2dNui}, \tag{17.2.6}$$

where $i = \sqrt{-1}$, and passing to the limit in N gives the identity

$$e^{-2iu} = 1, \tag{17.2.7}$$

with the convergence being uniform on compact u sets. If $0 < u < \frac{\pi}{2}$ and N is sufficiently large, then the left side of (17.2.6) has negative imaginary part, while the right side has positive imaginary part. Therefore, there is no solution in this range. Now, suppose that $u = u_N > 0$ is a solution of (17.2.6), and that $u_N \to u_\infty < \infty$ along a subsequence of N's. The limit u_∞ is necessarily $\geq \frac{\pi}{2}$ by the previous observation. Then u_∞ satisfies (17.2.7), and hence is a positive integer multiple of π. Therefore all positive solutions of (17.2.6) are asymptotically at least π.

To show that there is a solution of (17.2.6) that is asymptotically equal to π, take ϵ small and positive. Define the argument of a complex number z of modulus 1 in such a way that $-\pi < \arg z \leq \pi$, and let $L_N(u)$ and $R_N(u)$ be the left and right sides of (17.2.6) respectively. Then

$$\lim_{N\to\infty}\left[\arg L_N(\pi\pm\epsilon) - \arg R_N(\pi\pm\epsilon)\right] = \arg e^{-2i(\pi\pm\epsilon)} = \mp 2\epsilon.$$

It follows that for large N, $\arg L_N(\pi+\epsilon) - \arg R_N(\pi+\epsilon)$ and $\arg L_N(\pi-\epsilon) - \arg R_N(\pi-\epsilon)$ have opposite signs. Furthermore since ϵ is small, these arguments are close to zero. Therefore, the intermediate value theorem implies that for large N, there is a $u_N \in (\pi-\epsilon, \pi+\epsilon)$ so that $\arg L_N(u_N) - \arg R_N(u_N) = 0$. Since $L_N(u_N)$ and $R_N(u_N)$ have modulus 1, we conclude that $L_N(u_N) = R_N(u_N)$, i.e., u_N is a solution of (17.2.6).

We have now shown that the minimal positive solution of (17.2.6) converges to π as $N \to \infty$. Using (17.2.5), it follows that

$$4d\left(\lambda_c^N\right)^2 = 1 + \frac{\pi^2}{N^2} + o\left(\frac{1}{N^2}\right), \quad N \to \infty.$$

Taking square roots leads to the desired result.

17.3 Asymptotics of the Extinction Time

Here we prove Theorem 17.1.2. The main technique is to compute the eigenvalues and eigenvectors of A_N as explicitly as possible, and then write the mean number of particles at time t in terms of them. The hard work comes in determining the asymptotic behavior of the resulting expressions.

Suppose σ is an eigenvalue of C_N, so that $-1 + \lambda\sigma$ is an eigenvalue of A_N. We will show shortly that all the eigenvalues are real. Therefore, though it is not too important at this stage, we will write inequalities as though we knew this already. The largest eigenvalue σ^* satisfies $-1 + \lambda_c^N \sigma^* = 0$, so

$$\sigma \le \sigma^* = \frac{1}{\lambda_c^N} < 2\sqrt{d} \tag{17.3.1}$$

by Theorem 17.1.1. Let v and u be the left and right eigenvectors of C_N (and therefore also of A_N) respectively: $vC_N = \sigma v, C_N u = \sigma u$. Writing $v = (v(0), \ldots, v(N))$ and $u = (u(0), \ldots, u(N))$, this says that

$$\begin{aligned} v(1) &= \sigma v(0), \\ (d+1)v(0) + v(2) &= \sigma v(1), \\ dv(j-1) + v(j+1) &= \sigma v(j), \quad 2 \le j \le N-1, \\ dv(N-1) &= \sigma v(N) \end{aligned}$$

and

$$\begin{aligned} (d+1)u(1) &= \sigma u(0), \\ du(j+1) + u(j-1) &= \sigma u(j), \quad 1 \le j \le N-1, \\ u(N-1) &= \sigma u(N). \end{aligned}$$

The solutions of these difference equations are (up to constant multiples)

$$\begin{aligned} v(0) &= \beta - \alpha, \\ v(j) &= (d+1)\left(\alpha^{j-1} - \beta^{j-1}\right) + \sigma\left(\beta^j - \alpha^j\right), \quad 1 \le j \le N, \\ u(j) &= \frac{(d+1)}{d^j}\left(\alpha^{j-1} - \beta^{j-1}\right) + \frac{\sigma}{d^j}\left(\beta^j - \alpha^j\right), \quad 0 \le j \le N, \end{aligned} \tag{17.3.2}$$

where

$$\alpha = \alpha(\sigma) = \frac{\sigma - \sqrt{\sigma^2 - 4d}}{2}, \quad \beta = \beta(\sigma) = \frac{\sigma + \sqrt{\sigma^2 - 4d}}{2}.$$

One way to check (17.3.2) is to use the identity

$$\sigma\big(\beta^{j-1}-\alpha^{j-1}\big)=\big(\beta^{j}-\alpha^{j}\big)+d\big(\beta^{j-2}-\alpha^{j-2}\big).$$

Note that α and β are distinct and complex conjugates of each other by (17.3.1), and that $\alpha\beta = d$.

Making a change of variables in (17.2.1) and (17.2.3), we see that the characteristic polynomial $g_N(x)$ of C_N, which is a monic polynomial of degree $N+1$, satisfies the recursion

$$g_N(x) = xg_{N-1}(x) - dg_{N-2}(x), \quad N \geq 2, \tag{17.3.3}$$

and can be computed explicitly as

$$g_N(x)=\left[\frac{x}{2}+\frac{2+2d-x^2}{2\sqrt{x^2-4d}}\right]\big[\alpha(x)\big]^N+\left[\frac{x}{2}-\frac{2+2d-x^2}{2\sqrt{x^2-4d}}\right]\big[\beta(x)\big]^N. \tag{17.3.4}$$

It is easy to see from (17.3.3) that $g_N(x)$ is an even function of x if N is odd and an odd function of x if N is even. It is also easy to see from (17.3.3) that the eigenvalues of C_N are real and distinct. In fact, the $N+1$ zeros of g_N strictly interlace the N zeros of g_{N-1}. To see this, we argue inductively. Assume that g_{N-2} has $N-1$ distinct zeros, g_{N-1} has N distinct zeros, and the latter strictly interlace the former. By (17.3.3), the signs of $g_N(x)$ and $g_{N-2}(x)$ are strictly opposite at each zero of $g_{N-1}(x)$. Furthermore, since g_{N-2} is monic, it must be strictly positive at the largest zero σ of g_{N-1}. Therefore, g_N is strictly negative at σ. Since g_N is monic, it must have a zero strictly to the right of σ. Arguing in a similar fashion, g_N must have a zero strictly between successive zeros of g_{N-1}, and one strictly to the left of the smallest zero of g_{N-1}. This completes the induction step. We conclude that the eigenvalues $\sigma^* = \sigma_0 > \sigma_1 > \cdots > \sigma_N$ of C_N are real, distinct, and satisfy $\sigma_{N-k} = -\sigma_k$.

We will use subscripts to denote the eigenvectors u_k, v_k and the $\alpha_k = \alpha(\sigma_k), \beta_k = \beta(\sigma_k)$ corresponding to the eigenvalue σ_k. Then the mean matrix can be written as

$$M_N(t) = e^{A_N t},$$

and

$$M_N(t)u_k = e^{-t(1-\lambda\sigma_k)}u_k, \quad v_j M_N(t) = e^{-t(1-\lambda\sigma_j)}v_j.$$

It follows that

$$v_j M_N(t) u_k = e^{-t(1-\lambda\sigma_j)} v_j \cdot u_k = e^{-t(1-\lambda\sigma_k)} v_j \cdot u_k,$$

so that $v_j \cdot u_k = 0$ for $j \neq k$. This orthogonality is of course very useful in carrying out computations.

Fix a nonnegative integer m, and consider the branching random walk on $T_{d,N}$ in which initially, the number of particles a distance k from O is m^k for each $0 \leq k \leq N$. Then

$$E\sum_{x\in T_{d,N}} \zeta_t(x) = \big(1, m, m^2, \ldots, m^N\big)M_N(t)\big(1, \ldots, 1\big). \tag{17.3.5}$$

By the orthogonality mentioned above,

$$\left(1, m, m^2, \dots, m^N\right) = \sum_{j=0}^{N} c_j v_j, \quad \text{and} \quad \left(1, \dots, 1\right) = \sum_{j=0}^{N} d_j u_j,$$

where

$$c_j = \frac{\left(1, m, m^2, \dots, m^N\right) \cdot u_j}{v_j \cdot u_j}, \quad \text{and} \quad d_j = \frac{\left(1, \dots, 1\right) \cdot v_j}{v_j \cdot u_j}.$$

These coefficients can be computed by using (17.3.2) and summing the resulting finite geometric series. Here are the results:

$$\begin{aligned}\left(1, m, m^2, \dots, m^N\right) \cdot u_j &= G_N(\sigma_j),\\ \left(1, \dots, 1\right) \cdot v_j &= H_N(\sigma_j),\\ v_j \cdot u_j &= K_N(\sigma_j),\end{aligned}$$

where

$$\begin{aligned}G_N(\sigma) = {}& \frac{d\left[d+1-\sigma\alpha(\sigma)\right]}{\alpha(\sigma)\left[d-m\alpha(\sigma)\right]}\left[1-\left(\frac{m\alpha(\sigma)}{d}\right)^{N+1}\right]\\ &- \frac{d\left[d+1-\sigma\beta(\sigma)\right]}{\beta(\sigma)\left[d-m\beta(\sigma)\right]}\left[1-\left(\frac{m\beta(\sigma)}{d}\right)^{N+1}\right],\\ H_N(\sigma) = {}& \beta(\sigma)-\alpha(\sigma)+\left[d+1-\sigma\alpha(\sigma)\right]\frac{1-\alpha^N(\sigma)}{1-\alpha(\sigma)}\\ &- \left[d+1-\sigma\beta(\sigma)\right]\frac{1-\beta^N(\sigma)}{1-\beta(\sigma)},\end{aligned}$$

and

$$\begin{aligned}K_N(\sigma) = {}& \frac{d+1}{d}\left[\beta(\sigma)-\alpha(\sigma)\right]^2 + \frac{2N}{d}\left[\sigma^2-(d+1)^2\right]\\ &+ \left[d+1-\sigma\alpha(\sigma)\right]^2\frac{1-\alpha^{2N}(\sigma)d^{-N}}{d-\alpha^2(\sigma)}\\ &+ \left[d+1-\sigma\beta(\sigma)\right]^2\frac{1-\beta^{2N}(\sigma)d^{-N}}{d-\beta^2(\sigma)}.\end{aligned}$$

By (17.3.5),

$$\begin{aligned}E\sum_{x\in T_{d,N}} \zeta_t(x) &= \sum_{j=0}^{N} c_j d_j e^{-t(1-\lambda\sigma_j)} v_j \cdot u_j\\ &= \sum_{j=0}^{N} e^{-t(1-\lambda\sigma_j)}\frac{G_N(\sigma_j)H_N(\sigma_j)}{K_N(\sigma_j)}.\end{aligned} \tag{17.3.6}$$

Only the values of G_N, H_N and K_N at the eigenvalues of C_N appear in (17.3.6), and this fact makes it possible to simplify the above expressions. If σ is an eigenvalue of C_N, then it follows from (17.3.4) that

$$\frac{[\alpha(\sigma)]^N}{[\beta(\sigma)]^N} = \frac{2+2d-\sigma^2-\sigma\sqrt{\sigma^2-4d}}{2+2d-\sigma^2+\sigma\sqrt{\sigma^2-4d}}. \tag{17.3.7}$$

Using the fact that α and β are complex conjugates and have modulus $\sqrt{d}$, we see that

$$\alpha^{2N}(\sigma) = \frac{2+2d-\sigma^2-\sigma\sqrt{\sigma^2-4d}}{2+2d-\sigma^2+\sigma\sqrt{\sigma^2-4d}}d^N, \tag{17.3.8}$$

with a corresponding expression for β. Using (17.3.7) and (17.3.8) and a certain amount of algebra (aided by *Mathematica*), one can check that if σ is an eigenvalue of C_N, then

$$\frac{G_N(\sigma)H_N(\sigma)}{K_N(\sigma)} = \frac{m^N J(\sigma) \pm (\sqrt{d})^N L(\sigma)}{N\big[(d+1)^2-\sigma^2\big]-\sigma^2+2d(d+1)}, \tag{17.3.9}$$

where

$$J(\sigma) = \frac{m^2(4d-\sigma^2)\big[d+1+\sigma\big]}{2(d+m^2-m\sigma)},$$

$$L(\sigma) = \frac{(4d-\sigma^2)\sqrt{(d+1)^2-\sigma^2}\big[d^2+d-m\sigma\big]}{2(d+1-\sigma)(d+m^2-m\sigma)},$$

and the $\pm$ is the sign of

$$\beta^N(\sigma)\big(2+2d-\sigma^2-\sigma\sqrt{\sigma^2-4d}\big). \tag{17.3.10}$$

This quantity is real and nonzero. To see that it is real, crossmultiply in (17.3.7). The result is the statement that (17.3.10) is equal to its complex conjugate.

Next, we need to determine the locations of the eigenvalues of C_N with some precision. There are some parity issues that come up below, so we will assume from now on that N is even. This is no restriction. To see this, couple the processes on $T_{d,N}$ and $T_{d,N+1}$ in the natural way, so that the latter is always larger than the former. It follows that the left side of (17.3.6) is increasing in N for fixed t, so that our final conclusion (17.3.17) holds along the full sequence of N's provided it holds along the even subsequence.

As we have seen, the eigenvalues of C_N are the roots of (17.3.7) that lie in $(-2\sqrt{d}, 2\sqrt{d})$. Both sides of (17.3.7) are complex numbers of modulus 1. As σ traverses the interval $[-2\sqrt{d}, 2\sqrt{d}]$ from left to right, $\alpha(\sigma)/\beta(\sigma)$ goes around the unit circle in the complex plane once in the counterclockwise direction, so that the left side of (17.3.7) goes around the circle N times. It goes around once when $\alpha(\sigma)/\beta(\sigma)$ moves from one Nth root of unity to the next. The right side of (17.3.7), on the other hand, goes around the unit circle twice in the clockwise direction. Therefore, the σ_j's lie between successive x's for which

$$\frac{x}{2d}\sqrt{4d-x^2} = \sin\frac{2\pi k}{N}, \tag{17.3.11}$$

where k is an integer. Let $\{x_k, 0 \le k \le N\}$ be determined by the requirements that x_k solves (17.3.11) for that k, and

$$2\sqrt{d} = x_0 > x_1 > \cdots > 0 = x_{N/2} > \cdots > x_{N-1} > x_N = -2\sqrt{d}.$$

Saying this differently, put $x_k = \phi(k/N)$, where ϕ is the continuous decreasing function that maps $[0, 1]$ to $[-2\sqrt{d}, 2\sqrt{d}]$ with $\phi(0) = 2\sqrt{d}$, $\phi(1/2) = 0$, $\phi(1) = -2\sqrt{d}$ and

$$\frac{\phi(y)}{2d}\sqrt{4d - \phi^2(y)} = \sin 2\pi y, \quad 0 \le y \le 1.$$

More explicitly,

$$\phi(y) = \operatorname{sign}\left(\frac{1}{2} - y\right)\sqrt{2d\left[1 + \cos 2\pi y\right]}.$$

Then

$$\begin{aligned} x_{k+1} < \sigma_k < x_k \quad & \text{for } 0 \le k < \frac{N}{2} \\ \sigma_k = 0 \quad & \text{for } k = \frac{N}{2} \\ x_k < \sigma_k < x_{k-1} \quad & \text{for } \frac{N}{2} < k \le N. \end{aligned}$$

We also need to determine the sign that appears in (17.3.9) for a given σ_k. When the left side of (17.3.7) goes around the unit circle once, the numerator and denominator (being conjugates of each other) each goes around the circle of radius $\sqrt{d}$ half way. Therefore, the imaginary part of $\beta^N(\sigma_k)$ alternates in sign as k varies. In checking this, the reader may find it useful to note that since $\alpha(x_k)/\beta(x_k)$ are the successive Nth roots of unity,

$$\left[\frac{\beta(x_k)}{\sqrt{d}}\right]^N = (-1)^k.$$

It follows that the sign of (17.3.10) (evaluated at σ_k) alternates as well. The sign is $(-1)^{N/2}$ when $k = N/2$, so the sign of (17.3.10) at $\sigma = \sigma_k$ is $(-1)^k$. Combining these observations with (17.3.6) and (17.3.9) leads to

$$\begin{aligned} E \sum_{x \in T_{d,N}} \zeta_t(x) &= \sum_{j=0}^{N} e^{-t(1-\lambda\sigma_j)} \frac{m^N J(\sigma_j) + (-1)^j (\sqrt{d})^N L(\sigma_j)}{N\left[(d+1)^2 - \sigma_j^2\right] - \sigma_j^2 + 2d(d+1)} \\ &= M_J(m, t, N) + M_L(m, t, N), \end{aligned} \tag{17.3.12}$$

where M_J and M_L are defined to be the sums involving the J's and L's respectively.

As suggested by the decomposition above, in finding the asymptotics of (17.3.12), we will treat the parts involving J and those involving L separately. Noting that

$$d + m^2 - m\sigma = \left(\sqrt{d} - m\right)^2 + m\left(2\sqrt{d} - \sigma\right) \ge \left(\sqrt{d} - m\right)^2$$

for $|\sigma| \le 2\sqrt{d}$, we will assume now that $m \ne \sqrt{d}$ so that the factor $(d+m^2-m\sigma)$ in the denominators in the definitions of J and L is bounded away from zero. Begin with the part of (17.3.12) that involves J. Since $|\sigma_j| \le 2\sqrt{d}$ for all j, all the factors that occur in that sum are positive. Therefore, if we replace the denominator in (17.3.12) by $N\big[(d+1)^2-\sigma_j^2\big]$, the effect can be compensated by multiplying by a factor of $1+O(N^{-1})$, where the $O(\cdot)$ term is uniform in t. So, we need to consider

$$\sum_{j=0}^{N} e^{-t(1-\lambda\sigma_j)} \frac{J(\sigma_j)}{N\big[(d+1)^2-\sigma_j^2\big]}. \tag{17.3.13}$$

Excluding the middle term in this sum, this is just a Riemann sum corresponding to the partition $0, \frac{1}{N}, \frac{2}{N}, \ldots, 1$ for the integral

$$\int_0^1 e^{-t[1-\lambda\phi(y)]} \frac{J[\phi(y)]}{(d+1)^2-\phi^2(y)} dy. \tag{17.3.14}$$

To see how much error occurs in this approximation, recall that if f is any continuous function on [0, 1] and R is any Riemann sum corresponding to the uniform partition with spacing N^{-1}, then

$$\left| R - \int_0^1 f(y)dy \right| \le \frac{V(f)}{N},$$

where $V(f)$ is the total variation of f on [0, 1]. The total variation of the integrand in (17.3.14) is (up to a uniform constant multiple) $\le$ its maximum value. The middle term in (17.3.14) that was omitted above is also bounded by the maximum value of the integrand divided by N. The maximum value of the integrand is (up to a uniform constant multiple)

$$\le \max_{|\sigma|\le 2\sqrt{d}} \big[2\sqrt{d}-\sigma\big] e^{-t(1-\lambda\sigma)} \le \frac{e^{-1}}{t\lambda} e^{-t\left[1-2\lambda\sqrt{d}\right]}. \tag{17.3.15}$$

To determine the asymptotics of (17.3.14), note that the main contributions come from y near zero, since $\phi(y)$ is largest there. As $y \downarrow 0$,

$$\phi(y) \sim 2\sqrt{d} - \sqrt{d}\pi^2 y^2,$$

and hence

$$J\big[\phi(y)\big] \sim \frac{2d\pi^2 m^2 (\sqrt{d}+1)^2}{\big(m-\sqrt{d}\big)^2} y^2.$$

Therefore, the asymptotics of (17.3.14) are the same as the asymptotics of

$$\frac{2\pi^2 d m^2 \big(\sqrt{d}+1\big)^2}{(d-1)^2\big(m-\sqrt{d}\big)^2} e^{-t\left[1-2\lambda\sqrt{d}\right]} \int_0^1 y^2 e^{-t\lambda\sqrt{d}\pi^2 y^2} dy. \tag{17.3.16}$$

Making the change of variables $z = y^2 t$ in the integral in (17.3.16) and recalling the definition of the Gamma function, one sees that this integral is asymptotic to

$$\frac{1}{4\lambda^{3/2}\pi^{5/2}d^{3/4}t^{3/2}}$$

as $t \to \infty$, with an exponentially small error. Therefore,

$$\begin{aligned}\lim_{\substack{t\to\infty\\ t/N^2\to 0}} \quad & M_J(m,t,N)m^{-N}e^{t\left[1-2\lambda\sqrt{d}\right]}t^{3/2} \\ &= \frac{d^{1/4}m^2}{2(\sqrt{d}-1)^2(m-\sqrt{d})^2\lambda^{3/2}\sqrt{\pi}}.\end{aligned} \tag{17.3.17}$$

The requirement that $t/N^2 \to 0$ is what is needed to use the approximation of (17.3.13) by (17.3.14) using (17.3.15).

Thc analysis of the part of (17.3.12) that involves L is significantly more difficult because of the alternation of signs that appears there. In fact, for some choices of the parameters, the sum in $M_L(m,t,N)$ is much smaller than many of the summands, so very precise asymptotics are required – asymptotics that we will not carry out. Some additional comments will be made about this at the end of this section. Nevertheless, a simple trick allows us to prove Theorem 17.1.2 using only the asymptotics of $M_J(m,t,N)$ given in (17.3.17).

Let ζ_t be the branching random walk process on $T_{d,N}$ for which $\zeta_0(x) \equiv 1$. Then in the multi-type branching process terminology, initially there is one particle of type 0 and $(d+1)d^{n-1}$ particles of type n for each $1 \le n \le N$. Therefore,

$$\begin{aligned} dE\sum_{x\in T_{d,N}} \zeta_t(x) &= (d+1)\big[M_J(d,t,N) + M_L(d,t,N)\big] \\ &\quad -\big[M_J(0,t,N) + M_L(0,t,N)\big].\end{aligned}$$

But $M_J(0,t,N) = 0$ by (17.3.12), and $M_L(0,t,N) = (d+1)M_L(d,t,N)$ by (17.3.12) and the definition of $L(\sigma)$. It follows that the M_L terms exactly cancel above, leaving

$$dE\sum_{x\in T_{d,N}} \zeta_t(x) = (d+1)M_J(d,t,N). \tag{17.3.18}$$

Now make a change of variables

$$s = t\big(1-2\lambda\sqrt{d}\big) - N\log d + \frac{3}{2}\log N.$$

Letting $t \to \infty$, $N \to \infty$ so that the corresponding s remains fixed, we see from (17.3.17) and (17.3.18) that

$$E\sum_{x\in T_{d,N}} \zeta_t(x) \to \frac{(d+1)d^{\frac{1}{4}}}{2\big(\sqrt{d}-1\big)^4\sqrt{\pi}}\left[\frac{1-2\lambda\sqrt{d}}{\lambda\log d}\right]^{\frac{3}{2}} e^{-s}.$$

This proves Theorem 17.1.2 in the case $0 < \lambda < 1/2\sqrt{d}$.

Consider now the case $\lambda = 1/2\sqrt{d}$, which we will treat in less detail. In the previous case, the coefficient of t in the exponential in (17.3.13) did not approach zero. Now, when j is small, this coefficient is small:

$$1 - \lambda\sigma_0 \sim \frac{\pi^2}{2N^2}, \qquad N \to \infty$$

by Theorem 17.1.1. This is the main difference between the two cases. Since $J(\sigma_0)$ is of order N^{-2} (for $m \neq \sqrt{d}$) and there is an extra factor of m^N in (17.3.12) that was omitted in (17.3.13), in order for (17.3.12) to be bounded, it must be the case that the exponential factor in (17.3.13) tends to zero. In other words, the right regime to consider is $t/N^2 \to \infty$ rather than $t/N^2 \to 0$ as in (17.3.17). In the new regime, the estimates we have developed permit one to show that the $j = 0$ term in (17.3.13) dominates the rest of the sum, and the asymptotics of the $j = 0$ term can be determined by using Theorem 17.1.1 and the fact that $\sigma_0 = 1/\lambda_c^N$ (see (17.3.1). The conclusion is that

$$\lim_{\substack{N \to \infty \\ t/N^2 \to \infty}} \frac{N^3 \log m - 3N^2 \log N - N^2 \log M_J(m, t, N)}{t} = \frac{\pi^2}{2}.$$

Therefore, if we take the limit along the sequence $t = CN^3$ for some constant C it follows that

$$\lim_{N \to \infty, t = CN^3} \frac{1}{N} \log M_J(m, t, N) = \log m - \frac{\pi^2}{2} C. \tag{17.3.19}$$

This gives the final statement in Theorem 17.1.2. To see this, take $m = d$, use (17.3.18), and apply (17.3.19) with $C > 2 \log d/\pi^2$ and $C < 2 \log d/\pi^2$, concluding that the mean number of particles tends to zero in the first case and tends to infinity in the second. Therefore, t_N can be chosen so that the expected number of particles is a fixed value w, and

$$\frac{t_N}{N^3} \to \frac{2 \log d}{\pi^2}.$$

To see that a statement similar to Theorem 17.1.2 holds (with different constants) for initial configurations in which there are m^k particles at a distance k from O, with $m > \sqrt{d}$, argue as follows. Compute

$$\frac{L(\sigma)}{J(\sigma)} = \frac{d^2 + d - m\sigma}{m^2\sqrt{(d+1)^2 - \sigma^2}},$$

which is bounded for $|\sigma| \leq 2\sqrt{d}$. If $m > \sqrt{d}$, then it is clear from (17.3.12) that M_J dominates M_L as $N \to \infty$, so again the contributions of M_L can be neglected.

We conclude this section with some additional comments on the difficulties encountered in trying to carry out asymptotics on M_L. These would be required to obtain an analogue of Theorem 17.1.2 in the case $m < \sqrt{d}$. One way to proceed is to write an alternating sum in the form

$$\sum_{j=0}^{N}(-1)^j a_j = \sum_{j=0}^{N/2}\left[a_{2j} - a_{2j+1}\right]. \tag{17.3.20}$$

This is, after all, the simplest way to check the convergence of the alternating harmonic series, for example. The problem is that there is still a significant amount of cancellation that occurs even after this step is performed. Here is an example. Take $d = 4$ and $N = 30$. Write the sum for

$$e^t\left(\sqrt{d}\right)^{-N} M_L(0, t, N)$$

in the form (17.3.20). This depends on λ and t only through their product. Here are the approximate values of the 16 summands corresponding to the right side of (17.3.20) for two values of λt: For $\lambda t = 1$, they are

$$-.5, -.4, .07, .29, .25, .15, .07, .03, .01, .004,$$
$$.002, .0007, .0003, .0001, .00004, .000008.$$

Their sum is approximately .00000014. For $\lambda t = 5$, they are

$$-3 \times 10^6, 10^6, 10^6, 3 \times 10^5, 2 \times 10^4, 800, 20, .35, .006,$$
$$.0001, 2 \times 10^{-6}, 5 \times 10^{-8}, 2 \times 10^{-9}, 10^{-10}, 10^{-11}, 10^{-12}.$$

Their sum is approximately 67. The problem is not so much that the summands vary dramatically in size, but rather that the sum is much smaller than the larger summands, so very careful attention would have to be given to the cancellation that occurs.

Acknowledgments: This research was supported in part by NSF Grants # DMS 9703830 and DMS # DMS 9400644, and a Guggenheim Fellowship.

References

[1] K.B. Athreya and P.E. Ney. *Branching Processes*. Springer-Verlag, New York, 1972.

[2] C. Bezuidenhout and G. Grimmett. The critical contact process dies out. *Ann. Probab.*, **18** (1990), 1462–1482.

[3] C. Bezuidenhout and G. Grimmett. Exponential decay for subcritical contact and percolation processes. *Ann. Probab.*, **19** (1991), 984–1009.

[4] J.W. Chen. The contact process on a finite system in higher dimensions. *Chinese J. Contemp. Math.*, **15** (1994), 13–20.

[5] R. Durrett. The contact process, 1974–1989. In *Proceedings of the* 1989 *AMS Seminar on Random Media*, AMS Lectures in Applied Mathematics 27. AMS, Providence, RI, 1991, 1–18.

[6] R. Durrett and X. Liu. The contact process on a finite set. *Ann. Probab.*, **16** (1988a), 1158–1173.

[7] R. Durrett and R. Schonmann. The contact process on a finite set II. *Ann. Probab.*, **16** (1988b), 1570–1583.

[8] R. Durrett, R. Schonmann, and N. Tanaka. The contact process on a finite set III. The critical case. *Ann. Probab.*, **17** (1989), 1303–1321.
[9] S. Karlin and J. McGregor. Spectral theory of branching processes I. The case of discrete spectrum. *Z. Wahr. verw. Geb.*, **5** (1966), 6–33.
[10] T.M. Liggett. *Interacting Particle Systems*. Springer-Verlag, 1985.
[11] T.M. Liggett. Stochastic models of interacting systems. *Ann. Probab.*, **25** (1997), 1–29.
[12] T.M. Liggett. *Stochastic Interacting Systems: Contact, Voter and Exclusion Processes*. Springer-Verlag, Heidelberg, 1999.
[13] R. Pemantle and A.M. Stacey. The branching random walk and contact process on non-homogeneous and Galton-Watson trees, (1999).
[14] A.M. Stacey. The contact process on a finite tree, (1999).

Department of Mathematics
University of California
Los Angeles, CA 90095-1555
tml@math.ucla.edu

18

Toom's Stability Theorem in Continuous Time

Lawrence F. Gray

ABSTRACT This paper provides a continuous-time analogue of Toom's famous discrete-time stability criterion. Because of certain intrinsic differences between discrete and continuous time, simple analogues of Toom's result are not true, the main problem being that discrete-time models can undergo a spatial shift at each time step and continuous-time models cannot. In our main result, we show that once such shifts have been neutralized, the stability properties of a discrete-time model and its continuous-time analogue are the same. This result applies to a large class of models with finite range interactions in finite dimensions, including many for which the stability question was previously unanswered. Its proof uses an improved version of Toom's theorem that is found in [BG91]. We also obtain, as a byproduct of our analysis, an alternative criterion for stability in discrete time that is easy to check.

Keywords: Interacting particle systems, probability cellular automata, stability, critical phenomenon.

AMS Subject Classification: 60K35.

18.1 A Brief Description of Toom's Theorem

Throughout this paper, the setting will be the state space

$$\Xi = \text{ the collection of all subsets of the } d\text{-dimensional integer lattice } \mathbb{Z}^d.$$

The elements of the lattice $\mathbb{Z}^d$ are called *sites*. A common interpretation of a state $A \in \Xi$ is that the sites in A are *occupied*, while the sites in the complement of A are *vacant*.

The discrete-time models covered by Toom's stability result are defined in terms of an operator that maps the state space Ξ to itself. Let N be a finite subset of $\mathbb{Z}^d$. A mapping $T\colon \Xi \to \Xi$ is a *Toom operator* with *neighborhood* N if it satisfies each of the following for all $x \in \mathbb{Z}^d$ and $A, B \in \Xi$:

(i) Monotonicity: if $A \subseteq B$, then $T(A) \subseteq T(B)$;
(ii) Translation-invariance: $T(A + x) = T(A) + x$;
(iii) Neighborhood N: $x \in T(A)$ if and only if $x \in T(A \cap [N + x])$;
(iv) Two-traps: $T(\emptyset) = \emptyset$ and $T(\mathbb{Z}^d) = \mathbb{Z}^d$.

Let

$$B(r) = \{x \in \mathbb{R}^d : |x| \leq r\}$$

for $r > 0$, where $|\cdot|$ is the ordinary Euclidean norm. A Toom operator T with neighborhood N has *range* r if $N \subseteq B(r)$. Property (iii) is also called the *finite range condition* when we don't want to be specific about the neighborhood N.

Because of the two-traps condition, the states $\emptyset$ and $\mathbb{Z}^d$ are fixed points of any Toom operator. In [Too80], Toom investigated the stability of these fixed points under 'one-sided' random perturbations. Since the two fixed points can be treated in a completely analogous manner, we concentrate our attention on the stability of the state $\mathbb{Z}^d$.

For each Toom operator T, there is a corresponding family of discrete-time Markov processes on Ξ, parameterized by a real number $\varepsilon \in [0, 1]$ known as the *error probability*. We typically let η_t, $t = 0, 1, 2, \ldots$, denote such a process. The dynamics are simple to describe. Let η_0 be an initial state. For times $t \geq 0$ and finite sets $A \subseteq \mathbb{Z}^d$,

$$P(A \subseteq \eta_{t+1} \mid \eta_t) = \begin{cases} (1-\varepsilon)^{\sharp A} & \text{if } A \subseteq T(\eta_t); \\ 0 & \text{otherwise.} \end{cases}$$

(Here and elsewhere, the symbol $\sharp$ means "cardinality of".)

An equivalent description of (η_t) can be made in terms of random *error sets*. Let $E_1, E_2, E_3, \ldots$ be a sequence of random subsets of $\mathbb{Z}^d$, whose distribution is determined by the condition that the events $\{x \in E_t\}$, $x \in \mathbb{Z}^d$, $t = 1, 2, \ldots$, are independent and each have probability ε. Given an initial state η_0, define η_t inductively by

$$\eta_{t+1} = T(\eta_t) \setminus E_{t+1}, n = 1, 2, 3, \ldots. \tag{18.1.1}$$

We may informally describe the process (η_t) as follows. If, at some time t, the set of occupied sites is A, then, in the absence of errors, the set of occupied sites at time $t+1$ would be $T(A)$. But errors occur independently with probability ε at the sites in $T(A)$. As a result, the sites in $T(A) \cap E_{t+1}$ are vacant at time $t+1$, rather than occupied. The errors are *one-sided*, in the sense that the sites in $\mathbb{Z}^d \setminus T(A)$ are unaffected: a site in $\mathbb{Z}^d \setminus T(A)$ is vacant at time $t+1$, whether or not it happens to be part of the error set E_{t+1}.

The types of errors described in the preceding paragraph are those that were considered by Toom, and they suffice for the purposes of this section. In the proof of the main result in Section 18.4, we will use more general kinds of one-sided error sets E_t. Toom's results were extended to models with such error sets in [BG91].

We are interested in what happens when the initial state is $\eta_0 = \mathbb{Z}^d$ and ε is small but positive. There is a standard argument (found in [Lig85], for example) that shows that because of the monotonicity condition, the quantity $P[0 \in \eta_n]$ decreases in n, so the following limit exists:

$$\pi(\varepsilon) = \lim_{n \to \infty} P[0 \in \eta_n].$$

Furthermore, the monotonicity condition also implies that $\pi(\varepsilon)$ is monotonically decreasing as a function of ε. We say that T is *stable* at $\mathbb{Z}^d$ if $\lim_{\varepsilon \searrow 0} \pi(\varepsilon) = 1$. Note that the two-traps condition implies that $\pi(0) = 1$.

Toom proved that the following condition is necessary and sufficient for stability at $\mathbb{Z}^d$:

Toom's Eroder Condition

For every finite set $D \subseteq \mathbb{Z}^d$, there exists an n such that $T^n(D^c) = \mathbb{Z}^d$.

Any Toom operator T satisfying this condition is called an *eroder*. Toom's stability theorem says that T is stable at $\mathbb{Z}^d$ if and only if T is an eroder.

The eroder condition is quite beautiful, because of its simplicity. Nevertheless, in general it can be difficult to check, since it involves *every* finite set $D \subseteq \mathbb{Z}^d$. We will see (Theorem 18.2.1 and its corollary) that there are other more explicit criteria that involve analyzing Toom operators in terms of certain spatial shifts. By way of preparation, we conclude this section with a few definitions and general comments about the way in which spatial shifts affect Toom operators.

Given a Toom operator T and a point $z \in \mathbb{Z}^d$, the mapping $T + z$ is also a Toom operator, where $(T + z)(A) = T(A) + z$. The Toom operator $T + z$ is a *spatial shift* of T. It is obvious that the stability of a Toom operator is not affected by spatial shifts. It is also clear from Toom's eroder condition that a Toom operator T is stable if and only if the n-fold iterate T^n is stable for some (every) positive integer n. These facts motivate the following definition: we say that two Toom operators T_1 and T_2 are *shift-equivalent* if there exists a site $z \in \mathbb{Z}^d$ and positive integers m and n such that

$$(T_1)^m = (T_2)^n + z.$$

Note that shift-equivalence is an equivalence relation. In the next section, we will show that within each equivalence class, there is at least one Toom operator with particularly nice properties.

The main purpose of this paper is to prove a continuous-time analogue of Toom's stability theorem. As we will see in Section 18.3, for each Toom operator T there is a natural choice for a parameterized family of continuous-time Markov processes. However, there is an important difference between continuous time and discrete time. In discrete time, the model with $\varepsilon = 0$ is deterministic, while the analogous model in continuous time is not. One consequence is that if two different Toom operators are shift-equivalent, then the two models corresponding to them are trivially related to each other in discrete time, but not in continuous time. Indeed, we will see that there are examples in continuous time in which the two models do not even have the same stability properties.

18.2 The Eroder Condition and Spatial Shifts

One purpose of this section is to provide the context for our main result, Theorem 18.3.1; the concepts developed here appear in the hypotheses of Theorem 18.3.1, and the results given here make it clear that Theorem 18.3.1 is an appropriate analogue of Toom's stability theorem.

Another purpose of this section is to focus attention on certain aspects of the eroder condition that are often overlooked. In particular, Theorem 18.2.1 gives conditions that are equivalent to the eroder condition. These conditions are not as elegant as the eroder condition, but they have the advantage of being easier to verify and easier to apply. The essential idea behind one of these conditions, Equation (18.2.2) below, is due to Toom.

Throughout this discussion, we assume that we have been given a Toom operator T in d dimensions with finite range r.

In order to understand the role that spatial shifts play, it is natural to look at how T operates on half-spaces. For vectors $v, w \in \mathbb{R}^d$, let

$$\mathcal{H}(v, w) = \{x \in \mathbb{R}^d : \langle x - w, v\rangle < 0\},$$

where $\langle\cdot, \cdot\rangle$ denotes the usual Euclidean inner product. Thus $\mathcal{H}(v, w)$ is the open half-space in $\mathbb{R}^d$ that is uniquely determined by three conditions: (i) its boundary is oriented perpendicularly to v, (ii) its boundary contains the point w, (iii) v points towards the closed half-space $\mathcal{H}(v, w)^c$. It will also be useful to have some notation for the boundary of $\mathcal{H}(v, w)$,

$$\mathcal{P}(v, w) = \text{ the boundary of } \mathcal{H}(v, w),$$

and for the *lattice half-spaces*,

$$H(v, w) = \mathcal{H}(v, w) \cap \mathbb{Z}^d.$$

Let $\mathbf{0}$ denote the origin in $\mathbb{R}^d$, and let

$$R = \mathbb{R}^d \setminus \mathbf{0} \quad \text{and} \quad S = \mathbb{Z}^d \setminus \mathbf{0}.$$

In the following result, we introduce a function with domain R. This function is surprisingly useful, as it gives us a convenient formula (Equation (18.2.1)) for the way in which T operates on lattice half-spaces.

Proposition 18.2.1 *Let $\alpha: R \to \mathbb{R}$ be defined by the formula*

$$\alpha(v) = \sup\{a \in \mathbb{R} : \mathbf{0} \in T(H(v, -av))\}.$$

Then for $v \in R$ and $w \in \mathbb{R}^d$,

$$T(H(v, w)) = H(v, w + \alpha(v)v). \tag{18.2.1}$$

Furthermore, the function α has the following three properties: (i) $\alpha(cv) = \alpha(v)/c$ *for all $c > 0$ and all $v \in R$;* (ii) *α is uniformly continuous on any closed subset of R;* (iii) *α takes only rational values when restricted to S.*

Proof. We first prove (18.2.1). By the monotonicity of T and the fact that $\mathcal{H}(v, w)$ is open,

$$\mathbf{0} \in T(H(v, (-\alpha(v) + s)v)) \text{ iff } s > 0.$$

Since T has neighborhood N, it follows that the boundary of $\mathcal{H}(v, -\alpha(v)v)$ contains a (not necessarily unique) site $w(v) \in N$. So $\mathcal{H}(v, -\alpha(v)v) = \mathcal{H}(v, w(v))$. Using translation-invariance and monotonicity, it is now easy to first prove (18.2.1) for $w = -w(v)$, and then for all vectors $w \in \mathbb{Z}^d$. To extend to all $w \in \mathbb{R}^d$, use monotonicity, finiteness of range, and the simple fact that for every such w, there exists a sequence $w_1, w_2, \ldots \in \mathbb{Z}^d$ such that

$$\mathcal{H}(v, w) = \bigcup_{k=1}^{\infty} \mathcal{H}(v, w_k).$$

Property (i) is immediate from the definition of α. Property (ii) is an easy consequence of the finite range property of T. Finally, it is easy to see from the definitions that if $w(v)$ is as in the preceding paragraph, then $\alpha(v) = -\langle w(v), v\rangle$. Thus Property (iii) also holds. □

The function α is called the *speed coefficient* of T. We will eventually see how the speed coefficient can be used to determine whether T is an eroder.

In what follows, we will often be interested in how T acts on unions of half-spaces, so it will be convenient to have some special notation for such sets. Given a set V of vectors in R and a function $\varphi\colon V \to \mathbb{R}^d$, we let

$$\mathcal{H}(V, \varphi) = \bigcup_{v \in V} \mathcal{H}(V, \varphi(v)) \quad \text{and} \quad H(V, \varphi) = \mathcal{H}(V, \varphi) \cap \mathbb{Z}^d.$$

It will be common for our choice of the function φ to depend in some way on the identity function on R. We will use the notation

$$I = \text{ the identity function on } R.$$

For example, a common choice for φ will be the function αI. Note that (18.2.1) and monotonicity imply the following:

Proposition 18.2.2 *If $V \subseteq R$, then*

$$T(H(V, \varphi)) \supseteq H(V, \varphi + \alpha I),$$

for any function $\varphi\colon V \to \mathbb{R}^d$.

A typical application of the preceding result is as follows: let D be a finite subset of $\mathbb{Z}^d$, choose $c > 0$ so that $H(V, -cI) \subseteq D^c$. If there exists a positive integer n such that $H(V, n\alpha I - cI) = \mathbb{Z}^d$, then by monotonicity and Proposition 18.2.2, $T^n(D^c) = \mathbb{Z}^d$. So the proposition makes it possible to connect the eroder condition to the speed coefficient.

Of particular interest to us is the set $\mathcal{H}(R, \alpha I)$, which is the set obtained by shifting each half-space $\mathcal{H}(v, \mathbf{0})$ by the amount $\alpha(v)|v|$ in the v direction, and then taking the union over all vectors $v \in R$. We will see that T is an eroder if and only if $\mathcal{H}(R, \alpha I) = \mathbb{R}^d$. It will be convenient to be able to express this condition in terms of a nice subset of R. The next somewhat technical proposition and its corollary help us do just that. In preparation, we introduce the notation

$$L(v) = \text{ the linear span of } \mathcal{P}(v, \mathbf{0}) \cap B(2r) \cap \mathbb{Z}^d,$$

where we assume that T has range r, and that $r \geq 1$.

Lemma 18.2.1 *Let*

$$U = \{u \in R \colon L(u) \text{ has dimension } d - 1\}.$$

Then $\mathcal{H}(R, \alpha I) = \mathcal{H}(U, \alpha I)$.

PROOF. See Section 18.5. □

If $L(u)$ has dimension $d-1$, then u is the solution of a set of $d-1$ linear equations of the form $\langle u, v_i \rangle = 0$, where $v_1, \ldots, v_{d-1}$ are linearly independent vectors with integer coordinates. Thus, the method of Gaussian elimination makes it clear that some nonzero scalar multiple of u lies in S. By Property (i) in Proposition 18.2.1, $\mathcal{H}(cu, \alpha(cu)cu) = \mathcal{H}(u, \alpha(u)u)$ for all $c > 0$. Therefore, the following corollary is an immediate consequence of Lemma 18.2.1. It states that U may be replaced by a certain finite set W. Note that, like U, this set depends only on the dimension d and the range r.

Corollary 18.2.1 *Let*

$$W = \{w \in U \cap S \colon \text{the coordinates of } w \text{ have no common integer divisor}\}.$$

Then $\mathcal{H}(W, \alpha I) = \mathcal{H}(R, \alpha I)$.

We have one more technical result. It is a geometric result about unions and intersections of open half-spaces that does not really depend on the properties of Toom operators or their speed coefficients. However, to make the result easier to use in our context, the speed coefficient appears in its statement.

Lemma 18.2.2 *Suppose* $\mathcal{H}(W, \alpha I) = \mathbb{R}^d$, *and let* V *be a minimal subset of* W *such that* $\mathcal{H}(V, \alpha I) = \mathbb{R}^d$. *Then* $\mathcal{H}(V, -\alpha I)^c$ *contains at least one point with rational coordinates.*

PROOF. See Section 18.5. □

We are now ready to define two special classes of Toom operators. Let T be a Toom operator with speed coefficient α. If $\alpha(v) \leq 0$ for all vectors $v \in W$, then T is an *expander*. If there exists a set $V \subseteq W$ such that α is nonnegative on V and $\mathcal{H}(V, \alpha I) = \mathbb{R}^d$, then T is a *shrinker*. Like the term "eroder", the words "expander" and "shrinker" are intended to suggest the way in which T affects sets of vacant sites. Theorem 18.2.1 states that every Toom operator is shift-equivalent to either an expander or a shrinker, and that a Toom operator is an eroder if and only if it is shift-equivalent to a shrinker.

Theorem 18.2.1 *A Toom operator* T *with speed coefficient* α *is shift-equivalent to a shrinker if*

$$\mathcal{H}(W, \alpha I) = \mathbb{R}^d. \tag{18.2.2}$$

Otherwise, it is shift-equivalent to an expander. Furthermore, T *is an eroder if and only if* (18.2.2) *holds.*

PROOF. First assume that $\mathcal{H}(W, \alpha I) = \mathbb{R}^d$. By Lemma 18.2.2, there exists a finite set $V \subseteq W$ such that $\mathcal{H}(V, \alpha I) = \mathbb{R}^d$ and such that $\mathcal{H}(V, -\alpha I)^c$ contains a point x with rational coordinates. Let n be a positive integer such that nx has integer coordinates, and consider the Toom operator $T' = T^n - nx$. It is easy to see that $\mathbf{0} \in \mathcal{H}(V, -\alpha' I)^c$, where α' is the speed coefficient of T'. It follows that $\alpha' \geq 0$ on V. Since $\mathcal{H}(V, \alpha I) = \mathbb{R}^d$, it is obvious that $\mathcal{H}(V, \alpha' I) = \mathbb{R}^d$. So T' is a shrinker, as desired.

Next we show that if the set $A = \mathcal{H}(W, \alpha I)^c$ is nonempty, then T is shift-equivalent to an expander. By construction, A is closed and convex. Since T has finite range, A is bounded. Therefore, each extreme point of A equals the intersection of hyperplanes $\mathcal{P}(v, \alpha(v)v)$, taken over some set of vectors $v \in W$. As in the proof of Lemma 18.2.2, it follows that A must contain at least one point x with rational coordinates. Let n be a positive integer such that nx has integer coordinates, and consider the operator $T' = T^n - nx$. As in the previous step, it is easy to see that $\mathcal{H}(W, \alpha' I)^c$ contains the origin, so $\alpha' \leq 0$ on W, and T' is an expander, as desired.

Since the eroder property is preserved under shift-equivalence, it is enough to show that no expander is an eroder, and that every shrinker is an eroder. We first show that every shrinker is an eroder.

Let T be a shrinker, with speed coefficient α. By definition there exists a finite set V such that $\alpha \geq 0$ on V and $\mathcal{H}(V, \alpha I) = \mathbb{R}^d$, from which it follows that there exists a vector $v \in V$ such that $\alpha(v) > 0$. It is easy to see from these facts that for every $c > 0$, there exists a positive integer n such that $\mathcal{H}(V, (n\alpha - c)I) = \mathbb{R}^d$. By Propositions 18.2.1 and 18.2.2, $T^n(H(V, -cI)) = \mathbb{Z}^d$. It follows immediately from monotonicity that T is an eroder.

It remains to show that if T is an expander, then T is not an eroder. By Corollary 18.2.1 and the definition of expander, $\alpha(v) \leq 0$ for all $v \in R$. We now refer back to [Too80], where it is shown (using different terminology of course) that the speed coefficient of an eroder must take at least one positive value. Thus, T cannot be an eroder. □

As stated earlier, (18.2.2) is a condition that can be readily checked. We briefly describe here an explicit algorithm for doing so. First, determine the set W. One way to do this is to use brute force: look at all possible subsets of $B(2r) \cap S$ of size $d - 1$ to see which ones determine a $d - 1$-dimensional hyperplane through the origin, and for those that do, use algebra to find a vector with integer coordinates that is perpendicular to the hyperplane. Once W is determined, calculate $\alpha(w)$ for each $w \in W$ by comparing the two lattice half-spaces $H(w, \mathbf{0})$ and $T(H(w, \mathbf{0}))$. Then determine (using methods of linear programming) the solution set of the system of equations

$$\langle x - \alpha(w)w, w \rangle \geq 0, w \in W. \tag{18.2.3}$$

This solution set equals $\mathcal{H}(W, \alpha I)^c$, so (18.2.2) holds if and only if (18.2.3) has no solution. Thus, we have

Corollary 18.2.2 *A Toom operator T is stable if and only if (18.2.3) has no solution.*

When a solution to (18.2.3) exists, any member of the solution set with rational coefficients can be used, as in the proof of Theorem 18.2.1, to find an expander that is shift-equivalent to T. Otherwise, Lemma 18.2.2 and the proof of Theorem 18.2.1 indicate an explicit method for finding a shrinker that is shift-equivalent to T.

18.3 A Class of Continuous-Time Systems

In continuous time, we are concerned with certain Markov processes $(\xi_t, t \geq 0)$, with state space Ξ. They are defined in terms of *birth rates* and *death rates*. For each $x \in \mathbb{Z}^d$, let β_x denote the birth rate at the site x and δ_x denote the death rate at x. Each such birth or death rate is a function from the state space Ξ to the nonnegative real numbers. The following formulas give the meanings of these rates:

$$P(x \in \xi_{t+h} \mid \xi_t \text{ and } x \notin \xi_t) = \beta_x(\xi_t)h + o(h),$$

$$P(x \notin \xi_{t+h} \mid \xi_t \text{ and } x \in \xi_t) = \delta_x(\xi_t)h + o(h),$$

and

$$P(\text{a change occurs at both } x \text{ and } y \text{ during } (t, t+h] \mid \xi_t) = o(h).$$

Thus, the rates are instantaneous expected rates of change, given the current state of the system, with the birth rates being relevant at vacant sites and the death rates being relevant at occupied sites. Processes meeting the above description are sometimes called *Markovian spin-flip systems*. For the technical details of the construction of such processes, see a general reference such as [Lig85] or [Dur88].

Under very broad conditions (which are satisfied for all of the models considered here), the state $\mathbb{Z}^d$ is absorbing if and only if $\delta_x(\mathbb{Z}^d) = 0$ for all $x \in \mathbb{Z}^d$. In this case, the appropriate notion of stability at $\mathbb{Z}^d$ involves perturbing the death rates by adding a nonnegative constant ε. The parameter ε is called the *error rate*. In what follows, we will typically use the notation $(\xi_t, t \geq 0)$ for the parameterized Markovian particle system with birth rates β_x, death rates $\delta_x + \varepsilon$, and initial state $\mathbb{Z}^d$.

The formal definition of stability in continuous time is similar to that in discrete time: the state $\mathbb{Z}^d$ is *stable* if

$$\liminf_{\varepsilon \searrow 0} \pi(\varepsilon) = 1,$$

whcrc

$$\pi(\varepsilon) = \liminf_{t \to \infty} P[0 \in \xi_t].$$

We use the lim inf in these expressions to accommodate models with non-attractive rates. (Rates are *attractive* if they satisfy the following condition for all $x \in \mathbb{Z}^d$

and $A, B \subseteq \Xi$: if $A \subseteq B$, then $\beta_x(A) \leq \beta_x(B)$ and $\delta_x(A) \geq \delta_x(B)$.) If the rates are attractive, standard arguments ensure that the lim inf can be replaced by a limit in both of the expressions above involving $\pi(\varepsilon)$.

For the remainder of this paper, we will consistently make the following three assumptions about sets of birth and death rates (β_x, δ_x).

For all $A, B \in \Xi$ and $x, y \in \mathbb{Z}^d$:

(i) Weak attractiveness: if $A \subseteq B$, then $\beta_x(A) > 0 \Rightarrow \beta_x(B) > 0$ and $\delta_x(B) > 0 \Rightarrow \delta_x(A) > 0$;
(ii) Translation-invariance: $\beta_x(A) = \beta_{x+y}(A + y)$ and $\delta_x(A) = \delta_{x+y}(A + y)$;
(iii) Finite range property: there exists a finite set $N \subseteq \mathbb{Z}^d$ that does not depend on x or A such that $\beta_x(A) = \beta_x(A \cap (N + x))$ and $\delta_x(A) = \delta_x(A \cap (N + x))$.

The second and third assumptions guarantee sufficient regularity so that there are no difficulties with the construction of the corresponding Markovian interacting particle systems.

We are now ready to show how to relate continuous-time systems with discrete-time systems. Given a set of rates (β_x, δ_x) satisfying the above three conditions, define an operator $T: \Xi \to \Xi$ by the following formula:

$$T(A) = \{x \in A: \delta_x(A) = 0\} \cup \{x \in A^c: \beta_x(A) > 0\}. \tag{18.3.1}$$

The rates (β_x, δ_x) are called *Toom rates* if T is a Toom operator. If (β_x, δ_x) are Toom rates, then the operator defined in (18.3.1) is called the Toom operator *corresponding to* (β_x, δ_x). Toom rates are called *canonical* if they take only the values 0 and 1. Examples of models with Toom rates are given at the end of this section. These include d-dimensional versions of the ordinary contact process, the sexual contact process, and the majority vote model.

If we ignore the irrelevant values of $\delta_x(A)$ for $x \notin A$ and of $\beta_x(A)$ for $x \in A$, it is easy to see that (18.3.1) determines a one-to-one correspondence between Toom operators and canonical Toom rates. The "inverse" of (18.3.1) is

$$\beta_x(A) = \begin{cases} 1 & \text{if } x \in T(A \setminus \{x\}) \\ 0 & \text{otherwise,} \end{cases}$$

and

$$\delta_x(A) = \begin{cases} 1 & \text{if } x \notin T(A \cup \{x\}) \\ 0 & \text{otherwise,} \end{cases}$$

for $x \in \mathbb{Z}^d$ and $A \subseteq \mathbb{Z}^d$.

It is also easy to see from the four properties in the definition of Toom operator that (β_x, δ_x) are Toom rates if and only if they are weakly attractive, translation-invariant, finite range, and also satisfy the following two properties for $x \in \mathbb{Z}^d$ and $A, B \in \Xi$:

(iv) Two-traps: $\beta_x(\emptyset) = \delta_x(\mathbb{Z}^d) = 0$;
(v) Exclusiveness: $\beta_x(A \setminus \{x\})\delta_x(A \cup \{x\}) = 0$ for all $x \in \mathbb{Z}^d$ and $A \in \Xi$.

Just as in discrete time, Property (iv) implies that $\emptyset$ and $\mathbb{Z}^d$ are absorbing states. To see the reason for the Property (v), note that because of the monotonicity of T,

$$x \in T(A \setminus \{x\}) \quad \Rightarrow \quad x \in T(A \cup \{x\}).$$

This implication and its contrapositive immediately give Property (v). Note that canonical Toom rates are attractive, rather than just weakly attractive.

Here is the statement of our main result. The proof is given in Section 18.4.

Theorem 18.3.1 *Let (β_x, δ_x) be Toom rates, and let T be the corresponding Toom operator, as defined by (18.3.1). A continuous-time system with these rates is stable at $\mathbb{Z}^d$ if T is a shrinker. It is not stable at $\mathbb{Z}^d$ if T is an expander.*

A standard comparison argument using "coupling" (see [Lig85]) gives the following corollary:

Corollary 18.3.1 *Let (β_x, δ_x) be rates that satisfy the assumptions* (i), (ii), *and* (iii) *given above. Assume that $\delta_x(\mathbb{Z}^d) = 0$ for all $x \in \mathbb{Z}^d$. Let T be defined in terms of these rates as in (18.3.1). The continuous-time system with these rates is stable at $\mathbb{Z}^d$ if there exists a shrinker T' such that $T'(A) \subseteq T(A)$ for all $A \in \Xi$. It is not stable at $\mathbb{Z}^d$ if there exists an expander T' such that $T(A) \subseteq T'(A)$ for all $A \in \Xi$.*

We now give several examples. In each case, a class of models (or a specific model) is described in terms of the dimension d, a neighborhood set N, a formula for the birth rate $\beta_{\mathbf{0}}$, and a formula for the unperturbed death rate $\delta_{\mathbf{0}}$. The remaining rates β_x and δ_x can then be easily determined by applying translation-invariance.

Example 18.3.1 (Pure-birth models) A *pure-birth* model is one with weakly attractive, translation-invariant, finite-range rates (β_x, δ_x) such that $\delta_x \equiv 0$ for all x. Theorem 18.3.1 gives such a nice result in this case, that it is worth stating as a corollary. As a general stability criterion for pure-birth models, this result is new. After its statement and easy proof, we give several specific applications.

Corollary 18.3.2 *A pure-birth model with birth rates (β_x) and neighborhood N is stable at $\mathbb{Z}^d$ if and only if*

$$\beta_{\mathbf{0}}(A) > 0 \tag{18.3.2}$$

for some set $A \subseteq N$ whose convex hull in $\mathbb{R}^d$ does not contain the origin.

PROOF. The most important case is when the rates are Toom rates. Sincc the death rate is identically 0, the corresponding Toom operator has a nonnegative speed coefficient. So it is an expander if this speed coefficient is identically 0, and otherwise it is a shrinker. It is now easy to see that (18.3.2) is a necessary and sufficient condition for T to be a shrinker, and the desired result follows from Theorem 18.3.1.

Since the death rates are identically equal to 0, it is easy to see that the rates can only fail to be Toom rates if the birth rates are strictly positive, in which case (18.3.2) obviously holds. It is not hard to prove stability directly in this case, but stability is also an immediate consequence of Corollary 18.3.1 (just let the comparison operator T' be any shrinker). □

The most famous class of pure-birth models consists of the *contact processes*, for which $\beta_{\mathbf{0}}(A) > 1$ when $A \cap N$ is nonempty, and $\beta_{\mathbf{0}}(A) = 0$ otherwise. The two most common contact processes are defined by the birth rate formulas $\beta_{\mathbf{0}}(A) = \sharp(A \cap N)$ and $\beta_{\mathbf{0}}(A) = \min\{1, \sharp(A \cap N)\}$. The second formula gives canonical Toom rates. For any contact process with neighborhood N, the stability condition (18.3.2) is equivalent to the condition that $N \setminus \{\mathbf{0}\} \neq \emptyset$. This is a known result. For an introduction to the extensive body of research concerning contact processes, see [Lig85] and [Dur88].

A second important class of pure-birth models consists of the *sexual contact processes*. These are similar to the contact processes, except that $A \cap N$ must contain at least 2 sites in order for $\beta_{\mathbf{0}}(A)$ to be positive. In the literature, sexual contact processes are usually assumed to have canonical Toom rates, so that

$$\beta_{\mathbf{0}}(A) = \begin{cases} 1 & \text{if } \sharp(A \cap N) \geq 2 \\ 0 & \text{otherwise.} \end{cases}$$

The most studied example is the sexual contact process in 2 dimensions with the *NEC neighborhood* $N = \{(0, 0), (1, 0), (0, 1)\}$.

For the sexual contact process with arbitrary neighborhood N, the stability condition (18.3.2) is equivalent to the condition that $N \setminus \{\mathbf{0}\}$ contain at least two sites that do not lie on a straight line through the origin. This result is also known, but the proof, which is due to Durrett and Gray, is unpublished. Sexual contact processes have many interesting stability properties, which are discussed in [DG86], [Che92], and [Che94].

In order to find a completely new application for Corollary 18.3.2, we need to look at processes where a birth can only occur at x if more than three sites in $N + x$ are occupied. The simplest case is the following variation on the sexual contact process: let $d = 3$, $N = \{(0, 0, 0), (1, 0, 0), (0, 1, 0), (0, 0, 1)\}$, $\delta_x \equiv 0$, and

$$\beta_x(A) = \begin{cases} 1 & \text{if } N + x \subseteq A \cup \{x\} \\ 0 & \text{otherwise.} \end{cases}$$

This birth rate satisfies (18.3.2) with $A = N \setminus \{\mathbf{0}\}$. The stability at $\mathbb{Z}^3$ of the system with these rates was previously unknown. The reader can easily create other examples along these lines.

The remaining examples have nonzero death rates.

Example 18.3.2 (Majority vote models) For this class of examples, we assume that $\sharp N$ is odd. Let d be arbitrary. The birth rate $\beta_{\mathbf{0}}(A)$ is positive if $\sharp(A \cap N) > \sharp N/2$ and 0 otherwise. The unperturbed death rate $\delta_{\mathbf{0}}(A)$ is positive if $\sharp(A \cap N) < \sharp N/2$

and 0 otherwise. Usually, the rates are assumed to be canonical Toom rates, so that they equal 1 whenever they are not equal to 0. But there are good reasons to allow values other than 0 and 1 in these models (see the "double stability" example below).

If the neighborhood is symmetric, such as for the *nearest neighbor majority vote process*, for which $N = \{y: |y| \leq 1\}$, the corresponding Toom operator is an expander. Thus, majority vote models with symmetric neighborhoods are unstable at $\mathbb{Z}^d$. For many asymmetric neighborhoods, such as the NEC neighborhood in $\mathbb{Z}^2$ defined in the previous example, the corresponding Toom operator is a shrinker, so stability holds. Note that a majority vote process with the NEC neighborhood has the same birth rates as a sexual contact process with the NEC neighborhood (but not the same death rates).

The instability of symmetric neighborhood majority vote processes and the stability of NEC neighborhood majority vote processes are known. For many other choices of the neighborhood, Theorem 18.3.1 settles the stability question for the first time.

An instructive example for which Theorem 18.3.1 gives no information is the one-dimensional majority vote model with $N_x = \{x, x+1, x+2\}$. The corresponding Toom operator is neither a shrinker nor an expander (it is shift-equivalent to an expander). The stability properties of this model are unknown.

Example 18.3.3 (The double stability of the sexual contact process.) We have already explained that our main result implies directly that the sexual contact process with the NEC neighborhood in 2 dimensions is stable at $\mathbb{Z}^2$. As discussed in [DG86], this model is also stable at $\emptyset$ for any positive ε. This behavior contrasts with that of the ordinary contact process, which is not stable at $\emptyset$ for small positive $\varepsilon > 0$. (In an interacting particle system with $\emptyset$ as an absorbing state, stability at $\emptyset$ means that if the initial state is $\emptyset$ and a small positive quantity is added to the birth rate, then the probability that the origin is vacant at time t stays close to 1 as $t \to \infty$.) We explain here how the stability at $\emptyset$ of the sexual contact process follows from Corollary 18.3.1.

Let (ξ_t) be the sexual contact process with the NEC neighborhood in 2 dimensions, with error rate $\varepsilon > 0$. Consider the process (ζ_t) defined by $\zeta_t = (\xi_t)^c$. The death rates of this model are the death rates of an NEC majority vote model, and its birth rates equal the constant ε. This model does not have Toom rates. However, if T is defined as in (18.3.1), then it is easy to see that T satisfies the hypothesis of the first conclusion in Corollary 18.3.1, with T' being the Toom operator corresponding to an NEC majority vote model. So (ζ_t) is stable at $\mathbb{Z}^d$. Now transform back to see that (ξ_t) is stable at $\emptyset$. The original proof of this result, which is due to Durrett and Gray, is unpublished. See [DG86] for further details.

The next example illustrates a limitation of our results. Toom's stability theorem suffers from a similar limitation.

Example 18.3.4 (Stability without Toom rates) In [GG82], a necessary and sufficient stability condition is obtained for the class of models with $d = 1$ and

$N = \{-1, 0, 1\}$ whose rates are attractive, translation-invariant and satisfy the two-traps condition. But it is not assumed that the rates satisfy the exclusiveness condition, so that, for instance, both $\beta_{\mathbf{0}}(\{1\})$ and $\delta_{\mathbf{0}}(\{0, 1\})$ might be nonzero. When the rates are not exclusive, there is no corresponding Toom operator.

The necessary and sufficient condition for stability in [GG82] is that

$$\beta_{\mathbf{0}}(\{1\}) + \beta_{\mathbf{0}}(\{-1\}) > \delta_{\mathbf{0}}(\{0, 1\}) + \delta_{\mathbf{0}}(\{-1, 0\}).$$

Even when the rates are exclusive, the results in the present paper will not apply unless one of the two sides of this inequality equals 0, since otherwise, the corresponding Toom operator is neither a shrinker nor an expander.

We conclude this section with two examples showing that models with Toom rates need not have the same stability properties as their corresponding Toom operators, if these Toom operators are not shift-equivalent to shrinkers or expanders. These examples illustrate why shift-equivalence does not play the same role in continuous time as it does in discrete time. Because of space limitations, the reader will need to verify the assertions made in these two examples. (This would make a nice warm-up project for a talented graduate student.)

Example 18.3.5 (Unstable Toom rates corresponding to an eroder) Let $d = 2$. For each integer m, let $N^m = \{(m, k): k = -n, -n+1, \ldots, n-1, n\}$, where n is a sufficiently large positive integer. For $x \in \mathbb{Z}^d$ and $A \subseteq \mathbb{Z}^d$, define

$$\beta_x(A) = \begin{cases} 1 & \text{if } (N^{-2} \cup N^{-3} + x) \subseteq A \\ 0 & \text{otherwise,} \end{cases}$$

and

$$\delta_x(A) = \begin{cases} 1 & \text{if } \{x - (1, 0), x - (2, 0)\} \subseteq A^c \\ 0 & \text{otherwise.} \end{cases}$$

We leave it to the reader to check that we have defined a set of canonical Toom rates corresponding to a Toom operator T, whose speed coefficient satisfies $\alpha((1, 0)) = 2$, $\alpha((-1, 0)) = -1$, and $\alpha((0, 1)) = \alpha((0, -1)) = 0$. It is easy to see from these values of $\alpha(\cdot)$ that $T + (-1, 0)$ is a shrinker, so T itself is an eroder.

In discrete time, the half-plane $H((1, 0), (0, 0))$ moves to the right at speed 2, which is enough to make T into an eroder. In continuous time, the corresponding movement of occupied sites is much slower when n is large, due to the fact that it takes a long time for a large vertical column of vacant sites to become occupied. On the other hand, horizontal blocks of vacant sites spread relatively easily to the right. With a little work, it can be shown that the vacant sites "win out" when n is large, destroying stability at $\mathbb{Z}^2$.

The next example uses the main idea of the preceding example, but in reverse. For technical reasons, the rates are somewhat more complicated than in the preceding example.

Example 18.3.6 (Stable Toom rates corresponding to a noneroder) Let $d = 2$. For each integer m, let $N_+^m = \{(m, k): k = 0, 1, \ldots, n-1, n\}$, where n is a sufficiently large positive integer. For $x \in \mathbb{Z}^d$ and $A \subseteq \mathbb{Z}^d$, define

$$\delta_x(A) = \begin{cases} 1 & \text{if } (N_+^{-2} \cup N_+^{-3} + x - (0, j)) \subseteq A^c \text{ for some } j = 0, 1, \ldots, n \\ 0 & \text{otherwise,} \end{cases}$$

and

$$\beta_x(A) = \begin{cases} 1 & \text{if } \{x - (1,0), x - (2,0)\} \subseteq A \\ 0 & \text{otherwise.} \end{cases}$$

We leave it to the reader to check that we have defined a set of canonical Toom rates corresponding to a Toom operator T, whose speed coefficient satisfies $\alpha((1,0)) = 1, \alpha((-1,0)) = -2$, and $\alpha((0,1)) = \alpha((0,-1)) = 0$. Further investigation of the values of $\alpha(\cdot)$ shows that $T + (-1, 0)$ is an expander, so T is not an eroder.

It can be shown that the continuous-time system with unperturbed rates (β_x, δ_x) is stable at $\mathbb{Z}^d$ for sufficiently large n. The choice of n is made so that a vertical "wall" of vacant sites spreads very slowly to the right. Blocks of occupied sites spread to the right at speed 1 in the unperturbed system. For a sufficiently small noise rate $\varepsilon > 0$, these two facts can be used to show that the perturbed system with rates $(\beta_x, \delta_x + \varepsilon)$ prevents vacant regions from growing very large, so that the probability that the origin is occupied at time t stays close to 1 as $t \to \infty$. One way to prove this fact is to use the comparison techniques found in our proof of Theorem 18.3.1.

18.4 Proof of Theorem 18.3.1

The proof of our main theorem will be carried out by making a comparison between a continuous-time process with Toom rates and a discrete-time model whose Toom operator is related to the Toom operator determined by the rates. To define this discrete-time model, we will need some terminology.

Given a Toom operator T, we define two auxiliary operators T_β and T_δ:

$$T_\beta(A) = A \cup T(A) \text{ and } T_\delta(A) = A \cap T(A), \quad A \subseteq \mathbb{Z}^d.$$

Proposition 18.4.1 *Let T_β, T_δ be defined as above in terms of a Toom operator T. Then T_β and T_δ are themselves Toom operators, and they satisfy*

$$T_\beta(A) \supseteq A \quad \textit{and} \quad T_\delta(A) \subseteq A, \qquad A \in \Xi.$$

Furthermore, if α is the speed coefficient of T, then $\alpha \wedge 0$ is the speed coefficient of T_β and $\alpha \vee 0$ is the speed coefficient of T_δ. And finally, if T is a shrinker, then $T_\beta(T_\delta)^k$ is a shrinker for all positive integers k.

PROOF. Every conclusion in this result is routine to check, except possibly the last one needs a small hint. The hint is: the speed coefficient of $T_\beta(T_\delta)^k$ is $(\alpha \wedge 0) + k(\alpha \vee 0)$. □

We now turn to the proof of Theorem 18.3.1. Before giving the formal proof, we first sketch the main idea.

The hard part is to show stability when T is a shrinker. One would like to make some sort of direct comparison between the continuous-time process (ξ_t) with rates $(\beta_x, \delta_x + \varepsilon)$, and the discrete-time process (η_t) with Toom operator T and error probability ε. To make such a comparison, the two processes (ξ_t) and (η_t) need to be defined jointly on the same probability space. Then, to get the stability of the process (ξ_t) from the stability of (η_t), it would be sufficient to find some positive integer J such that $\xi_{Jt} \supseteq \eta_t$ for all $t = 0, 1, 2, \ldots$.

This is clearly asking for too much. There are three main problems. The first is that, in continuous time, it is possible for a vacant site x to stay vacant for a long time, even when the birth rate stays large and there are no "errors". Since we want to compare ξ_{Jt} with η_t, this problem is solved by letting J be large, so that if the birth rate at a site x is positive during some time interval $(J(t-1), Jt]$, then a birth at that site is highly likely. The rare "stubborn" vacant sites are included among the errors. With this expanded notion of errors, things can be arranged so that the occurrence of a birth at x at time t in the discrete-time system implies a birth at x during the time interval $(J(t-1), Jt]$ in the continuous-time system.

The second problem is more difficult. It arises from the fact that, in continuous time, it is possible for many related deaths to occur over a short time interval. Such "death-clusters" are unlikely, but arbitrarily large ones do occur in any non-trivial system, and they must be reckoned with. This problem cannot be handled in the same way as the first problem, since the integer J appears in the wrong place in the comparison. We would need to compare $\xi_{t/J}$ with η_t if we wanted to make the probability of a death-cluster small during a single discrete time step.

The solution is to replace T by a different Toom operator, namely $T_\beta(T_\delta)^k$ (see Proposition 18.4.1). We will take k to be proportional to J^2. This "speeds up" the deaths in the discrete-time process, essentially achieving the desired comparison of the preceding paragraph.

The third problem is also due to "death-clusters". If we want to include them among the errors, we need to generalize the types of error sets that we use to construct our discrete-time comparison process, since "death-clusters" involve more than one site. Processes with these kinds of multi-site errors are studied in [BG91]. We will need the estimates in [BG91] to finish off the proof.

We are now ready for the proof of our main result.

Proof of Theorem 18.3.1. We begin with the easy half. Assume that T is an expander. We will first show that there exists a finite set $D \subseteq \mathbb{Z}^d$ such that $T(D^c) \subseteq D^c$. A simple comparison argument shows that we may assume that $\alpha(w) = 0$ for all $w \in W$. By Theorem 18.2.1, T is not an eroder, so there exists a finite set $B \subseteq \mathbb{Z}^d$ such that $T^n(B^c) \neq \mathbb{Z}^d$ for all positive integers n. Let $D_n = T^n(B^c)^c$. Since $\alpha(w) = 0$ for all W, it is easy to see that the sets D^n all lie inside a single bounded subset of $\mathbb{Z}^d$. It follows that there exists a positive integer k such that, for all sufficiently large n, $D_n = D_{n+k}$. Fix such an n, and let $D = D_n \cup D_{n+1} \cup \ldots \cup D_{n+k-1}$. Clearly, D is nonempty and finite. By monotonicity, $T(D^c) = D^c$.

Now we can complete the proof of the easy half. It follows from the monotonicity of T and the definition of Toom rates that $\beta_x(A) = 0$ for all $x \in D$ if $A \subseteq D^c$. Standard properties of interacting particle systems imply that if the perturbed system with rates $(\beta_x, \delta_x + \varepsilon)$ reaches a state in which all of the sites in D are vacant, then all of the sites in D will remain vacant forever after. Since the death rates of a perturbed system are bounded away from 0, it is easy to see that such a system will almost surely reach such a state. Lack of stability at $\mathbb{Z}^d$ is an easy consequence of this fact and translation-invariance.

We turn our attention to the hard part. Assume that T is a shrinker. By the finite range condition, the positive values of the birth rates are bounded away from 0, and the values of the death rates are bounded above. Because of these facts, standard comparison arguments show that it is enough to prove stability at $\mathbb{Z}^d$ in the case for which the birth rates only take the two values 0 and 1 and the death rates only take the two values 0 and M for some $M \geq 1$. We will make these simplifying assumptions throughout the rest of the proof.

It is now quite easy to describe a useful 'graphical' construction of the interacting particle system with rates $(\beta_x, \delta_x + \varepsilon)$. Let

$$B_x, C_x, D_x, \quad x \in \mathbb{Z}^d,$$

be independent Poisson point processes on the time line $[0, \infty)$. For each x, the intensity measures of B_x, C_x, D_x, are respectively λ, $\varepsilon\lambda$, $M\lambda$, where λ is Lebesgue measure on $[0, \infty)$. It can be shown that there exists a Markov process $(\xi_t, t \geq 0)$ with state space Ξ, defined on the same probability space as the Poisson point processes $B_x, C_x, D_x, x \in \mathbb{Z}^d$, such that $\xi_0 = \mathbb{Z}^d$ and for all sites x, $\xi_t(x) = \xi_{t-}(x)$ unless either

$$\begin{aligned} t \in B_x \text{ and } \beta_x(\xi_{t-}) = 1, &\quad \text{in which case } \xi_t(x) = \xi_{t-}(x) \cup \{x\}, \\ \text{or} \quad t \in C_x, &\quad \text{in which case } \xi_t(x) = \xi_{t-}(x) \setminus \{x\}, \\ \text{or} \quad t \in D_x \text{ and } \delta_x(\xi_{t-}) = M, &\quad \text{in which case } \xi_t(x) = \xi_{t-}(x) \setminus \{x\}. \end{aligned}$$

Furthermore, these conditions uniquely determine the process $(\xi_t, t \geq 0)$ almost surely, and this process is an interacting particle system with rates $(\beta_x, \delta_x + \varepsilon)$ and initial state $\mathbb{Z}^d$.

Informally speaking, the conditions say that (i) a birth can occur at a site x only at a time in B_x, and then only if the birth rate at x at time $t-$ is 1, and (ii) a death can occur at a site x only at a time in C_x or D_x, and in the latter case, only if the unperturbed death rate at x at time $t-$ is M. The deaths at x that occur at times in C_x are interpreted as errors at x. They happen at rate ε.

This graphical construction of the perturbed interacting particle system will make it possible for us to define a discrete-time model that can be directly compared with the continuous-time process. Our discrete-time model will be defined with an inductive definition that is similar to (18.1.1), with the error sets E_t being more general than in Section 18.1. These error sets will be determined by the graphical construction described above.

Let J be a positive integer, and let $\varepsilon = 1/J^{250d}$, so that (ξ_t) now depends on J instead of ε. This parameter J will be used to define a discrete-time model (η_t) that will dominate (ξ_t) in a certain sense. The time interval $((t-1)J, tJ]$ of our continuous-time model (ξ_t) will correspond to the tth time step in our discrete-time model (η_t), and we will eventually see that $\eta_t \supseteq \xi_{Jt}$ for $t = 0, 1, 2, \ldots$.

Our discrete-time model will be defined in terms of error sets E_t that are more general than the ones in Section 18.1. These error sets will consist of three types of errors. The first two types are easy to define. For each $t = 1, 2, 3, \ldots$, let

$$E_t^{(1)} = \{x \in \mathbb{Z}^d \colon C_x \cap ((t-1)J, tJ] \neq \emptyset\}$$
$$E_t^{(2)} = \{x \in \mathbb{Z}^d \colon B_x \cap ((t-1)J, tJ] = \emptyset\}.$$

As indicated by the set $E_t^{(1)}$, an error of the first type occurs at x during the tth time step if an ordinary error occurs at x during $((t-1)J, tJ]$ in the continuous-time model. An error of the second type occurs at x during the tth time step if a birth at x during $((t-1)J, tJ]$ is impossible in the continuous-time model because of the fact that $B_x \cap ((t-1)J, tJ] = \emptyset$.

Errors of the third type are slightly more complicated. These errors indicate places in the continuous-time model where a rapid succession of related deaths is possible. Some terminology will be useful. As usual, let r be the range of the Toom operator T. Given an integer $k \geq 2$, a time interval $\mathcal{I}$, and a site $x \in \mathbb{Z}^d$, we say that a sequence $((x_j, t_j), j = 1, 2, \ldots, k)$ is a *k-chain in $\mathcal{I}$ with head x* if (i) $x = x_1$, (ii) $t_1 \in D_x \cap \mathcal{I}$, and (iii) for all $j = 2, \ldots, k$, we have

$$t_j \in D_{x_j} \cap \mathcal{I}, \quad |x_j - x_{j-1}| \leq r, \quad \text{and} \quad t_j > t_{j-1}.$$

The set $\{x_1, \ldots, x_k\}$ is called the *body* of the k-chain.

We now define the third type of error for our discrete-time model. For positive times t and sites x, let

$$E_t^{(3,x)} = \bigcup_{\ell=0}^{J^2-1} \{y \colon y \text{ is in the body of a } 250d\text{-chain}$$
$$\text{in } \big((t-1)J + \tfrac{\ell}{J},\ (t-1)J + \tfrac{\ell+1}{J}\big] \text{ with head } x\big\},$$

and

$$E_t^{(3)} = \bigcup_{x \in \mathbb{Z}^d} E_t^{(3,x)}.$$

The full error set corresponding to the tth time step is

$$E_t = E_t^{(1)} \cup E_t^{(2)} \cup E_t^{(3)}.$$

Note that the sets $E_1, E_2, E_3, \ldots$ are iid.

We now define a discrete-time Markov process $(\eta_t, t = 0, 1, 2, \ldots)$ inductively as follows. Let $\eta_0 = \mathbb{Z}^d$, and for $t = 0, 1, 2, \ldots$, let

$$\eta_{t+1} = \Big(T_\beta\, (T_\delta)^{250dJ^2}\, (\eta_t \setminus E_{t+1})\Big) \setminus E_{t+1}.$$

Since T is assumed to be a shrinker, Proposition 18.4.1 implies that the Toom operator $T_\beta(T_\delta)^{250dJ^2}$ is a shrinker for every J.

We claim that for all $t = 0, 1, 2, \ldots,$

$$\xi_{Jt} \supseteq \eta_t. \tag{18.4.1}$$

The case $t = 0$ is obvious. To prove the claim for $t > 1$, we proceed inductively. So assume that $\xi_{Jt} \supseteq \eta_t$, and let x be in the set η_{t+1}. We must show that $x \in \xi_{J(t+1)}$.

Since the definition of η_{t+1} implies that $x \notin E^{(1)}_{t+1} \cup E^{(2)}_{t+1}$, we know that no error occurs at x in the continuous-time system during $(Jt, J(t+1)]$, and we also know that $B_x \cap (Jt, J(t+1)] \neq \emptyset$. Thus, to prove that $x \in \xi_{J(t+1)}$, it is sufficient to show that $\beta_x(\xi_s) = 1$ and $\delta_x(\xi_s) = 0$ for all $s \in (Jt, J(t+1)]$. By the exclusiveness of the birth and death rates, it is enough to show that

$$\beta_x(\xi_s) = 1, \quad \text{for all } s \in (Jt, J(t+1)]. \tag{18.4.2}$$

For the purposes of obtaining a contradiction, assume that $\beta_x(\xi_s) = 0$ for some $s \in (Jt, J(t+1)]$. For $j = 0, 1, \ldots, 250dJ^2$, let

$$\eta^{(j)} = T_\delta^{250dJ^2-j}(\eta_t \setminus E_{t+1}). \tag{18.4.3}$$

By Proposition 18.4.1,

$$x \in \eta_{t+1} \subseteq T_\beta(\eta^{(0)}) \subseteq T_\beta T_\delta(\eta^{(1)}).$$

It follows from the definitions of T_β, T_δ that $\beta_x(\eta^{(1)}) = 1$. Since $\eta^{(1)} \subseteq \eta_t \subseteq \xi_{Jt}$, weak attractiveness implies that a death must occur at some site $y_1 \in \eta^{(1)} \cap (N+x)$ at some time $s_1 \in (Jt, J(t+1)]$, since otherwise we could not have $\beta_x(\xi_s) = 0$ for some $s \in (Jt, J(t+1)]$. Since $\eta^{(1)}$ does not include any sites in E_{t+1}, no error could occur at y_1 at time s_1, so it must be that $\delta_{y_1}(\xi_{s_1^-}) = M$ and $s_1 \in D_{y_1}$.

Now we proceed inductively. Let (y_1, s_1) be as in the preceding paragraph, and suppose that for some $j = 1, \ldots, 250dJ^2 - 1$, there exist space-time points $(y_1, s_1), \ldots, (y_j, s_j)$ such that for $i = 2, \ldots, j$,

$$s_i \in D_{y_i} \cap (Jt, s_{i-1}), \quad \delta_{y_i}(\xi_{s_i^-}) = M, \quad \text{and} \quad y_i \in \eta^{(i)} \cap (y_{i-1} + N).$$

Since $y_j \in \eta^{(j)} = T_\delta(\eta^{(j+1)})$, we know that $\delta_{y_j}(\eta^{(j+1)}) = 0$. As before, it follows that a death must occur at some site $y_{j+1} \in \eta^{(j+1)} \cap (N + y_j)$ at some time $s_{j+1} \in (Jt, s_j)$. Since $\eta^{(j+1)}$ does not include any sites in E_{t+1}, this death cannot be due to an error at y_{j+1}, so $s_{j+1} \in D_{y_{j+1}}$ and $\delta_{y_{j+1}}(\xi_{s_{j+1}^-}) = M$.

Our inductive argument provides us with a sequence

$$(y_1, s_1), \ldots, (y_{250dJ^2}, s_{250dJ^2})$$

having certain properties. These properties imply that the reversed sequence is a $250dJ^2$-chain in $(Jt, J(t+1)]$ whose head is a site in η_t, and none of the sites involved in the chain lie in the set E_{t+1}. Such a chain must contain a $250d$-chain in some interval of the form $(Jt + \ell/J, Jt + (\ell+1)/J]$ for some $\ell = 0, \ldots, J^2 - 1$. Since E_{t+1} contains all of the sites that could be involved in such a chain, we have a contradiction.

We have shown that (18.4.1) holds for all $t = 0, 1, 2, \ldots$. To complete the proof, we will need some estimates that are found in [BG91]. Some work will be required to move from our present context into the context of [BG91].

We will first define a discrete-time process (ζ_t) that is equivalent to a process found in [BG91]. Let $\zeta_0 = \mathbb{Z}^d$, and then define ζ_t inductively by

$$\zeta_{t+1} = T_\beta (T_\delta)^{250dJ^2}(\zeta_t) \setminus E_{t+1,J}, \quad t = 0, 1, 2, \ldots,$$

where $E_{t+1,J}$ denotes the set $\{x + y : x \in E_{t+1},\ y \in B(250dJ^2 r) \cap \mathbb{Z}^d\}$. This differs from the definition of (η_t) in two ways: (i) the error set has been "thickened" by the amount $250dJ^2 r$, and (ii) the error set only appears once in the formula, in contrast to (18.4.3). The thickening of the error set is intended to compensate for the fact that it only appears once in the formula. Indeed, it is easy to check that, since T_δ has range r, the compensation is more than adequate, so $\zeta_t \subseteq \eta_t$ for all times t.

The process (ζ_t) is precisely of the type considered in [BG91], where a comparison is made between discrete-time processes like (ζ_t) and certain auxiliary processes that are denoted in [BG91] by (A_t). The comparison is that $A_t^c \subseteq \zeta_t$, provided the parameters for the process A_t are chosen appropriately. The parameters for (A_t) are denoted in [BG91] as $\alpha, \beta, \varepsilon, \theta$. As we shall see, the appropriate choices for these parameters in our context all depend on J, except for α.

An appropriate choice for α is any positive value of the speed coefficient of the Toom operator $T_\beta (T_\delta)^{250dJ^2}$. The positive part of this speed coefficient equals the speed coefficient of T_β. Since T_β is a shrinker, its speed coefficient does have at least one positive value, and we let α be this value.

An appropriate choice for β is any lower bound on the speed coefficient of $T_\beta (T_\delta)^{250dJ^2}$. This speed coefficient is not less than $-250dJ^2 r$. So we let $\beta = \beta(J) = -250dJ^2 r$.

The parameter ε of the process (A_t) in [BG91] is an error probability. It is sufficient to let it equal

$$P(x \in E_t^{(1)} \cup E_t^{(2)} \cup E_t^{(3,x)}) = 1 - P(x \notin E_t^{(1)}) P(x \notin E_t^{(2)}) P(E_t^{(3,x)} = \emptyset)$$

(which does not depend on t or x). We denote this quantity by $\varepsilon(J)$.

We need an upper bound on $\varepsilon(J)$. It is easy to check that $P(x \notin E_t^{(1)}) = \exp(-\varepsilon J) \geq \exp(-1/J^{249d})$ and that $P(x \notin E_t^{(2)}) = 1 - \exp(-J)$. To obtain an adequate lower bound on $P(E_t^{(3,x)} = \emptyset)$, we first note that there are $\leq (2r)^{dk}$ choices for the body of a k-chain whose head is at a particular site x. Thus,

$$\begin{aligned} P\left(E_t^{(3,x)} = \emptyset\right) &\geq \left(1 - \sum_{k=250d}^{\infty} P\,(x \text{ is the head of a } k\text{-chain in } (0, 1/J])\right)^{J^2} \\ &\geq \left(1 - \sum_{k=250d}^{\infty} \frac{(2r)^{dk} \exp(-M/J)(M/J)^k}{k!}\right)^{J^2}. \end{aligned} \tag{18.4.4}$$

A straightforward estimate shows that the right side is bounded below by $(1 - A/J^{250d})^{J^2}$ for some positive constant A. Combining these three bounds, we have

$$\varepsilon(J) \le 1 - \exp(-1/J^{249d})(1 - \exp(-J))(1 - A/J^{250d})^{J^2}.$$

A further simple estimate gives

$$\lim_{J\to\infty} J^{240d}\,\varepsilon(J) = 0. \tag{18.4.5}$$

Finally, we have the parameter θ from [BG91], which we denote here by $\theta(J)$. Define a random variable X by the formula

$$X = \inf\{c \ge 1: E_1^{(3,\mathbf{0})} \subseteq B(c)\}.$$

The parameter $\theta(J)$ depends on the distribution of X, so it measures, in some sense, the size of our error sets. It is sufficient to let $\theta(J)$ be any positive number such that

$$C(J) = \sup_k \exp\left(\theta(J)(k + 750drJ^2)\right) P(k-1 < X \le k) < \infty.$$

The appearance of the term $750drJ^2$ in this expression comes from the "thickening" that was done to our error sets and the fact that the operator $T_\beta(T_\delta)^{250dJ^2}$ has range $250drJ^2$ (see the definitions of μ and Δ in [BG91, Theorem 2]). We will need to choose $\theta(J)$ so that $C(J)$ stays bounded as $J \to \infty$. By making estimates similar to those used for the terms in (18.4.4), we see that $\sup_k \exp(k)P(k-1 < X \le k)$ stays bounded as $J \to \infty$. So we may take $\theta(J) = 1/(750dJ^2r) = 1/(3|\beta(J)|)$.

With these choices of the parameters $\alpha, \beta(J), \varepsilon(J), \theta(J)$, it is shown in the proof of [BG91, Theorem 2] that $A_t^c \subseteq \zeta_t$, so we have

$$A_t^c \subseteq \xi_{Jt}, \quad t = 0, 1, 2, \ldots.$$

So it is enough to check that the estimates in [BG91] imply that

$$\lim_{J\to\infty} P(\mathbf{0} \in A_t) = 0. \tag{18.4.6}$$

This is precisely the kind of result that appears in the proof of [BG91, Theorem 1]. So we only need to check that the parameters $\alpha, \beta(J), \varepsilon(J), \theta(J)$ defined above satisfy the conditions that appear in the proof of [BG91, Theorem 1].

Here is a brief description of those conditions. The reader will need to refer to the proof of [BG91, Theorem 1] for further details. In the proof of [BG91, Theorem 1], there is a constant K_3 that depends on the parameters $\alpha, \beta(J), \theta(J)$. Since α is constant, and since $\theta(J)$ is proportional to $\beta(J)$, the relationships given in (2-10) in [BG91] imply that, in our context, K_3 grows no faster than $|\beta(J)|^{12d}$. Given our choice of $\beta(J)$, this means that it is sufficient to let K_3 be on the order of J^{24d}. In that same proof, another constant K_2 is chosen, subject to several conditions, the critical one being that $K_2 \ge (K_3)^2$. This means that, in our context, K_2 can be chosen to grow no faster than J^{48d}.

Now we come to the key requirement. Condition (2-12) in the proof of [BG91, Theorem 1] requires that $\varepsilon(J)$ go to 0 faster than $1/(K_2)^5$. This last condition is guaranteed in our context by (18.4.5). For our particular application, all remaining conditions in the proof of Theorem 1 are weaker than the ones just stated. So the estimates given there apply to our situation. In particular, the results labeled (2-21), (2-22), and (2-23) in [BG91] all hold, and these immediately imply our Equation (18.4.6), as desired. □

18.5 Proofs of Lemmas 18.2.1 and 18.2.2

PROOF OF LEMMA 18.2.1. Note that $\mathcal{H}(U, \alpha I) \subseteq \mathcal{H}(R, \alpha I)$, so it is enough to show that if $x \in \mathcal{H}(R, \alpha I)$, then there exists a vector $u \in U$ such that $x \in \mathcal{H}(u, \alpha(u)u)$.

Let x be in $\mathcal{H}(R, \alpha I)$. Then there exists a vector v such that $x \in \mathcal{H}(v, \alpha(v)v)$. If $v \in U$, we are done, so assume $v \notin U$. Thus, $L(v)$ has dimension less than $d - 1$. It follows that the hyperplane $\mathcal{P}(v, \mathbf{0})$ can be rotated about the space $L(v)$. Pick a particular $(d-2)$-dimensional 'axis' of rotation that contains the space $L(v)$, and rotate the hyperplane $\mathcal{P}(v, \mathbf{0})$ continuously about that axis. Parameterize the resulting family of rotations with a real parameter s, and let v_s be the result of rotating v with the rotation corresponding to s. We assume that (i) $v_0 = v$; (ii) v_s is a continuous function of s; (iii) the rotations corresponding to positive s are in the opposite direction from the rotations corresponding to negative s; (iv) if $s \in (-1, 1)$, then $\mathcal{P}(v_s, \mathbf{0})$ does not contain any of the sites in $B(2r) \cap \mathbb{Z}^d$ except for those in $L(v)$; (v) both hyperplanes $\mathcal{P}(v_{\pm 1}, \mathbf{0})$ contain a site in $B(2r) \cap \mathbb{Z}^d$ that is not in $L(v)$.

Now consider the family of hyperplanes $\mathcal{P}(v_s, \alpha(v_s)v_s)$, $s \in [-1, 1]$. The definition of α and the fact that T has range r imply that for $s \in (-1, 1)$, these hyperplanes all contain the same sites in $B(r) \cap \mathbb{Z}^d$. Letting $w(v)$ be as in the proof of Proposition 18.2.1, it follows that each of these hyperplanes contains the affine linear space $L(v) + w(v)$. In other words, as the hyperplanes $\mathcal{P}(v_s, \mathbf{0})$ rotate continuously about the space $L(v)$, the corresponding shifted hyperplanes $\mathcal{P}(v_s, \alpha(v_s)v_s)$ rotate continuously about $L(v) + w(v)$. By continuity, the hyperplanes $\mathcal{P}(v(\pm 1), \alpha(v_{\pm 1})v_{\pm 1})$ contain $L(v) + w(v)$. It is easy to see that at least one of the two half-spaces $\mathcal{H}(v_{\pm 1}, \alpha(v_{\pm 1})v_{\pm 1})$ also contains the point x. We may assume that the parameterization has been chosen so that $x \in \mathcal{H}(v_1, \alpha(v_1)v_1)$. Note that our construction ensures that the dimension of $L(v_1)$ is at least one more than the dimension of $L(v)$.

Repeating the above procedure if necessary, we eventually obtain a vector u such that $x \in \mathcal{H}(u, \alpha(u)u)$ and the dimension of $L(u)$ is $d - 1$, as desired. □

PROOF OF LEMMA 18.2.2. To streamline notation in this proof somewhat, we will write

$$A(V') = \mathcal{H}(V', \alpha I)^c$$

for any set $V' \subseteq W$. Note that each such $A(V')$ is the intersection of finitely many closed half-spaces in $\mathbb{R}^d$, and hence, is a closed convex set. Our hypothesis is that $A(W) = \emptyset$ and V is a minimal subset of W such that $A(V) = \emptyset$.

Fix a vector $v \in V$ and let $V' = V \setminus \{v\}$. Let X be the set of points $x \in A(V')$ that minimize the quantity

$$d(x) = \inf\{|x - z| : z \in \mathcal{P}(v, \alpha(v)v)\}.$$

Since $A(V') \subseteq \mathcal{H}(v, \alpha(v)v)$, the set X consists of extreme points of $A(V')$, and $d(x)$ is a positive constant for $x \in X$. The minimality of V implies that X is an affine linear subspace of $\mathbb{R}^d$ (possibly consisting of a single point), and X is parallel to $\mathcal{P}(v, \alpha(v)v)$ by construction. Thus, there exists a set $B \subseteq V'$ such that

$$X = \bigcap_{v \in B} \mathcal{P}(v, \alpha(v)v). \tag{18.5.1}$$

Assume that B is the maximal set with this property. We will show that $B = V'$.

Suppose that $B \neq V'$. Since $X \subseteq A(V')$, the maximality of B and the definition of X imply that X lies in the interior of $A(V' \setminus B)$. Thus, for any $x \in X$ and $y \in A(B)$, a part of the line segment from x to y also lies in the interior of $A(V' \setminus B)$. By the choice of x, every point on the line segment connecting y to x lies at least as far from $\mathcal{P}(v, \alpha(v)v)$ as does x itself. Therefore, $\langle y - x, v\rangle \leq 0$ for all such y. Since $X \subseteq \mathcal{H}(v, \alpha(v)v)$, it follows that the ray starting at x and pointing in the direction of $y - x$ is entirely contained in $\mathcal{H}(v, \alpha(v)v)$. Thus, $A(B)$ is entirely contained in $\mathcal{H}(v, \alpha(v)v)$, so that $\mathcal{H}(B \cup \{v\}, \alpha I) = \mathbb{R}^d$. The minimality of V now implies that $B = V'$.

Again let x be a point in X. Since $B = V'$, we have $\langle x - \alpha(v')v', v'\rangle = 0$ for all $v' \in V'$. We also have $\langle x - \alpha(v)v, v\rangle < 0$. So x lies in the closure of each of the sets $\mathcal{H}(w, \alpha(w)w)$, $w \in V$. Each such closure can be written as $\mathcal{H}(-w, \alpha(w)w)^c$. Reflecting through the origin, we find that $-x$ lies in $\mathcal{H}(V, -\alpha I)^c$. It follows that $-X \subseteq \mathcal{H}(V, -\alpha I)^c$.

To complete the proof, it suffices to show that X contains at least one point with rational coordinates. By definition, each vector in V has integer coordinates, and by Proposition 18.2.1, $\alpha(w)$ is rational for each $w \in V$. These facts and (18.5.1) show that X can be expressed as the solution set of a system of linear equations with integer coefficients. The desired conclusion now follows from the method of Gaussian elimination. □

REFERENCES

[BG91] M. Bramson and L. Gray, *A useful renormalization argument*, Random Walks, Brownian Motion and Interacting Particle Systems (R. Durrett and H. Kesten, eds.), Birkhäuser, Boston, 1991, pp. 113–152.

[Che92] H.-W. Chen, *On the stability of a population growth model with sexual reproduction on* $\mathbb{Z}^2$, Ann. Probab. **20** (1992), 232–285.

[Che94] H.-W. Chen, *On the stability of a population growth model with sexual reproduction on* $\mathbb{Z}^d$ $(d \geq 2)$, Ann. Probab. **22** (1994), 1195–1226.

[DG86] R. Durrett and L. Gray, *Some peculiar properties of a particle system with sexual reproduction*, Stochastic Spatial Processes (P. Tautu, ed.), Lecture Notes in Mathematics, vol. 1212, Springer-Verlag, New York, 1986, pp. 106–111.

[Dur88] R. Durrett, *Lecture Notes on Particle Systems and Percolation*, Wadsworth & Brooks/Cole Advanced Books and Software, Pacific Grove, California, 1988.

[GG82] L. Gray and D. Griffeath, *A stability criterion for attractive nearest-neighbor spin systems on* $\mathbb{Z}$, Ann. Probab. **10** (1982), 67–85.

[Lig85] T.M. Liggett, *Interacting Particle Systems*, Grundlehren der mathematischen Wissenschaften, vol. 276, Springer-Verlag, New York-Berlin-Heidelberg-Tokyo, 1985.

[Too80] A. Toom, *Stable and attractive trajectories in mulicomponent systems*, Adv. Probab. **6** (1980), 549–575.

School of Mathematics
University of Minnesota
Minneapolis, MN 55455
gray@math.umn.edu

19

The Role of Explicit Space in Plant Competition Models

Claudia Neuhauser

ABSTRACT Most natural plant communities show a considerable amount of diversity. Classical ecological theory predicts that the number of stably coexisting species is limited by the number of resources. Since plants rely on a relatively small number of resources, this poses a dilemma. A current topic of plant ecology is thus to explain patterns of natural plant communities. In this article we focus on the effect of an explicit spatial dimension on plant community structure. This is done within the framework of a stochastic spatial model.

Keywords: Interacting particle systems, plant competition models, spatially explicit models, contact process, voter model, Lotka-Volterra model, intraspecific competition, interspecific competition.

AMS Subject Classifications: 60K35, 92B05.

19.1 Introduction

Early naturalists, such as Alexander von Humboldt or Charles Darwin (Humboldt 1814, Darwin 1836) were deeply impressed by the diversity of life they encountered on their voyages, in particular the diversity of terrestrial plants as expressed by Darwin in *The Voyage of the Beagle* when he wrote 'a traveler should be a botanist, for in all views plants form the chief embellishment.'

This interest in plant diversity has continued until today. In the beginning the study of plant diversity was purely descriptive. The second half of the 19th century saw an analytic study of plant communities that went well beyond the purely descriptive methods of earlier naturalists. Instead of simply describing plant communities, the focus changed towards asking about the forces that shape plant communities and this has remained an important question in plant ecology.

Darwin's *Origin of Species* (Darwin, 1859) proved to be extremely influential on how we think about species interactions. Prior to Darwin's work it was commonly acknowledged that interspecific competition, that is, competition between different species, is an important force that shapes communities. Darwin emphasized the importance of intraspecific competition, that is, competition between individuals of the same species, as the principal mechanism for evolutionary change. It is now generally accepted that both intraspecific and interspecific competition play an important role in shaping plant communities.

Early plant ecology investigated the correlation between the type of plant community and the environment. Today, plant ecology focuses on species richness, species abundance and spatial and temporal patterns (Crawley, 1997). To find the underlying causes, both experimental and theoretical methods are employed. Experimental methods include field observations, field manipulations and laboratory experiments; theoretical methods frequently rely on the analysis of simple mathematical models.

The use of mathematical models to illuminate the effects of competition goes back to Vito Volterra (1860-1940), a physicist, and Alfred James Lotka (1880-1949), a mathematician and demographer (Kingsland, 1991). The Lotka-Volterra model of interspecific competition (Volterra 1926; Lotka 1932) describes the competitive interactions between a finite number of species. In the two species formulation, the model is given by

$$\frac{dN_1}{dt} = r_1 N_1 \left(1 - \frac{N_1}{K_1} - \alpha_{12}\frac{N_2}{K_1}\right) \tag{19.1.1}$$

$$\frac{dN_2}{dt} = r_2 N_2 \left(1 - \frac{N_2}{K_2} - \alpha_{21}\frac{N_1}{K_2}\right) \tag{19.1.2}$$

where $N_i = N_i(t)$ denotes the number of individuals of species i at time t. The constants r_1 and r_2 are the respective intrinsic rates of increase, that is, the per capita rate of population growth at low densities. The constants K_1 and K_2 are the respective carrying capacities, that is, the respective equilibrium abundance of each species in the absence of the other species. The term N_i/K_i measures the effect of intraspecific competition; the term $\alpha_{ij}N_j/K_i$ measures the effect of interspecific competition; in particular, the coefficient α_{ij} measures the strength of the effect *of* species j *on* species i.

It is easy to analyze this model. It exhibits three basic types of behavior, namely coexistence, competitive exclusion and founder control. We will discuss this in more detail below but wish to point out here that one of its conclusions is that coexistence is only possible if intraspecific competition is more important than interspecific competition (i.e., $\alpha_{12}K_2 < K_1$ and $\alpha_{21}K_1 < K_2$).

Lotka and Volterra's work led Georgii F. Gause (1935) to test some of the conclusions of this theoretical work in laboratory experiments. His experiments on two species of *Paramecium*, a protozoan, clearly demonstrated that one species can competitively exclude the other. He found that in order for species to coexist, they need to partition their resources. This is known as *Gause's principle of competitive exclusion*. This principle has been demonstrated in numerous field and laboratory experiments by other researchers as well, for instance, Tansley (1917), Connell (1961) and Tilman (1976, 1977) (see also, Tilman 1997 and references therein). The common theme in all these experiments is that coexistence between competing species requires niche differentiation.

Many of the simple competition models share some basic assumptions, namely, (i) a spatially and temporally homogeneous environment, (ii) global dispersal and competition, and (iii) the system is in equilibrium. These assumptions allow one to formulate the dynamics as a system of ordinary differential equations which keep track of the number of individuals of each species. One of the principal conclusions of these models is that the number of stably coexisting species is limited by the number of available niches or resources.

This limit on the number of coexisting species poses a dilemma for plants. Plants require a limited number of resources, namely water, light and a few mineral nutrients. The theoretical models therefore predict that plant communities contain only a few species. However, many natural plant communities are species rich. For instance, in fields of the Cedar Creek Natural History Area, Minnesota, one can find tens to hundreds of species of vascular plants that stably coexist (Tilman, 1997). This discrepancy between theory and observations was first pointed out by G.E. Hutchinson (1959, 1961). Since then it has become clear that by relaxing the basic assumptions in the mathematical models one can formulate models in which an arbitrarily large number of species can stably coexist (see Tilman and Pacala, 1993, for an overview).

We will be concerned with the role of explicit space in competitive interactions. Our approach will be to compare three basic types of models: (1) spatially unstructured models; (2) spatially implicit models in which individuals or local populations occupy discrete patches and all patches are connected with each other; (3) spatially explicit models in which individuals or local populations occupy discrete patches and only nearby patches influence the dynamics of any given patch. The main difference between cases (2) and (3) is thus that interactions are global in case (2) and local in case (3).

One of the most obvious mechanisms that can facilitate coexistence and generate spatial patterns, is spatial heterogeneity. Indeed, as Crawley (1997) pointed out, "[the] traditional approach in plant ecology was to view spatial patterns as a reflection of underlying heterogeneity in soil conditions and microclimate." But even in the absence of spatial heterogeneities it has been shown that simple, spatially explicit models can produce complex large scale behavior. Many such examples can be found in a recent book by J. Bascompte and R.Solé (1998), who compiled a comprehensive overview on this subject. Most of these results in the biological literature are based on computer simulations; more recently, analytic approximation methods have been used to analyze spatially explicit plant competition models (Pacala 1997; Pacala and Levin 1997; Bolker and Pacala 1998). One of the conclusions that emerged from the study of spatially explicit models can be found in the article by Hartway et al. (1998) where the authors stated that "[one] of the most general results of spatial models is that the addition of a spatial dimension (or "patchiness") provides opportunities for competitors, or for predators and prey, to coexist—opportunities that would be absent in a nonspatial world without population subdivision."

In this article we will review some of the mathematically rigorous results that pertain to the importance of explicit space on the outcome of competitive interactions in simple plant competition models. This will be done in the framework of *interacting particle systems*, a class of spatially explicit stochastic processes where local populations or individuals are arranged on a grid and local interactions (i.e., interactions between close neighbors on the grid) define the dynamics.

There is by now an extensive body of work in the mathematical literature on interacting particle systems (see, e.g., Liggett 1985, Durrett 1988 and 1995); they provide an ideal framework for modeling plant populations or plant communities. In Section 19.2 we begin with single species models that will serve as the building blocks for multispecies models. Section 19.3 is devoted to a spatially explicit Lotka-Volterra model with interspecific competition. In Section 19.4 we discuss a spatially explicit hierarchical model in which species are ordered according to competitiveness. Conclusions and suggestions for future research can be found in Section 19.5.

19.2 Single Species Models

We begin our discussion with single species models. The first mathematical model that addressed the effects of intraspecific competition, was the logistic model. It was originally formulated by Pierre-François Verhulst (1838) who called it *logistic model.* This model did not receive much attention until Raymond Pearl, a statistician, independently came up with the same model (Pearl and Reed, 1920). Pearl then discovered that Verhulst had used this model previously. Pearl adopted Verhulst's name for the model and publicized it. (See Kingsland, 1985, for the history of this model.) This model takes into account the decrease in the per capita growth rate as the population density increases. The equation describing this model is given by

$$\frac{dN}{dt} = rN\left(1 - \frac{N}{K}\right) \tag{19.2.1}$$

where $N = N(t)$ is the size of the population at time t, r is the intrinsic rate of growth and K is the carrying capacity. If $r > 0$, then the population has a globally stable nontrivial equilibrium, namely $N^\star = K$, the carrying capacity.

In 1969, Levins introduced so called *metapopulation* models (Levins 1969, Levins and Culver 1971). The conceptual advance of the metapopulation modeling approach is that instead of keeping track of the density of the population, they considered the population as occupying discrete patches and kept track of the proportion of occupied patches. Each patch is either vacant or occupied by a subpopulation; the precise dynamics in each patch were not specified further. Instead, one simply assumes that occupied patches go extinct at rate e and vacant patches become occupied at rate c times the fraction of occupied patches. The colonization of vacant patches is thus global in the sense that a vacant patch can receive migrants from *any* other patch. Dividing the habitat into discrete patches introduces a spatial component into the model; such metapopulation models are referred to as *spatially implicit* models since all patches are linked together through migration. (In contrast, in spatially explicit models local neighborhoods are defined and interactions are only possible between nearby neighbors.)

The number of patches is assumed to be large so that a deterministic model for the fraction of occupied patches provides a good approximation. If $q = q(t)$ denotes the fraction of occupied patches at time t, then the dynamics are given by

$$\frac{dq}{dt} = cq(1 - q) - mq. \tag{19.2.2}$$

Levins' model (19.2.2) predicts the existence of a phase transition. If $m/c \geq 1$, then there is only one equilibrium, namely the trivial equilibrium $q^\star = 0$. If $m/c < 1$, then, in addition to the trivial equilibrium, there is a nontrivial equilibrium, namely

$$q^\star = 1 - \frac{m}{c}.$$

That is, the colonization rate c must exceed the extinction rate m in order to maintain the population.

Levins' model can also be interpreted as an individual based model by stipulating that each patch is occupied by at most one individual, instead of entire subpopulations (see, for instance, Tilman et al. 1997). The parameters c and m are then interpreted as the fecundity and the mortality of an individual plant. We will use this interpretation in the following.

Individual based models for plant populations typically have three components, fecundity, establishment and mortality. Each component may be density dependent. In the models we consider below, fecundity and establishment will be closely linked. Fecundity refers to the ability to produce offspring, whereas establishment refers to whether the offspring are able to survive and become adult plants. We assume in all the models below that there is at most one individual per site. Consequently, establishment will always be density dependent.

The spatially explicit analogue of Levins' model is the *basic contact process*. This model is an interacting particle system and was introduced by Harris (1974) to study the spread of an epidemic in a spatial habitat. The model can also be interpreted as a single species plant competition model. It is a continuous time Markov process whose state at time t is a configuration $\eta_t \in \{0, 1\}^{\mathbf{Z}^d}$. That is, each site $x \in \mathbf{Z}^d$ is either occupied (i.e., $\eta_t(x) = 1$) or vacant (i.e., $\eta_t(x) = 0$). For each site $x \in \mathbf{Z}^d$ we define a local neighborhood $\mathcal{N}_x = x + \mathcal{N}$ with $\mathcal{N} = \{z : 0 < \|z\|_p \leq R\}$ where R denotes the radius of the neighborhood. Here $\|\cdot\|_p$ denotes the p norm; frequently we choose $p = \infty$, that is, $\|z\|_\infty = \sup_i z_i$ for $z = (z_1, \ldots, z_d)$. When we say that something happens at rate κ, we mean that the interarrival times between events are exponentially distributed with mean $1/\kappa$. The dynamics are then as follows:

(i) If x is occupied, it becomes vacant at rate 1.
(ii) If x is vacant, it becomes occupied at rate

$$\lambda|\{y : \eta_t(y) = 1 \text{ for } y \in \mathcal{N}_x\}|/|\mathcal{N}|.$$

Mortality is density independent. Fecundity is expressed by the parameter λ which gives the rate at which offspring are produced; this is density independent. Since only vacant sites can become occupied, i.e., births onto occupied sites are suppressed, establishment is density dependent.

Equation (19.2.2) can be viewed as the *mean field* version of the contact process. In a mean field model one assumes that all sites are independent.

The contact process shows a phase transition similar to Levins' model. There exists a critical value λ_c, so that if $\lambda \leq \lambda_c$, then the only stationary distribution is the point mass at the all empty configuration; whereas if $\lambda > \lambda_c$, in addition to the trivial stationary distribution, there exists a nontrivial translation invariant stationary distribution with positive density of 1's. The existence of a phase transition was shown by Harris (1974). The fact that the contact process dies out at the critical value was established by Bezuidenhout and Grimmett (1990). There are rigorous bounds available for the critical value. Furthermore, computer simulations show that with nearest neighbor interactions (i.e., $R = 1$ and $p = 1$), the one-dimensional critical value is $\lambda_c = 3.2988$ and the two-dimensional critical value is $\lambda_c = 1.648$.

The radius of the neighborhood, R, is called the dispersal range. It is known that the contact process has positive correlations, that is, the spatial pattern in equilibrium shows clumping. This is more pronounced for smaller values of the dispersal range R. Bramson et al. (1989) showed that the critical value converges to 1 as the range R tends to infinity and they gave rates of convergence. The critical value in the long range case thus agrees with the critical value for Levins' model. In addition, as the range R tends to infinity, the nontrivial equilibrium (if it exists) is close to product measure with density $1 - 1/\lambda$; that is, spatial correlations vanish and the global density agrees with the corresponding value for Levins' model ($m = 1, c = \lambda$).

Another model that is used as a building block for plant competition models has density dependent death. The metapopulation formulation is

$$\frac{dq}{dt} = cq(1-q) - \alpha q^2 \tag{19.2.3}$$

where $q = q(t)$ is again the fraction of occupied patches at time t. That is, individuals give birth at rate c and establishment is only possible in vacant patches; individuals die at a rate which is proportional to the fraction of occupied patches, hence mortality is density dependent. This model has a trivial equilibrium $q^\star = 0$ and a nontrivial equilibrium

$$q^\star = \frac{1}{1 + \alpha/c}.$$

The critical value for the existence of a nontrivial equilibrium is $c = 0$; that is, survival is possible for any $c > 0$ provided $\alpha < \infty$.

The spatially explicit analogue of this model is the biased annihilating branching process (BABP). The unbiased case was studied by Bramson and Gray (1985) and Bramson et al. (1991). The biased case was investigated by Neuhauser and

Sudbury (1993) and Sudbury (1997). This is again a continuous time Markov process whose state at time t is a configuration $\eta_t \in \{0,1\}^{\mathbf{Z}^d}$. We define the local neighborhood as before, namely, for $x \in \mathbf{Z}^d$, we set $\mathcal{N}_x = x + \mathcal{N}$ with $\mathcal{N} = \{z : 0 < \|z\|_p \leq R\}$ where R denotes the radius of the neighborhood. We define the dynamics as follows.

(i) If x is occupied, then at rate β the particle at x gives birth and the offspring chooses one of the sites in $\mathcal{N}_x$ at random. If the chosen site is vacant, the offspring can establish itself. If the chosen site is occupied, the offspring perishes.

(ii) If x is occupied, then the particle at x dies at rate

$$\frac{|\{y : \eta(y) = 1 \text{ for } y \in \mathcal{N}_x\}|}{|\mathcal{N}|}.$$

Neuhauser and Sudbury (1993) showed that the only equilibrium measures for the BABP on $\mathbf{Z}$ are the product measure with density $\beta/(1+\beta)$ and the measure that concentrates on the all empty configuration. They also showed that if $\beta > 1/3$, the probability of a site being occupied was bounded away from 0. Mountford (1993) showed that if $\beta > 1/3$ then the distribution is product measure with density $\beta/(1+\beta)$ if the initial configuration is finite. Sudbury (1997) extended these results to $\beta > 0$ and showed that starting from any product measure with positive density, product measure with density $\beta/(1+\beta)$ is the equilibrium.

Remarkable from a plant competition point of view is that sites are uncorrelated at equilibrium. This is due to the fact that both the competition neighborhood (the neighborhood in (ii)) and the dispersal neighborhood (the neighborhood in (i)) are the same. When the competition neighborhood is much larger than the dispersal neighborhood, the process resembles the contact process and thus shows clumping. Biologically more realistic is the case when the competition neighborhood is much smaller than the dispersal neighborhood; this results in a more even spatial distribution of occupied sites.

The two models introduced in this section can be used as building blocks for multispecies models. In addition to the dynamics of each species in the absence of all other species, one needs to specify interactions between the different species. The next two sections discuss two such models.

19.3 The Lotka-Volterra Model

The earliest mathematical model that included both intraspecific and interspecific competition between two species was formulated by Volterra (1926) and, independently, by Lotka (1932). We introduced this model in equations (19.1.1) and (19.1.2) in the introduction. Comparing (19.1.1) and (19.1.2) with (19.2.1), we see that the Lotka-Volterra model of interspecific competition is an extension of the logistic equation. Now, both species contribute to density dependent competition; N in (19.2.1) is replaced by $N_1 + \alpha_{12}N_2$ in (19.1.1) where the first term (N_1) signifies the effect of species 1 on itself and the second term ($\alpha_{12}N_2$) the effect of species 2 on species 1 (multiplying N_2 by α_{12} and using the same denominator, namely K_1 allows one to interpret $\alpha_{12}N_2$ as "N_1-equivalents"). A similar interpretation holds for (19.1.2). It is straightforward to analyze the equilibria of this system. The following holds.

(i) When $K_1 > K_2\alpha_{12}$ and $K_2 > K_1\alpha_{21}$, that is, when interspecific competition is less important than intraspecific competition, *coexistence* is possible; i.e., there exists a nontrivial stable equilibrium in which both species have positive densities.
(ii) When $K_1 < K_2\alpha_{12}$ and $K_2 > K_1\alpha_{21}$, then species 2 takes over and species 1 dies out. That is, a strong interspecific competitor outcompetes a weak interspecific competitor. This is called *competitive exclusion.* (If we reverse the inequalities, then species 1 takes over and species 2 dies out.)
(iii) When $K_1 < K_2\alpha_{12}$ and $K_2 < K_1\alpha_{21}$, that is, when interspecific competition is more important than intraspecific competition, the two species cannot coexist. The outcome of the competition between the two species depends on initial densities. This is called *founder control.*

As in the case of the logistic equation, the Lotka-Volterra model of interspecific competition tracks population sizes and not individuals. It is easy to reformulate this model as a metapopulation model. To do this, we introduce a spatially implicit model analogous to the metapopulation model (19.2.2) but interpret it as an individual based model. We denote by u_i the fraction of patches that are occupied by an individual of species i. We formulate the model in the following way.

$$\frac{du_1}{dt} = \beta_1 u_1(1 - u_1 - u_2) - u_1^2 - \alpha_{12}u_1u_2 \tag{19.3.1}$$

$$\frac{du_2}{dt} = \beta_2 u_2(1 - u_1 - u_2) - u_2^2 - \alpha_{21}u_1u_2. \tag{19.3.2}$$

Looking at (19.3.1) and (19.3.2) we see that fecundity is density independent, but establishment and mortality are density dependent. Mortality has two components, one due to intraspecific competition (u_i^2), the other due to interspecific competition ($\alpha_{ij}u_iu_j$). The equilibrium behavior of this model is the same as for the Lotka-Volterra model of interspecific competition and is summarized in Figure 19.1.

Following the approach in (19.3.1) and (19.3.2), we can formulate the analogous spatially explicit model. Again, this will be a continuous time Markov process with local interactions (or interacting particle system) on $\mathbf{Z}^d$. The state at time t is a configuration $\eta_t \in \{0, 1, 2\}^{\mathbf{Z}^d}$. That is, each site $x \in \mathbf{Z}^d$ is either vacant (i.e., $\eta_t(x) = 0$), occupied by an individual of species 1 (i.e., $\eta_t(x) = 1$), or occupied by an individual of species 2 (i.e., $\eta_t(x) = 2$). We define a local neighborhood as before, namely, for $x \in \mathbf{Z}^d$, we set $\mathcal{N}_x = x + \mathcal{N}$ with $\mathcal{N} = \{z : 0 < \|z\|_p \leq R\}$ where R denotes the radius of the neighborhood. We set

$$f_i(x) = |\{y \in \mathcal{N}_x : \eta(y) = i\}|/|\mathcal{N}|.$$

Then the dynamics are as follows.

(i) If x is occupied by an individual of species i, then the site becomes vacant at rate $f_i(x) + \alpha_{ij} f_j(x)$.

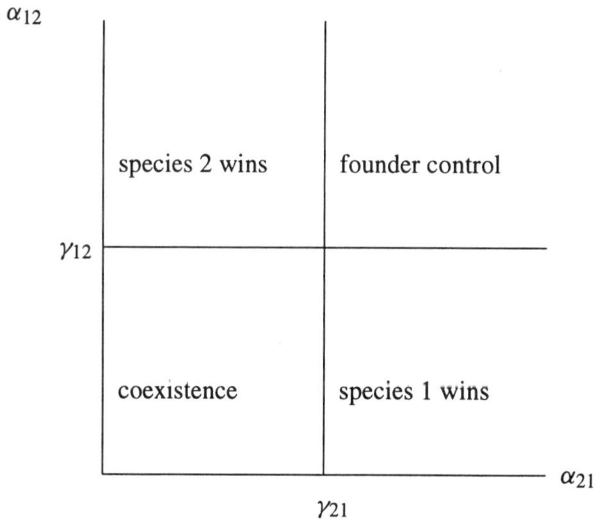

FIGURE 19.1. Phase diagram for the Lotka-Volterra model in equations (19.1.1) and (19.1.2) where $\gamma_{ij} = K_i/K_j$ and the metapopulation model in equations (19.3.1) and (19.3.2) where $\gamma_{ij} = 1$.

(ii) If x is vacant, it becomes occupied by an individual of species i at rate $\beta_i f_i(x)$.

Equations (19.3.1) and (19.3.2) can then also be viewed as the mean field version of this spatially explicit model. This model assumes that offspring can only land on vacant sites. (In the next section we will discuss a model in which individuals can displace each other.)

Before we discuss the effects of explicit space on the outcome of competition, we wish to comment on the relationship between spatially implicit and spatially explicit models. A detailed discussion of this can be found in Durrett and Levin (1994). Broadly speaking the spatially explicit model is well approximated by the spatially implicit model when the neighborhood is sufficiently large. [Durrett and Levin (1994) discussed the case when individuals are allowed to move about at a rate which is much faster than the population dynamics. But their conclusions can be extended to the case of long range dispersal.] That is, if we fix all parameters in the spatially explicit model and then let the range R go to infinity, the spatially explicit model shows the same behavior except in the case of founder control. When the spatially implicit model predicts founder control, the two monoculture equilibria compete with each other in the spatially explicit model and the stronger one wins regardless of the initial state provided both types are initially present at positive densities.

Using results in Durrett and Neuhauser (1994) and Neuhauser (1994) the above comparison can be made rigorous. The main conclusion from this approach is that if the competition/dispersal neighborhoods are large compared to the scale of individual plants, a spatially explicit model will not add much beyond a spatially implicit model. Unless we can demonstrate that there are conditions where the spatially explicit model behaves qualitatively different from the corresponding spatially implicit model, there would be no need for spatially explicit models.

In Neuhauser and Pacala (1999) we studied the spatially explicit Lotka-Volterra model with interspecific competition with the specific goal to understand the effects of short-range interactions on the outcome of competition. We were able to demonstrate that the spatially explicit model with short range interactions exhibits behavior that differs from the behavior of the spatially implicit model. We investigated the case when

$$\beta_1, \beta_2 \to \infty \qquad \text{with} \qquad \frac{\beta_1}{\beta_2} \to \lambda.$$

In this limit vacant sites are filled instantly, that is, the state space reduces to $\{1, 2\}^{\mathbf{Z}^d}$. We therefore call this the *high density limit*. In the following discussion we will restrict ourselves to the case $\lambda = 1$. The dynamics then simplify to

(i) If $\eta_t(x) = 1$, it becomes a 2 at rate $f_2(f_1 + \alpha_{12} f_2)$.
(ii) If $\eta_t(x) = 2$, it becomes a 1 at rate $f_1(f_2 + \alpha_{21} f_1)$.

We exclude the one-dimensional nearest neighbor case from the following discussion. (The one-dimensional nearest neighbor case exhibits clustering for all choices of parameters and is therefore different from the other cases.) Three major conclusions emerged from our investigation of the effects of short range interactions. First, we demonstrated that spatial segregation occurs in $d \leq 2$ when $\alpha \equiv \alpha_{12} = \alpha_{21}$ and $\alpha \uparrow 1$. The case $\alpha = 1$ is the voter model (Clifford and Sudbury 1973, and Holley and Liggett 1975); the voter model exhibits clustering in $d \leq 2$. We conjecture that coexistence is possible in $d \leq 2$ when $\alpha < 1$. More precisely, we believe that the average cluster size stays finite as long as $\alpha < 1$ but is increasing in α. The spatial segregation for $\alpha < 1$ is more pronounced when R, the radius of the interaction neighborhood, is small. In fact, if we fix $\alpha < 1$ and increase R, the clusters break up.

Second, we demonstrated that short range interactions reduce the size of the coexistence region. That is, there are parameter values such that the spatially implicit model predicts coexistence whereas in the spatially explicit model one species competitively excludes the other species.

Third, we demonstrated a reduction in the size of the founder control region which is more pronounced for short range interactions. That is, there are parameter values such that the spatially implicit model predicts founder control whereas in the spatially explicit model one species competitively excludes the other species. We conjecture that there is no founder control in the spatially explicit model; instead, if $\alpha_{21} > \alpha_{12} > 1$, then species 1 competitively excludes species 2, and if $\alpha_{12} > \alpha_{21} > 1$, then species 2 competitively excludes species 1. We call the strategy that results in an expansion of the regions where one species competitively excludes the other species, a *phalanx strategy*.

Our results clearly demonstrate that the general belief in plant ecology, namely an additional spatial dimension facilitates coexistence, is not true in general.

Bolker and Pacala (1998) investigated a version of the Lotka-Volterra model with interspecific competition. Their model is a point process. Using moment equations and neglecting all central moments of order 3 and higher, they were able to identify two mechanisms that facilitate coexistence.

The first mechanism is the well-known *competition-colonization trade-off*. Here, a strong competitor, say species 1, outcompetes the other species in the spatially implicit version. If species 1 has positive correlations in the spatially explicit model, then the global density of species 1 decreases as the dispersal range decreases. This allows one to choose parameter values for species 2 so that if species 2 has long range dispersal, it may be able to invade species 1 provided the dispersal range of species 1 is sufficiently small.

The second mechanism is called *successional niche* and was first demonstrated in a metapopulation model in Pacala and Rees (1998). It relies on a short range advantage of the weaker competitor. If the stronger competitor leaves open corridors, the weaker competitor might be able to utilize these corridors if it disperses over short distances.

Both mechanisms are also present in our grid based model on $\{0, 1, 2\}^{\mathbf{Z}^d}$ (i.e., when β_1 and β_2 are both finite).

19.4 Hierarchical Models

In the model in the previous section we stipulated that offspring could only land on vacant sites. We now assume that species are hierarchically ordered and that better competitors can replace weaker competitors. This model was originally formulated by Hastings (1980) and Nee and May (1992) and then later extended and extensively studied by Tilman and collaborators (e.g., Tilman 1994, Lehman and Tilman 1997). It can be thought of as the metapopulation analogue of Tilman's resource models (Tilman 1977, 1982) which predict that if species are hierarchically ordered and compete for a single resource, then eventually the best competitor, that is, the species which reduces the resource level to its lowest level, will dominate and exclude all other species. We denote by $p_i(t)$ the fraction of patches that are occupied by species i for $i = 1$ and 2. The spatially implicit version is given by

$$\frac{dp_i}{dt} = c_i p_i \left(1 - \sum_{j=1}^{i} p_j\right) - m_i p_i - \sum_{j=1}^{i-1} c_j p_i p_j. \tag{19.4.1}$$

Species 1 is the better competitor and can colonize vacant sites and, in addition, replace the weaker competitor 2. Competitor 2 can only colonize vacant sites. This model can easily be extended to more than two species in which case species i can replace species j provided $j > i$. Tilman (1994) gave conditions for coexistence.

He showed that any number of species can stably coexist provided weaker competitors have sufficiently high colonization rates. In the case of just two species, the conditions for coexistence are as follows.

$$\frac{c_1}{m_1} > 1 \qquad \text{and} \qquad \frac{m_2}{c_2} < \frac{m_1}{c_1} - \frac{c_1 - m_1}{c_2}. \tag{19.4.2}$$

The ratio c_i/m_i can be interpreted as the average lifetime fecundity of an individual of species i. The second condition in (19.4.2) implies that the lifetime fecundity of the weaker competitor must exceed the lifetime fecundity of the stronger competitor by a positive amount in order to be able to coexist with the stronger competitor.

The mechanism at work here is the *competition-colonization trade-off* we encountered in the previous section. The behavior of this model is in contrast to Tilman's original resource models. In his model for one limiting resource the species that reduced the limiting resource to the lowest level, eventually dominated and excluded all other species. The reason for this difference lies in the fact that the spatially implicit model allows for empty patches which can be colonized by the weaker competitors, so-called fugitive species, provided they are sufficiently good colonizers. (Horn and MacArthur 1972, Lehman and Tilman (1997) and references therein).

A spatially explicit version of this model was formulated by Durrett and Swindle (1991) and Durrett and Schinazi (1993). It is a continuous time Markov model whose configuration at time t is $\eta_t \in \{0, 1, 2\}^{\mathbf{Z}^d}$. As before we define a local neighborhood $\mathcal{N}_x$ with range R, $\mathcal{N}_x = x + \{z : 0 < \|z\|_p \leq R\}$ and denote by $f_i(x)$ the fraction of neighbors of x of type i. The dynamics are as follows.

(i) If $\eta_t(x) = 1$ ($\eta_t(x) = 2$, respectively), then the site becomes vacant (i.e., 0) at rate δ_1 (δ_2, respectively).
(ii) If $\eta_t(x) = 0$ or 2, it becomes a 1 at rate $\lambda_1 f_1(x)$.
(iii) If $\eta_t(x) = 0$, it becomes a 2 at rate $\lambda_2 f_2(x)$.

(The spatially implicit version (19.4.1) is the mean field version of this spatially explicit model with $m_i = \delta_i$ and $c_i = \lambda_i$.) Durrett and Swindle (1991) investigated the spatially explicit model in the case of long range dispersal and showed that if the conditions in (19.4.2) hold, then coexistence is possible in the spatially explicit model for sufficiently large dispersal range. Durrett and Schinazi (1993) showed that if we reverse the inequality in the second condition in (19.4.2), then coexistence is not possible if the dispersal range is sufficiently large. Durrett (1995) conjectured that except for the one-dimensional nearest neighbor case we can find parameter values so that coexistence is possible. Simulations in $d = 2$ with nearest neighbor dispersal show that coexistence is possible but that the set of parameter values where coexistence occurs, is rather small (Durrett, 1995).

As mentioned above the existence of implicit space allows for an arbitrary number of species to coexist. This will likely have to be modified in a spatially explicit model. Being a better colonizer in a spatially implicit model simply means that better colonizers have higher fecundity. In a spatially explicit model, however, there are two parts of reproduction, namely the production of seeds and their dispersal. If being a better colonizer simply means a higher production of seeds, then competitors sufficiently far down in the hierarchy will likely be recruitment limited: If their density falls below the reciprocal of the neighborhood size, survival will be difficult if not impossible. That is, in order to have an arbitrary number of species, weaker competitors must not only rely on a larger number of seeds but also increase their dispersal range. Consequently, a weaker competitor not only needs to invest more energy into seed production but also needs to produce seeds that are more easily dispersed over large distances in order to take advantage of the competition-colonization trade-off.

From a biological point of view it is not very realistic to assume that a better competitor can instantly replace a weaker competitor. A spatially implicit model in which a time delay was allowed, was introduced in Pacala and Rees (1998). In their model patches that are occupied by the superior competitor are initially susceptible to invasion by the inferior competitor. Once the superior competitor resided in a patch long enough, it can successfully exclude the weaker competitor from its patch. Besides the competition-colonization trade-off they found a new mechanism, the successional niche, which we discussed earlier. The successional niche mechanism operates when both species are not recruitment limited, for instance, if both species have an unlimited number of colonists available.

Durrett and Neuhauser (1997) introduced a spatially explicit competition model in which both species were able to coexist simultaneously at the same site, however, the death rates of the two species depended on whether an individual was alone in a patch or together with an individual of the other type. This model allows for three extreme cases. (1) Neither species is affected by the presence of the other species. In this case both species evolve like two independent contact processes. (2) Neither species can invade a site that is already occupied. This is the multitype contact process (Neuhauser 1992). (3) Only one of the two species is allowed to invade a site that is already occupied by the other species and kills the other species upon invasion. This is the hierarchical model introduced above.

In case (1) coexistence can be trivially achieved as long as both species are able to survive. In case (2) Neuhauser (1992) showed that coexistence is impossible in $d \leq 2$ when both species die at the same rate. The species with the higher birth rate excludes the species with the lower birth rate. If both birth rates are the same, then clustering occurs in $d \leq 2$ as in the case of the voter model we discussed earlier. In case (3) we saw that the weaker competitor needs a colonization rate that is sufficiently large to be able to coexist.

19.5 Conclusions and Outlook

The discrepancy between predictions from earlier competition models and observations in natural plant communities calls for the investigation of models that relax some of the assumptions made in the early competition models. In this article, we summarized some of the more recent mathematical results in theoretical plant ecology that address the role of explicit space on the outcome of competition in spatially and temporally homogeneous environments. As already pointed out by Hutchinson (1961) explicit space might provide a mechanism that allows for many more species to stably coexist than there are limiting resources.

That space will not always facilitate coexistence was demonstrated, for instance, in Neuhauser and Pacala (1999). There it was shown that local competitive interactions between plants can hinder coexistence by generating patterns, such as spatial segregation, or by employing the phalanx strategy which qualitatively changed the outcome of competition.

On the other hand, we also summarized results that clearly demonstrated that space can serve as an additional niche and thus facilitate coexistence. The two mechanisms that are important in this context, are the competition colonization trade-off and the successional niche.

We are still at the beginning of understanding the role of explicit space on the composition of natural plant communities. Simple mathematical models can help to identify basic mechanisms that shape communities. Results that have been obtained so far, are model specific. Clearly, no natural system behaves exactly like any of the theoretical models. It would therefore be important to investigate how robust these results are under changing the dynamics of the simple mathematical models. In addition, there is a need to interpret theoretical results in light of empirical observations and to suggest field experiments that would allow field biologists to test predictions from theoretical models in an experimental setting.

Models that have been rigorously studied are largely confined to two-species models. Already spatially implicit models for three or more species can show complex behavior. The investigation of spatially explicit, individual based models for three or more species should provide important insight into the question of how many species can stably coexist in natural communities.

Acknowledgment: This research was partially supported by NSF grant DMS-97-03694.

REFERENCES

[1] Bascompte, J. and R.V. Solé (Eds.) (1998). *Modeling Spatiotemporal Dynamics in Ecology*. Springer-Verlag, New York.

[2] Bezuidenhout, C. and G. Grimmett (1990). The critical contact process dies out. *Ann. Probab.* 18:1462–1482.

[3] Bolker, B. and S.W. Pacala (1998). Spatial moment equations for plant competition: understanding spatial strategies and the advantage of short dispersal. Submitted to *American Naturalist*.

[4] Bramson, M., Ding Wan-ding, and R. Durrett (1991). Annihilating branching processes. *Stoch. Proc. and their Appl.* 37:1–17.

[5] Bramson, M., R. Durrett and G. Swindle (1989). Statistical mechanics of crabgrass. *Ann. Probab.* 17:444–481.

[6] Bramson, M. and L. Gray (1985). The survival of branching annihilating random walk. *Z. für Wahr.* 68:447–460

[7] Clifford, P. and A. Sudbury (1973). A model for spatial conflict. *Biometrika* 60:581–588.

[8] Connell, J.H. (1961). The influence of interspecific competition and other factors on the distribution of the barnacle *Chthamalus stellatus*. *Ecology* 42:710–723.

[9] Crawley, M.J. (1997). The structure of plant communities. Pages 475–531. In *Plant Ecology*, edited by Michael J. Crawley, second edition. Blackwell Science Ltd.

[10] Darwin, C. (1836). *The Voyage of the Beagle*. Published by Nal Penguin Inc., New York, 1988.

[11] Darwin, C. (1859). *The Origin of Species*. Published in Penguin Books, New York, 1985.

[12] Durrett, R. (1988). *Lecture Notes on Particle Systems and Percolation*. Wadsworth, Belmont, CA.

[13] Durrett, R. (1995). Ten Lectures on Particle Systems. In *Lecture Notes on Probability Theory. Ecole d'Eté de Probabilités de Saint-Flour XXIII-1993*, P. Bernard (ed.), pages 97–201. Springer-Verlag, New York.

[14] Durrett, R. and S.A. Levin (1994). The importance of being discrete (and spatial). *Theor. Pop. Biol.* 46:363–394.

[15] Durrett, R. and C. Neuhauser (1994). Particle systems and reaction diffusion equations. *Ann. Probab.* 22:289–333.

[16] Durrett, R. and C. Neuhauser (1997). Coexistence results for some competition models. *Ann. Appl. Probab.* 7:10–45.

[17] Durrett, R. and R. Schinazi (1993). Asymptotic critical value for a competition model. *Ann. Appl. Probab.* 3:1047–1066.

[18] Durrett, R. and G. Swindle (1991). Are there bushes in the forest? *Stoch. Proc. Appl.* 37:19–31.

[19] Gause, G.F. (1935). *The Struggle for Existence*. Williams & Wilkins, Baltimore.

[20] Harris, T.E. (1974). Contact interaction on a lattice. *Ann. Probab.* 2:969–988.

[21] Hartway, C., M. Ruckelshaus, and P. Kareiva (1998). The challenge of applying spatially explicit models to a world of sparse and messy data. In *Modeling Spatiotemporal Dynamics in Ecology*, Bascompte, J. and Solé R.V. (eds.), pages 215–223. Springer-Verlag, New York.

[22] Hastings, A. (1980). Disturbance, coexistence, history, and competition for space. *Theor. Popul. Biol.* 18:363–373.

[23] Holley, R. and T.M. Liggett (1975). Ergodic theorems for weakly interacting particle systems. *Adv. Math.* 32:149–174.

[24] Horn, H.S. and R.H. MacArthur (1972). Competition among fugitive species in a harlequin environment. *Ecology* 53:749–752.

[25] Hutchinson, G.E. (1959). Homage to Santa Rosalia; or why are there so many kinds of animals. *The American Naturalist* 93:145–159.

[26] Hutchinson, G.E. (1961). The paradox of the plankton. *The American Naturalist* 95:137–147.

[27] Humboldt, A. von (1814). *Personal Narrative of a Journey to the Equinoctial Regions of the New Continent*. Abridged and translated with an Introduction by Jason Wilson. Published in Penguin Classics, New York, 1995.

[28] Kingsland, S.E. (1985). Modeling nature: Episodes in the history of population ecology. University of Chicago Press, Chicago.

[29] Kingsland, S.E. (1991). Defining ecology as a science. Pages 1–13. In *Foundations of Ecology*. Edited by Leslie A. Real and James H. Brown. The University of Chicago Press, Chicago.

[30] Lehman, C. and D. Tilman (1997). Competition in spatial habitats. In *Spatial ecology: The role of space in population dynamics and interspecific interactions*. Edited by David Tilman and Peter Kareiva. Princeton University Press, Princeton.

[31] Levins, R. (1969). Some demographic and genetic consequences of environmental heterogeneity for biological control. *Bulletin of the Entomological Society of America* 15:237–240.

[32] Levins, R. and D. Culver (1971). Regional coexistence of species and competition between rare species. *Proc. Nat. Acad. Sci. USA* 68:1246–1248.

[33] Liggett, T.M. (1985). *Interacting Particle Systems*. Springer-Verlag, New York.

[34] Lotka, A.J. (1932). The growth of mixed populations: two species competing for a common food supply. *Journal of the Washington Academy of Sciences* 22:461–469.

[35] Mountford, T. (1993). A coupling of finite particle systems. *J. Appl. Probab.* 30:258–262.

[36] Nee, S. and R.M. May (1992). Dynamics of metapopulations: habitat destruction and competitive coexistence. *J. Anim. Ecol.* 61:37–40.

[37] Neuhauser, C. (1994). A long range sexual reproduction process. *Stoch. Proc. Appl.* 53:193–220.

[38] Neuhauser, C. (1992). Ergodic theorems for the multitype contact process. *Probab. Theory Related Fields* 91:467–506.

[39] Neuhauser, C. and S.W. Pacala (1999). An explicitly spatial version of the Lotka-Volterra model with interspecific competition. To appear in *Ann. Appl. Probab.*

[40] Neuhauser, C. and A. Sudbury (1993). The biased annihilating branching process. *Stoch. Proc. Appl.* 53:193–220.

[41] Pacala, S.W. (1997). Dynamics of plant communities. In *Plant Ecology*, edited by Michael J. Crawley, second edition. Blackwell Science Ltd.

[42] Pacala, S.W. and S.A. Levin (1997). Biologically generated spatial pattern and the coexistence of competing species. In *Spatial ecology: The role of space in population dynamics and interspecific interactions*. Edited by David Tilman and Peter Kareiva. Princeton University Press, Princeton.

[43] Pacala, S.W. and M. Rees (1998). Models suggesting field experiments to test two hypotheses explaining successional diversity. *American Naturalist* 152(5):729–737.

[44] Pearl, R. and L.J. Reed (1920). On the rate of growth of the population of the United States since 1790 and its mathematical representation. *Proc. Nat. Acad. Sci. USA* 6:275–288.

[45] Sudbury, A. (1997). The convergence of the biased annihilating branching process and the double-flipping process on $\mathbf{Z}^d$. *Stoch. Proc. Appl.* 68:255–264.

[46] Tansley, A.G. (1917). On competition between *Galium saxatile* L. (G. *hercynicum* Weig.) and *Galium sylvestre* Poll. (G. *asperum* Schreb.) on different types of soil. *Journal of Ecology* 5:173–179.

[47] Tilman, D. (1976). Ecological competition between algae: experimental confirmation of resource-based competition theory. *Science* 192:463–465.

[48] Tilman, D. (1977). Resource competition between planktonic algae: an experimental and theoretical approach. *Ecology* 58:338-348.

[49] Tilman, D. (1982). *Resource Competition and Community Structure*. Princeton University Press, Princeton.

[50] Tilman, D. (1994). Competition and biodiversity in spatially structured habitats. *Ecology* 75:2–16.
[51] Tilman, D. (1997). Mechanisms of plant competition. Pages 239–261. In *Plant Ecology*, edited by Michael J. Crawley, second edition. Blackwell Science Ltd.
[52] Tilman, D., C.L. Lehman, and P. Kareiva (1997). Population dynamics in spatial habitats. Pages 3–20. In *Spatial ecology: The role of space in population dynamics and interspecific interactions*. Edited by David Tilman and Peter Kareiva. Princeton University Press, Princeton.
[53] Tilman, D. and S.W. Pacala (1993). The maintenance of species richness in plant communities. Pages 13–25. In *Species Diversity in Ecological Communities*. Edited by Robert E. Ricklefs and Dolph Schluter. Chicago University Press, Chicago.
[54] Verhulst, P.F. (1838). Notice sur la loi que la population suit dans son accroissement. *Correspondences Math. Phys.* 10:113–121.
[55] Volterra, V. (1926). Variations and fluctuations of the number of individuals in an animal species living together. (Reprinted in 1931. In R.N. Chapman, *Animal Ecology*. Mc Graw Hill, New York.)

Department of Mathematics
University of Minnesota
127 Vincent Hall
206 Church St. SE
Minneapolis MN 55455
nhauser@math.umn.edu

20

Large Deviations for Interacting Particle Systems

S.R.S. Varadhan

ABSTRACT We consider large systems of interacting particles. We obtain a large deviation principle for the empirical process viewed as a random measure on the path space. The precise rate function is obtained. While the formula is fairly general, the explicit dependence on the model is seen through quantities like bulk and self diffusion coefficients.

Keywords: Large deviations, interacting particle systems.

AMS Subject Classifications: 60K35, 60F10.

20.1 Introduction

We will consider large systems of interacting particles, i.e., a large number of particles interacting with each other and moving randomly in R^d or Z^d. The interactions will be local in the sense that the motion of each particle will be affected only by presence and positions of a few nearby particles that happen to be within a fixed finite range of it. We will be studying space-time rescalings of such systems, with a diffusive or parabolic relation between the space and time scales. For technical reasons, we will start with periodic boundary conditions on a finite but large cube, so that after rescaling, the space can be considered to be a fixed torus or some finely imbedded lattice inside one. We will give several classes of examples.

One important feature of all these examples is the validity of a hydrodynamical limit under diffusive scaling. We are rescaling systems with conserved quantities, like the total number of particles in our examples. The conserved quantities change slowly while the system rapidly equilibrates to a local equilibrium that is determined by the local average of the conserved quantity. Consequently in our examples, for large values of the scaling parameter, the system is essentially fully described by the density as a function of rescaled space and time. The density itself will evolve according to some (usually) nonlinear heat

---

NOTE: The present article contains a summary of results obtained jointly with J. Quastel and F. Rezakhnalou. The details can be found in an article that is to appear in a forthcoming issue of Probability Theory and Related Fields.

equation. One can use the hydrodynamical scaling principle to reduce many questions, including the macroscopic large deviation behavior and the diffusive motion of a tracer particle in nonequilibrium, to considerations of density alone.

20.2 Some Classes of Examples

The scaling parameter is a large number N that represents the "size" of the system. In physical terms, $\frac{1}{N}$ is the typical interparticle distance and this requires the number of particles to be of order N^d. The actual number of particles k_N will be assumed to be asymptotically equal to $\bar{\rho}N^d$ for some fixed constant $\bar{\rho}$.

Model 1 The scaled interacting diffusion model in $\mathbf{T}^d$ consists of a system of k_N stochastic differential equations for the positions $\{x_i(t)\}$ at time t of the k_N particles in $\mathbf{T}^d$.

$$dx_i(t) = d\beta_i(t) - N\sum_j V'(N(x_i(t) - x_j(t)))dt.$$

Here β_i are independent d-dimensional Brownian motions and the interaction consists of a drift term which is the gradient of the potential

$$V_N = \sum_{i<j} V(N(x_i - x_j))$$

where $V(x)$ is a repulsive pair potential with compact support.

Model 2 Zero Range Processes describe the motion of k_N particles that are located on the sites of a sublattice with width $h = \frac{1}{N}$. Each particle waits for a random exponential time and jumps to a new site. The jumps themselves are random and mutually independent, with probability $p(z)$ for a jump of size z in the unscaled lattice or of size hz in the rescaled one. It is natural to assume that $\sum_z zp(z) = 0$ or more strongly that $p(z) \equiv p(-z)$. The interaction is introduced through the jump rate that is allowed to depend on the number of particles that share the same site as the particle that is to jump. The model then is fully specified by $\{p(z)\}$ and a function $c(k)$. The jump rate of the i-th particle is $N^2c(k(x_i))$. Here x_i is the loction of the i-th particle and $k(x_i)$, is the total number of particles at that site. The factor N^2 is the effect of rescaling time. The case where $c(k) \equiv 1$ is the non-interacting case of scaled independent random walks.

Model 3 The simple exclusion process is a slight modification of the second model. No more than one particle is allowed to be at any given site at any given time. The jump rates are all N^2, except that if there is already a particle present at the selected site where the particle is to jump, the jump is disallowed and the particle is forced to wait for a new exponential waiting time. The effective jump rate then is $N^2\sum' p(z)$ where the summation is restricted to jumps that lead to empty sites.

20.3 The Problem

In all of these problems, for each fixed time T, we have the collection of k_N trajectories $\{x_1(\cdot), x_2(\cdot), \ldots, x_{k_N}(\cdot)\}$ defined on the interval $[0, T]$. The empirical process, which is a measure on the path space $D[[0, T]; \mathbf{T}^d]$, is defined by

$$R_N = \frac{1}{N^d} \sum_{i=1}^{k_N} \delta_{x_i(\cdot)}.$$

It is a random measure and its distribution is determined by the model and the initial configuration $(x_1, \ldots, x_{k_N})$ of the k_N particles.

The basic problem is the asymptotic behavior of R_N. It is natural to assume that $(x_1, \ldots, x_{k_N})$ is such that the initial configuration has an asymptotic empirical distribution. For simplicity we could assume that the initial conditions are deterministic and that

$$\lim_{N\to\infty} \frac{1}{N^d} \sum_{i=1}^{k_N} \delta_{x_i} = \nu_0(dx) = \rho_0(x)dx \tag{20.3.1}$$

exists. One expects the asymptotic behavior of R_N to be governed for the most part by $\rho_0(\cdot)$. We will be concerned with the laws of large numbers asserting that

$$\lim_{N\to\infty} R_N = Q \quad \text{in probability}$$

for a Q that is to be determined, as well as calculating the rates for large deviations from it. One interesting aspect is a degree of universality in the form of the large deviation rate function, depending only on a few physically meaningful objects, like bulk and self diffusion coefficients, that are to be calculated specifically for each model.

We will describe the results in the context of simple exclusion processes, the third set of models that we described. At the end we will provide a brief summary of what is known for the other models.

20.4 Laws of Large Numbers

The first step in these problems is to look at the laws of large numbers for the family of empirical densities

$$\nu_t^N(dx) = \frac{1}{N^d} \sum_i \delta_{x_i(t)} \tag{20.4.1}$$

viewed as random maps from $[0, T]$ into the space $\mathcal{M}(T^d)$ of measures on T^d. In the context of the symmetric simple exclusion process it is quite elementary to establish the following law of large numbers. If at time $t = 0$, the initial conditions are such that (20.3.1) holds in probability, then

$$\lim_{n\to\infty} \nu_t^N(dx) = \nu_t(dx) = \rho(t,x)dx$$

exists in probability, where $\rho(t,x)$ is the solution of the linear heat equation

$$\frac{\partial\rho}{\partial t} = \frac{1}{2}\nabla D\nabla\rho \quad \text{for } t > 0 \tag{20.4.2}$$

with the initial condition

$$\rho(0,x) = \rho_0(x).$$

Here D is the diffusion matrix, which is just the covariance matrix of the jump distribution

$$\langle Du, u\rangle = \sum_z p(z)\langle u, z\rangle^2.$$

Although in other models the bulk diffusion matrix D is a function of the density ρ, for the symmetric simple exclusion model it turns out not to depend on ρ. To see this we make the following simple calculation that is almost a proof. The generator can be written in terms of the basic exchange operators

$$(\nabla_{x,y}f)(\eta) = f(\eta^{x,y}) - f(\eta)$$

where $\eta^{x,y}$ is the new configuration obtained by making an exchange between sites x and y.

$$\eta^{x,y}(z) = \begin{cases} \eta(z) & if \ z \neq x, y \\ \eta(y) & if \ z = x \\ \eta(x) & if \ z = y. \end{cases}$$

The generator is written as

$$\mathcal{L}_N f = N^2 \sum_x \eta(x) \sum_z (1 - \eta(x+z))p(z)\nabla_{x,x+z} f$$

which can be rewritten using the symmetry of $p(\cdot)$ as

$$\mathcal{L}_N f = \frac{N^2}{2}\sum_{x,y}[\eta(x)(1-\eta(y)) + \eta(y)(1-\eta(x)]p(y-x)\nabla_{x,y} f.$$

Because $\nabla_{x,y} f(\eta) = 0$ unless $\eta(x)(1-\eta(y)) + \eta(y)(1-\eta(x)) = 1$, we can simplify

$$\mathcal{L}_N f = \frac{N^2}{2}\sum_{x,y} p(y-x)\nabla_{x,y} f$$

and it looks quite noninteracting. In fact, if we now compute

$$\mathcal{L}_N \frac{1}{N^d}\sum J(\frac{x}{N})\eta(x),$$

it is not hard to see that

$$\mathcal{L}_N \frac{1}{N^d} \sum J(\frac{x}{N})\eta(x) \simeq \frac{1}{2} \sum_x (\nabla D \nabla J)(\frac{x}{N})\eta(x).$$

To complete the proof we only have to show that the martingale term in

$$d \frac{1}{N^d} \sum J(\frac{x}{N})\eta(x,t)$$

goes to zero with N.

The next step is to investigate the behavior of a tagged particle. In other words, at time zero one particular particle is tagged and its motion is followed. In equilibrium we expect the tagged particle to diffuse like a Brownian motion. In fact for a wide class of models this is now a well understood phenomenon [KV]. The covariance matrix for the limiting Brownian motion will in general depend on the equilibrium that is used. In our model on Z^d, the equilibria are the various Bernoulli or product measures μ_ρ with particle density

$$\rho = \mu_\rho\{\eta(x) = 1\}.$$

We will denote by $S(\rho)$ the limiting covariance usually referred to as the self-diffusion matrix. In the noninteracting case it will not depend on ρ. But for symmetric simple exclusion, if we exclude the very special case of nearest neighbor model in one dimension, it is not surprising that at low density, i.e., near $\rho = 0$, $S(\rho) \simeq D$ because the interaction is negligible and near high density, i.e., $\rho = 1$, $S(\rho) \simeq 0$ essentially due to the gridlock that develops when nearly every site is filled.

The behavior of a tagged particle in nonequilibrium is far from obvious. The density ρ is no longer a constant, but a function $\rho(t,x)$ of t and x, given by the solution of equation (20.4.2). One might guess that a tagged particle located at $x(t)$ at time t, being myopic, will only see its immediate neighborhood and behave as if it is in equlibrium at density $\rho = \rho(t,x)$. As a first approximation we could expect $x(t)$ to behave like a time dependent diffusion with generator

$$\frac{1}{2}\nabla S(\rho(t,x))\nabla,$$

but in reality it is more likely that the generator needs to be modified by a first order term and takes the form

$$\frac{1}{2}\nabla S(\rho(t,x))\nabla + b(t,x)\nabla$$

for some drift term $b(\cdot,\cdot)$. Because we have an extra as yet undetermined first order term, it does not matter if we write the higher order term in divergence form or not. We have chosen the divergence form for convenience.

If we picked initially a particle at random and tagged it, its distribution is given, subject to normalization, by a density evolving according to the forward (adjoint) equation corresponding to (20.4.2).

$$\frac{\partial\rho}{\partial t}(t,x) = \frac{1}{2}\nabla S(\rho(t,x))\nabla\rho(t,x) - \nabla\cdot(b(t,x)\rho(t,x) \quad \text{for } t>0 \quad (20.4.3)$$

with the initial condition

$$\rho(0,x) = \rho_0(x).$$

Tagging is only an exercise of the mind. The particles are totally oblivious to it. Since the empirical density is the same for the tagged or untagged system, the solutions of (20.4.2) and (20.4.3) must coincide. This yields

$$\frac{1}{2}\nabla S(\rho(t,x))\nabla\rho(t,x) - \nabla\cdot(b(t,x)\rho(t,x)) = \frac{1}{2}\nabla D\nabla\rho(t,x).$$

While it does not follow, it is quite likely that

$$\frac{1}{2}S(\rho(t,x))\nabla\rho(t,x) - b(t,x)\rho(t,x) = \frac{1}{2}D\nabla\rho(t,x)$$

which can be solved to yield

$$b(t,x) = \frac{1}{2\rho(t,x)}[S(\rho(t,x)) - D]\nabla\rho(t,x).$$

If we now denote by Q a Markov process (with total mass $\bar{\rho}$) having initial density $\rho_0(x)$ and time dependent (backward) generator

$$L_t = \frac{1}{2}\nabla S(\rho(t,x))\nabla + \frac{1}{2}\frac{[S(\rho(t,x)) - D]\nabla\rho(t,x)}{\rho(t,x)}\cdot\nabla, \quad (20.4.4)$$

then one expects

$$\lim_{N\to\infty} R_N = Q \quad (20.4.5)$$

in probability. For simple exclusion processes (20.4.5) was indeed proved by Rezakhanlou [R]. Once the distribution of the tagged particle is shown to converge to a diffusion with generator L_t given by (20.4.4), is not very hard to prove (20.4.5). One needs to show that the motion of two different tagged particles are asymptotically independent. However if we start with an initial configuration that is far away from equilibrium, like an arbitrary deterministic initial condition, to prove the limit theorem when an *arbitrary* particle in the initial configuration is tagged is very hard. On the other hand proving (20.4.5) requires only that the limit theorem be valid for *nearly all* particles in the initial configuration.

The proof of (20.4.5) in [R] uses earlier results in [Q] that study a situation that mediates between tagged and untagged versions. Let us divide the particles into a finite number of types that we will think of as colors. Colors do not affect the motion which is as before. We simply keep track of the colors. At any time t, we have the empirical distribution of each one of the k colors. In the scaled limit we will have k densities $\tilde{\rho}(t,x) = \{\rho_j(t,x) : 1 \le j \le k\}$. They should evolve as a nonlinear system

$$\frac{\partial \tilde{\rho}(t,x)}{\partial t} = \nabla \mathbf{B}(\tilde{\rho}(t,x)) \nabla \tilde{\rho}(t,x) \tag{20.4.6}$$

with suitable initial conditions. The system (20.4.6) is equivalent to solving first the linear equation (20.4.2) for the total density

$$\rho(t,x) = \sum_{j=1}^{k} \rho_j(t,x)$$

and then at the next step solving a set of (identical) equations that are again linear with the correct initial conditions for the densities of individual colors

$$\frac{\partial \rho_j(t,x)}{\partial t} = L_t^* \rho_j(t,x),$$

which is just the forward equation for the probability density of the position of the tagged particle.

In [Q] the multicolor system was viewed in its own right and the law of large numbers under hydrodynamical scaling was established for it. For the simple exclusion model, this turns out to be a nongradient system and the analysis becomes considerably harder. To see this let us consider k different colors. Denote the configuration by $\zeta(x) = \{\xi_j(x)\}$. The variable $\xi_j(x)$ is 1 if the site x contains a particle of color j. Exclusion still applies so that $\eta(x) = \sum_j \xi_j(x) \leq 1$. The generator $\mathcal{L}_N$ now takes the form

$$(\mathcal{L}_N f)(\zeta) = N^2 \sum_{x,y} \eta(x)(1-\eta(y))p(y-x)[f(\zeta^{x,y}) - f(\zeta)].$$

If we compute

$$\mathcal{L}_N \frac{1}{N^d} \sum_x J(\frac{x}{N}) \xi_j(x)$$

we get

$$\frac{N^2}{N^d} \sum_{x,y} \xi_j(x)(1-\eta(y))p(y-x)[J(\frac{y}{N}) - J(\frac{x}{N})]$$

which is nearly equal to

$$\frac{N}{N^d} \sum_{x,y} \xi_j(x)(1-\zeta(y))p(y-x)\langle \nabla J(\frac{x}{N}), y-x\rangle.$$

Although $p(\cdot)$ is symmetric, the possibility of having the sites x, y filled by different colors makes the situation asymmetric, and we cannot simplify the expression above and do summation by parts once more, to neutralize the factor N. The microscopic currents for the different colors in the different coordinate directions

$$W_x^{j,i} = N \sum_y \xi_j(x)(1-\zeta(x+z))p(z)\langle e_i , z\rangle$$

can be replaced only asymptotically by linear combinations of density gradients of the form

$$W_x^{j,i} = -N\mathbf{B}_{r,s}^{j,i}(\mathbf{q})[\xi_r(x+e_s) - \xi_r(x)]$$

where $\mathbf{q} = \{q_1, \ldots, q_k\}$ is the empirical local density. This is the hard part in the analysis of nongradient models. Once this is done (20.4.6) is easy to establish.

The transition from the multicolor system to (20.4.5) is essentially a question of compactness. Let $\mathcal{P} = \{A_1, \ldots, A_p\}$ be an arbitrary but fixed partition of the $\mathbf{T}^d$ into nice sets. If E is a subset of the form

$$E = \{x : x(t_1) \in A_{i_1}, \ldots, x(t_n) \in A_{i_n}\}$$

and the convergence in probability of $R_N(E)$ to $Q(E)$ can be proved by induction on n. For $n = 1$, this just involves the empirical density and a single color. If we assume the result for n time points, we have p^n possible past histories and we code them by p^n distinct colors. The empirical density of the different colors at the next time t_{n+1} is sufficient to yield information on subsets E at stage $n+1$. Given the law of large numbers for the multicolor system, the induction proceeds smoothly. Since the partition is arbitrary, if we have compactness, we can pass from finite dimensional distributions to weak convergence of the processes R_N. This was carried out in [R].

20.5 Large Deviations

We now turn to the question of Large Deviation rates. We examine first the empirical densities ν_t^N.

The probability distribution of the empirical density satisfies a large deviation principle with the following rate functional $I(\rho(\cdot,\cdot))$ on the space of weakly continuous maps ρ from $[0, T]$ into $L_1(T^d)$ that satisfy $0 \le \rho(t, x) \le 1$.

$$I(\rho(\cdot,\cdot)) = I_0(\rho(0,\cdot) + \frac{1}{2}\int_0^T ||\frac{\partial \rho}{\partial t} - \frac{1}{2}\nabla D \nabla \rho||^2_{-1,\rho(t,\cdot)(1-\rho(t,\cdot))} dt. \quad (20.5.1)$$

The first term I_0 is the rate function for deviations of the initial profile from its appropriate limit. For deterministic initial conditions it is 0 for the true value and ∞ for all others. Otherwise it depends on the statistics of the initial configuration. The second term, which is the dynamical part, measures in some way by how much ρ fails to satisfy the heat equation (20.4.2). For any $\rho(\cdot)$ the norm $||\ ||_{-1,\rho(t,\cdot)(1-\rho(t,\cdot))}$ refers to the dual of the weighted Sobolev norm

$$||f||^2_{1,\rho(\cdot)} = \int_{T^d} \langle D\nabla f(x), \nabla f(x)\rangle \rho(x)(1-\rho(x)dx.$$

This was proved in [KOV] and it is necessary for us to understand how these large deviations arise. One can introduce a small bias in the system by perturbing the symmetric jump rates by a weak asymmetric term, i.e., the probabilities of transition from x to $x+z$ at time t are changed from $p(z)$, by a small perturbation, to $p(z) + \frac{1}{N}q(t, \frac{x}{N}, z)$. This has an entropy cost as well as a macroscopic effect. To calculate these quantities we need to know that the system is locally close to an equilibrium (some Bernoulli measure) and the empirical density dictates which equilibrium we have to pick. Quantities not explicitly dependent on the empirical densities are replaced by their expectations in equilibrium which means by suitable functions of the empirical density.

Macroscopically, with weak asymmetry, the equation (20.4.2) gets replaced by (20.5.2) below.

$$\frac{\partial \rho}{\partial t} = \frac{1}{2}\nabla D \nabla \rho - \nabla \cdot b\rho(1-\rho) \quad \text{for } t > 0 \tag{20.5.2}$$

with the initial condition

$$\rho(0, x) = \rho_0(x).$$

Here

$$b(t, x) = \sum_z z q(t, x, z)$$

and the entropy cost (after optimizing over $q(\cdot,\cdot,\cdot)$ for a given $b(\cdot,\cdot)$ and normalization by a factor of N^{-d}) is given by

$$\mathcal{E}(b(\cdot,\cdot)) = \frac{1}{2}\int_0^T \int_{T^d} \langle Db(t,x), b(t,x)\rangle \rho(t,x)(1-\rho(t,x))dxdt.$$

For a given $\rho(\cdot,\cdot)$ we consider

$$\mathcal{B}_{\rho(\cdot,\cdot)} = \{b(\cdot,\cdot) : (20.5.2) \text{ holds}\}.$$

The dynamical term in (20.5.1) is the result of optimizing $\mathcal{E}(b(\cdot,\cdot))$ over $b(\cdot,\cdot) \in \mathcal{B}_{\rho(\cdot,\cdot)}$.

The weak asymmetry has an effect on the behavior of a tagged particle which can be described through $b(\cdot,\cdot)$. The generator L_t gets replaced by

$$L_{t,b(t,\cdot)} = L_t + (1-\rho(t,x))b(t,x)\cdot\nabla$$

and the tagged particle process with the new generator will be denoted by Q_b. As long as $b(\cdot,\cdot) \in \mathcal{B}_{\rho(\cdot,\cdot)}$, Q_b will have $\rho(t,\cdot)$ for its marginal density at time t. Suppose Q is a general stochastic process on $C[[0,T]; T^d]$ with total mass $\bar{\rho}$. Its marginal densities can be denoted by $q(t,\cdot)$. The 'current' is the collection of expected values of the stochastic integrals

$$E^Q\Big[\int_0^T \langle (t, x(t)), dx(t)\rangle\Big]$$

and we can try to find $b_Q \in \mathcal{B}(q(\cdot,\cdot))$ such that the marginals as well as the 'current' of Q match those of the tagged particle distribution Q_{b_Q} that we will obtain under our perturbation. The measure Q will then determine b_Q and Q_{b_Q}. The large deviation rate function for R_N is the functional

$$\mathcal{I}(Q) = I_0(q(0,\cdot)) + \mathcal{E}(b_Q(\cdot,\cdot)) + H(Q; Q_{b_Q}). \tag{20.5.3}$$

The important point here is the interpretation of the terms. The first term is clearly the contribution of the statistics of the initial condition. The second term depends on the hydrodynamical scaling behavior of the model and depends on the objects that describe them. The last term is pure noise, and says that once the correct macroscopic behavior of density and current are assured the tagged particle motions are nearly independent.

The proof of this theorem again proceeds through the intermediate step of multicolor systems. A large deviation principle is established for them with a rate function that is similar to the one for single color.

$$I(\rho(\cdot,\cdot)) = I_0(\rho(0,\cdot) + \frac{1}{2}\int_0^T ||\frac{\partial \rho}{\partial t} - \nabla \mathbf{B}(\rho)\nabla\rho||^2_{-1,Q((\rho(t,\cdot))}dt. \quad (20.5.4)$$

Here $\rho = \{\rho_j\}$ are the densities of the various colors and Q is a matrix of diffusion coeffecients closely related to $\mathbf{B}(\rho)$ and are determined in [Q]. We then partition T^d into a finite number p of cells and consider where the individual labeled particles fall at a finite number of specified times $0 \le t_1 < t_2 < \cdots < t_n \le T$. The large deviation rates for these p^n relative frequencies can be calculated by induction on n. The induction step depends on the large deviation property of a multicolor system. We have to let p and n go to ∞, and a careful calculation with the rate functions through the limiting procedure yields formula (20.5.3). The details can be found in the article [QRV].

20.6 Related Work

While we have described in some detail the work related to symmetric simple exclusion models, there are partial results concerning models 1 and 2.

In the case of interacting Brownian motions, there is the possibility of phase transitions in dimensions larger than one. In dimension one and at densities below phase transition in higher dimensions, there are results concerning laws of large numbers. There are general results in the one dimensional case [V] and the methods of [Y] are applicable in any dimension when the density is restricted to a nice region free of phase transition . For a special model in one dimension, with a $V(x)$ that is like a δ-function, the distribution of any tagged particle in nonequilibrium has been worked out in [G]. Little else is known regarding tagged particles in nonequilibrium.

A lot is known about Zero Range Processes, especially regarding laws of large numbers and tagged particles. See for instance [GJL] and a forthcoming book [KL]. In some sense this is a simpler class of models because the multicolor system is still a gradient system. The analysis is therefore simpler and the results that we have described for the empirical processes of simple exclusion models should have analogs here.

Acknowledgments: This research was supported by NSF grant DMS-9803140 and ARO grant DAAH04-95-1-0666.

REFERENCES

[G] I. Grigorescu, New York University Thesis, 1997.

[GJL] D. Gabrielli, G. Jona-Lasinio, and C. Landim, Onsager reciprocity relations without microscopic reversibility, *Phys. Rev. Lett.* **77**, no. 7, 1202–1205 (1996).

[KL] C. Kipnis and C. Landim, *Scaling Limits of Interacting Particle Systems*, Springer-Verlag, Berlin, Heidelberg, 1999.

[KOV] C. Kipnis, S. Olla, and S.R.S. Varadhan, Hydrodynamics and large deviations for simple exclusion processes, *Comm. Pure Appl. Math.* **42**, 115–137 (1989).

[KV] C. Kipnis and S.R.S. Varadhan, Central limit theorem for additive functionals of reversible Markov processes and applications to simple exclusion, *Comm. Math. Phys.* **106**, 1–19.

[Q] J. Quastel, Diffusion of color in the simple exclusion process, *Comm. Pure Appl. Math.* **45**, 623–679 (1992).

[QRV] J. Quastel, F. Rezakhanlou, and S.R.S. Varadhan, Large deviations for the symmetric simple exclusion process in dimensions $d \geq 3$, *Probab. Theory Relat. Fields* **113**, 1–84 (1999).

[R] F. Rezakhanlou, Propagation of chaos for symmetric simple exclusion, *Comm. Pure Appl. Math.* **117**, 943–957 (1994).

[V] S.R.S. Varadhan, Scaling limit for interacting diffusions, *Comm. Math. Phys.* **135**, 313–353 (1991).

[Y] Yau, H.T., Relative entropy and the hydrodynamics of Ginzburg–Landau models, *Lett. Math. Phys.* **22**, 63–80 (1991).

Courant Institute of Mathematical Science
Department of Mathematics
New York University
251 Mercer Street
New York, NY 10012
varadhan@cims.nyu.edu

21

The Gibbs Conditioning Principle for Markov Chains

Ana Meda and Peter Ney

ABSTRACT Let $X_1, X_2, \ldots$ be an irreducible Markov chain taking values in a measurable space $(S, \mathcal{S})$, $u : S^2 \to \mathbb{R}^d$, $U_n = \sum_{i=1}^{n} u(X_i, X_{i+1})$, $C \subset \mathbb{R}^d$ open and convex. Then conditioned on $\{U_n \in nC\}$ (and under some hypotheses on $\{X_n\}$), it is shown that $\{X_n\}$ converges to a Markov chain, whose transition mechanism is specified.

Keywords: Markov chains, large deviations, Gibbs conditioning.

AMS Subject Classifications: 60F10, 60J10.

21.1 Some Background on Independent Random Variables

The study of conditioned limit laws goes back to Gibbs [G] (1902). He considered a system taking values in a finite state space $\{1, \ldots, k\}$ with probabilities $\{p_1, \ldots, p_k\}$, and with $\{a_1, \ldots, a_k\}$ denoting "energy levels" associated with the states. He proposed the idea of a "canonical distribution" as that which maximized the entropy $H(p) = -\sum p_i \log p_i$, subject to the constraint that the average energy be constant ($\Sigma a_i p_i = c$). Later it was argued (e.g., Jaynes [J] (1967)) that in large physical systems, empirical distributions subject to certain constraints, should approximate maximum entropy distributions.

This was formalized by Vasicek [V] (1980) in the form of a conditional law of large numbers for empirical distributions, as follows. Let $X_1, X_2, \ldots$ be a sequence of independent identically distributed random variables (i.i.d.r.v.'s) taking values in a finite set $S = \{x_1, \ldots, x_k\}$, with $P\{X_1 = x_i\} = q_i > 0$, $i = 1, \ldots, k$, where $Q = (q_1, \ldots, q_k) \in \mathcal{P} =$ the set of d.f.'s on S. Let $\widetilde{P}_n = (\widetilde{p}_{n1}, \ldots, \widetilde{p}_{nk})$, with

$$\widetilde{p}_{ni} = \frac{1}{n} \sum_{j=1}^{n} \mathbf{1}_{x_i}(X_j), \qquad i = 1, \ldots, k,$$

be the empirical distribution of $(X_1, \ldots, X_n)$, where $\mathbf{1}_x(\cdot)$ is the indicator function. Suppose a "constraint set" is given more generally by

$$D_\delta = \left\{ P \in \mathcal{P} : \left| \sum_{i=1}^{k} a_{ji} p_i - c_j \right| \leq \delta, \quad j = 1, \ldots, r \right\}.$$

Assume $D_\delta \neq \phi$, and define the entropy of P relative to Q (for $P, Q \in \mathcal{P}$) by

$$\mathcal{H}(P; Q) = -\sum_{i=1}^{k} p_i \log(p_i/q_i). \tag{21.1.1}$$

Let $V = (v_1, \ldots, v_k) \in \mathcal{P}$ be the maximum entropy distribution, namely

$$\max_{P \in D_0} \mathcal{H}(P; Q) = \mathcal{H}(V; Q).$$

Then Vasicek proved that given any $\epsilon > 0$ there is a $\delta > 0$ such that

$$P\{\|\widetilde{P}_n - V\| > \epsilon \mid \widetilde{P}_n \in D_\delta\} \to 0 \quad \text{as } n \to \infty. \tag{21.1.2}$$

More generally, suppose $X_1, X_2, \ldots$ is a sequence of i.i.d.r.v.'s taking values in a measurable space $(S, \mathcal{S})$, having probability measure μ; let $\{\widetilde{P}_n\}$ be its empirical measure, and let M be a set of probability measures. There have been extensive studies (under various conditions on μ and M) on conditional measures of the form

$$\mathbb{P}\{(X_1, \ldots, X_k) \in \cdot \mid \widetilde{P}_n \in M\} \tag{21.1.3}$$

and of

$$\mathbb{P}\{\widetilde{P}_n \in \cdot \mid \widetilde{P}_n \in M\}. \tag{21.1.4}$$

Csiszar [Cs] (1984) showed that conditioned on $\widetilde{P}_n \in M$, with M satisfying a convexity condition, $(X_1, \ldots, X_n)$ have a property which he calls "asymptotically quasi-independence". This implies that the conditional distribution of $(X_1, \ldots, X_k)$ converges in total variation norm to the k'th product of a measure P^* that he called the I-projection of μ. In Dembo and Zeitouni [D-Z1] (1996), [D-Z2] (1998) and Dembo and Kuelbs [D-K] (1997) this result is extended by allowing $k = k(n) \to \infty$. Stroock and Zeitouni [St-Z] (1991) make an extensive study of (21.1.4) and some generalizations and applications.

Another formulation was made by Lehtonen and Nummelin [Le-Nu2] (1990). Let $X_1, X_2, \ldots \in S$ be as above, $g : S \to \mathbb{R}^{d_1}$, $u : S \to \mathbb{R}^{d_2}$, $G_n = \sum_{i=1}^n g(X_i)$, $U_n = \sum_{i=1}^n u(X_i)$, and let $C \subset \mathbb{R}^{d_2}$ be convex with $C^0 =$ interior of $C \neq \phi$. Say that a sequence of r.v.'s $\{Z_i\}$ converges exponentially to z_0, written $Z_i \xrightarrow{\exp} z_0$, with respect to $\mathbb{P}_n$, if for any $\epsilon > 0$ there is an $a > 0$ such that

$$\mathbb{P}_n\{|Z_n - z_0| > \epsilon\} \leq e^{-an}, \quad n = 1, 2, \ldots. \tag{21.1.5}$$

Then they showed that for g bounded

$$\frac{G_n}{n} \xrightarrow{\exp} v_0 \in \mathbb{R}^{d_1} \tag{21.1.6}$$

with respect to $\mathbb{P}\{\cdot \mid \frac{U_n}{n} \in C\}$, where v_0 is a point that will be identified below.

From (21.1.6) it is argued in [Le-Nu2] that as $n \to \infty$, $\mathbb{P}\{X_1 \in \cdot \mid \frac{U_n}{n} \in C\}$ b-converges (in the sense of integration against bounded functions) to an exponential transform of μ. A similar limit (called a Gibbs measure) may result in (21.1.4).

(See, e.g., [D-Z2, section 7.3] and references therein.) A special case of the above result is in an earlier paper of Van Campenhout and Cover [Va-Co] (1981).

Some papers that contain surveys of previous literature on the subject and on statistical mechanics applications are, for example, Stroock and Zeitouni [St-Z] (1991), Martin-Löf [ML] (1979), Ellis [E] (1985) and Lehtonen and Nummelin [Le-Nu2] (1990). Also, see the book by Dembo and Zeitouni [D-Z2], and Zabell [Za1, Za2] (1980, 1993) for some relevant material, and Bartfai [B] (1972) for an early special case of (21.1.2).

In this note we extend some of the results surveyed above to the case when $X_1, X_2, \ldots$ is a Markov chain. The exact statement is Theorem 21.3.1 below.

We call the point v_0 in (21.1.6) a *dominating point*. This expression was introduced by the second author in [N] (1983) in the context of convexity theory, to describe some properties of a point at which the convex conjugate of a function achieves its infimum over a set. In the present context, the large deviation rate functions are convex conjugates, and the dominating points have the above extremum characteristics, as well as certain additional separation properties that are needed in the analysis. These are used in [Le-Nu2] and again below (e.g., Lemma 21.2.2) to obtain explicit representations of the conditioned limits. Kuelbs [K] (1998) has extended this construction to infinite dimensional spaces. See a similar construction, called an *exposed point* in [D-Z2, section 2.3], as well as the minimizing set "K_F" in Bolthausen, Deuschel and Tamura [Bo-Deu-T, p. 237] (1995). The latter consider multiple minimizing points (the elements of their K_F) and their mixtures, to prove large deviation results in a Markov context. For the i.i.d. case see previous papers by Bolthausen [Bo1] (1986) and [Bo2] (1987).

Other related work is Deuschel, Stroock and Zessin [Deu-St-Ze] (1991) from the perspective of random fields, and Dobrushin and Tirozzi [Do-Ti] (1977).

21.2 The Markov Conditioned Case: Background

Conditioned limit results for Markov chains are less complete than the i.i.d. case. Consider a Markov chain $X_1, X_2, \ldots$ taking values in a measurable space $(S, \mathcal{S})$. Csiszar, Cover, and Choi [Cs-Co-Ch] (1987) studied the case when S is a finite set. They showed that (under suitable hypothesis)

$$\lim_{n\to\infty} P\{X_2 = x_2, \ldots, X_k = x_k \mid \widetilde{P}_n^2 \in \Pi, X_1 = x_1\} = \prod_{i=1}^{k-1} P^*(x_{i+1} \mid x_i), \tag{21.2.1}$$

where $\widetilde{P}_n^2$ is the (2nd order) empirical measure of $\{X_n\}$, Π is a set of measures on $S \times S$, and the measures $P^*(\,\cdot \mid \cdot\,)$ are determined by a measure $P^*(\cdot, \cdot)$ on $S \times S$, at which a relative entropy is minimized over Π (assuming this minimum is achieved).

The most definitive results to date for Markov chains on a general state space are by Schroeder [Sc] (1993). She proves a weak convergence result for conditional expectations of empirical measures [Sc, Theorem 3.2] which is similar to our Theorem 21.3.1. Our proofs are quite different, and our use of the dominating point construction allows more general conditioning sets. Under some restrictions on the M.C. $\{X_n\}$ she also proves an "asymptotically quasi-Markov" property which is an analog of Csiszar's result for the i.i.d. case.

In this section we summarize (without proof) some known background results on Markov chains that we will need. Lemmas 21.2.3 and 21.2.4 below are taken from our paper Meda and Ney [M-N] (1998).

We take $X_0, X_1, \dots$ to be a M.C. taking values in $(S, \mathcal{S})$, irreducible with respect to a measure φ on $(S, \mathcal{S})$, with transition function

$$P = \{p(x, A)\} = \{\mathbb{P}(X_{n+1} \in A \mid X_n = x); \;\; x \in S, \; A \in \mathcal{S}\}. \qquad (21.2.2)$$

Let $f : S \times S \to \mathbb{R}^d$ be a bounded, measurable function, $S_n = \Sigma_{i=0}^{n-1} f(X_i, X_{i+1})$. A central role will be played by the "transform kernel"

$$\hat{P}(\alpha) = \{\hat{p}_\alpha(x, A)\} = \left\{ \int_A e^{\langle \alpha, f(x,y) \rangle} p(x, dy), \; A \in \mathcal{S}, \; x \in S \right\}, \quad \alpha \in \mathbb{R}^d.$$

We need to make a hypothesis under which we can assert the existence of eigenvalues and invariant functions for $\hat{P}(\alpha)$. To that end, define the following condition:

For some $0 < a \leq b < \infty$ and some measure ν on $(S, \mathcal{S})$

$$a\nu(A) \leq p(x, A) \leq b\nu(A) \quad \text{for all } x \in S, A \in \mathcal{S}. \qquad (21.2.3)$$

Later we will add the further restriction

$$\nu(x_0) = \delta > 0 \;\; \text{for some } x_0 \in S. \qquad (21.2.4)$$

Lemma 21.2.1 *Assume* (21.2.3). *Then*

(i) *For every* $\alpha \in \mathbb{R}^d$ *there exists a (maximal real) eigenvalue* $\lambda(\alpha) < \infty$ *of* $\hat{P}(\alpha)$, *with (right) eigenfunction* $\{r_\alpha(x); \;\; x \in S\}$ *and (left) eigenmeasure* $\{\ell_\alpha(A); \;\; A \in \mathcal{S}\}$.

(ii) *Let* $\Lambda(\alpha) = \log \lambda(\alpha)$. Λ *is convex and essentially smooth.*

The proof is an immediate consequence [I-N-Nu, Lemma 3.1 and (3.4)].

For other conditions under which the conclusions of the lemma hold see, e.g., [N-Nu].

Let Λ^* = the convex conjugate of Λ. Namely

$$\Lambda^*(v) = \sup_{\alpha \in \mathbb{R}^d} [\langle \alpha, v \rangle - \Lambda(\alpha)], \quad v \in \mathbb{R}^d,$$

and let $\mathcal{D}(\Lambda^*) = \{v : \Lambda^*(v) < \infty\}$. Let ∇ denote gradient. For $B \subset \mathbb{R}^d$, let B^0 = interior of B, $\bar{B}$ = closure of B, ∂B = boundary of B.

Lemma 21.2.2 *Assume* (21.2.3), *and let* $B \subset \mathbb{R}^d$ *be open and convex with* $[B \cap \mathcal{D}(\Lambda^*)]^0 \neq \phi$. *Then*

(i) $\inf[\Lambda^*(v) : v \in B]$ *is achieved at a unique (dominating) point* $v_B \in \bar{B} \cap \mathcal{D}^0(\Lambda^*)$.

(ii) *The equation*

$$\nabla\Lambda(\alpha) = v_B \tag{21.2.5}$$

has a solution $\alpha_B \in \mathbb{R}^d$.

This lemma is [N, Lemma, parts (i) and (ii)]. Lemma 21.2.1(ii) also implies that the level sets of Λ^*, namely $L_a(\Lambda^*) = \{v : \Lambda^*(v) \leq a\}$, are compact.

Recall $S_n = \sum_{i=0}^{n-1} f(X_i, X_{i+1})$. Then we have

Lemma 21.2.3 *Assume* (21.2.3) *and again let* B *be open and convex with* $[B \cap \mathcal{D}(\Lambda^*)]^0 \neq \phi$. *Then*

$$\lim \frac{1}{n} \log \mathbb{P}_x\left\{\frac{S_n}{n} \in B\right\} = -\Lambda^*(v_B), \tag{21.2.6}$$

and

$$\frac{S_n}{n} \overset{\text{exp}}{\longrightarrow} v_B \tag{21.2.7}$$

with respect to the measures $\mathbb{P}_x\left\{ \cdot \mid \frac{S_n}{n} \in B\right\}$.

The large deviation result (21.2.6) is a special case of [N-Nu, Theorem 5.1]. The conditional limit (21.2.7) is a special case of [M-N, Lemma 2.3]. We will need an alternative identification of v_B. To that end define the stochastic kernel

$$Q(\alpha) = \{q_\alpha(x, A)\} = \left\{(e^{\Lambda(\alpha)} r_\alpha(x))^{-1} \int_A e^{\langle \alpha, f(x,y)\rangle} r_\alpha(y) p(x, dy)\right\}. \tag{21.2.8}$$

Then one can check that

$$\pi_\alpha(A) = \int_A r_\alpha(x) \ell_\alpha(dx),$$

normalized so that $\pi_\alpha(S) = 1$, is an invariant probability measure for $Q(\alpha)$. By [N-Nu, Lemma 5.3]

$$\nabla\Lambda(\alpha) = E_{\pi_\alpha}^{Q(\alpha)} S_1 = \int_{S\times S}\int f(x, y) \pi_\alpha(dx) q_\alpha(x, dy).$$

If we take $\alpha = \alpha_B$ then $\nabla\Lambda(\alpha_B) = v_B$ by (21.2.5), and

$$v_B = \int_{S\times S}\int f(x, y) \pi_{\alpha_B}(dx) q_{\alpha_B}(x, dy). \tag{21.2.9}$$

Remark The conditions on B, that it be open and convex with $[B \cap \mathcal{D}(\Lambda^*)]^o \neq \phi$ are not necessary. A sufficient hypothesis for Lemma 21.2.3 is that there exists a unique dominating point for (Λ, B). By Lemma 21.2.2, the above conditions on B are sufficient to assure this. A similar remark applies to the set C in Lemma 21.2.4, and in Theorem 21.3.1.

Now (as in [Le-Nu2]) split the function $f = (g, u)$, where $g : S^2 \to \mathbb{R}^{d_1}$ and $u : S^2 \to \mathbb{R}^{d_2}$; $d_1 + d_2 = d$, g and u bounded. Let

$$G_n = \sum_{i=0}^{n-1} g(X_i, X_{i+1}), \quad U_n = \sum_{i=0}^{n-1} u(X_i, X_{i+1}). \tag{21.2.10}$$

For any $v \in \mathbb{R}^d$, write $v = (v_1, v_2)$ with $v_i \in \mathbb{R}^{d_i}$. Take $\alpha_0 = (O, \beta) \in \mathbb{R}^d$, with $O \in \mathbb{R}^{d_1}$, $\beta \in \mathbb{R}^{d_2}$, and define

$$\hat{P}(\alpha_0) = \hat{P}_u(\beta), \quad \Lambda(\alpha_0) = \Lambda_u(\beta); \tag{21.2.11}$$

$$r_{\alpha_0}(x) = r_{u,\beta}(x), \; x \in S, \quad \pi_{\alpha_0}(A) = \pi_{u,\beta}(A), \; A \in \mathcal{S}. \tag{21.2.12}$$

Note that the functions in (21.2.11) and (21.2.12) depend on u, but are independent of g.

Let q_α be as defined in (21.2.8) and define $Q_u(\beta) = \{q_{u,\beta}(x, A);\; x \in S, A \in \mathcal{S}\}$, where

$$q_{u,\beta}(x, A) = q_{\alpha_0}(x, A), \quad \text{with } \alpha_0 = (O, \beta). \tag{21.2.13}$$

Then we have

Lemma 21.2.4 *Assume* (21.2.3). *Let $C \subset \mathbb{R}^{d_2}$ be open and convex with $[C \cap \mathcal{D}(\Lambda_u^*)]^o \neq \phi$. Then*

(i) $\inf_{v \in C} \Lambda_u^*(v)$ *is achieved at a unique (dominating) point v_C, and*

$$\nabla \Lambda_u(\beta) = v_C \;\; \text{has solution } \beta_C \in \mathbb{R}^{d_2}. \tag{21.2.14}$$

(ii)

$$\mathbb{P}_x\left\{\left\|\frac{G_n}{n} - v_1\right\| > \epsilon \;\middle|\; \frac{U_n}{n} \in C\right\} \xrightarrow{\exp} 0,$$

where

$$v_1 = (\nabla \Lambda(0, \beta_C))_1 = \int_{S^2}\int g(x, y)\pi_{u,\beta_C}(dx) q_{u,\beta_C}(x, dy). \tag{21.2.15}$$

Part (i) follows from Lemma 21.2.2, part (ii) follows from a translation of [M-N, Theorem 2] to the present setting, and from (21.2.9).

21.3 Convergence to the Markov Limit

For any measure μ on $(S, \mathcal{S})$ and measurable function $f : S \to \mathbb{R}^d$, writc $f\mu = \int_S f(x)\mu(dx)$, and for mcasures μ_n write $\mu_n \xrightarrow{b} \mu$ on $\mathbb{R}^d$ if $f\mu_n \to f\mu$ for all bounded measurable $f : S \to \mathbb{R}^d$. (This is convergence in the τ-topology.)

Definition Let $\{\widetilde{\mu}_n(\cdot, \omega),\; \omega \in \Omega$ with σ-field $\mathcal{F}$, $n = 1, 2, \ldots\}$ be random measures on $(S, \mathcal{S})$, μ be a non-random measure, and $\{\mathbb{P}_n; n = 1, 2, \ldots\}$ be probability measures on $(\Omega, \mathcal{F})$. Write

$$\widetilde{\mu}_n \xrightarrow{\exp} \mu \text{ on } \mathbb{R}^d \text{ with respect to } \{\mathbb{P}_n\} \tag{21.3.1}$$

if $f\widetilde{\mu}_n \stackrel{\exp}{\longrightarrow} f\mu$ with respect to $\{\mathbb{P}_n\}$ for all bounded measurable $f : S \to \mathbb{R}^d$.

Lemma 21.3.1 ([Le-Nu1]) *Let $\widetilde{\mu}_n$ and μ be as above. If $\widetilde{\mu}_n \stackrel{\exp}{\longrightarrow} \mu$ on $\mathbb{R}^d$ with respect to $\{\mathbb{P}_n\}$ then*

$$E_{\mathbb{P}_n}(\widetilde{\mu}_n) \stackrel{b}{\longrightarrow} \mu \quad on\ \mathbb{R}^d. \tag{21.3.2}$$

For $x_i \in S$, $A_i \in \mathcal{S}$, $i = 1, 2, \ldots$ adopt the notation

$$(x_k, \ldots, x_\ell) = x_k^\ell, \quad A_k \times \cdots \times A_\ell = A_k^\ell, \quad 0 \le k \le \ell < \infty,$$

and write

$$(x_k^\ell, x_{\ell+1}^m) = x_k^m, \quad k \le \ell \le m.$$

For example

$$(x_1^k, x_{k+1}) = x_1^{k+1}.$$

Let $\mathbf{1}_x(A) = 1$ if $x \in A$, $= 0$ otherwise.

Define the k'th order empirical measure of $\{X_n\}$ on $(S^k, \mathcal{S}^k)$ by

$$\widetilde{P}_n^{(k)}(\cdot) = \frac{1}{n}\sum_{i=0}^{n-1} \mathbf{1}_{X_{i+1}^{i+k}}(\cdot), \quad n = 1, 2, \ldots.$$

We can now state

Theorem 21.3.1 *Let $\{X_n; n = 0, 1, \ldots\}$ satisfy* (21.2.3) *and* (21.2.4), *let $0 < d_1, d_2 < \infty$ be any integers, and $C \subset \mathbb{R}^{d_2}$ be open and convex. Let u and U_n be as defined in* (21.2.10). *Let $e^{\Lambda_u(\beta)}$, $\beta \in \mathbb{R}^{d_2}$, be the eigenvalue of $\hat{P}_u(\beta)$ whose existence was asserted in Lemma* 21.2.1 *and* (21.2.11). *Assume that $[C \cap \mathcal{D}(\Lambda_u^*)]^o \ne \phi$. Then*

$$\mathbb{E}_x\left\{\widetilde{P}_n^{(k)} \;\Big|\; \frac{U_n}{n} \in C\right\} \stackrel{b}{\longrightarrow} \rho^{(k)} \quad on\ \mathbb{R}^{d_1}, \tag{21.3.3}$$

where

$$\rho^{(k)}(A) = \int_A \cdots \int \pi^*(dx_1) \prod_{i=1}^{k-1} q^*(x_i, dx_{i+1}), \tag{21.3.4}$$

and where $q^*(\cdot,\cdot)$ is a stochastic transition kernel on $(S, \mathcal{S})$ with invariant probability measure π^*, which is identified in (21.3.16).

Thus

$$\frac{1}{n}\sum_{i=0}^{n-1} \mathbb{P}_x\left\{X_{i+1}^{i+k} \in \cdot \;\Big|\; \frac{U_n}{n} \in C\right\} \stackrel{b}{\to} \rho^{(k)}(\cdot). \tag{21.3.5}$$

(See also the remark after Lemma 21.2.3.)

Proof. Consider the M.C. $\{Y_n\} = \{X_{n+1}^{n+k}\} \in S^k$, $n = 0, 1, \ldots$, where $\{X_n\} \subset S$ is the M.C. on S defined in section 2. Denote the transition function of $\{Y_n\}$ by

$$P^{(k)} = \{p^{(k)}(x_1^k, A),\ x_1^k \in S^k,\ A \in \mathcal{S}^k\}.$$

Then

$$p^{(k)}(x_1^k, A_1^k) = P\{Y_{n+1} \in A_1^k \mid Y_n = x_1^k\} \qquad (21.3.6)$$
$$= \mathbf{1}_{x_2^k}(A_1^{k-1}) p(x_k, A_k).$$

Note that $\{Y_n\}$ is irreducible with respect to the measure ν^k in (21.2.3).

Now consider $f^{(2k)}(x_1^k, y_1^k) : S^2 \to \mathbb{R}^d$, and split $f^{(2k)} = (g^{(2k)}, u^{(2k)})$, $g^{(2k)} : S^{2k} \to \mathbb{R}^{d_1}$, $u^{(2k)} : S^{2k} \to \mathbb{R}^{d_2}$. Take $u^{(2k)}$ to be of the special form

$$u^{(2k)}(x_1^k, y_1^k) = u(x_k, y_k) : S^2 \to \mathbb{R}^{d_2}.$$

When $y_1^k = x_2^{k+1}$, define the abbreviation

$$f^{(2k)}(x_1^k, x_2^{k+1}) = \bar{f}^{(k+1)}(x_1^{k+1}).$$

Thus $\bar{f}^{(k+1)} : S^{k+1} \to \mathbb{R}^d$. Again define the split $\bar{f}^{(k+1)} = (\bar{g}^{(k+1}, \bar{u}^{(k+1)})$, with $\bar{g}^{(k+1)} : S^{k+1} \to \mathbb{R}^{d_1}$, $\bar{u}^{(k+1)} : S^{(k+1)} \to \mathbb{R}^{d_2}$, and with $\bar{u}^{(k+1)}$ now taking the special form

$$\bar{u}^{(k+1)}(x_1^{k+1}) = u(x_k, x_{k+1}).$$

Now define the kernel

$$\hat{P}^{(k)}(\alpha) = \{\hat{P}_a(x_1^k, A_1^k)\} = \left\{ \int_{A_1^k} \cdots \int e^{\langle \alpha, f^{(2k)}(x_1^k, y_1^k)\rangle} p^{(k)}(x_1^k, dy_1^k);\ A_1^k \in \mathcal{S}^k \right\}$$
$$= \left\{ \mathbf{1}_{x_2^k}(A_1^{k-1}) \int_{A_k} e^{\langle \alpha, \bar{f}(x_1^{k+1})\rangle} p(x_k, dx_{k+1}) \right\}, \quad \alpha \in \mathbb{R}^d. \qquad (21.3.7)$$

We will need

Lemma 21.3.2 *Assume* (21.2.3) *and* (21.2.4), *then*

(i) $P^{(k)}(\alpha)$ *has a (maximal, Perron-Frobenius (P–F)) eigenvalue* $e^{\Lambda^{(k)}(\alpha)}$ *with associated right eigenfunction* $r_\alpha^{(k)}(x_1^k)$ *and left eigenmeasure* $\ell_\alpha^{(k)}(A)$, $x_1^k \in S^k$, $A \in \mathcal{S}^k$.

(ii) $\Lambda^{(k)}(\alpha)$ *is strictly convex and essentially smooth.*

The idea of the proof of this lemma is as follows. It consists of showing first that $(P_\alpha^{(k)})^k$ (= the k-fold matrix product of $P_\alpha^{(k)}$) satisfies (21.2.3) for suitable a, b and ν. This implies that it has a convergence parameter $R_k^{(k)}(\alpha)$ and is $R_k^{(k)}(\alpha)$-recurrent. One can then argue from this that $P_\alpha^{(k)}$ itself has a convergence parameter $R^{(k)}(\alpha)$ and is $R^{(k)}(\alpha)$-recurrent; also that $R_k^{(k)}(\alpha) = (R^{(k)}(\alpha))^k$. Now condition (21.2.4) is invoked (for the first and only time) to show that $P^{(k)}(\alpha)$ has a minorization, namely

$$\widetilde{h}(x_1^k)\widetilde{\nu}(A_1^k) \le p_\alpha^{(k)}(x_1^k, A_1^k)$$

for suitable $\widetilde{h}$, $\widetilde{\nu}$ such that $\int \widetilde{\nu}(dx_1^k) h(x_1^k) > 0$. This, together with its $R^{(k)}(\alpha)$-recurrence, implies that $(R^{(k)}(\alpha))^{-1}$ is an eigenvalue. (Let $\Lambda^{(k)}(\alpha) = \log(R^{(k)}(\alpha))^{-1}$.) Also there are associated (right) eigenvector $r_\alpha^{(k)}(x_1^k)$ and (left) eigenmeasure $\ell_\alpha^{(k)}(A)$, $A \in \mathcal{S}^k$. This follows, e.g., from [N-Nu, (2.7)–(2.10)]. Finally one can show that $\Lambda(\alpha) < \infty$ for all $\alpha \in \mathbb{R}^d$, and then (ii) follows from [N-Nu, Lemma 3.4].

As in (21.2.11), (21.2.12) and (21.2.13) with $\alpha = \alpha_0 = (0, \beta)$, we define $\hat{P}^{(k)}(\alpha_0) = \hat{P}_u^{(k)}(\beta)$ and similarly $\Lambda_u^{(k)}(\beta)$, $r_{u,\beta}^{(k)}$, $\ell_{u,\beta}^{(k)}$. Then one can again define

$$q_\alpha^{(k)}(x_1^k, A_1^k) = \frac{e^{-\Lambda^{(k)}(\alpha)}}{r_\alpha^{(k)}(x_1^k)} \int_{A_1^k} \cdots \int e^{\langle \alpha, f^{(2k)}(x_1^k, y_1^k)\rangle} r_\alpha^{(k)}(y_1^k) p^{(k)}(x_1^k, dy_1^k) \quad (21.3.8)$$

which by (21.3.4)

$$= \frac{e^{-\Lambda^{(k)}(\alpha)}}{r_\alpha^k(x_1^k)} \int_{A_k} e^{\langle \alpha, \bar{f}^{(k+1)}(x_1^{k+1})\rangle} \mathbf{1}_{x_2^k}(A_1^{k-1}) r_\alpha^{(k)}(x_2^{k+1}) p(x_k, dx_{k+1}). \quad (21.3.9)$$

Now with $\alpha = \alpha_0 = (0, \beta)$ this defines

$$\begin{aligned} &q_{u,\beta}^{(k)}(x_1^k, A_1^k) \quad (21.3.10) \\ &= \frac{e^{-\Lambda_u^{(k)}(\beta)}}{r_{u,\beta}^{(k)}(x_1^k)} \mathbf{1}_{x_2^k}(A_1^{k-1}) \int_{A_k} e^{\langle \beta, u(x_k, x_{k+1})\rangle} r_{u,\beta}^{(k)}(x_2^{k+1}) p(x_k, dx_{k+1}). \end{aligned}$$

We now apply Lemma 21.2.4 to the M.C. $\{Y_n\}$ with

$$\begin{aligned} G_n &= \sum_{i=0}^{n-1} g^{(2k)}(Y_i, Y_{i+1}) = \sum_{i=0}^{n-1} g^{(2k)}(X_{i+1}^{i+k}, X_{i+2}^{i+k+1}) \quad (21.3.11) \\ &= \sum_{i=0}^{n-1} \bar{g}^{(k+1)}(X_{i+1}^{i+k+1}), \end{aligned}$$

and

$$U_n = \sum_{i=0}^{n-1} u^{(2k)}(Y_i, Y_{i+1}) = \sum_{i=0}^{n-1} u(X_{i+k}, X_{i+k+1}).$$

This implies that

$$\frac{G_n}{n} \xrightarrow{\exp} v_1^{(k)} \quad (21.3.12)$$

with respect to

$$\mathbb{P}_n = \mathbb{P}_x\left\{ \cdot \,\Big|\, \frac{U_n}{n} \in C \right\}. \quad (21.3.13)$$

where $v_1^{(k)}$ is identified in (21.3.18) below.

Now recall the eigenvalue $e^{\Lambda_u(\beta)}$ and eigenfunction $r_{u,\beta}(\cdot)$ of $\hat{P}_u(\beta)$, as defined in (21.2.11) and (21.2.12). We can check that $e^{\Lambda_u(\beta)}$ and $r_{u,\beta}(\cdot)$ are also an eigenvalue and eigenfunction of $\hat{P}_u^{(k)}(\beta)$ for all k. Namely

$$\int_{S^k}\cdots\int \hat{P}^{(k)}_{u,\beta}(x_1^k, dy_1^k) r_{u,\beta}(y_k) \tag{21.3.14}$$
$$= \int_{S^k}\cdots\int e^{\langle\beta, u(x_k, y_k)\rangle}\mathbf{1}_{x_2^k}(dy_1^{k-1}) r_{u,\beta}(y_k)p(x_k, dy_k)$$
$$= \int_S e^{\langle\beta, u(x_k, y_k)\rangle} r_{u,\beta}(y_k)p(x_k, dy_k)$$
$$= e^{\Lambda_u(\beta)} r_{u,\beta}(x_k).$$

Thus we can take $r^{(k)}_{u,\beta}(x_1^k) = r_{u,\beta}(x_k)$ and $\Lambda_u^{(k)}(\beta) = \Lambda_u(\beta)$ in (21.3.10), to obtain

$$q^{(k)}_{u,\beta}(x_1^k, A_1^k) = \frac{e^{-\Lambda_u(\beta)}}{r_{u,\beta}(x_k)}\mathbf{1}_{x_2^k}(A_1^{k-1})\int_{A_k} e^{\langle\beta, u(x_k, x_{k+1})\rangle} r_{u,\beta}(x_{k+1})p(x_k, dx_{k+1})$$
$$= \mathbf{1}_{x_2^k}(A_1^{k-1}) q_{u,\beta}(x_k, A_k), \tag{21.3.15}$$

where $q_{u,\beta}$ is defined by (21.2.8) and (21.2.13).

We can now identify the kernel q^* in the theorem, namely we show that

$$q^* = q_{u,\beta_C}, \tag{21.3.16}$$

where β_C is defined in (21.2.14).

From (21.2.15) we can identify

$$v_1^{(k)} = \int_{S^{2k}}\cdots\int g^{(2k)}(x_1^k, y_1^k)\pi^{(k)}_{u,\beta_C}(dx_1^k) q^{(k)}_{u,\beta_C}(x_1^k, dy_1^k). \tag{21.3.17}$$

(Note $\nabla\Lambda_u^{(k)}(\beta) = \nabla\Lambda_u(\beta) = v_C$ has solution β_C.) Substituting for $q^{(k)}_{u,\beta_C}$ from (21.3.15), this

$$= \int_{S^{k+1}} \bar{g}^{(k+1)}(x_1^{k+1})\pi^{(k)}_{u,\beta_C}(dx_1^k) q_{u,\beta_C}(x_k, dx_{k+1}). \tag{21.3.18}$$

Abbreviating $\pi^{(k)}_{\mu,\beta_C}(dx_1^k) q_{u,\beta_C}(x_k, dx_{k+1}) = \rho^{(k+1)}(dx_1^{k+1})$,

$$v_1^{(k)} = \int \bar{g}^{(k+1)}(x_1^{k+1})\rho^{(k+1)}(dx_1^{k+1}).$$

Hence we see by Lemma 21.2.4 that

$$\frac{1}{n}\sum_{i=0}^{n-1}\bar{g}^{(k+1)}(X_{i+1}^{i+k+1}) \xrightarrow{\text{exp}} \int_{S^{k+1}}\cdots\int \bar{g}(x_1^{k+1})\rho^{(k+1)}(dx_1^{k+1})$$

with respect to $\mathbb{P}_n = \mathbb{P}_x\{\cdot \mid U_n \in nC\}$. Define the k'th order empirical measure of $\{X_n\}$ by

$$\tilde{P}_n^{(k)}(A) = \frac{1}{n}\sum_{i=0}^{n-1}\mathbf{1}_A(X_{i+1}^{i+k}),\ A \in \mathcal{S}^k.$$

Then

$$\frac{G_n}{n} = \int_{S^{k+1}} \bar{g}(x_1^{k+1})\widetilde{P}_n^{(k+1)}(dx_1^{k+1}) \xrightarrow{\text{exp}} \int \bar{g}(x_1^{k+1})\rho^{(k+1)}(dx_1^{k+1}) \tag{21.3.19}$$

with respect to $\{\mathbb{P}_n\}$, or

$$\widetilde{P}^{(k+1)} \xrightarrow{\exp} \rho^{(k+1)} \text{ on } \mathbb{R}^{d_1}$$

with respect to $\mathbb{P}_n$.

Now by Lemma 21.3.1

$$E_{\mathbb{P}_n}\widetilde{P}_n^{(k+1)}(\cdot) = \frac{1}{n}\sum_{i=0}^{n-1} P(X_{i+1}^{i+k+1} \in \cdot | \frac{U_n}{n} \in C) \xrightarrow{b} \rho^{(k+1)}(\cdot) \tag{21.3.20}$$

$$= \int_{(\cdot)} \int \pi_{u,\beta_C}^{(k)}(dx_1^k) q_{u,\beta_C}(x_k, dx_{k+1}).$$

Integrating over the last coordinate (x_{k+1}) this says that

$$\frac{1}{n}\sum_{i=0}^{n-1} \mathbb{P}_x(X_{i+1}^{i+k} \in \cdot | \frac{U_n}{n} \in C) \xrightarrow{b} \int_{(\cdot)} \cdots \int \pi_{u,\beta_C}^{(k)}(dx_1^k) = \pi_{u,\beta_C}^{(k)}(\cdot). \tag{21.3.21}$$

But applying (21.3.20) with k replaced by $k-1$, we also have

$$\frac{1}{n}\sum_{i=0}^{n-1} \mathbb{P}_x(X_{i+1}^{i+k} \in \cdot | \frac{U_n}{n} \in C) \xrightarrow{b} \int_{(\cdot)} \cdots \int \pi_{u,\beta_C}^{(k-1)}(dx_1^{k-1}) q_{u,\beta_C}(x_{k-1}, dx_k). \tag{21.3.22}$$

Thus, comparing (21.3.21) and (21.3.22), we see that

$$\pi_{(u,\beta_C)}^{(k)}(dx_1^k) = \pi_{u,\beta_C}^{(k-1)}(dx_1^{(k-1)}) q_{u,\beta_C}(x_{k-1}, dx_k). \tag{21.3.23}$$

Iterating, this

$$= \pi_{u,\beta_C}^{(k-2)}(dx_1^{k-2}) q_{u,\beta_C}(x_{k-2}, dx_{k-1}) q_{u,\beta_C}(x_{k-1}, dx_k) \tag{21.3.24}$$

$$= \cdots = \pi_{u,\beta_C}(dx_1) q_{u,\beta_C}(x_1, dx_2) q_{u,\beta_C}(x_2, dx_3) \cdots q_{u,\beta_C}(x_{k-1}, x_k).$$

Substituting this in (21.3.20) and replacing $k+1$ by k yields the theorem. □

21.4 Further Remarks on the I.I.D. Case

Return to the case when $X_1, X_2, \ldots$ are i.i.d. $\subset (S, \mathcal{S})$ with law μ. We have commented in section 1 on several kinds of conditioned limit laws for $\{X_n\}$. In particular, there was the result (stated more precisely) that

Lemma 21.4.1 ([Le-Nu2]) *If $u : S \to \mathbb{R}^{d_2}$, $C \subset \mathbb{R}^{d_2}$ is open and convex with $[C \cap \mathcal{D}(\Lambda_u^*)]^o \neq \phi$ then*

$$\mathbb{P}\Big\{X_1 \in \cdot \,\Big|\, \frac{1}{n}\sum_{i=1}^{n} u(X_i) \in C\Big\} \xrightarrow{b} \mu_{\beta_C}(\cdot) \quad as\ n \to \infty, \tag{21.4.1}$$

where $\mu_\beta(\Gamma) = e^{\Lambda_u(\beta)} \int_\Gamma e^{\langle \beta, u(x)\rangle} \mu(dx)$, $e^{\Lambda_u(\beta)} = Ee^{\langle \beta, u(X_1)\rangle}$, and β_C is the solution of $\nabla\Lambda_u(\beta) = v_C$; v_C being the dominating point of (Λ_u, C).

Let μ^k = the k-fold product measure of μ. From this lemma it follows that

Corollary 21.4.1 *Under the hypothesis of Lemma* 21.4.1

$$\mathbb{P}\Big\{X_1^k \in \cdot \,\Big|\, \frac{1}{n}\sum_{i=1}^{nk} u(X_i) \in C\Big\} \xrightarrow{b} \mu_{\beta_C}^k(\cdot) \text{ as } n \to \infty. \tag{21.4.2}$$

PROOF. Let $Z_n = (X_{kn+1}, \ldots, X_{k(n+1)})$, $u^{(k)}(\cdot) : S^k \to \mathbb{R}^{d_2}$, with $u^{(k)}(x_1^k) = u(x_1) + \cdots + u(x_k)$. Then

$$\frac{e^{\langle \beta, u^{(k)}(x_1^k)\rangle} dx_1^k}{Ee^{\langle \beta, u^{(k)}(X_1^k)\rangle}} = \prod_{i=1}^{k} \frac{e^{\langle \beta, u(x_i)\rangle} dx_i}{e^{\Lambda_u(\beta)}} = \mu_\beta^k(dx_1^k),$$

and

$$\mathbb{P}\Big\{Z_1 \in \cdot \,\Big|\, \frac{1}{n}\sum_{i=1}^{n} u^{(k)}(Z_i) \in C\Big\} \xrightarrow{b} \mu_{\beta_c}^k(\cdot),$$

implying (21.4.2). □

Dembo and Zeitouni [D-Z2, section 7.3] observe that Sznitman [Sz] has a result very similar to this corollary. Their approach is more general but the Lehtonen-Nummelin argument shows the limiting measure μ_{β_c} in terms of the dominating point.

Both proofs are based on the exchangeability of the r.v.'s $X_1, X_2, \ldots$ and it is exactly what we below call symmetry right after (21.4.5).

If u is taken to be a function on S^2 (instead of S) and the conditioning in (21.4.2) is replaced by

$$\Big\{\frac{1}{n}\sum_{i=1}^{n} u(X_i, X_{i+1}) \in C\Big\}, \tag{21.4.3}$$

then (as observed above, and in [Cs-Co-Ch]) the limit random variables are no longer independent. The transition kernel simplifies to $p(x, A) = \mu(A)$ for all x, but there is no essential simplification in the transform kernel $\hat{P}_u(\beta)$, which becomes

$$\hat{p}_\beta(x, A) = \int_A e^{\langle \beta, u(x,y)\rangle} p(dy), \quad \beta \in \mathbb{R}^{d_2}. \tag{21.4.4}$$

Thus we have the same conclusion as Theorem 21.3.1 with $\pi_{u,\beta}$ and $q_{u,\beta}$ obtained from $\hat{p}_\beta$ in (21.4.4).

We note that the proof of [Le-Nu2, (4.1)] first showed that

$$\frac{1}{n}\sum_{j=1}^{n} \mathbb{P}\Big\{X_j \in A \,\Big|\, \frac{1}{n}\sum_{i=1}^{n} u(X_i) \in C\Big\} \to \mu_{\beta_c}(A). \tag{21.4.5}$$

By symmetry, the quantities in (21.4.5) are equal for $j = 1, \ldots, n$, and hence (21.4.1) followed. When $u(X_i)$ is replaced by $u(X_i, X_{i+1})$ in the conditioning, this symmetry no longer holds. It appears to be an open problem how to go from the Cesaro convergence to termwise convergence.

Acknowledgments: We thank the referee for noting several references which we had not previously mentioned.

We appreciate the opportunity to participate in this volume in honor of Harry Kesten, and to thank him for the help and influence that he and his work have had on our work and that of our students.

REFERENCES

[B] Bartfai, P. (1972), On a conditional limit theorem, *Coll. Math. Soc. János Bolyai.* 9 European Meeting of Statisticians, Budapest, 85–91.

[Bo1] Bolthausen, E. (1986). Laplace approximations for sums of independent random vectors. *Probab. Theory Related Fields* (72) 305–318.

[Bo2] Bolthausen, E. (1987). Laplace approximations for sums of independent random vectors. Part II. *Probab. Theory Related Fields* (76) 167–205.

[Bo-Deu-T] Bolthausen, E., Deuschel, J.D., and Tamura, Y. (1995). Laplace approximations for large deviations of nonreversible Markov processes. The nondegenerate case. *Ann. Prob.* (23) 236–267.

[Cs] Csiszar, I. (1984). Sanov property, generalized I-projection and a conditional limit theorem. *Ann. Prob.* (12) 768–793.

[Cs-Co-Ch] Csiszar, I., Cover, T.M., and Choi, B.S. (1987). Conditional limit theorems under Markov conditioning. *IEEE Trans. Inf. Theory* (IT-33) 788–801.

[D-K] Dembo, A. and Kuelbs, J. (1997). A Gibbs conditioning principle for certain infinite dimensional statistics. *University of Wisconsin Technical Report.*

[D-Z1] Dembo, A. and Zeitouni, O. (1996). Refinements of the Gibbs conditioning principle. *Prob. Th. Relat. Fields* (104) 1–14.

[D-Z2] Dembo, A. and Zeitouni, O. (1998). *Large Deviation Techniques and Applications*, 2nd Edn., Springer-Verlag, New York.

[Deu-St-Ze] Deuschel, J.D., Stroock, D.W., and Zessin, H. (1991). Microcanonical distributions for lattice gases. *Commun. Math. Phys.* (139) 83–101.

[Do-Ti] Dobrushin, R.L., and Tirozzi, B. (1977). The central limit theorem and the problem of equivalence of ensembles. *Commun. Math. Phys.* (54) 173–192.

[E] Ellis, R.S. (1985). *Entropy, Large Deviations and Statistical Mechanics*. Springer-Verlag, Berlin.

[G] Gibbs, J.W. (1902). *Principles in Statistical Mechanics*. Yale Univ. Press, New Haven, Conn.

[I-N-Nu] Iscoe, I., Ney, P., and Nummelin, E. (1985). Large deviations of uniformly recurrent Markov additive processes. *Adv. in Appl. Math.* (6) 373–412.

[J] Jaynes, E.T. (1967). Foundations of Probability Theory and Statistical Mechanics. In *Delaware Seminary in Foundation of Physics*. Springer-Verlag, Berlin.

[K] Kuelbs, J. (1998). Large deviation probabilities and dominating points for open, convex sets: non-logarithmic behavior. *University of Wisconsin Technical Report.*

[Le-Nu1] Lehtonen, T. and Nummelin, E. (1988). On the convergence of empirical distributions under partial observations. *Ann. Acad. Sc. Fennicae. Ser. A.I.* (13) 219–223.

[Le-Nu2] Lehtonen, T. and Nummelin, E. (1990). Level I theory of large deviations in the ideal gas. *Int'l J. of Theoret. Phys.* (29) 621–635.

[ML] Martin-Löf, A. (1979). *Statistical Mechanics and the Foundations of Thermodynamics*. Springer-Verlag, Berlin.

[M-N] Meda, A. and Ney, P. (1998). A conditioned law of large numbers for Markov additive chains. *Studia Sc. Math. Hungarica* (34) 305–316.

[Nev] Neveu, J. (1963). Sur le Théorème ergodique de Chung-Erdös, *C.R. Aca. Sci. Paris*, (257) 2953–2955.

[N] Ney, P. (1983). Dominating points and the asymptotics of large deviations on $\mathbb{R}^d$. *Ann. Prob.* (11) 158–167.

[N-Nu] Ney, P. and Nummelin, E. (1987). Markov additive processes I: Eigenvalue properties and limit theorems. *Ann. Prob.* (15) 561–592.

[Nu] Nummelin, E. (1984). General irreducible Markov chains and non-negative operators. *Cambridge University Press*, Cambridge, UK.

[Sc] Schroeder, C. (1993). I-projection and limit theorems for discrete parameter Markov chains. *Ann. Prob.* (21) 721–758.

[St-Z] Stroock, D.W. and Zeitouni, O. (1991). *Microcanonical distributions, Gibbs states, and the equivalence of ensembles.* In *F. Spitzer Festschrift*, R. Durrett and H. Kesten Eds., Birkhäuser, Boston.

[Sz] Sznitman, A. (1991). *Topics in propagation of chaos*. Lecture Notes in Mathematics 1464. Springer-Verlag, New York.

[Va-Co] Van Campenhout, J.M. and Cover, T.M. (1981). Maximum entropy and conditional probability. *IEEE Trans. Inf. Theory*. IT–27, 483–489.

[V] Vasicek, A.O. (1980). A conditional law of large numbers. *Ann. Prob.* (8) 142–147.

[Za1] Zabell, S. (1980). Rates of convergence for conditional expectations. *Ann. Prob.* 5(8) 928–941.

[Za2] Zabell, S. (1993). A limit theorem for expectations conditional on a sum. *J. Th. Prob.* 2(6) 267–283.

Department of Mathematics
University of Wisconsin
Madison, WI 53706
ney@math.wisc.edu